Hot Cracking Phenomena in Welds III

John Lippold · Thomas Böllinghaus · Carl E. Cross
Editors

Hot Cracking Phenomena in Welds III

Springer

Editors
Prof. Dr. John Lippold
The Ohio State University
Dept. of Industrial, Welding,
and Systems Engineering
Arthur E. Adams Drive 1248
43221 Columbus, Ohio
USA
lippold.1@osu.edu

Prof. Dr.-Ing. Thomas Böllinghaus
Bundesanstalt für Materialforschung
und -prüfung (BAM)
Unter den Eichen 87
12205 Berlin
Germany
Thomas.Boellinghaus@bam.de

Prof. Dr. Carl E. Cross
Bundesanstalt für Materialforschung
und -prüfung (BAM)
Unter den Eichen 87
12205 Berlin
Germany
cecross@bajabb.com

ISBN 978-3-642-16863-5 e-ISBN 978-3-642-16864-2
DOI 10.1007/978-3-642-16864-2
Springer Heidelberg Dordrecht London New York

Library of Congress Control Number: 2011921710

Cover design: deblik, Berlin

Printed on acid-free paper

Springer is part of Springer Science+Business Media (www.springer.com)

This conference and the proceedings are dedicated in memory of Prof. Dr.-Ing. habil. Dr. E. h. Horst Herold, formerly of Magdeburg University in Germany. He was the "founding father" of the inaugural Hot Cracking Workshop and a co-editor of the first two proceedings.

Prof. Herold was born in Leipzig in 1945. He graduated from Rostock University in 1969 as a Shipbuilding Engineer and in 1976 earned his Dr.-Ing. degree. From 1988 until his death in 2008, he was on the faculty of the University of Magdeburg in Germany. From 1988 to 2005 he served as the Director of the Institute of Joining and Beam Technology, and from 2006 to 2007 the Director of the Institute of Materials Joining Technology.

Prof. Herold will be remembered as a kind and gentle person who had a passion for his profession. He has touched the lives of many and left us all the better for it. We will miss him as a friend and colleague.

Preface

The term "hot cracking" is generally used to describe cracking that occurs at elevated temperature and, in the context of materials joining, generally constitutes weld solidification cracking, and weld metal and heat-affected zone (HAZ) liquation cracking. Ductility-dip cracking can also be considered a "hot cracking" phenomenon, even though DDC occurs in the solid-state but only slightly below the solidus temperature of the material. Hot cracking is a major weldability issue with many structural materials including aluminum alloys, steels, stainless steels and nickel-base alloys.

This conference was the third in a series of workshops that started in 2004. The first workshop was organized by Prof. Horst Herold and Prof. Thomas Böllinghaus and held in Berlin in March 2004. The proceedings from that workshop, entitled *Hot Cracking Phenomena in Welds* (ISBN 3-540-22332-0) were published in January 2005. The second workshop in March 2007 was organized by Prof. Böllinghaus, Prof. Herold, Prof. Carl Cross, and Prof. John Lippold and also held in Berlin. The proceedings from that workshop, *Hot Cracking Phenomena in Welds II* (ISBN 978-3-540-78627-6), were published in May 2008. These first two proceedings constitute 42 papers with contributions from over 10 countries.

The third workshop was held in Columbus, Ohio in March 2010 and hosted by The Ohio State University and Edison Welding Institute. There were over 80 participants and 22 presentations from 8 countries (USA, Germany, France, United Kingdom, Finland, Japan, Sweden, and Brazil). Paper topics included weld solidification cracking, liquation cracking, ductility dip cracking, weldability testing, and modeling. The conference was organized into three major sessions based on material type: aluminum and magnesium alloys, steels and stainless steels, and nickel-base alloys. Each of these sessions was started with a keynote presentation. Keynote speakers at this workshop were Prof. Sindo Kou (Univ. of Wisconsin – USA), Prof. Thomas Böllinghaus (BAM – Germany), and Prof. John DuPont (Lehigh University – USA).

The papers collected here together with the 42 papers from the previous workshops represent the best compilation of information on the topic of hot cracking of welds that has ever been collected. Among these papers, readers can find information on hot cracking mechanisms, characterization, weldability testing approaches, and modeling techniques. The authors are among the leading experts in the field and have been provided the opportunity (without page limitations) to present their results and express their opinions.

The editors wish to thanks all the authors for their excellent contributions and for keeping to a schedule that allowed publication of these proceedings in a timely manner. We would also like to thank the Edison Welding Institute for providing the venue for this workshop and, in particular, to Mr. Nate Ames and Mr. Brad Hudson for assisting with the organization and advertising of the workshop.

Columbus, OH — John Lippold
Berlin, Germany — Thomas Böllinghaus
Berlin, Germany — Carl E. Cross
January 2011

Photo of Presenters

Front row (from left): C. Cross, S. Kou, J. DuPont, T. Böllinghaus, J. Lippold
Back row (from left): A. Ramirez, K. Nishimoto, Y-P. Yang, E. Giraud, D. Keil, M. Karhu, M. Gittos, M. Gallagher, J. Andersson, V. Karkhin, J. Sowards, B. Alexandrov, S. McCracken, A. Niel

Contents

Contributors

J. Adrien MATEIS, INSA Lyon, CNRS UMR5510, Lyon, France, jerome.adrien@insa-lyon.fr

B.T. Alexandrov Welding Engineering Program, The Ohio State University, Columbus, OH, USA, alexandrov.1@osu.edu

T.D. Anderson ExxonMobil Upstream Research Company, Houston, TX, USA

J. Andersson Volvo Aero Corporation, Trollhättan, Sweden; Chalmers University of Technology, Gothenburg, Sweden, joel.andersson@volvo.com

S.S. Babu The Ohio State University, Columbus, OH, USA, babu.13@osu.edu

C. Bordreuil Mechanics and Civil Engineering Laboratory LMGC UMR 5508, University Montpellier2, Montpellier, France, bordreuil@iut-nimes.fr

S.M.I. Birch Chesterfield College, Chesterfield, UK

Thomas Böllinghaus BAM Federal Institute for Materials Research and Testing, D-12205 Berlin, Germany, Thomas.Boellinghaus@bam.de

Fisseha Brhane Laboratory of Welding Technology and Laser Processing, Lappeenranta University of Technology, Lappeenranta, Finland; Department of Mechanical Engineering, Lappeenranta University of Technology, Lappeenranta, Finland, fisseha.brhane@lut.fi

F.W. Brust Engineering Mechanics Corporation of Columbus, Upper Arlington, OH, USA, bbrust@emc-sq.com

N. Coniglio University of Adelaide, Adelaide, Australia, nicolas.coniglio@adelaide.edu.au

M. Coret LAMCoS, INSA Lyon, CNRS UMR5514, Lyon, France, Michel.Coret@insa-lyon.fr

Carl E. Cross Federal Institute for Materials Research and Testing, Berlin, Germany, cecross@bajabb.com

F. Deschaux-Beaume Mechanics and Civil Engineering Laboratory LMGC UMR 5508, University Montpellier2, Montpellier, France, deschaux@iut-nimes.fr

J.-M. Drezet Computational Materials Laboratory, Ecole Polytechnique Federale Lausanne, Lausanne, Switzerland, jean-marie.drezet@epfl.ch

John N. DuPont Department of Materials Science and Engineering, Lehigh University, Bethlehem, PA, USA, jnd1@lehigh.edu

V. Firouzdor Department of Materials Science and Engineering, University of Wisconsin, Madison, WI, USA, firouzdor@wisc.edu

G.S. Frankel The Ohio State University, Columbus, OH, USA, frankel.10@osu.edu

G. Fras Mechanics and Civil Engineering Laboratory LMGC UMR 5508, University Montpellier2, Montpellier, France, fras@iut-nimes.fr

M.L. Gallagher Edison Welding Institute, Columbus, OH, USA, mgallagher@ewi.org

E. Giraud SIMaP, Universités de Grenoble, CNRS UMR5266, Saint Martin d'Hères Cedex, France, eliane.giraud@simap-grenoble-inp.fr

M.F. Gittos TWI Ltd, Cambridge, UK, mike.gittos@twi.co.uk

A. Gumenyuk BAM Federal Institute for Materials Research and Testing, D-12205, Berlin, Germany, andrey.gumenyuk@bam.de

H. Hänninen Aalto University School of Science and Technology, Espoo, Finland

I.W. Haygood Department of Materials Science and Engineering, University of Wisconsin, Madison, WI, USA

Esa Hiltunen Laboratory of Welding Technology and Laser Processing, Lappeenranta University of Technology, Lappeenranta, Finland; Department of Mechanical Engineering, Lappeenranta University of Technology, Lappeenranta, Finland, esa.hiltunen@lut.fi

P.W. Hochanadel Los Alamos National Laboratory, Los Alamos, NM, USA, phoch@lanl.gov

S. Ivanov Department of Welding and Laser Technologies, St. Petersburg State Polytechnic University, Saint Petersburg, Russia

M.Q. Johnson Los Alamos National Laboratory, Los Alamos, NM, USA, mqj@lanl.gov

Paul Kah Laboratory of Welding Technology and Laser Processing, Lappeenranta University of Technology, Lappeenranta, Finland, Paul.Kah@lut.fi

Miikka Karhu VTT-Technical Research Centre of Finland, FI-53851 Lappeenranta, Finland, mikka.karhu@vtt.fi

Victor Karkhin Department of Welding and Laser Technologies, St. Petersburg State Polytechnic University, Saint Petersburg, Russia, victor.karkhin@mail.ru

D. Keil Institute of Materials and Joining Technology, University of Magdeburg, Magdeburg, Germany, daniel.keil@ovgu.de

J.M. Kikel The Babcock & Wilcox Company, Barberton, OH, USA, jmkikel@bacock.com

K. Kiuchi Japan Atomic Energy Agency, Naka, Japan

S. Kou Department of Materials Science and Engineering, University of Wisconsin, Madison, WI, USA, kou@engr.wisc.edu

Veli Kujanpää VTT-Technical Research Centre of Finland, FI-53851 Lappeenranta, Finland; Lappeenranta University of Technology, FI-53851 Lappeenranta, Finland, veli.kujanpaa@lut.fi

T.J. Lienert Los Alamos National Laboratory, Los Alamos, NM, USA, lienert@lanl.gov

John Lippold Welding Engineering Program, The Ohio State University, Columbus, OH, USA, lippold.1@osu.edu

E. Maire MATEIS, INSA Lyon, CNRS UMR5510, Lyon, France, eric.maire@insa-lyon.fr

Jukka Martikainen Department of Mechanical Engineering, Lappeenranta University of Technology, Lappeenranta, Finland, jukka.martikainen@lut.fi

J.N. Martinez Los Alamos National Laboratory, Los Alamos, NM, USA

R.J. Martinez Los Alamos National Laboratory, Los Alamos, NM, USA

Steven L. McCracken Electric Power Research Institute, Charlotte, NC, USA, smccracken@epri.com

M. Mousavi Høgskolen i Buskerud, Konigsberg, Norway, mehdi.gebreil.mousavi@hibu.no

J. Nakayama Kobe Steel, Ltd., Kobe, Japan

A. Niel Mechanics and Civil Engineering Laboratory LMGC UMR 5508, University Montpellier2, Montpellier, France, aurelie.niel@iut-nimes.fr

K. Nishimoto Osaka University, Suita, Osaka, Japan, nishimoto@mapse.eng.osaka-u.ac.jp

R.J. Pargeter TWI Ltd, Cambridge, UK, richard.pargeter@twi.co.uk

H. Pries Institute of Joining and Welding, Braunschweig University of Technology, Brunswick, Germany, h.pries@tu-braunschweig.de

V. Quiroz BAM Federal Institute for Materials Research and Testing, D-12205 Berlin, Germany, Vanessa.Quiroz@bam.de

A.J. Ramirez Brazilian Synchrotron Light Laboratory (LNLS), Campinas, Brazil, ramirez@lnls.br

K. Saida Osaka University, Suita, Japan, saida@mapse.eng.osaka-u.ac.jp

P. Schempp Federal Institute for Materials Research and Testing, Berlin, Germany, Philipp.Schempp@bam.de

G. Sjöberg Volvo Aero Corporation, Trollhättan, Sweden; Chalmers University of Technology, Gothenburg, Sweden, sjobergg@chalmers.se

Richard E. Smith Structural Integrity Associates, Inc., Huntersville, NC, USA, Rsmith@Structint.com

J.W. Sowards The Ohio State University, Columbus, OH, USA; National Institute of Standards and Technology, Boulder, CO, USA, jeffrey.sowards@nist.gov

M. Suéry SIMaP, Universités de Grenoble, CNRS UMR5266, Saint Martin d'Hères Cedex, France, michel.suery@simap-grenoble-inp.fr

E.A. Torres Brazilian Synchrotron Light Laboratory (LNLS), Campinas, Brazil; School of Mechanical Engineering, State University of Campinas, Campinas, Brazil

J.S. Unfried Brazilian Synchrotron Light Laboratory (LNLS), Campinas, Brazil; School of Mechanical Engineering, State University of Campinas, Campinas, Brazil, junfried@lnls.br

E.M. Westin Outokumpu Stainless, Avesta Research Center, Avesta, Sweden, elin.westin@bsga.ata

Y.P. Yang Edison Welding Institute, Columbus, OH, USA, yyang@ewi.org

M. Zinke Institute of Materials and Joining Technology, University of Magdeburg, Magdeburg, Germany, zinke@mb.uni-magdeburg.de

Part I
Aluminum and Magnesium Alloys

Hot Cracking in Welds of Aluminum and Magnesium Alloys

S. Kou, V. Firouzdor, and I.W. Haygood

Introduction

Aluminum (Al) alloys are known to be susceptible to hot cracking during fusion welding, including solidification cracking and liquation cracking. Figure 1 shows these two types of hot cracking in a bead-on-plate weld made by gas-metal arc welding of a 6.4 mm-thick 6,061 Al plate with a 4,043 Al filler metal (Fig. 1a). Both types of hot cracking are intergranular. Solidification cracking occurs in the fusion zone during solidification of the weld pool and is thus characterized by a dendritic fracture surface (Fig. 1b). Liquation cracking, on the other hand, occurs in the partially melted zone during solidification of the grain-boundary liquid and is thus characterized by an essentially smooth fracture surface (Fig. 1c). More details about hot cracking can be found elsewhere [1].

Magnesium (Mg) alloys, similar to Al alloys in having a relatively wide freezing temperature range as compared to the melting point, are also susceptible to hot cracking especially liquation cracking. In view of the growing interest in using Mg alloys to reduce the vehicle weight and energy consumption in the automotive industry [2], the hot cracking of Mg as well as Al welds will be discussed.

Liquation and solidification cracking in fusion welding of Al and Mg alloys, including arc and resistance spot welding, will be discussed in the present study. Liquation cracking in friction stir welding, which is considered as a solid-state welding process, will also be discussed.

Liquation Cracking

Liquation Cracking in Arc Welding

Liquation and liquation cracking in Al arc welding have been studied extensively, e.g., by Kou and coworkers [3–14] and Rao et al. [15–17]. Figure 2 shows

S. Kou (✉)
Department of Materials Science and Engineering, University of Wisconsin, Madison, WI, USA
e-mail: kou@engr.wisc.edu

J. Lippold et al. (eds.), *Hot Cracking Phenomena in Welds III*,
DOI 10.1007/978-3-642-16864-2_1, © Springer-Verlag Berlin Heidelberg 2011

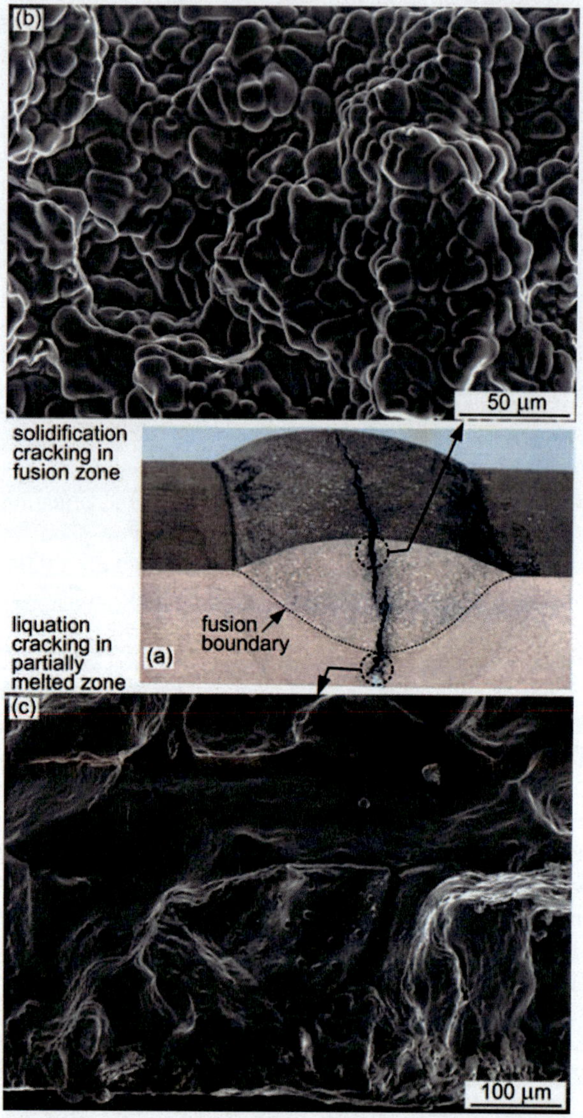

Fig. 1 Hot cracking in a bead-on-plate, partial-penetration weld of 6,061 Al: (**a**) transverse weld macrograph; (**b**) dendritic fracture surface typical of solidification cracking; (**c**) smooth fracture surface typical of liquation cracking

schematically the weld pool and its surrounding area during arc welding, where the cooling rate is usually not too high and undercooling is usually negligible. A pure metal melts and solidifies at its melting point T_m, so the weld pool boundary is solid S (Fig. 2a). However, an alloy melts and solidifies over a temperature range. The upper bound of the temperature range is the liquidus temperature T_L. Most Al and Mg alloys contain eutectic-forming constituents, and the lower bound is often the eutectic temperature T_E. During welding of an alloy, a semisolid region exists all around the weld pool [1, 7, 10, 18] (Fig. 2b). The semisolid behind the pool (the

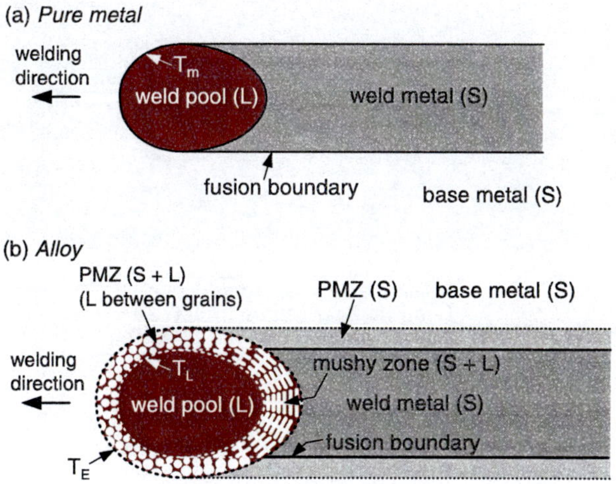

Fig. 2 Weld pool and its surrounding area: (**a**) pure metal; (**b**) alloy. PMZ: partially melted zone

mushy zone) solidifies into the fusion zone or the weld metal. The semisolid S+L on the sides of the pool solidifies into the partially melted zone (PMZ) immediately outside the weld metal.

During welding liquid exists between partially melted grains (as well as in the grain interior). Liquid formation (i.e., liquation) can be initiated by the reaction between the matrix and the eutectic-forming constituent at T_E upon rapid heating during welding. This is known as "constitutional liquation" [19, 20]. Other liquation mechanisms have been discussed by Kou and coworkers [1, 3, 7]. Regardless of the mechanism by which liquation is initiated, once the semisolid region (S+L) of the phase diagram is reached, the fraction of liquid can increase dramatically. For Al and Mg alloys, the semisolid region tends to be rather wide as compared to the liquidus temperature. Furthermore, according to Fourier's law of conduction, the heat flux $q = -k\nabla T$ where k is the thermal conductivity and ∇T the temperature gradient [21]. Under the same heat flux, a high thermal conductivity such as that of Al and Mg favors a low temperature gradient. Thus, arc welds of Al and Mg alloys can have a rather wide PMZ, in which a wide band of liquated material can exist along the grain boundary.

Huang et al. [4–7] discovered that the grain boundary liquid solidifies directionally, that is, upward and toward the heat source. This is illustrated schematically in Fig. 3a, and verified by the transverse micrographs of the partial-penetration Al-4.5Cu weld shown in Fig. 3b, c. Furthermore, the grain boundary liquid solidifies with the planar mode, resulting in first a soft solute-depleted band and then a brittle solute-rich eutectic film along the grain boundary, as shown by Fig. 3d. This explains the severe weakening of high-strength Al alloys in the PMZ. Such grain boundary segregation is sometimes mistaken as the result of solid-state diffusion during welding.

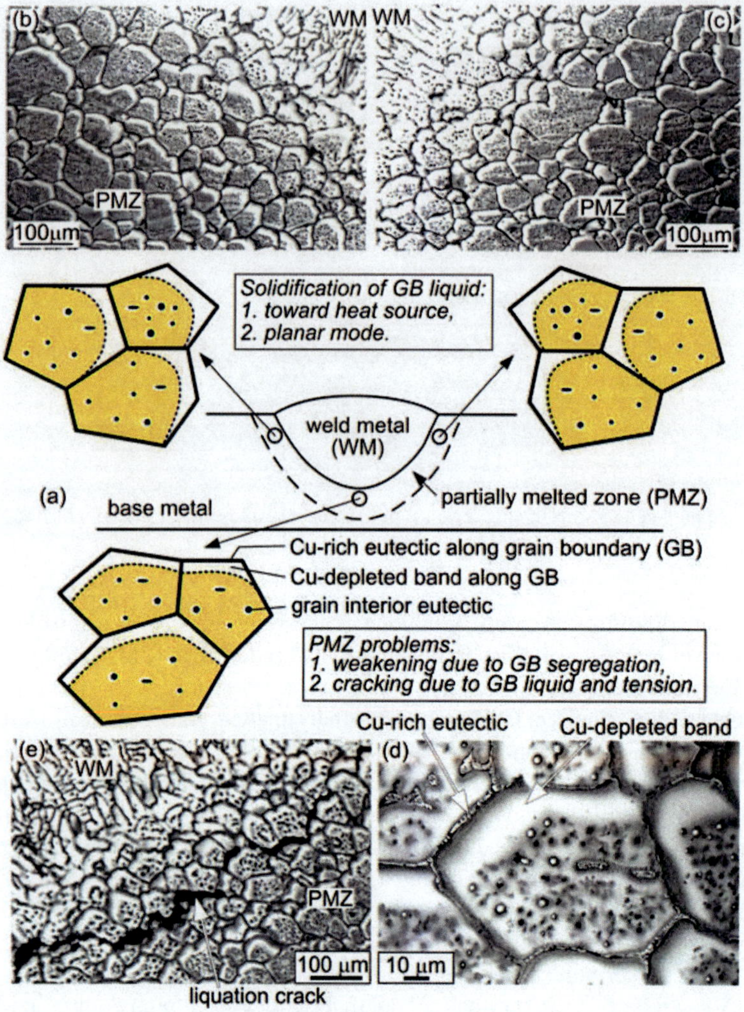

Fig. 3 Solidification of grain-boundary (GB) liquid in PMZ of partial-penetration Al-4.5Cu weld: (**a**) schematic; (**b**) (**c**) transverse micrographs; (**d**) GB segregation; (**e**) liquation cracking along GBs [6]

Similar grain-boundary (GB) solidification in the PMZ has been observed in Mg arc welds. The micrographs in Fig. 4 show the top views of an autogenous weld made on a 1.6 mm-thick AZ31B Mg sheet by gas-tungsten arc welding. Grain growth is evident in the heat-affected zone (HAZ) (Fig. 4a). Directional planar solidification of the GB liquid is evident in the PMZ (Fig. 4b, c).

Since the density of solid Al is greater than that of liquid Al, the weld pool contracts and pulls the solidifying PMZ nearby during solidification. The tension so induced in the PMZ can cause liquation cracking. Consider the semisolid at the

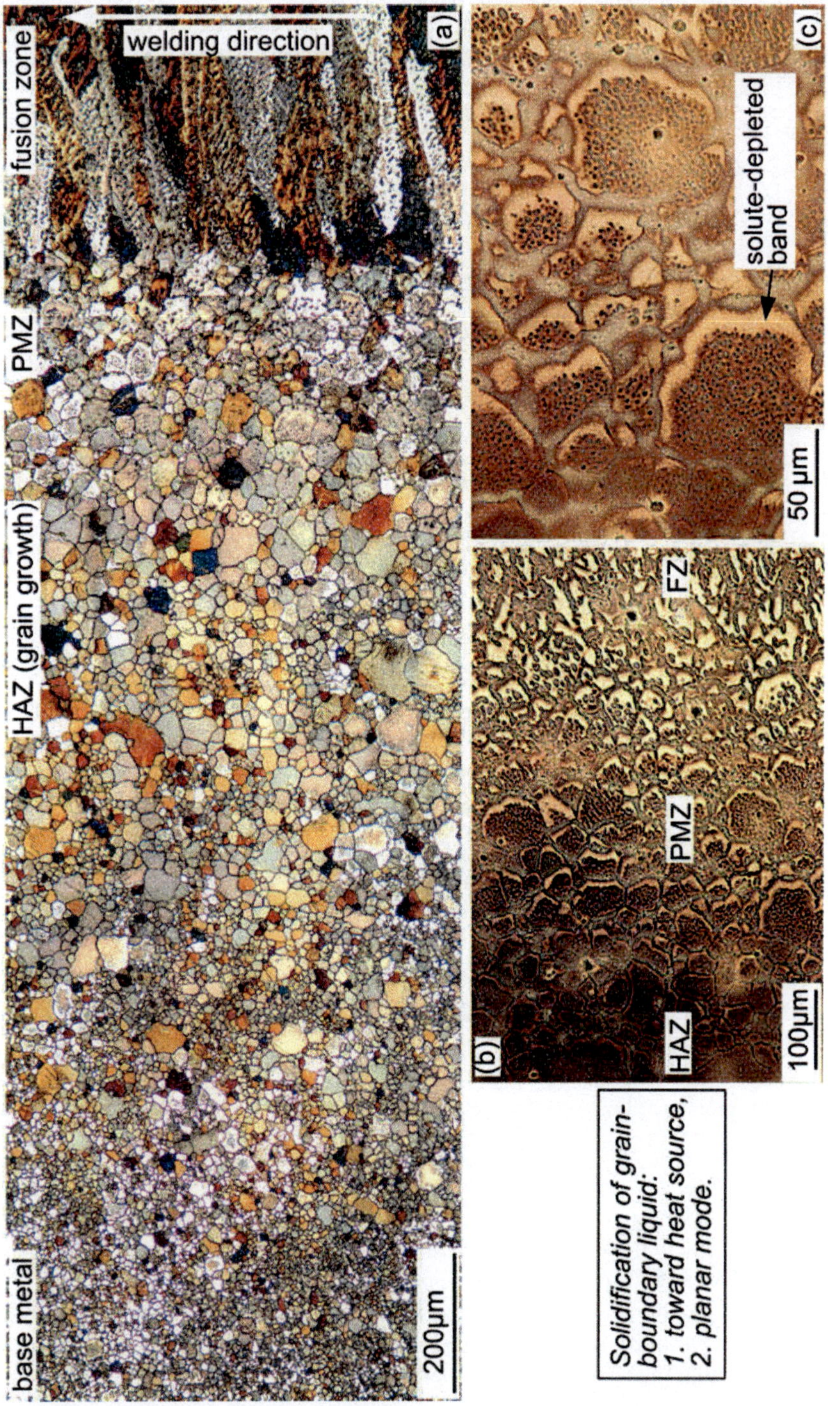

Fig. 4 Top surface of AZ31 Mg weld: **(a)** coarse grains in heat-affected zone (HAZ) and columnar grains in fusion zone (FZ); **(b)** solidification of grain-boundary in partially melted zone (PMZ); **(c)** enlarged **(b)**

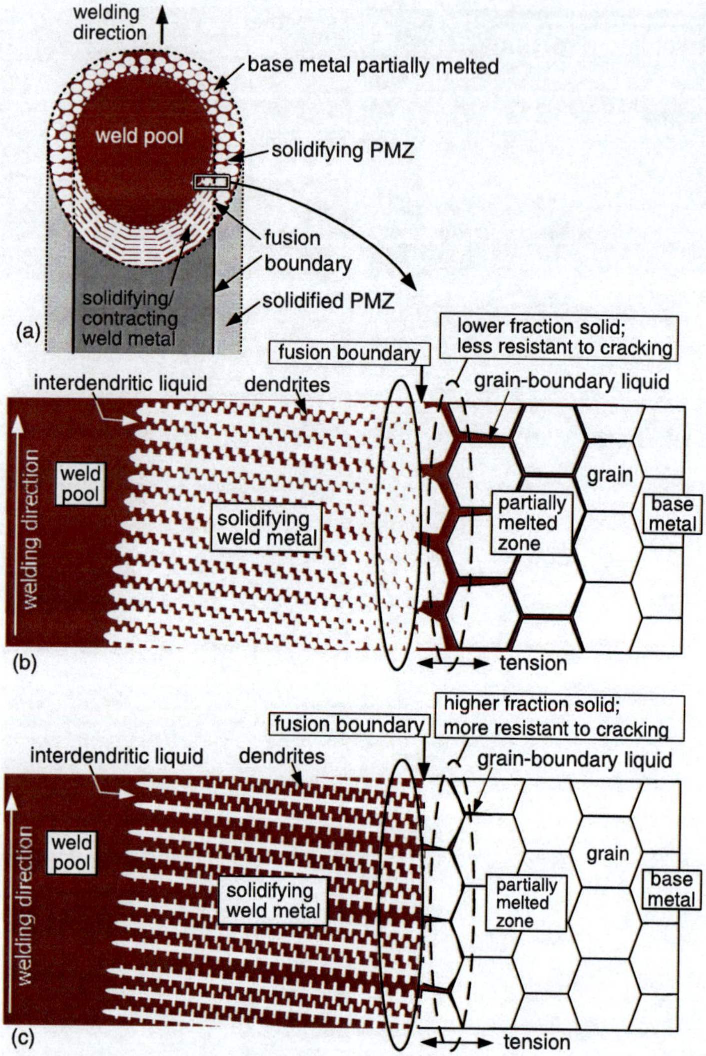

Fig. 5 Liquation cracking susceptibility: (**a**) top view of weld pool; (**b**) PMZ susceptible to liquation cracking when weld-metal f_S > PMZ f_S; (**c**) not susceptible when weld-metal f_S < PMZ f_S

fusion boundary in the boxed area in Fig. 5a. Since the strength of a semisolid increases with its fraction solid, it seems reasonable to assume that the resistance to cracking increases with increasing fraction solid. As illustrated in Fig. 5b, the PMZ is lower in fraction solid than the weld metal. The PMZ may not be able to resist pulling by the solidifying/contracting weld metal and can thus be susceptible to liquation cracking. The opposite is true, as illustrated in Fig. 5c. Due to its small amount of liquid, the solidifying PMZ does not induce significant tension in the weld metal to cause solidification cracking. If solidification cracking does occur, it is caused by the tension in the solidifying/contracting weld metal itself.

Based on the difference between the solidifying weld metal (WM) and the solidifying PMZ in fraction solid f_S, Kou and coworkers [9–12] proposed the following criterion for the liquation-cracking susceptibility of full-penetration Al arc welds:

$$\ldots\ldots \textit{Not susceptible to liquation cracking if WM } f_s < \textit{PMZ } f_s \qquad (1)$$

Since a semisolid develops significant strength only after f_S reaches 0.2–0.4 [21], the comparison is more meaningful after f_S becomes about 0.3. Even curves of T-f_S (temperature vs. fraction solid) calculated based on the simple Scheil solidification model work well. The Scheil model is based on three assumptions: 1. equilibrium at the S/L interface (no undercooling), 2. no back diffusion of solutes into the solid phase, and 3. uniform liquid composition [22]. In normal arc welding, undercooling is negligible. Furthermore, since solidification takes only seconds in normal arc welding, solid-state diffusion is also negligible.

This simple criterion has been confirmed by available experimental data from both UW-Madison and the literature. It has been used successfully to eliminate PMZ cracking from Al welds, including welds of highly crack-susceptible Al alloys such as 7,075 Al and 2,024 Al, by changing the composition and hence T-f_S curve of the weld metal with a proper filler wire [9–12].

Figure 6 shows a weld made on 6,061 Al (3.2 mm thick, full-penetration) with filler 5,356 Al at 64% dilution can be susceptible to liquation cracking as predicted by Scheil T-f_S curves. Because of the presence of the grain boundary liquid in the PMZ during welding, liquation cracking is intergranular (as also shown in Figs. 1c

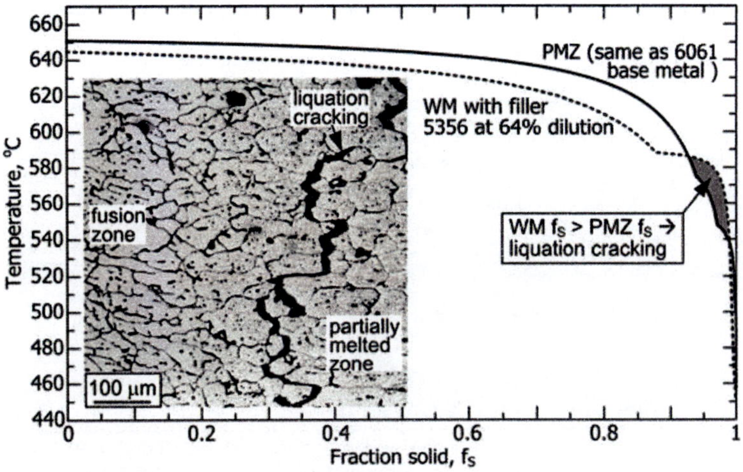

Fig. 6 Liquation cracking in 6,061 Al weld made with filler 5,356 Al at 64% dilution [12]. T-f_S curves calculated using Pandat of CompuTherm LLC [23]

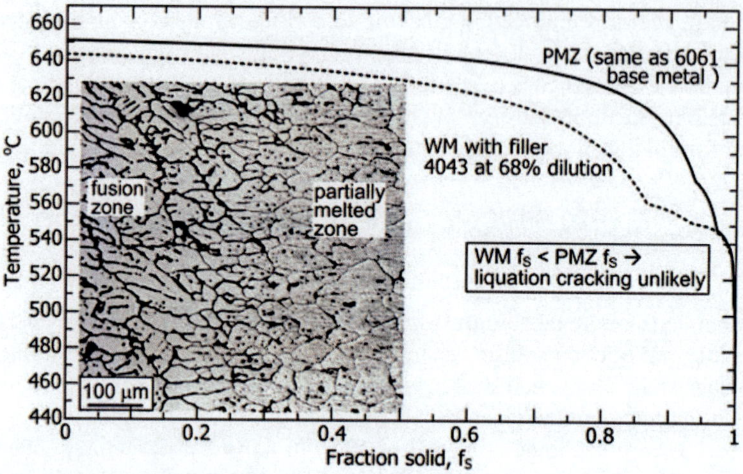

Fig. 7 No liquation cracking in 6,061 Al weld made with filler 4,043 Al at 68% dilution [12]. T-f_S curves calculated using Pandat of CompuTherm LLC [23]

and 3e). Figure 7, on the other hand, shows welding the same 6,061 Al with filler 4,043 Al at 68% dilution is not susceptible to liquation cracking.

Liquation cracking has also been observed in Mg welds. Figure 8 shows an SEM image of liquation cracking in gas-metal arc welding of AZ91 Mg with filler AZ61 Mg [24]. Susceptibility to liquation cracking is likely as shown by the Scheil T-f_S curves. The dilution ratio was assumed 40% since it was not specified. The condition of weld-metal f_S > PMZ f_S can still exist even if the ratio is changed, e.g., to 50 or 60%.

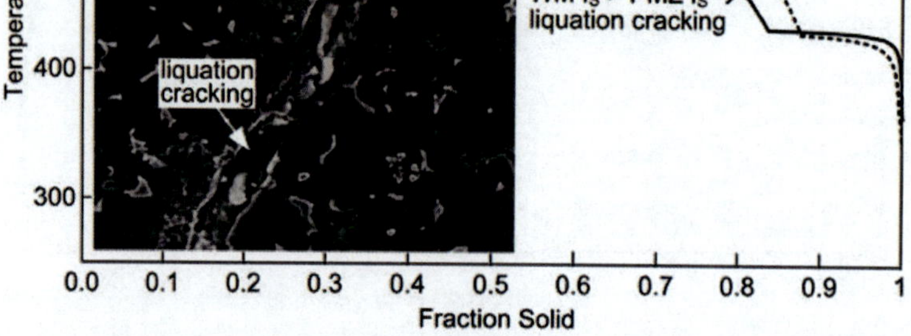

Fig. 8 Liquation cracking in AZ91 Mg weld made with filler AZ61 Mg at 40% dilution (assumed). T-f_S curves calculated using Pandat of CompuTherm LLC [22]. SEM image from Sun et al. [25]

Liquation cracking has been observed in bead-on-plate welding of Al and Mg alloys with nearly complete penetration. Both thermal contraction and solidification shrinkage can induce high tension in the region of the PMZ between the bottom of the weld and the bottom surface of the workpiece. The tendency for the workpiece to develop angular distortion during welding can be considered as an indication of such tension. If the bottom surface is within the PMZ, liquation cracking may occur regardless of the filler metal used. This may explain the liquation cracking in the 6,061 Al weld in Fig. 1a (which was made with filler 4,043 Al at 67% dilution). Similar liquation cracking was observed in bead-on-plate welding (but not in butt welding) of AZ91 Mg by the gas-tungsten arc welding process with partial-penetration and no filler metal [25].

Liquation Cracking in Resistance Spot Welding

Liquation cracking has been observed in resistance spot welding of Al and Mg alloys. Figure 9 is a micrograph showing evidence of liquation in resistance spot welding of AZ31 Mg alloy [26]. The light-etching bands appear to be along the lower and left boundaries of the grains. This seems to suggest planar directional solidification of the grain-boundary liquid although the position of the nugget relative to the micrograph was not specified by the authors.

Figure 10 shows liquation cracking in resistance spot welds of 5,754 Al alloy [27] and AZ31 Mg alloy [26]. During resistance spot welding, liquid can be present along grain boundaries in the region of the PMZ that is under tension, such as that shown in Fig. 10a. The coexistence of tension and grain-boundary liquid in the PMZ causes liquation cracking [27].

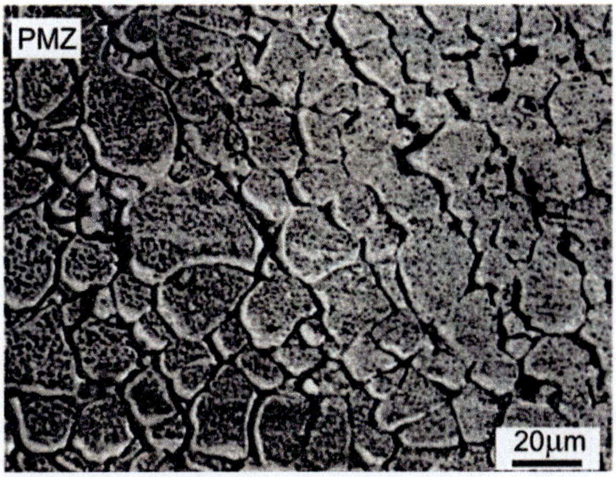

Fig. 9 Evidence of grain-boundary liquation in the partially melted zone of a resistance spot weld of AZ31 Mg. From Lang et al. [26]

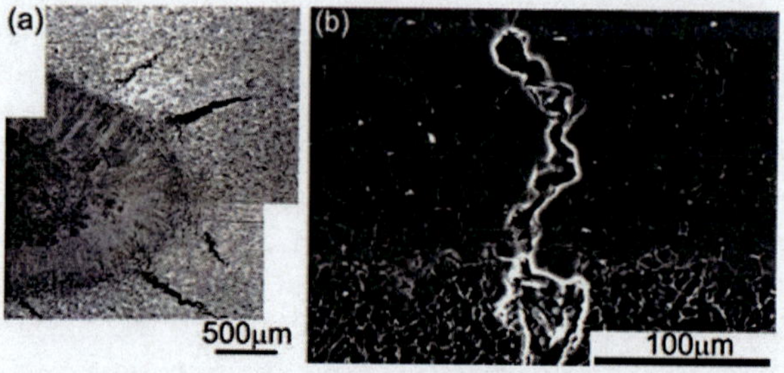

Fig. 10 Liquation cracking in resistance spot welds: (**a**) 5,754 Al, from Senkara et al. [27]; (**b**) AZ31 Mg, from Lang et al. [26]

Liquation Cracking in Friction Stir Welding

Friction stir welding (FSW) is a solid-state welding process invented by The Welding Institute in 1991 [28]. It has been widely used for welding soft materials such as Al and Mg alloys since then. In FSW the pin at the bottom of a rotating cylindrical tool is plunged into a rigidly held workpiece and traversed along the joint to be welded. Welding is achieved by plastic flow of frictionally heated material from ahead of the pin to behind it. In friction stir spot welding (FSSW), the rotating tool is also plunged into the workpiece but without being traversed along any direction. Welding is achieved by plastic flow of frictionally heated material around the pin. A keyhole is left in the workpiece after the tool is withdrawn at the end of FSW or FSSW.

Liquation cracking has been observed in FSSW of Mg and Al alloys [29–34]. Figure 11a shows the vertical cross-section of a spot weld made on AZ91B Mg by FSSW. The tool for welding was a H13-steel tool, with a shoulder of 15 mm diameter and a threaded pin of 5.5 mm diameter and 5.1 mm length beyond the bottom of the shoulder. The tool rotation speed was 1,000 rpm. The plunge rate was 0.15 mm/s and the dwell time 4 s. As shown in the optical micrograph in Fig. 11b, a crack is visible in the stir zone. The evidence of liquation is shown by the SEM image in Fig. 11c. The γ phase in the workpiece before welding is visible in the lower right corner of the image, which is essentially in the base metal. The closer to the keyhole, that is, the tool, the more the γ phase reacts with the Mg-rich matrix α to form liquid and solidifies as the α/γ eutectic. The stir zone cracked along liquid films along the α-grain boundaries under severe shearing of the rotating tool.

Most Mg alloys (e.g., AZ91 Mg, AM60 Mg and AZ31 Mg) have Al as a major alloying element and they are often used in the as-cast condition. Thus, there can be plenty of γ (a $Mg_{17}Al_{12}$ compound containing other alloying elements, e.g., Zn) to react with the Mg-rich phase α (Mg) and cause liquation by the eutectic reaction $\alpha + \gamma \rightarrow$ L upon heating to a eutectic temperature T_E (e.g., $\leq$ 437°C). This temperature is relatively low and easily achieved during FSW. The susceptibility of these Mg alloys to liquation in FSSW is promoted by: (i) a liquation-causing constituent γ and a low T_E, and (ii) the heat input dwelling at one spot (instead of spreading along the joint path as in FSW). Although Mg is very similar to Al in being light, soft and weldable by FSSW with an alloy-steel tool, a welding schedule good for a

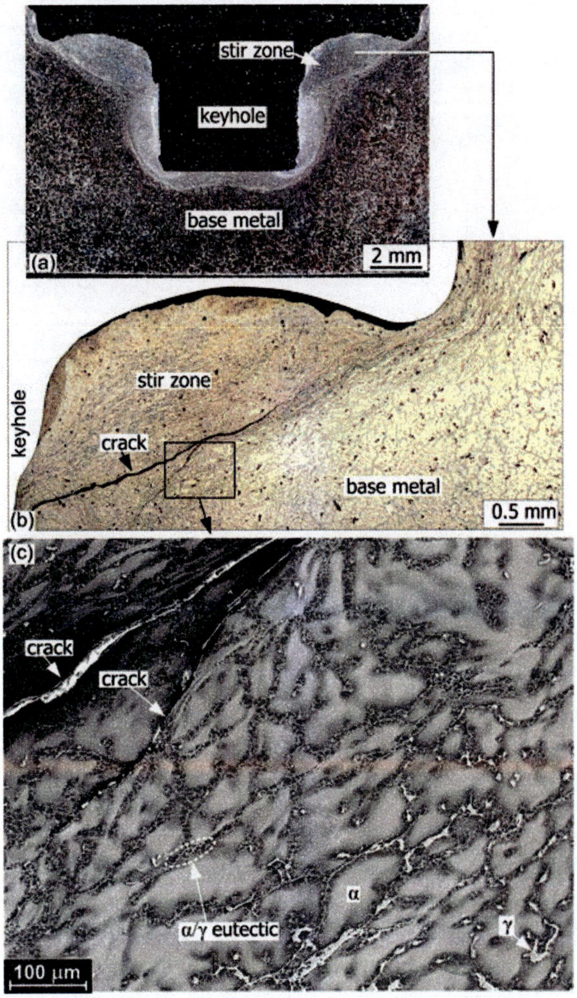

Fig. 11 Liquation cracking in friction stir spot welding of AZ91 Mg: (**a**) vertical weld cross-section; (**b**) optical micrograph showing cracking; (**c**) SEM image showing liquated microstructure and cracking [33]

very weldable Al alloy such as 6,061 Al may, in fact, cause liquation in Mg alloys. It is essential to keep the heat input as low as possible, e.g., by using a small tool at a low rotation speed.

Liquation cracking can easily occur in FSW of Al alloys to Mg alloys [35–37]. Figure 12a shows the transverse optical micrograph of a lap weld made between 6,061 Al at the top and AZ31 Mg at the bottom. The travel speed was 38 mm/min (1.5 ipm) and the rotation speed 1,400 rpm. The shoulder of the tool was 10 mm in diameter and the pin 4 mm in diameter and 1.5 mm in length. A crack is visible on the retreating side of the pin tip, where the pin rotated in the direction opposite to the travel direction. The particle inside the crack is shown by the SEM image in Fig. 12b, c. As shown in Fig. 12c, the particle exhibits a laminar α/γ eutectic

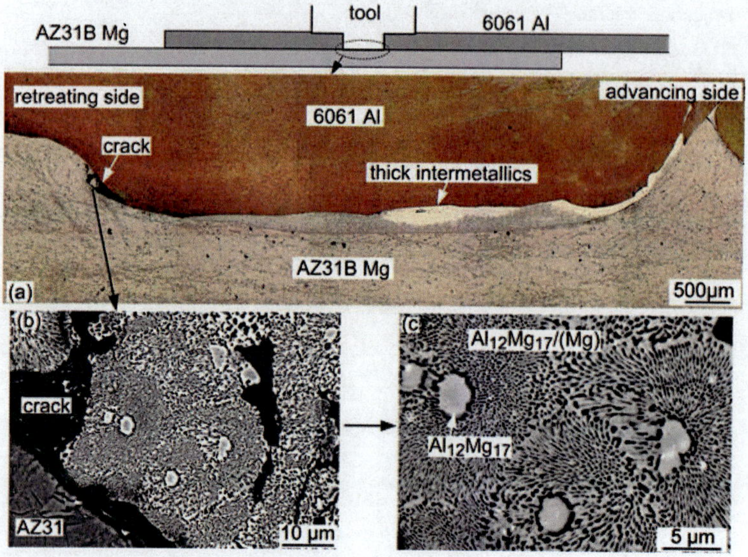

Fig. 12 Transverse cross-section of an Al-to-Mg lap weld: (**a**) optical micrograph; (**b**) (**c**) SEM images showing eutectic solidification structure of the γ-phase $Al_{12}Mg_{17}$ and the Mg-rich phase (Mg) [37]

structure (and some γ particles). This solidification structure is clear evidence of liquation during FSW. It has been demonstrated that liquation and cracking can be reduced by putting AZ31 Mg on the top [35].

Figure 13a shows a lap weld made along the edge of the upper sheet (6,061 Al) with a pin of 2.3 mm length at 38 mm/min (1.5 ipm) and 2,200 rpm. Solidified droplets of the liquid squeezed out of the stir zone (through the gap between the two sheets) during welding are present along the edge. These droplets are direct evidence of liquation during welding, even though FSW is considered as a solid-state welding process.

Microstructural examination showed that the droplets contained Al_3Mg_2 or $Al_{12}Mg_{17}$ or both. This indicates that the composition varied from droplet to droplet even along the same weld. Figure 13b shows the microstructure of a droplet that contained $Al_{12}Mg_{17}$ and no Al_3Mg_2. The compositions measured by EPMA at the marked locations are listed in Table 1. At points 4 through 6 the local mean compositions were measured by enlarging the beam diameter to 10 μm. The dendrites at points 1 through 3 contain about 58 wt% Mg. Based on the binary phase diagram in Fig. 14 as an approximation, $Al_{12}Mg_{17}$ (γ) contains about 58 wt% Mg. The composite-like structure at points 4 through 6 between the dendrites, on the other hand, contains about 67 wt% Mg. Based on the phase diagram, the γ/(Mg) eutectic contains about 67 wt% Mg. Thus, the dendrites are primary $Al_{12}Mg_{17}$ (γ) and the interdendritic regions are the γ/(Mg) eutectic. The absence of (Mg) dendrites suggests that the overall composition of the droplet is between points "d" and "e" in Fig. 14.

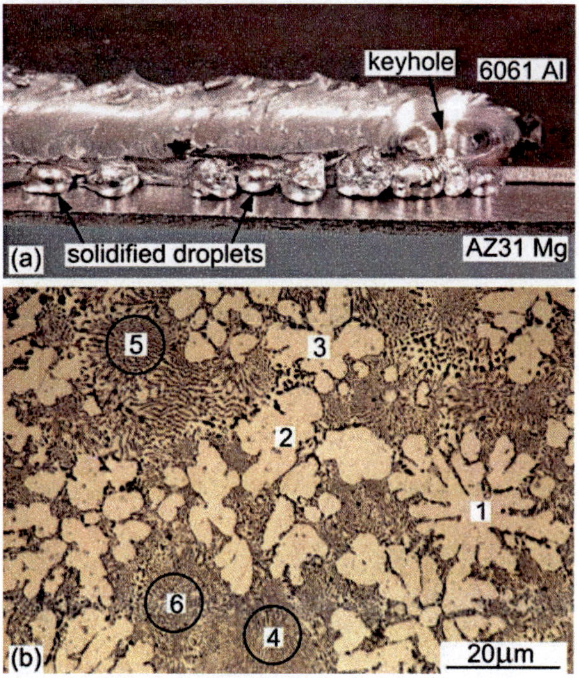

Fig. 13 Direct evidence of liquation in Al-to-Mg lap welding: (**a**) solidified droplets of liquid squeezed out of the stir zone during welding; (**b**) solidification microstructure inside a droplet [37]

Table 1 Compositions (in wt%) by EPMA at locations shown in Fig. 13b [37]

Location	Mg	Al	Si	Cu	Zn	Mn	Phase
1	57.94	38.41	0.06	0.03	0.51	0.11	γ
2	58.01	38.28	0.05	0.05	0.51	0.08	γ
3	58.00	38.1	0.04	0.02	0.45	0.09	γ
4	66.91	29.82	0.08	0.17	1	0.06	γ/(Mg)
5	67.36	29.11	0.08	0.13	0.92	0.07	γ/(Mg)
6	67.41	29.41	0.09	0.01	0.65	0.13	γ/(Mg)

Solidification Cracking

Solidification Cracking in Arc Welding

Solidification cracking in Al welds has been studied extensively, including several papers in the previous two conferences on hot cracking in welds [39, 40]. Solidification cracking in Mg welds, on the other hand, has not been studied much so far. Thus, only Mg welds will be covered in the following discussion as solidification cracking in Al welds will be covered by several papers in the present conference.

Figure 15a shows solidification cracking in gas-metal arc welding of AZ91D Mg alloy with filler AZ61 Mg [25]. Figure 15b shows $Al_{12}Mg_{17}$ (γ) along the

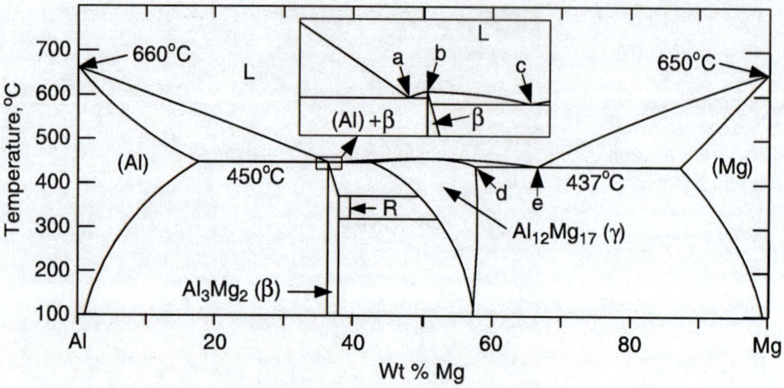

Fig. 14 Al-Mg phase diagram [37, 38]

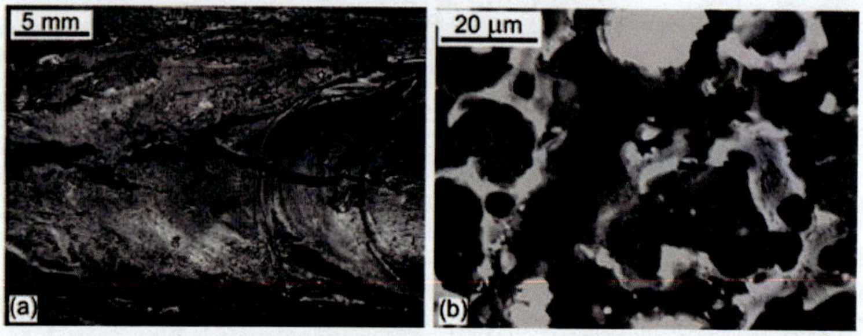

Fig. 15 Solidification cracking in gas-metal arc welding of AZ91D Mg alloy with filler AZ61 Mg: (**a**) macrograph; (**b**) optical micrograph showing intergranular cracking. From Sun et al. [25]

crack. This suggests that the γ/(Mg) eutectic formed when the grain boundary liquid solidified.

Recently, new creep-resistant Mg alloys have been developed. The susceptibility of these alloys to solidification cracking in casting, called hot tearing, is an important subject because it affects their castability. Kou and coworkers [41, 42] have used constrained-rod casting in a steel mold to evaluate the hot tearing susceptibility of Mg alloys. The experimental data of the hot tearing susceptibility of Al alloys have been widely used to study solidification cracking in Al welds [1]. Thus, hot tearing data of Mg alloys can be useful for studying solidification cracking in Mg welds.

Figure 16a shows the design of a steel mold for constrained-rod casting [41, 42]. As shown, each rod expands suddenly at its two ends, that is, the sprue end and the ball end, which act as anchors to keep the rod from contracting during solidification. Consequently, tension develops in the rod during solidification and cracks can initiate at the stress concentrations at the sprue end and the ball end. The cast pin tear test for evaluating solidification cracking in Ni-base alloys [43] is based on the same principle except it is smaller in scale. Figure 16b shows an example of Mg rods prepared by constrained rod casting. Cracks at the sprue ends of the rods are

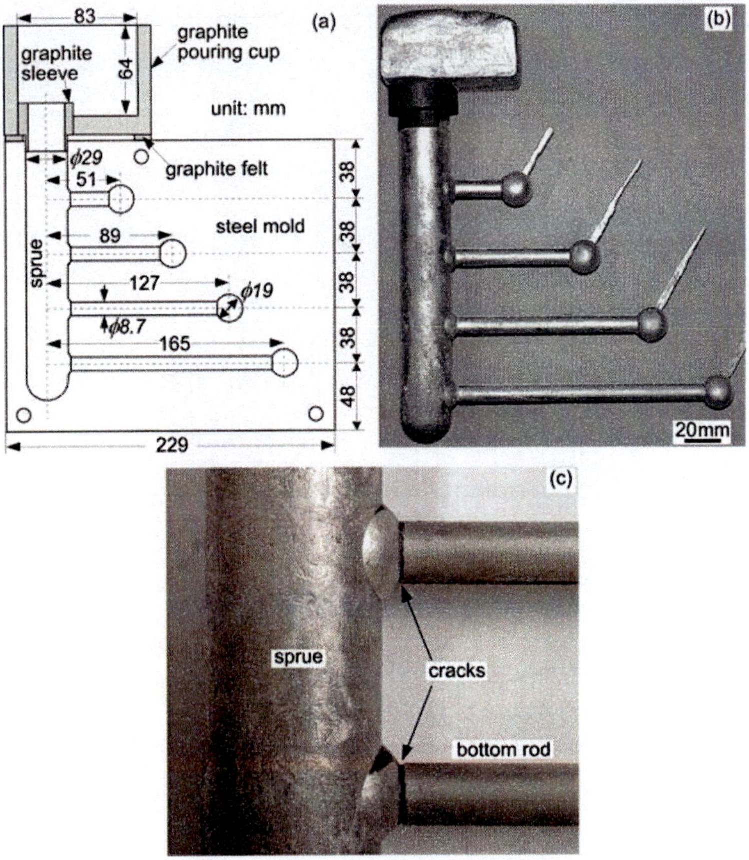

Fig. 16 Constrained rod casting for evaluating the crack susceptibility during solidification of Mg castings: (**a**) design of steel mold; (**b**) cast rods of a Mg alloy; (**c**) cracks near the sprue ends of rods [41, 42]

visible in Fig. 16c. The hot tearing susceptibility is evaluated based on the widths and locations of the cracks [41, 42].

Figure 17 shows the hot tearing susceptibility of new creep-resistant Mg-Al-Ca and Mg-Al-Sr alloys determined by constrained rod casting [41, 42]. For comparison, a readily castable commercial Mg casting alloy, that is, AZ91E Mg, is also included. For the Mg-xAl-yCa alloys, increasing the Ca content from 0.5 to 3.5 wt% in the Mg-4Al-yCa alloys reduces hot tearing. Increasing the Al content in the Mg-xAl-2.5Ca from 4 to 6 wt%, however, does not affect hot tearing significantly. As for the Mg-xAl-ySr alloys, increasing the Sr content from 1.5 to 3 wt% reduces hot tearing in the Mg-4Al-ySr, Mg-6Al-ySr and Mg-8Al-ySr alloys. Increasing the Al content from 4 to 8 wt% in the Mg-xAl-1.5Sr alloys reduces hot tearing significantly but less significantly in the Mg-xAl-3Sr alloys. These results are consistent with the hot tearing experience in casting the Mg-Al-Ca and Mg-Al-Sr alloys.

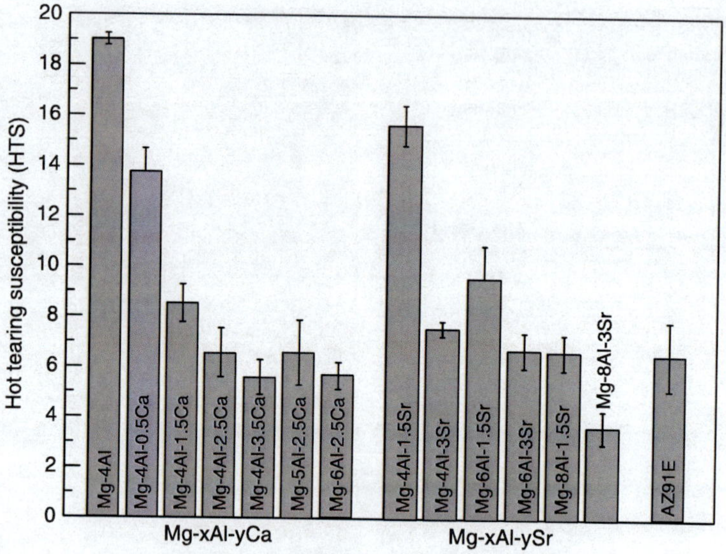

Fig. 17 Hot tearing susceptibility of new creep-resistant Mg-Al-Ca and Mg-Al-Sr alloys as compared to the readily castable alloy AZ91E Mg [41, 42]

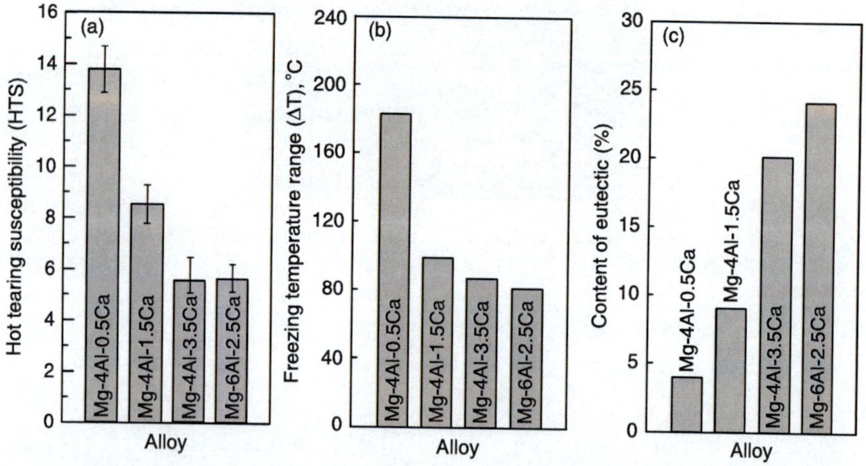

Fig. 18 Correlations between hot tearing susceptibility and freezing temperature range and amount of eutectic: (**a**) hot tearing susceptibility, (**b**) freezing temperature range, and (**c**) amount of eutectic [41]

Figure 18 shows the freezing temperature range based on Suzuki's liquidus projection [44] and the eutectic content in alloys Mg-4Al-0.5Ca, Mg-4Al-1.5Ca, Mg-4Al-3.5Ca and Mg-6Al-2.5Ca, along with their hot tearing susceptibility taken from Fig. 17. The freezing range first decreased sharply from Mg-4Al-0.5Ca to Mg-4Al-1.5Ca and then more mildly to Mg-4Al-3.5Ca and Mg-6Al-2.5Ca. In general,

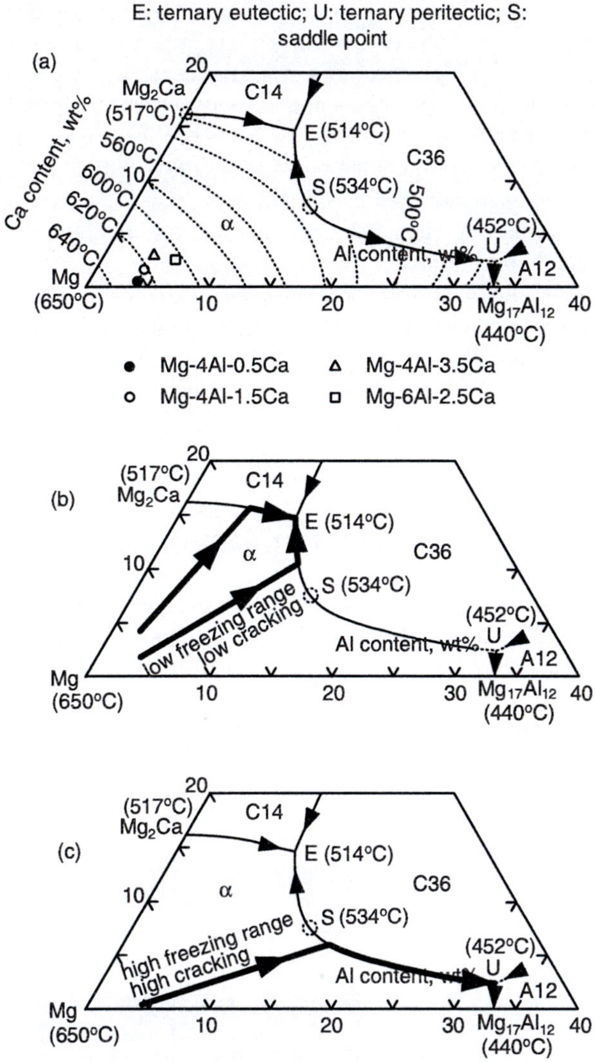

Fig. 19 Ternary Mg-Al-Ca: (**a**) liquidus projection determined by Suzuki et. a. [44]; (**b**) solidification paths with a narrower freezing range ending at a higher-temperature ternary eutectic (E at 514°C); (**c**) path with a wider freezing range ending at lower-temperature ternary peritectic point (U at 452°C) or binary eutectic (B at 440°C), that is, with A12 formation [41]

the wider the freezing range, the longer the time a semisolid is vulnerable to hot tearing. Furthermore, the more eutectic liquid is present, the more likely the solidifying alloy can accommodate the rising tension and cracks can be healed. Thus, the severe hot tearing susceptibility of alloy Mg-4Al-0.5Ca could have resulted from the combination of a very wide freezing range and a very low eutectic content.

Figure 19 shows the effect of the solidification path on the freezing temperature range and the hot tearing susceptibility for Mg-Al-Ca ternary alloys in the Mg-rich corner. The liquidus projection according to Suzuki et al. [44] is shown in Fig. 19a. Alloys with a solidification path ending at the ternary eutectic point (E) tend to have a narrower freezing range because of the ternary eutectic temperature is relatively high, that is, 514°C (Fig. 19b). On the other hand, alloys with a solidification path reaching the ternary peritectic point (U) and ending at the binary eutectic point ($Mg_{17}Al_{12}$) tend to have a wider freezing range because the binary eutectic point ($Mg_{17}Al_{12}$) is much lower, that is, 440°C (Fig. 19c).

For the Mg-xAl-1.5Sr alloys, increasing the Al content from 4 to 8 wt% decreased the freezing temperature range and increased the fraction of eutectic [42]. For the Mg-xAl-3Sr alloys, a similar trend was observed. Thus, the hot tearing susceptibility of the Mg-Al-Sr alloys shown in Fig. 17 can be explained in a way similar to that of the Mg-Al-Ca alloys.

Solidification Cracking in Resistance Spot Welding

Solidification cracking has also been observed in resistance spot welding of Al and Mg alloys. Figure 20a shows a large crack in a resistance spot weld of AZ31 Mg alloy [45]. The crack is intergranular as shown at a higher magnification in Fig. 20b. It was reported that the weld nugget had a relatively high hot cracking susceptibility. The cracks appeared in the nuggets when the welding current was higher than 15 kA,

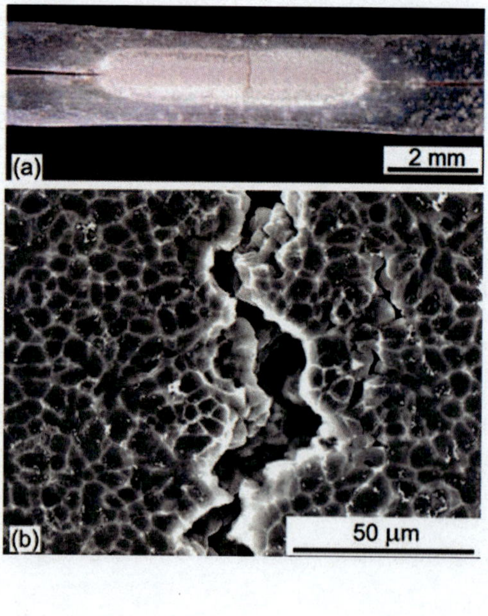

Fig. 20 Solidification cracking in resistance spot welding of AZ31 Mg alloy: (**a**) macrograph; (**b**) micrograph. From Sun et al. [45]

for instance, 23 kA (Fig. 20a). Increasing the welding current increased the cracking tendency because of the increased tensile stresses developed during cooling [45].

Conclusions

1. Liquation cracking can occur in fusion and solid-state welding of Al alloys, Mg alloys, and Al alloys to Mg alloys, including arc welding, resistance-spot welding and friction stir welding.
2. The T-f_S curves of the weld metal and the PMZ (same as the base metal) can be used to predict the liquation cracking susceptibility of full-penetration arc welds of Al alloys and, most likely, Mg alloys as well. They can be used to select or develop filler metals to eliminate liquation cracking.
3. If the workpiece bottom surface is within the PMZ in bead-on-plate, partial-penetration welding of Al and Mg alloys, thermal contraction and solidification shrinkage can induce high tension in the PMZ to cause liquation cracking regardless of the filler metal used.
4. Hot tearing of new Mg alloys during solidification in casting has been investigated recently, and the data can be useful for studying solidification cracking in welds of these Mg alloys.

Acknowledgements The research was supported by the National Science Foundation under Grants DMR-9803589, DMR-0098776, DMR-0309468 and DMR-0605662 and by the Wisconsin Alumni Research Foundation of the University of Wisconsin-Madison. The authors would like to thank Professor Y. Austin Chang of the University of Wisconsin at Madison, WI for using the software package Pandat, and Professor Rainer Schmid-Fetzer of Clausthal University of Technology at Clausthal-Zellerfeld, Germany for using the database PanMg. They would also like to thank former graduate students C. Huang, C. Cao and Y. K. Yang for their research work on hot cracking.

References

1. Kou S (2003) Welding Metallurgy, 2nd edition. John Wiley, New York.
2. Kulekci MK (2008) Magnesium and Its Alloys Applications in Automotive Industry, International Journal of Advanced Manufacturing Technology 39:851–865.
3. Huang C and Kou S (2002) Liquation Mechanisms in Multicomponent Aluminum Alloys during Welding. Welding Journal 81:211s–222s.
4. Huang C and Kou S (2001) Partially Melted Zone in Aluminum Welds – Planar and Celluar Solidification. Welding Journal 80:46s–53s.
5. Huang C and Kou S (2001) Partially Melted Zone in Aluminum Welds – Solute Segregation and Mechanical Behavior. Welding Journal 80:9s–17s.
6. Huang C, Kou S and Purins JR (2000) Liquation, Solidification, Segregation and Hot Cracking Adjacent to the Fusion Boundary of Al-4.5wt%Cu Welds, Proceedings of Merton C. Flemings Symposium, June 28–30, 2000, MIT, Cambridge, MA, TMS, Warrendale, PA, 2001, pp. 229–234.
7. Huang C and Kou S (2000) Partially Melted Zone in Aluminum Welds – Liquation Mechanism and Directional Solidification. Welding Journal 79:113s–120s.

8. Huang C, Cao G and Kou S (2007) Liquation Cracking in Aluminum Welds. Materials Science Forum. 539–543:4036–4041.
9. Cao G and Kou S (2006) Predicting and Reducing Liquation-Cracking Susceptibility Based on Temperature vs. Fraction Solid. Welding Journal 85:9s–18s.
10. Cao G and Kou S (2005) Liquation Cracking in Full-Penetration Al-Si Welds. Welding Journal 84:63s–71s.
11. Huang C and Kou S (2004) Liquation Cracking in Full-Penetration Al-Cu Welds. Welding Journal 83:50s–58s.
12. Huang C and Kou S (2004) Liquation Cracking in Full-Penetration Al-Mg-Si Welds. Welding Journal 83:111s–122s.
13. Huang C, Cao G and Kou S (2004) Liquation Cracking in Partial-Penetration Aluminum Welds: Assessing Tendencies to Liquate, Crack and Backfill. Science and Technology of Welding and Joining 9:1–9.
14. Huang C and Kou S (2003) Liquation Cracking in Partial-Penetration Aluminum Welds: Effect of Penetration Oscillation and Backfilling. Welding Journal 82:184s–194s.
15. Rao KS, Reddy GM and Rao KP (2005) Studies on partially melted zone in aluminum-copper alloy welds – effect of techniques and prior thermal temper. Materials Science and Engineering A 403:69–76.
16. Rao KS and Rao KP (2005). Partially melted zone in Al-Mg-Si alloy gas tungsten arc welds: effect of techniques and prior thermal temper. Materials Science and Technology 21(9): 955–1002.
17. Rao KP, Ramanaiah N and Viswanathan N (2008) Partially melted zone cracking in AA6061 welds. Materials and Design 29:179–186.
18. Kou S and Le Y (1986) Nucleation Mechanisms and Grain Refining of Weld Metal – A Fundamental Study. Welding Journal 65:305s.
19. Pepe JJ and Savage WF (1967) Effects of Constitutional Liquation in 18-Ni Maraging Steel Weldment. Welding Journal 46(9):411s–422s.
20. Pepe JJ and Savage WF (1970) Weld Heat-affected Zone of The 18Ni Maraging Steels. Welding Journal 49(12):545s–553s.
21. Kou S (1996) Transport Phenomena and Materials Processing. Wiley, New York, NY, p. 119.
22. Flemings MC (1974) Solidification Processing. McGraw-Hill, New York, NY, p. 256.
23. Pandat – Phase Diagram Calculation software package for Multicomponent Systems, CompuTherm LLC, Madison, WI 53719, 2001.
24. Zhou W, Long TZ and Mark CK (2007) Hot cracking in tungsten inert gas welding of magnesium alloy AZ91D. Materials Science and Technology 23(11):1294–1299.
25. Sun DX, Sun DQ, Gui XY and Xuan ZZ (2009) Hot Cracking of Metal Inert Gas Arc Welded Magnesium Alloy AZ91D. ISIJ International 49(2):270–274.
26. Lang B, Sun DQ, Xuan ZZ and Qin XF (2008) Hot Cracking of Resistance Spot Welded Magnesium Alloy. ISIJ International 48(1):77–82.
27. Senkara J and Zhang H (2000) Cracking in spot welding aluminum alloy AA5754. Welding Journal 79:194s–201s.
28. Thomas WM et al. Friction Stir Butt Welding, International Patent Application No. PCT/GB92, Patent Application No. 9125978.8, December 6, 1991.
29. Gerlich A, Su P and North TH (2005) Peak temperatures and microstructures in aluminum and magnesium alloy friction stir spot welds. Science and Technology of Welding and Joining 10(6):647–652.
30. Yamamoto M, Gerlich A, North TH and Shinozaki K (2007) Mechanism of cracking in AZ91 friction stir spot welds. Science and Technology of Welding and Joining 12(3):208–216.
31. Yamamoto M, Gerlich A, North TH and Shinozaki K (2007) Cracking in the stir zones of Mg-alloy Friction Stir Spot Welds. Journal of Materials Science 42:7657–7666.
32. Gerlich A, Yamamoto M and North TH (2007) Local melting and cracking in Al 7075 and Al 2024-T3 friction stir spot welds. Science and Technology of Welding and Joining 12(9): 472–480.

33. Yang YK, Dong H, Cao H, Chang YA and Kou S (2008) Liquation of Mg Alloys in Friction-Stir Spot Welding. Welding Journal 87:167s–177s.
34. Yang YK, Dong H and Kou S (2008, August) Liquation Tendency and Liquid-Film Formation in Friction-Stir Spot Welding. Welding Journal 87:202s–211s.
35. Firouzdor V and Kou S (2009) Al-to-Mg Friction Stir Welding: Effect of Positions of Al and Mg with Respect to the Welding Tool. Welding Journal 88:213s–224s.
36. Firouzdor V and Kou S (in press) Al-to-Mg Friction Stir Butt Welding: Effect of Material Position, Travel Speed and Rotation Speed. Metallurgical and Materials Transactions A.
37. Firouzdor V and Kou S (in press) Formation of Liquid and Intermetallics in Al-to-Mg Friction Stir Welding. Metallurgical and Materials Transactions A.
38. American Society for Metals (1986) Binary Alloy Phase Diagrams vol. 1: p. 106, Metals Park, O.H., American Society for Metals.
39. Bollinghaus T and Herold H (2005) Hot Cracking Phenomena in Welds. Springer, Berlin.
40. Bollinghaus T, Herold H, Cross CE and Lippold JC (2008) Hot Cracking Phenomena in Welds II. Springer, Berlin.
41. Cao G and Kou S (2006) Hot Tearing in Ternary Mg-Al-Ca Alloy Castings. Metallurgical and Materials Transactions A 37A:3647–3663.
42. Cao G, Zhang C, Cao H, Chang YA and Kou S (2010) Hot Tearing Susceptibility of Ternary Mg-Al-Sr Alloy Castings. Metallurgical and Materials Transactions A 41A:706–716.
43. Alexandrov BT and Lippold JC (2010) The cast pin tear test for evaluating solidification cracking in Ni-base alloys, 3rd International Hot Cracking Workshop, Columbus, OH, March 15–16, 2010.
44. Suzuki A, Saddock ND, Jones JW and Pollock TM (2005) Solidification Paths and Intermetallic Phases in Mg-Al-Ca Ternary Alloys. Acta Materialia 53(9):2823–2834.
45. Sun DQ, Lang B, Sun DX and Li JB (2007) Microstructures and Mechanical Properties of Resistance Spot Welded Magnesium Alloy Joints. Materials Science and Engineering A 460–461:494–498.

Critical Conditions for Weld Solidification Crack Growth

Carl E. Cross, N. Coniglio, P. Schempp, and M. Mousavi

Introduction

The occurrence of solidification cracking during welding remains a little understood phenomenon, in spite of extensive studies and tests performed to evaluate and compare the relative weldability of many different alloys. From an approach often adopted in the welding community attributed to Prokhorov [1], solidification cracks are believed to form when a critical tensile strain is exceeded, specific to the alloy, assuming that the mushy-zone has limited ductility. Tensile stresses and strains normally form behind a moving weld pool as a result of solidification shrinkage and thermal contraction, as influenced by welding parameters and the degree of restraint [2–4].

The apparent loss of mushy zone ductility that results in solidification cracking is normally observed to occur over a specific temperature range that is bounded in the extreme by liquidus and solidus temperatures. The term brittle temperature range (BTR) has often been applied to this interval, reflecting upon the dramatic loss in apparent ductility. The upper bound to the BTR does not normally extend all the way to the liquidus. Nevertheless, alloys with a large solidification range will usually, but not always, display higher susceptibility to cracking. It has been argued that alloys with a large BTR will be exposed to a greater accumulated strain during cooling and thus exceed the ductility limit [5]. It has also been argued that rapid rates of straining facilitate exceeding the ductility limit [6].

The problem with applying the Prokhorov limited ductility approach to solidification cracking is that it cannot be physically linked to any cracking mechanism involving liquid rupture. Liquid rupture mechanisms have instead centered on cavitation and the conditions necessary to achieve a liquid pressure drop. Of particular interest in this regard has been the Rappaz-Drezet-Gremaud (RDG) interdendritic pressure drop model [7], where pressure drop has been related to solidification

C.E. Cross (✉)
Federal Institute for Materials Research and Testing, Berlin, Germany
e-mail: cecross@bajabb.com

J. Lippold et al. (eds.), *Hot Cracking Phenomena in Welds III*,
DOI 10.1007/978-3-642-16864-2_2, © Springer-Verlag Berlin Heidelberg 2011

shrinkage and the rate of transverse thermal strain. From RDG analysis it is found that strain rate, and not strain, is the key parameter controlling rupture.

Recent application of RDG analysis to aluminum welding has suggested that conditions for cavitation are not easily or normally achieved [8]. Alternative mechanisms for crack nucleation may involve gas pore formation, which is also initiated due to interdendritic pressure drop. In particular, gas pore formation in Al alloys requires the presence of dissolved hydrogen at levels high enough to overcome nucleation barriers. However, pre-existing pore nuclei in the form of oxide films may circumvent the need for pore nucleation altogether. None of the mechanisms for crack nucleation are very well understood at this point.

Following nucleation of a solidification crack, conditions must likewise favor crack growth if a continuous crack is to persist. An interesting outcome of recent analyses [8, 9] is that steady-state crack growth also requires a minimum transverse strain rate, as determined from a simple mass balance. Exploring the details of this mechanism, developing a model, and applying it to commercial Al alloys is the subject of this paper.

The fact that solidification cracking often manifests itself in the form of continuous centerline cracks tells us that the requirements for both crack nucleation and growth have been satisfied. But it does not tell us which event (i.e. nucleation or growth) is limiting the occurrence of cracking. For the sake of argument, it will be assumed in this analysis that crack nucleation is an easily occurring (i.e. non-limiting) event and that crack growth is the controlling factor. Determining whether or not this is assumption is valid must be the subject of future work, e.g. comparing critical strain rates needed for crack nucleation versus crack growth.

Coherency

Important to solidification cracking and models given below is the concept of coherency. Coherency is where adjacent dendrites have interconnected, either mechanically or through atomic bonding between dendrite arms. This allows for the transmission of transverse stress, needed for the initiation and growth of a crack. The coherency temperature (i.e. where coherency begins) defines the location in the mushy zone where cracks may exist. Above the coherency temperature, the liquid experiences only hydrostatic stress which may lead to porosity, but not cracks. This temperature is a unique property of the alloy and solidification conditions, and must be determined experimentally [10]. Likewise, the solid fraction interval over which coherency exists will determine the amount of solidification shrinkage experienced.

Crack Growth Model

Formulation of the model presented here is based upon the steady-state nature of weld solidification and the continuous nature of centerline crack growth. If it is assumed that a solidification crack tip grows in the mushy zone at the same speed

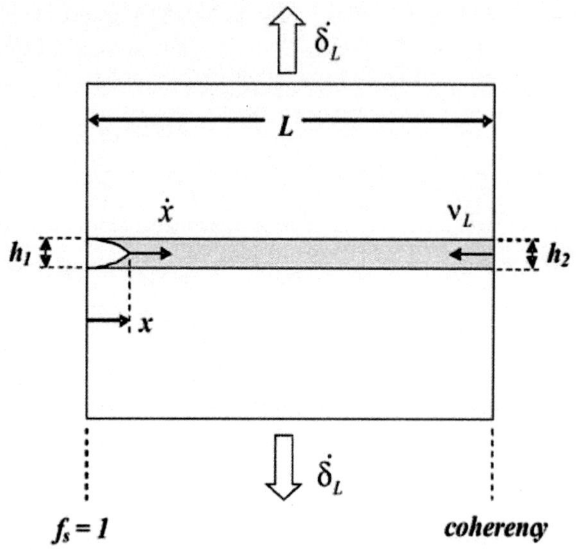

Fig. 1 Schematic showing liquid film (*shaded*) located at grain boundary in coherent region of weld mushy zone

that the weld pool advances (i.e. weld torch speed), then it follows that steady-state conditions of mass flow must apply. As demonstrated in Fig. 1, the transverse displacement of a mushy zone grain boundary (δ_L) must be compensated by either advancement of the crack tip (x) or back-feeding of liquid (v_L) over an interval of time. Accordingly, the following mass balance is proposed:

$$\dot{\delta}_L(L - x) = \dot{x}h_1 + v_L h_2 \tag{1}$$

$\dot{\delta}_L$ is transverse displacement rate at the grain boundary (i.e. transverse to the direction of welding), L is length of the coherent zone (distance between coherency and solidus temperatures), x is location of the crack tip (relative to the solidus temperature), $\dot{x}$ is crack growth rate, h is liquid film thickness, and v_L is liquid flow rate.

When considering boundary conditions for the model represented by Eq. 1, the coherent zone length is found from the difference between coherent and solidus temperatures:

$$L = (T_c - T_s)/G \tag{2}$$

where G is temperature gradient. Both of these temperatures (T_c and T_s) can be measured experimentally (e.g. from thermal analysis of castings), and the weld temperature gradient can be measured from an implanted thermocouple. The position of the crack tip (x), for purposes of calculating a critical displacement rate, will be assumed small (i.e. at a fixed solid fraction, $f_s = 0.98$). In other words, we need the (L-x) term to be as large as possible to find the smallest (i.e. critical) $\dot{\delta}_L$. Crack tip velocity is set equal to the weld travel speed [11], which is a pre-set welding parameter. Liquid film thickness is found from liquid fraction and primary dendrite spacing ($f_L \cdot \lambda$). Liquid flow rate is determined using Darcy's Equation:

$$v_L = \frac{-K}{\mu \cdot f_L} \frac{d(\Delta P)}{dx} \tag{3}$$

where μ is viscosity and K is permeability expressed as a function of dendrite spacing [12]:

$$K = \frac{\lambda^2 (1 - fs)^2}{8\pi} \tag{4}$$

Permeability is a measure of the openness of the dendritic structure, reflecting upon the ability of liquid to flow and feed shrinkage. ΔP represents the liquid pressure drop, with contributions from both thermal strain (ΔP_ε) and solidification shrinkage (ΔP_{sh}) calculated using the RDG model [7]:

$$\Delta P = \Delta P_\varepsilon + \Delta P_{sh} = (1+\beta)\mu \int_{fs=0.98}^{fs^{coh}} \frac{E(x)}{K} dx + v\beta\mu \int_{fs=0.98}^{fs^{coh}} \frac{(1-fs)}{K} dx \tag{5}$$

where β represents solidification shrinkage and v is the isotherm velocity. The integral is evaluated over the specific interval of interest (i.e. *L-x*). Required here is knowledge of the relation between temperature and distance (provided by *G*) and also temperature and solid fraction (obtained from experimental measurement or thermodynamic software). The function *E(x)* contains a transverse strain rate term (transverse to the plane of the liquid film), which is expressed in terms of a displacement rate normalized to the dendrite spacing (gage length):

$$E(x) = \int fs \dot{\varepsilon}_L dx = \int fs \left(\frac{\dot{\delta}_L}{\lambda} \right) dx \tag{6}$$

Thus, solutions for $\dot{\delta}_L$ can be found through an iterative process, assuming an initial value to estimate *E(x)* and then using this to calculate ΔP and v_L (Eqs. 5 and 3), leading to a refined value for $\dot{\delta}_L$ (Eq. 1). Repeated iterations lead to convergence and a unique value for $\dot{\delta}_L$.

Strain Rate Distribution Model

Of considerable practical importance is the ability to relate grain boundary liquid displacement rate $\dot{\delta}_L$ (from model above) to the transverse displacement rate across the mushy-zone $\dot{\delta}$. This requires an understanding of how strain and strain rate are distributed between the grains in the mushy-zone. Establishing this relationship is important because $\dot{\delta}$ can be easily and accurately measured with extensometers [13], whereas $\dot{\delta}_L$ cannot be easily determined (e.g. see, however, MISO technique [14]). Thus, this distribution model becomes the link between experiment and theory. Likewise, displacement in the HAZ near the fusion line can be readily simulated with FE analysis through continuum mechanics, whereas displacement in the mushy-zone cannot.

A particularly important aspect of this problem is to understand how mushy-zone grain size and shape affects strain rate distribution at liquid grain boundaries. The susceptibility to solidification cracking is known to be significantly reduced when large columnar grains are refined (i.e. smaller, equiaxed grains) [15]. It has been proposed for castings that this reduction in cracking susceptibility is the result of a reduced strain experienced at each grain boundary [16].

For purposes of approximation in this work, a very simple 2-D strain rate distribution model is proposed as represented in Fig. 2. Here grains are assumed to be free from curvature, arranged in a parallel array, and aligned in the direction of welding. It is assumed that the displacement rate across the mushy-zone $\dot{\delta}$ is distributed equally among N grains, partitioned between the grain boundary liquid $\dot{\delta}_L$ and the grain itself $\dot{\delta}_G$:

$$\frac{\dot{\delta}}{N} = \dot{\delta}_L + \dot{\delta}_G \tag{7}$$

The term $\dot{\delta}_G$ is comprised primarily of solidification shrinkage (~6 vol.% for Al alloys), large in comparison to any liquid pressure driven displacement as demonstrated in Ref. [8]. While it is clear that the grains themselves also contain liquid films, the grain boundaries are assumed to be coherency-free and hence will experience a higher portion of liquid displacement [17]. This also explains why solidification cracking is typically observed along solidification grain boundaries.

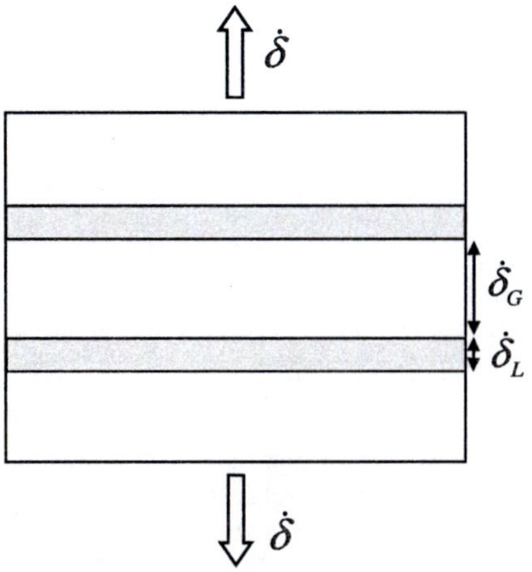

Fig. 2 Schematic showing distribution of transverse deformation rate between grains and grain boundaries in weld mushy zone

Experimental

Material

The aluminum alloys selected for examination in this study included 2219, 2014, and 2024 (aluminum association alloy designations) with compositions compared in Table 1. While all three alloys are copper bearing, they represent a wide spectrum in resistance to solidification cracking. Alloy 2219 is essentially a commercial grade Al-Cu binary alloy, high in Cu content and low in Mg, developed specifically to have good weldability in welded assemblies (e.g. Space Shuttle External Tank). Alloys 2014 and 2024 are both considered to be Al-Cu-Mg alloys, in that they can attain high strength through formation of S'-$CuAl_2Mg$ precipitates. Both alloys are lower in Cu and higher in Mg than Alloy 2219 and both are generally known to have poor weldability [18], although Alloy 2014 can be welded successfully (e.g. Titan Missile casing) using a 4043 filler metal. It should also be noted that both Alloys 2014 and 2024 are much higher in Mn, Si, and Fe content than Alloy 2219 (Table 1).

Welds and Castings

Autogenous, bead-on-plate, partial penetration, gas-tungsten arc welds were made on 7 mm thick plate using the welding parameters given in Table 2. Coupon size was 23 mm wide × 140 mm long. Weld bead dimensions varied slightly with alloy and were approximately 8 mm wide × 4 mm deep. Weld metal cooling rate was measured by plunging a type-K thermocouple (0.5 mm dia.) into the moving weld pool.

Castings were made by melting 50 g samples of wrought plate, taken from the same plate used to make weld coupons. Samples were melted in a furnace held

Table 1 Measured aluminum alloy compositions for alloys examined in this study (in wt%)

Alloy	Cu	Mg	Mn	Si	Fe
2219	6.42	0.02	0.23	0.06	0.17
2014	5.13	0.46	0.74	0.90	0.24
2024	4.66	1.42	0.69	0.09	0.18

Table 2 Gas-tungsten arc welding parameters used to make autogenous, partial penetration, bead-on-plate welds for microstructure and thermal analysis

Arc polarity	Direct current, electrode negative (DCEN)
Arc current	160 A
Arc voltage	18 V
Torch travel speed	4.0 mm/s
Shielding gas	Ar-He (50/50)

at 750°C and then cast into a cylindrical graphite mold (19 mm dia. × 60 mm length) coated with boron-nitride. Metal was poured into the top of the cylinder, while at the bottom of the cylinder there was a heat sink made from stainless steel (80 mm dia. × 30 mm thick) used to promote directional solidification. Cooling curves and temperature difference measurements were obtained from type-K thermocouples (0.25 mm dia.) pre-placed along the cylinder centerline, and calibrated with 99.999% pure Al. One thermocouple was placed at the bottom of the cylinder, and another placed 3 mm directly above it.

Both welds and castings were cross-sectioned and prepared for metallographic examination. Specimens were mounted, ground, and polished to 1 micron and examined at 500× and 1000 × magnification in the un-etched state.

Results and Discussion

Solidification Analysis

Cast microstructures for the three different alloys are compared in Fig. 3. Based upon the observation of phases in Fig. 3 in comparison with previous analyses of cast structure [19, 20], the predominant interdendritic eutectic phase for Alloy 2219 is taken to be Al_2Cu. Whereas for Alloys 2014 and 2024 there are three distinct phases present: Al_2Cu (light grey), Mg_2Si (black), and $Al_{15}(CuFeMn)_3Si_2$ (dark grey). While these are the constituents that predominate, it is possible that there may

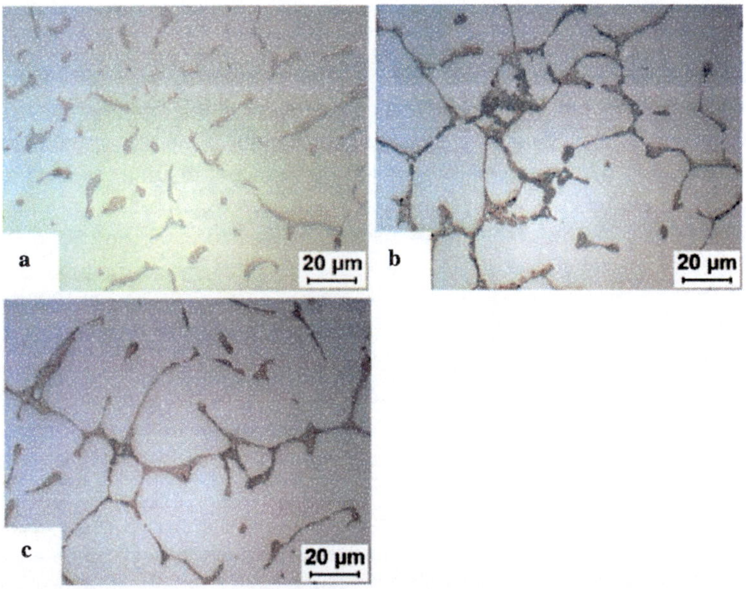

Fig. 3 Solidification microstructure for cast alloys (**a**) 2219, (**b**) 2014, and (**c**) 2024

Table 3 Comparison of solidification phase reactions for Alloys 2219 [19] and 2024 [20]

Alloy	Reaction	Temperature (°C)
2219		
	$L \rightarrow \alpha\,(Al)$	648
	$L \rightarrow \alpha\,(Al) + Al_2Cu$	547
2024		
	$L \rightarrow \alpha\,(Al)$	637
	$L \rightarrow \alpha\,(Al) + Al_{15}(CuFeMn)_3Si_2$	613
	$L \rightarrow \alpha\,(Al) + Al_{15}(CuFeMn)_3Si_2 + Al_{20}Cu_2Mn_3$	544
	$L + Al_{20}Cu_2Mn_3 \rightarrow \alpha\,(Al) + Al_2Cu + Al_{15}(CuFeMn)_3Si_2$	
	$L \rightarrow \alpha\,(Al) + Mg_2Si + Al_2Cu$	480
	$L \rightarrow \alpha\,(Al) + Al_2Cu + Mg_2Si + Al_2CuMg$	

be smaller amounts of other phases present (e.g. S-phase or $Al_{20}Cu_2Mn_3$-phase). It is important to note that it is the minor and impurity elements (Mn, Si, Fe) that have a major effect on phase formation in Alloys 2014 and 2024.

In comparison with the simple Al-Cu eutectic for Alloy 2219, the sequence of solidification phase reactions for Alloy 2024 is shown in Table 3, with corresponding reaction temperatures, taken from Bäckerud et al. [20]. While the liquidus temperatures are similar ($T_{L/2219} = 648°C$, $T_{L/2024} = 637°C$), the terminal solidus is significantly lower for Alloy 2024 ($T_{S/2219} = 547°C$, $T_{S/2024} = 480°C$) and hence the solidification range is larger for this alloy. This is further demonstrated by comparing the temperature-vs-solid fraction curves for these two alloys shown in Fig. 4, taken from published thermodynamic software predictions [21, 22]. Here the terminal eutectic plateau for Alloy 2219 occurs near 547°C in agreement with binary equilibrium, whereas for Alloy 2024 the plateau is some 20 degrees higher than Bäckerud's measurement. It should be noted that there is more terminal eutectic generated for Alloy 2219 (13% solid fraction) than for Alloy 2024 (9% solid fraction). This, in itself, would be expected to have significant effect in aiding the feeding of shrinkage in Alloy 2219.

Shown in Fig. 5 are the results from thermal analysis of cast Alloy 2024. These curves together with similar data for Alloys 2219 and 2014 were used in the identification of the liquidus, coherency, and temperatures as summarized in Table 4. The measured liquidus and solidus temperatures for Alloys 2219 and 2024 are in approximate agreement with Fig. 4. The solidus value for Alloy 2014 is found to lie between those for Alloys 2219 and 2024. Accordingly, the solidification range ($T_L - T_S$) is seen to increase in the order 2219, 2014, 2024.

Coherency temperatures were determined using a method outlined in Ref. [23] and demonstrated in Fig. 5. Coherency is assumed to begin at the end of the α-phase plateau as indicated by the dip in the temperature difference curve (at 5 s) due to a change in heat transfer. Comparing coherency start temperatures in Table 4 shows a similar value for all three alloys (~629°C). The range of temperature over which coherency occurs (i.e. $T_C - T_S$) is thus found to increase in the order 2,219, 2,014,

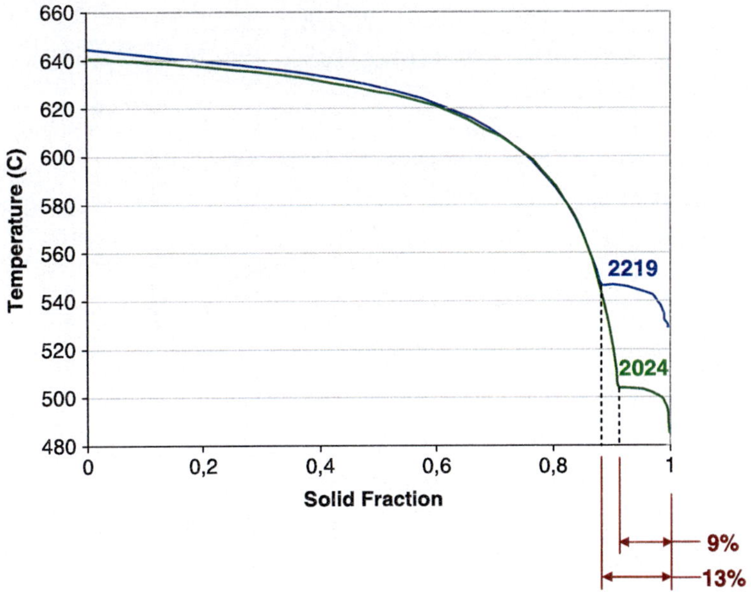

Fig. 4 Solidification curves for Alloys 2219 and 2024 based upon thermodynamic predictions, with data replotted from [21, 22]

2,024. However, when expressed in terms of solid fraction with the aid of Fig. 4, the coherency range for Alloy 2219 is actually found to be the same or larger than for Alloys 2014 and 2024 (see $1–f_s^c$ in Table 4).

Application of Proposed Models

The crack growth model, as represented by Eqs. 1–6, was applied using the physical constants shown in Table 5 together with the thermal and solid fraction data of Figs. 4 and 5. Figure 6 gives a comparison of the resulting critical grain boundary deformation rates $\dot{\delta}_L$ shown as a function of crack tip location x for the three different alloys. If for purposes of comparison the crack tip is assumed to be located at a solid fraction of 0.98, the critical deformation rates $\dot{\delta}_L$ are seen to be similar for all three alloys (between 0.18 and 0.22 μm/s) as compared in Table 6, with Alloy 2219 having the highest value.

When these $\dot{\delta}_L$ values at $f_s = 0.98$ are substituted into the deformation rate distribution model (Eq. 7), one is able to obtain the corresponding critical deformation rate $\dot{\delta}$ across the mushy zone assuming a variety of different grain sizes and corresponding N values. These values can in turn be converted to strain rate $\dot{\varepsilon}$ (i.e. dividing $\dot{\delta}$ by bead width) as represented by the curves in Fig. 7. Clearly these curves do not reflect the established behavior of these alloys, where Alloy 2219 should be

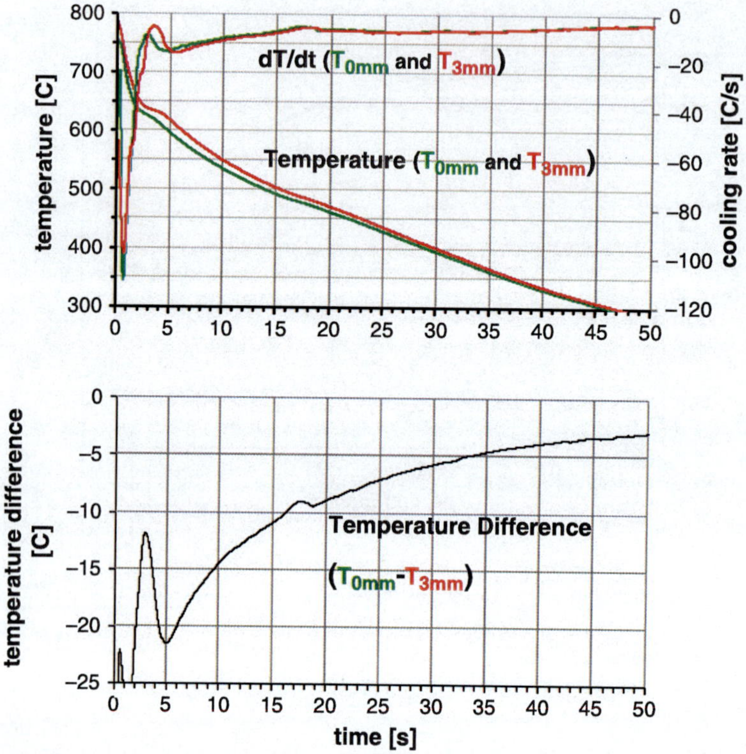

Fig. 5 Differential thermal analysis of Alloy 2219 cast specimen

Table 4 Summary of critical (liquidus, coherent, and solidus) temperatures measured from thermal analysis of castings (in °C). Also compared are solidification temperature range (ΔT_S), coherency temperature range (ΔT_C), and coherency solid fraction range ($1 - f_s^c$)

Alloy	T_L	T_C	T_S	ΔT_S	ΔT_C	$(1 - f_s^c)$
2219	641	629	543	98	86	0.50
2014	645	629	501	134	128	0.50
2024	637	622	487	150	135	0.40

Table 5 Physical constants pertaining to aluminum used in making calculations for crack growth model

Parameter	Symbol	Value	Reference
Shrinkage factor	ß	0.06	[24]
Liquid viscosity	μ	1×10^{-3} Pa·s	[24]
Dendrite arm spacing	λ	10 μ	–

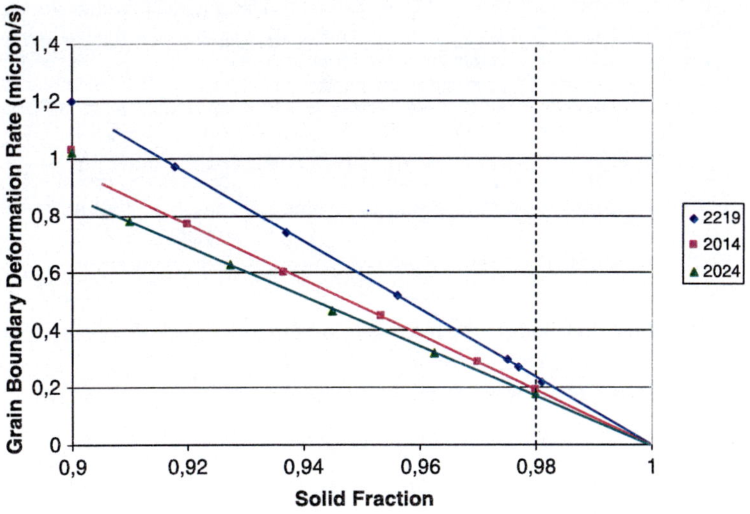

Fig. 6 Critical grain boundary deformation rates obtained from crack growth model, assuming 10 μm dendrite spacing for each alloy

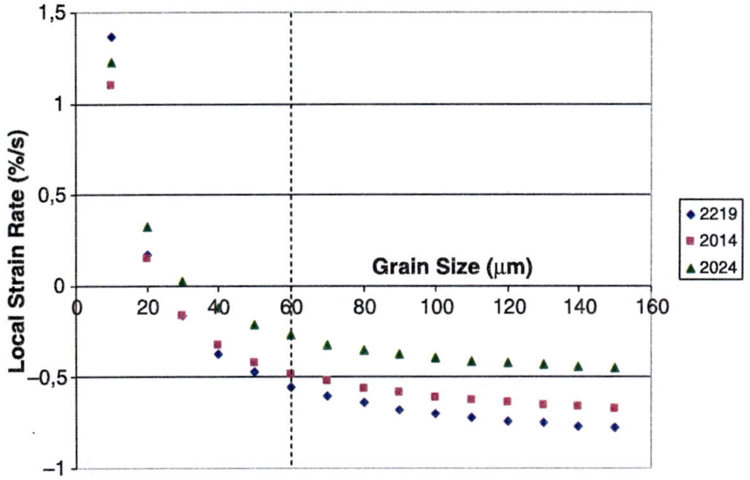

Fig. 7 Critical strain rate across mushy zone required for crack growth based upon strain rate distribution model and data from Fig. 6

expected to have the best weldability and the highest critical $\dot{\varepsilon}$ values. When comparing alloys at a normal grain size (e.g. 60 μm diameter) in Fig. 7 and Table 6, the $\dot{\varepsilon}$ value for 2219 is found to be negative which is also not to be expected. From previous experience, a $\dot{\varepsilon}$ value on the order of +0.5%/s is expected for weld metal exhibiting good weldability [25].

Table 6 Summary of critical displacement and strain rate values for solidification crack growth taken from Figs. 6–9 (assuming crack tip at $f_s = 0.98$ and grain size = 60 μm). Two sets of data are shown, one for constant λ (10 μm) and one for variable λ

Alloy	λ (μm)	$\dot{\delta}_L$ (μm/s)	$\dot{\varepsilon}$ (%/s)	λ (μm)	$\dot{\delta}_L$ (μm/s)	$\dot{\varepsilon}$ (%/s)
2219	10	0.22	−0.55	39	0.86	0.5
2014	10	0.19	−0.48	29	0.54	0.10
2024	10	0.18	−0.27	1	0.28	−0.11

In order to adjust our input to the models to obtain more realistic results, the relative permeability of the three alloys was adjusted accordingly. To achieve the desired result, dendrite arm spacing was arbitrarily selected as follows: 2219-39 μm, 2014-29 μm, 2024-16 μm. Using these adjusted values as input to the crack growth model produced $\dot{\delta}_L$ curves shown in Fig. 8 with a greater difference now observed between alloys. Specifically, when compared at $f_s = 0.98$, $\dot{\delta}_L$ is seen to vary between 0.28 and 0.86 μm/s (see Table 6). When converted to strain rate values in Fig. 9, it is observed that Alloy 2219 now demonstrates the highest weldability and has a realistic positive $\dot{\varepsilon}$ value at 60 μm grain size. Likewise, Alloy 2024 has a negative $\dot{\varepsilon}$ value at 60 μm grain size, indicative of very poor weldability.

It must be clarified that the dendrite spacing numbers arbitrarily selected above have no relation to actual values. Dendrite spacing is determined principally by cooling rate (~100°C/s for arc welds) and is typically on the order of 10 μm for most alloy arc welds. This suggests that the problem lies with the permeability formulation (Eq. 4), an observation also noted and debated by others [26]. Alloys 2014 and 2024 both contain coarse constituents, not found in Alloy 2219, that likely

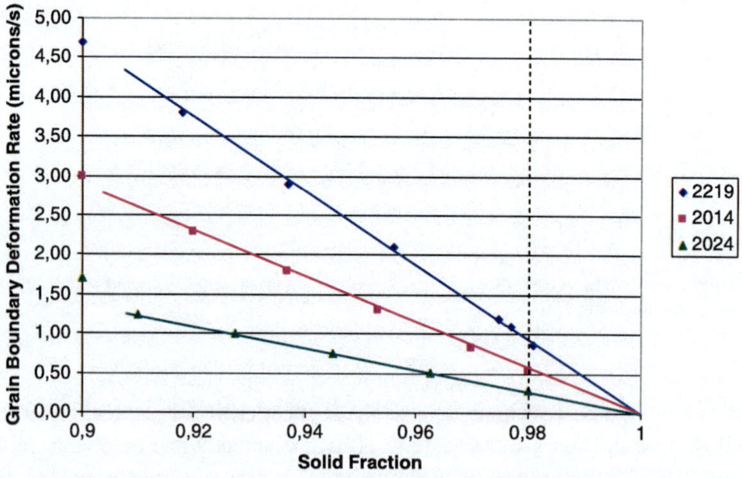

Fig. 8 Critical grain boundary deformation rates obtained from crack growth model, assuming different dendrite spacings for each alloy (2219-39 μm, 2014-29 μm, 2024-16 μm)

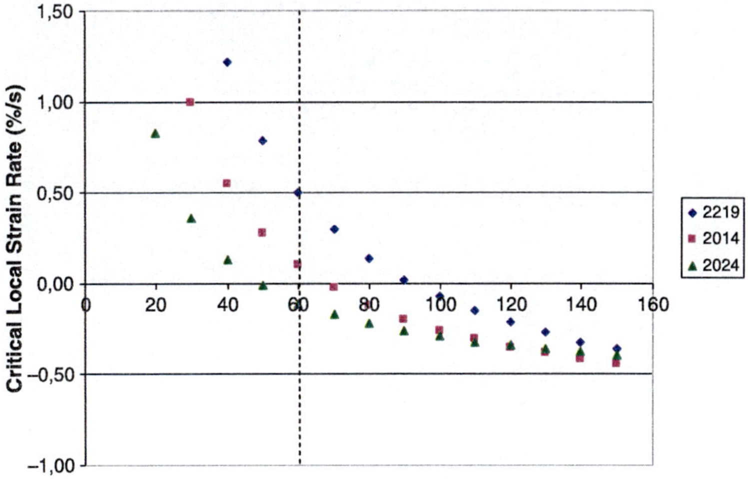

Fig. 9 Critical strain rate across mushy zone required for crack growth based upon strain rate distribution model and data from Fig. 8

serve to impede fluid flow and reduce permeability. Such intricate differences in microstructure are not accounted for in Eq. (4).

The curves of Fig. 9 suggest a significant influence of grain size on weldability. In particular, if Alloy 2219 were to be welded without filler at high heat input to get course grains of 100 μm, it could potentially experience weldability problems. Also, if Alloy 2024 could be grain refined to 30 μm grain size through use of sufficient grain refiner or current pulsing techniques, it could be made weldable according to these predictions. This fine 30 μm grain size can be achieved in arc welded Al alloys, e.g. through the use of abnormally high $TiAl_3$ additions up to 0.3 wt% Ti [15]. Likewise, the requirement to use a Si containing 4043 filler alloy in order to weld Alloy 2014 may be linked to the grain refinement attributed to Si. Si is known to produce high constitutional undercooling in Al alloy solidification favoring grain nucleation [27], which is believed to be the major factor contributing to improved weldability when using a 4043 filler [25].

Weld Microstructure

Thermal analysis and phase identification for this study was made using castings cooled at a rate of 10°C/s. This was done because identification of coarser phases is easier to accomplish and the resolution of thermal arrests on cooling curves is improved. The cooling rate for a weld, in comparison, was measured to be 110°C/s at 550°C, which results in much finer microstructures as shown in Fig. 10. Questions arise as to how much cooling rate might affect the critical temperature values in Table 4, and how this will affect model predictions. Although the thermal and phase

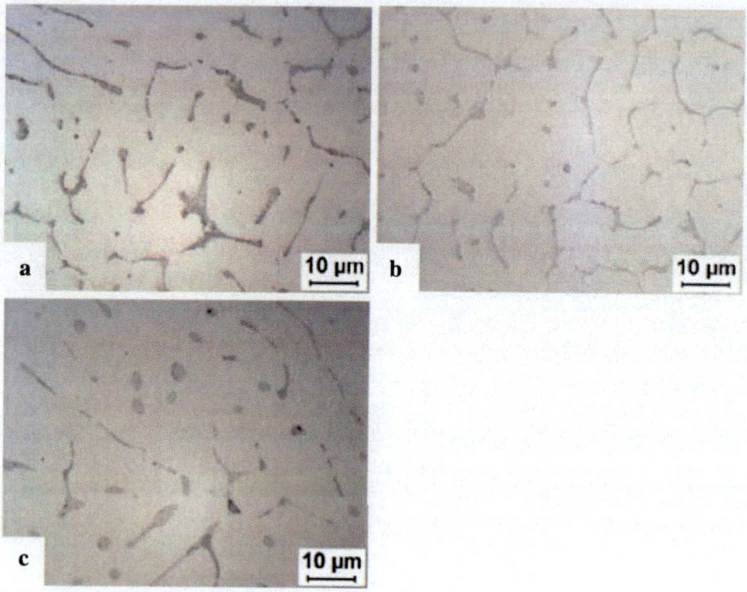

Fig. 10 Weld metal microstructures for alloys (**a**) 2219, (**b**) 2014, and (**c**) 2024

analysis of welds provides much more a challenge, work is progressing in this area [28] using single sensor thermal analysis (SSTA) specifically developed for welding [29].

Conclusion

The critical strain rate across the mushy zone needed to sustain continuous crack growth has been taken as a measure of susceptibility to solidification cracking. Two models have been proposed; one model defining the critical grain boundary displacement rate based upon a mass balance, and one model relating grain boundary displacement rate to the strain rate across the mushy zone (a measureable quantity). The former takes into account the ability to feed thermal and solidification shrinkage as controlled by permeability. Both models have been applied to three different Al-Cu (-Mg) alloys each having markedly different weldability (Alloys 2219, 2014, and 2024).

When the inherent permeability of each alloy is assumed to be the same, the models predict that Alloy 2024 should have the best weldability, which goes against common knowledge and experience. Although the coherent temperature range is greater for Alloy 2024, the corresponding solid fraction range is smaller, meaning that this alloy will experience less shrinkage over this critical solidification interval, thus leading to this result. However, it has been demonstrated that if the permeability of the dendrite structure is adjusted to reflect the better feedability of Alloy 2219 (i.e.

less constituent phases blocking dendrite channels), model predictions can be made to coincide with real-world experience. Thus, it appears to be the coarse Si, Fe, and Mn- containing constituents in Alloys 2024 and 2014 that reduce their weldability in relation to Alloy 2219.

The models have also demonstrated a strong dependence of weldability on grain size. The results suggest that grain refinement in arc welds (down to 30 μm grain size) may allow difficult alloys (e.g. Alloy 2024) to be welded free from cracking. This together with data collected previously [15] provides a compelling argument supporting the need for grain refinement in aluminum weld metal.

Considering in greater depth the application of the simple models presented here, one must concede that the problem of solidification cracking is much more complicated and multi-faceted than what has been assumed. Even so, the simple ideas presented here are congruent with observed behavior. The phenomenon of crack "healing" by backfilling represents a case in point. This is a condition often observed in weldability testing when severe strain is applied during welding (e.g. varestraint test) where pools of eutectic are observed in the form of lens-shaped features. This phenomenon may in fact not represent the healing of cracks as is usually assumed, but rather the feeding of strain and shrinkage. In fact, it is difficult to conceive how a crack could possibly form and later be backfilled in a continuous manner behind a moving weld pool (although this is quite a different matter for HAZ liquation cracks).

Another case in point is the typical smooth appearance of solidification cracks and the absence of dendrite tearing. The absence of any sign of dendrite tearing does not mean that dendrite coherency is not present. On the contrary, coherency is an absolute requirement in order to generate the transverse stress needed for cracking. The growth of cracks depends upon this transverse stress, but does not necessarily involve the tearing of dendrites. The proposed crack growth model incorporates the continuous separation of a liquid film, thus allowing for a smooth, undulating fracture surface as is normally observed.

Acknowledgements The authors are grateful for material donated by Th. Seefeld at BIAS in Bremen. Also acknowledged is technical support at BAM, including metallography performed by R. Breu and weld set-up by K. Babu. Funding for this work was provided through a German DVS/AiF industry research grant.

References

1. Prokhorov NN (1956) The problem of the strength of metals while solidifying during welding. Svarochnoe Proizvodstvo 6:5–11.
2. Chihoski RA (1972) The character of stress field around a weld pool moving on aluminum sheet. Welding Journal 51:9s–18s.
3. Feng Z (1994) A computational analysis of thermal and mechanical conditions for weld metal solidification cracking. Welding in the World 33:340–347.
4. Zacharia T (1994) Dynamic stresses in weld metal hot cracking. Welding Journal 73: 164s–172s.

5. Pumphrey WI and Jennings PH (1948) A consideration of the nature of brittleness and temperature above the solidus in castings and welds in aluminium alloys. Journal Institute of Metals 75:235–256.
6. Arata Y, Matsuda F, Nakata K and Shinozaki K (1977) Solidification crack susceptibility of aluminum alloy weld metals (Report II)- Effect of straining rate on cracking threshold in weld metal during solidification. Trans JWRI 6:91–104.
7. Rappaz M, Drezet J-M and Gremaud M (1999) A new hot-tearing criterion. Metal lurgical and Materials Transactions 30A:449–455.
8. Coniglio N and Cross CE (2009) Mechanisms for solidification crack initiation and growth in aluminum welding. Metallurgical and Materials Transactions 40A:2718–2728.
9. Coniglio N and Cross CE (2008) Weld parameter and minor element effects on so lidification crack initiation in aluminum. In: Hot Cracking Phenomena in Welds II. Springer, pp. 277–310.
10. Dahle K and Arnberg L (1997) Development of strength in solidifying aluminum alloys. Acta Materialia 45:547–559.
11. Robino CV, Reece M, Knorovsky GA, DuPont JN and Feng Z (2005) Prediction of maximum crack length in longitudinal varestraint testing. In: Proceedings of the 7th International Conference Trends in Welding Research, ASM International, pp. 313–318.
12. Ganesan S, Chan CL and Poirier DR (1992) Permeability for flow parallel to primary dendrite arms. Materials Science and Engineering-A 151:97–105.
13. Kannengiesser T, McInerney T, Florian W, Böllinghaus T and Cross, CE (2002) The influence of local weld deformation on hot cracking susceptibility. In: Mathematical Modeling of Weld Phenomena-6, Institute of Materials, pp. 803–818.
14. Matsuda F, Nakagawa H, Nakata K, Kohmoto H and Honda Y (1983) Quantitative evaluation of solidification brittleness of weld metal during solidification by means of in-situ observation and measurement (Report 1)- Development of the MISO technique. Trans JWRI 12:65–72.
15. Mousavi MG, Cross CE and Grong Ø (1999) Effect of scandium and titanium-boron on grain refinement and hot cracking of aluminium alloy 7108. Science and Technology of Welding and Joining 4:381–388.
16. Spittle JA and Cushway AA (1983) Influences of superheat and grain structure on hot tearing susceptibility of Al-Cu alloy castings. Metals Technology 10:6–13.
17. Drezet J-M and Allehaux D (2008) Application of the Rappaz Drezet Gremaud hot tearing criterion to welding of aluminum alloys. In: Hot Cracking Phenomena in Welds II. Springer, pp. 19–38.
18. Cross CE, Kramer LS, Tack WT and Loechel LW (1990) Aluminum weldability and hot cracking theory. In: Weldability of Materials, ASM International, pp. 275–282.
19. Mondolfo LF (1976) Aluminum Alloys- Structure and Properties. Butter worths, London.
20. Bäckerud L, Król E and Tamminen J (1986) Solidification Characteristics of Aluminium Alloys, Vol. 1. Skanaluminium, Oslo.
21. Cao G and Kou S (2005) Liquation cracking in full-penetration Al-Si welds. Welding Journal 84:63s–71s.
22. Cao G and Kou S (2006) Predicting and reducing liquation cracking susceptibility based on temperature vs. fraction solid. Welding Journal 85:9s–18s.
23. Arnberg L, Bäckerud L and Chai G (1996), Solidification Characteristics of Aluminium Alloys, Vol. 3. Skanaluminium, Oslo.
24. Van Horn KR (1967) Aluminum, Vol. 1: Properties, Physical Metallurgy, and Phase Diagrams, ASM International.
25. Coniglio N, Cross CE, Michael Th and Lammers M (2008) Defining a critical weld dilution to avoid solidification cracking in aluminum. Welding Journal 87:237s–247s.
26. Carlson KD and Beckermann C (2009) Authors' Reply to Discussion of "Prediction of shrinkage pore volume fraction using a dimensionless Niyama Criterion". Metallurgical and Materials Transactions 40A:3054–3055.
27. Easton MA and StJohn DH (2001) A model of grain refinement incorporating alloy constitution and potency of heterogeneous nucleant particles. Acta Materialia 49:1867–1878.

28. Coniglio N, Cross CE, Dörfel I and Österle W (2009) Phase formation in 6060/4043 aluminum weld solidification. Materials Science and Engineering A 517:321–327.
29. Alexandrov BT and Lippold JC (2006) A new methodology for studying phase transformations in high strength steel weld metal. In: Proceedings of the 7th International Conference Trends in Welding Research, ASM International, pp. 975–980.

Hot Tearing Test for TIG Welding of Aluminum Alloys: Application of a Stress Parallel to the Fusion Line

A. Niel, F. Deschaux-Beaume, C. Bordreuil, G. Fras, and J.-M. Drezet

Introduction

Welding processes are extensively used to assemble components in many manufacturing industries, such as aeronautics, construction, energy and automotive. To increase their productivity, constructors try to reduce manufacturing time. This involves for welding operations an increase of welding speed. However, various defects such as hot tearing might limit this increase. Figure 1 illustrates the presence of a crack created during arc welding on a 6061 aluminum alloy.

Hot tearing appears at the solidification end of an alloy, and is commonly observed in welding. During welding, components are subjected to high thermal gradients around the melting zone due to localized heat input. The solidification zone, where microstructure forms, is located at the rear of the melting zone and is bordered by two isothermal surfaces corresponding to liquidus and solidus temperatures. In welding, due to high thermal gradients, solidification by epitaxial growth of columnar dendrites is generally observed from the border of the melting zone (Fig. 2). However, equiaxed dendritic grains can also form in the center area of the melting zone.

Solidification can be divided in several stages (Fig. 2) [1]. Nucleation is the first step. Solid particles nucleate in the liquid and are free to move. The material behaves

Fig. 1 Hot tearing in a GTAW weld seam

A. Niel (✉)
Mechanics and Civil Engineering Laboratory LMGC UMR 5508,
University Montpellier2, Montpellier, France
e-mail: aurelie.niel@iut-nimes.fr

J. Lippold et al. (eds.), *Hot Cracking Phenomena in Welds III*,
DOI 10.1007/978-3-642-16864-2_3, © Springer-Verlag Berlin Heidelberg 2011

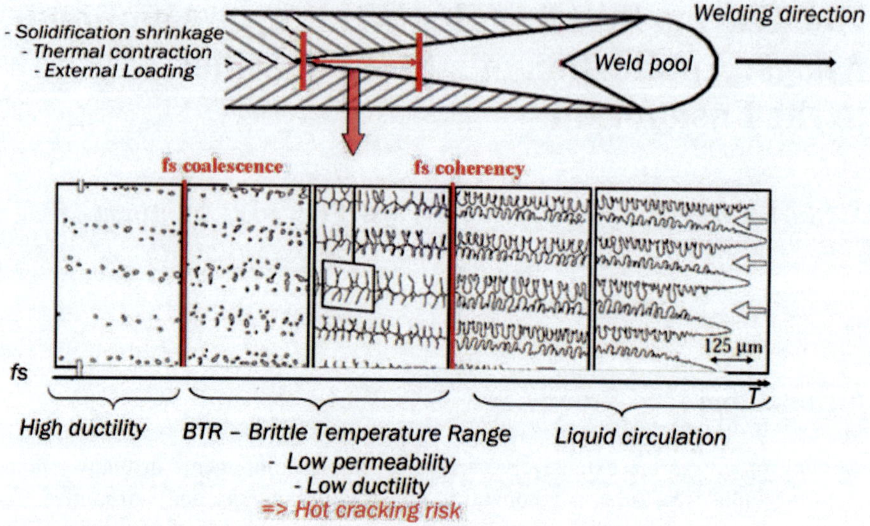

Fig. 2 Schematic representation of hot tearing formation in a columnar structure (from [1])

like a viscous fluid with a very low tensile strength. Then, grains grow. During welding, the solidification rate is high, and the solidification front (solid-liquid interface) is dendritic. The solid fraction increases gradually and the temperature approaches the solidus. The coherency solid fraction is then reached and corresponds to first solid bridge bonding. The dendritic grains are in contact with each other and form a coherent solid skeleton. A sharp drop in permeability is then observed, due to the compact network formation. Thin liquid films between the solid grains remain and are subjected to high strains, induced by solidification shrinkage and thermal contraction of the solid. During this stage, the strain to fracture of the alloy is very weak, because the liquid can no longer flow to accommodate deformation and the solid network is not resistant enough to avoid cracking. Finally, the last stage begins when the coalescence solid fraction is reached. The last isolated liquid pockets solidify, and the solid fraction tends toward one. The ductile solid can then resist to strain. The stress and strain to tensile failure is rising rapidly.

Brittle Temperature Range (BTR) is defined as the temperature interval corresponding to the solid fractions where the microstructure is in critical configuration. Hot tearing risk is maximal in this temperature range. The BTR corresponds to the interval between the coherency solid fraction, where the liquid does not easily circulate because of the low permeability, and the coalescence solid fraction, where the solid opposes mechanical resistance due to rise of the number of solid bridges formed.

The hot cracking susceptibility is highly correlated to the solidification path, which depends on the chemical composition of the alloy and the solidification rate.

These factors affect the quantity of residual elements with low melting point favoring the survival of residual liquid films, but also the grain size and shape, which alter the permeability of the mushy zone.

From a mechanical point of view, hot cracking in welding is due to thermal and mechanical loading of the weld pool during solidification, which cause liquid films to debond and/or solid bridges to break [2]. The welding process leads to a non-uniform temperature distribution, inducing thermal stresses and localized plastic deformations. The thermally induced deformation, combining solidification shrinkage and thermal contraction, is the main factor of mechanical loading in the BTR. An additional loading due to samples clamping or self clamping can also promote hot cracking. Resulting stress and strain fields in the BTR are very complex to predict. Chihoski [3] was the first to try to qualitatively explain the stress distribution around a moving localized heat source on thin sheets of aluminum alloy. However, the differential shrinkage induced by temperature gradient and the thermal dependent mechanical behavior of alloys require numerical modeling to determine the stress and strain distributions.

Welding conditions, i.e. process parameters (especially welding power and welding speed) and sample geometry and clamping act on the cooling rate (then on the solidification microstructure) and on the thermal gradient (then on the mechanical loading). They have then a great influence on hot cracking.

To assess hot cracking sensitivity of an alloy, various tests have been developed [4]. In Fig. 3, existing tests have been classified into two categories: crack initiation tests (Varestraint test type) and crack propagation tests (Joining and Welding Research Institute test type). The principle of all these tests is to apply a mechanical loading to the weld, produced by an external loading device or by clamping or self clamping of the sample. These tests are generally difficult to interpret and implement in an industrial environment. Moreover, it is difficult for the tests based on camped or self clamped samples, to distinguish the sample geometry effects and the process effects, because cracks are the result of complex interactions between these factors.

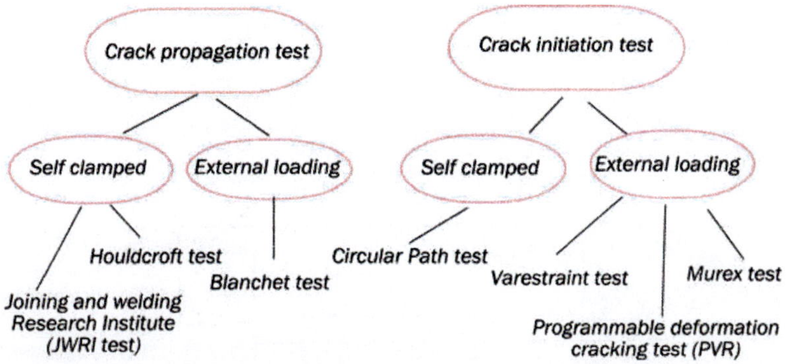

Fig. 3 Classification of the main hot cracking tests

The purpose of this research is to explore ways for optimizing the welding parameters with respect to hot tearing. For that, a new hot tearing test has been developed to better understand the influence of the process parameters on the sensitivity to crack initiation, according to the study of the interactions process – mechanical loading and process – solidification microstructure. The other originality of this study is that we combine experimental and numerical results to study these interactions.

This paper first presents the original test developed, and experimental results obtained in welding on 6061 aluminum alloy. Next, a thermo-mechanical modeling of the welding operation is developed, in order to evaluate local loading of the solidification zone. Finally, the analysis of the microstructure influence on susceptibility to hot cracking is discussed in order to propose improvements to the welding process.

The Experimental System

A simple original hot tearing test for thin sheets has been developed in this study. It consists to impose on a sheet sample, before welding without additional filler material, an external controlled mechanical loading promoting cracking, to study the welding process and material influence on hot cracking susceptibility. This test allows the initiation of a hot tear under controlled experimental conditions.

The loading of the sample promoting hot cracking is a constant uniaxial loading, with an enforced tensile stress or enforced displacement. Preliminary tests revealed that an enforced displacement or stress transverse to the welding direction doesn't favor hot cracking for our welding conditions, whereas a longitudinal tensile stress promote it (Fig. 4).

During welding, the specimen is clamped on both sides between two jaws. The process used is the Gas Tungsten Arc Welding (GTAW). This process uses a tungsten refractory electrode to create an electric arc, and an inert gas, generally argon, to protect metal against oxidation. A fusion line is made on a parallelepiped sheet in the longitudinal direction. Hot tearing test is placed on a mobile two axes (X, Y) table, as depicted in Fig. 5. During the test, welding arc is in a fixed position, and the sample is translated in the longitudinal direction at a constant speed. The used sheets are thin (2.3 mm) and the samples, cut with a water jet machine, are 265 × 50 mm^2 in size.

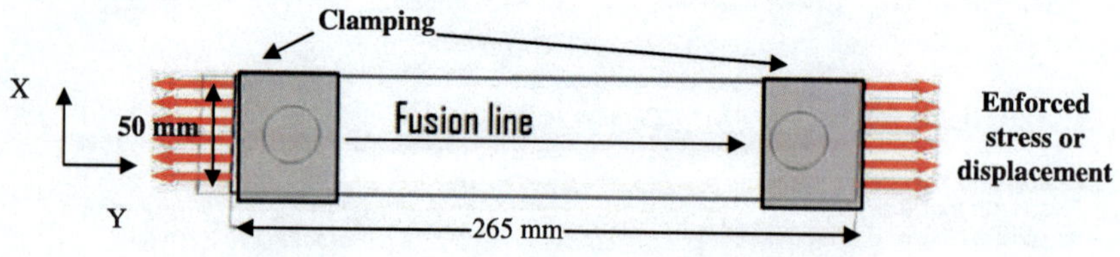

Fig. 4 Longitudinal tensile specimen with GTAW fusion line

Fig. 5 Two axes table with hot tearing test

The test advantage, compared to other hot tearing tests such as the Varestraint test, is its simplicity which is interesting for welds qualifications in industrial environment. It is also more representative of clamping conditions applied for welding structural parts than existing tests with rising loading, like Varestraint, Controlled Tensile Weldability, or PVR tests [4]. In addition, the simple sample geometry (thin sheet with full penetration weld) and boundary conditions make the 2D numerical simulation of the test easy. Indeed, measurements using strain gauges have shown that flexion of the sample was negligible, so the stress state can be supposed planar.

This test allows varying mechanical loading of the solidification zone by changing the enforced stress or displacement, and/or the welding parameters. Solidification microstructure change is also possible by adjusting thermal cycle, depending on welding power, speed and sample size. To follow the formation of solidification microstructures, the back part of the weld pool is observed using a high speed camera. The camera recording is facilitated by the static position of the weld pool, the fusion line in the sheet being produced by the table translation.

The present experimental campaign was achieved with an enforced initial displacement of 0.8 mm, corresponding to an initial tensile stress in the sheet of about 200 MPa. Six welding speeds, between 5 and 20 mm/s, were retained. Note that these speeds are rather high for GTAW welding process. Four tests were conducted for each speed by varying the welding current in the range [110 A, 260 A]. Welding parameters are resumed in Fig. 6. The material tested is a 6061-T6 aluminum alloy. The alternative current in GTAW is necessary on aluminum. At each cycle, there is a polarity reversal that breaks the alumina layer formed on the surface. The ratio welding current/welding speed is set to have full penetration. A 3 mm arc length is imposed for all the welding tests, which corresponds to a welding voltage of about 10 V. In total, 24 samples have been studied with varying welding parameters and constant enforced displacement. Note that the weld pool starts rather far from the jaws to have a uniaxial initial stress state.

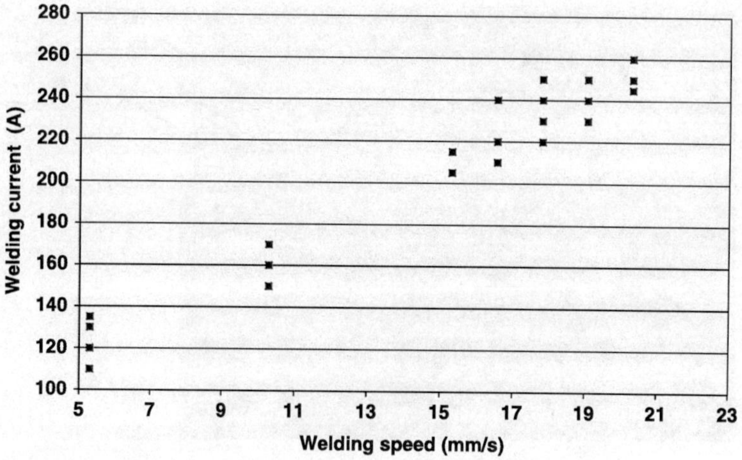

Fig. 6 Welding conditions retained for the hot cracking test campaign (U~10 V)

This test allows the crack initiation in the solidification zone, in the quasi-stationary thermal state. For all tested conditions, the formed cracks initiate transversely to the welding direction, near the border of the weld seam. For some tests, these cracks propagate longitudinally in the weld seam center as shown in Fig. 1. In the crack initiation zone, grains have a columnar morphology, and cracks are formed into the intergranular spaces.

The aim of our study is to explain the correlation between the transverse initiation of hot cracking, observed on the edges of the weld seam, and the process parameters and the enforced displacement. The coupling between these parameters being complex, a numerical simulation using Finite Element Method has been achieved to access strain and stress fields into the solidification zone.

Numerical Modeling

A numerical simulation of hot cracking test described in the previous section has been achieved using Sysweld finite element software. Due to the quasi-planar configuration of the test, a 2D modeling of the sample, using plane stress quadrilateral finite elements has been retained. A finer mesh is used in the region of high thermal gradient, in and around the weld seam zone. A 0.8 mm displacement along Y axis is imposed as initial condition to the sample before welding (Fig. 7). The heat flux provided by the electrical arc to the sample is modeled by a double Gaussian power distribution (in the X,Y plane), in translation along Y axis at the 15 mm/s speed. The heat source model parameters have been adjusted to obtain the same weld pool size than experimentally observed. Convection and radiation heat fluxes are prescribed as boundary conditions to the sample surface.

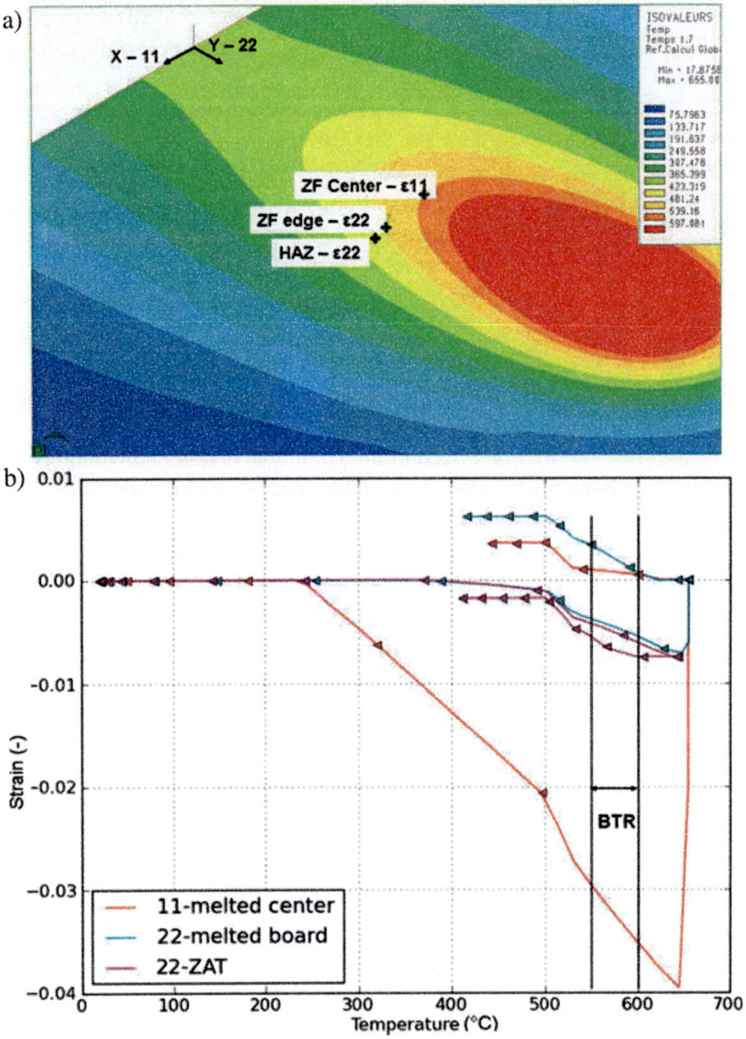

Fig. 7 (**a**) Localization of studied characteristic points, (**b**) Strain histories during welding test

The thermal dependence of density, thermal conductivity, specific heat Young's modulus, thermal expansion coefficient and yield strength is taken into account. The solidification shrinkage is modeled by the increase of the thermal expansion coefficient in the solidification temperature range. The evolution of the yield strength is modeled using a standard thermoplastic law, with a mixed hardening. Preliminary simulations with various elasto-visco-plastic laws having shown only few differences with elasto-plastic laws, viscosity has been neglected. In addition, the dissolution of precipitates in the molten zone and the hardening restoration in the heat affected zone have been taken into account. All the data concerning high temperature mechanical behavior of 6061-T6 aluminum alloy were adapted from [5].

Figure 7b shows the evolution of strain tensor components during welding and cooling versus temperature in three points of the sheet (Fig. 7a): the first one in the weld seam center, the second one at the fusion boundary, and the last one in the heat-affected zone. The position of these points and the strain component for each one were chosen in agreement with experimental observations of cracks location and orientation. The normal strain transverse to the fusion line (ε11) is shown in the weld center, and the longitudinal normal strain (ε22) is chosen for two other points.

In the BTR, represented by two vertical black lines in Fig. 7b, the strain evolution calculated promotes hot cracking. Indeed, the solidification zone at the fusion boundary is submitted to positive and growing longitudinal strain, mainly due to shrinkage, thus facilitating crack initiation. The transverse strain in the weld seam center is also positive and growing at solidification end, but twice lower than the longitudinal strain at the fusion boundary, which can explain why cracks initiation is always transverse, at the fusion boundary. Positive transverse strain in the weld seam center however can facilitate crack propagation, which is in accordance with experimental observations. The risk of liquation cracking is limited, as the point situated in the heat affected zone exhibit a negative strain.

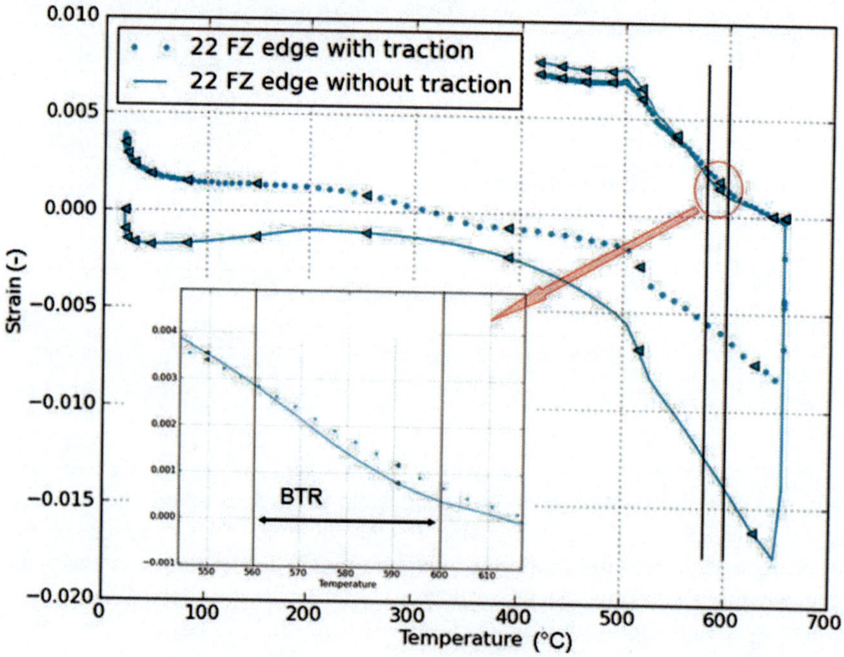

Fig. 8 Influence of an initial tensile stress on the longitudinal strain at the fusion boundary

Simulation has been used also to show the influence of an enforced initial displacement on the loading of the solidification zone. The strain histories of this zone in samples with or without enforced displacement as initial condition have been compared. Longitudinal strain evolutions at the fusion boundary for both cases are presented in Fig. 8. In the BTR, the strain is about 30% greater for the case with enforced displacement than without. An initial tensile stress then exacerbates the tensile loading at the weld edge in the welding direction, and consequently could promote hot cracking initiation.

However, mechanical loading of the sheet is not the only parameter affecting hot cracking initiation. Characteristics of the solidification microstructure also play an important role and should be considered.

Experimental Results

In Situ Observation

High speed camera recordings of the solidification zone during welding were carried out, in order to observe crack initiation. Solidification zone observation is complicated by the formation of oxide at the weld pool surface. With an improved gas protection, the weld pool boundary appears, as can be seen in Fig. 9. In-situ observations could permit a better understanding of the hot tearing initiation, by the identification of solid fractions and dendrite size and morphology at the crack initiation.

Two areas with distinct granular structure can be observed on the in-situ recording (Fig. 9), in the re-solidified zone: a columnar dendritic zone on the edge of the weld seam, and an equiaxed dendritic zone in the center. However, at the lowest welding speed, the weld seam is composed entirely of columnar dendritic grains. Theoretical considerations can explain these results. The formation in weld pool of

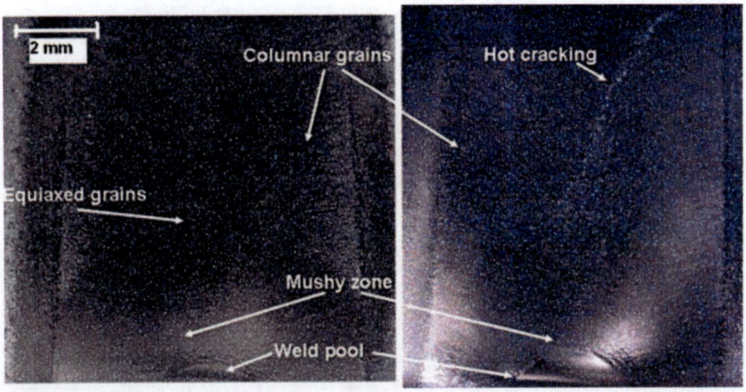

Fig. 9 Video image recording during welding showing the back part of the weld pool

equiaxed dendritic grains, instead of columnar dendrites, is promoted by undercooling in the liquid in front of the solid/liquid interface. Undercooling is favored by high solidification rate and low thermal gradient [6]. The thermal gradient in the solidification range is lower in the weld seam center, especially for high welding speed, than at the fusion boundary, at the opposite of the solidification rate, which increases from the fusion boundary to the weld seam center. That could explain the nucleation of equiaxed grains in the central zone of the weld pool.

Observation with high speed camera confirms that cracks initiate in the columnar zone, in a direction transverse to the welding direction (Fig. 9). Depending on welding parameters, crack can then stop in the equiaxed grains zone or propagate in the longitudinal direction into this zone.

Post-Mortem Study of Crack Morphology

After welding, the samples were cut out to obtain cross section views. The microstructure is studied using a combination of optical microscopy and scanning electron microscopy, to complete in-situ observations.

Microscopic observations of cracked areas (Fig. 10b) clearly show intergranular cracks generated by a debonding of liquid films. Indeed, the coalescence solid fraction, corresponding to the lower boundary of the BTR, is first reached between two dendrites of the same grain, and then between two grains, which promotes the initiation of intergranular cracks rather than intragranular. The presence of liquid films at the end of solidification is visible on the fracture surfaces using scanning electron microscopy (Fig. 10c).

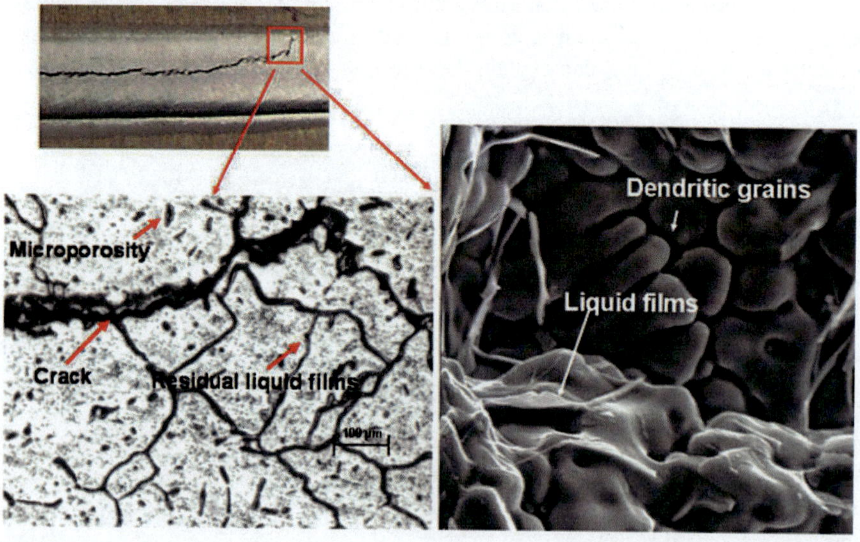

Fig. 10 Observation of cracks: (**a**) macrograph top view, (**b**) micrograph in transverse section, (**c**) SEM fracture pattern

The microstructure observation confirms the characteristic grain structure, with peripheral columnar grains oriented in a direction perpendicular to the welding direction, and central equiaxed grains, which is very typical of weld structures [7,8]. The width of each zone depends on changing welding parameters, i.e. welding current and welding speed.

It is admitted in the literature that columnar dendritic morphology, with dendrites growing in the thermal gradient direction, generally observed for rather low solidification rate and/or high thermal gradient, is more sensitive than equiaxed dendritic morphology [2]. This can be understood by the fact that liquid feeding between the equiaxed grains is facilitated even at the solidification end, as well as crack healing by liquid feeding. It has been confirmed with these observations that for most of the samples, the crack initiation occurs between the columnar dendritic grains, near the fusion boundary. This is not a surprising result, because this area combines an unfavorable microstructure, and the higher strain level, as shown in previous section. The application of an initial longitudinal tensile stress also favors transverse cracking between columnar grains.

Some other morphological characteristics of the weld pool structure can influence the hot cracking sensitivity. For instance, the emergence of a centerline grain boundary, corresponding to the junction of columnar grains from each side of the weld pool, increases hot cracking sensibility [9]. If the thermal conditions leading to the formation of various grain morphologies are well known, the relationship between welding parameters and microstructure is still not fully explained, because of the complex relationships between these parameters and the temperature field. In order to try to correlate process parameters, weld pool structure and hot cracking sensitivity, the morphology of the fusion zone of all samples has been characterized by quantitative parameters, such as grain shape and size, growth direction, and disorientation between grains (Fig. 11).

For our tests, the mean width of the columnar dendritic zone formed on each side of the weld seam is about 1.5 mm, and the mean width of the equiaxed dendritic zone is about 2 mm. However, the proportion of each zone is a function of welding parameters. A correlation between the relative size of these zones and the crack propagation can be seen. When the equiaxed zone width is small compared to the total width of the weld, crack will tend to propagate more easily in a large

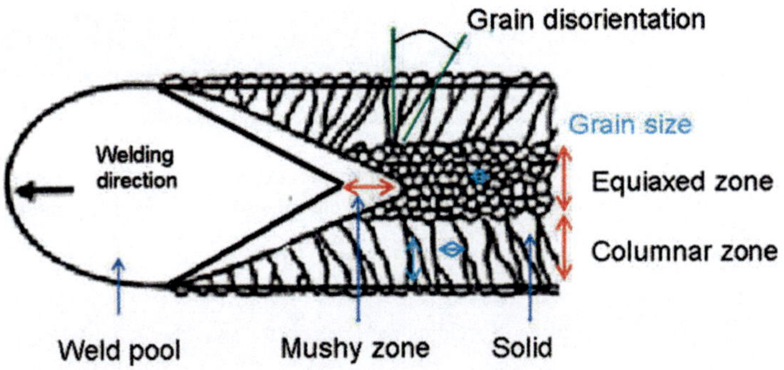

Fig. 11 Morphological characteristics of the weld seam structure

longitudinal crack. Moreover, when the weld pool is shifted closer to one sample edge, the crack always initiates from that side. That can be correlated to the creation of asymmetrical heat dissipation from the weld pool, inducing asymmetrical weld seam structure, but also asymmetrical strain and stress fields. Any clear correlation has been found between welding parameters and relative size of columnar and equiaxed zones in the fusion line.

The columnar grains growth direction in our samples is characterized by an angle of about 80 degrees relative to the welding direction. Grain orientation, which can be evaluated by the dendrite axes directions into the grain, also affects the ability of the liquid to feed interdendritic spaces. Wang showed the importance of disorientation between two grains on the hot cracking phenomenon [10]. This disorientation also limits coalescence and possible crack healing. The maximum disorientation found in our samples is about 22 degrees. However, crack initiation does not occur where the grains are highly disoriented between them in our case. The distance between arc initiation and first crack initiation appears rather constant, between 2 and 4 cm.

Influence of Process Parameters

Welding parameters were classified as cracking or not cracking conditions for a given enforced initial displacement. Figure 12 shows the crack sensitivity of 6061 alloy as a function of welding parameters. After welding, the cracks were revealed

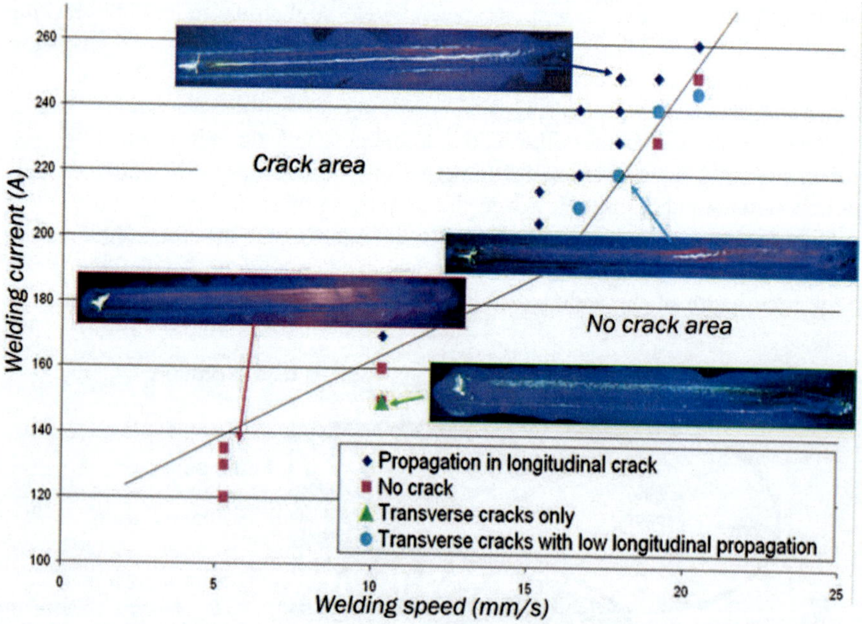

Fig. 12 Hot cracking map according to welding parameters (for an enforced initial displacement of 0.8 mm)

by dye penetrant testing. The results highlight several failure modes, depending on process parameters. For high welding speeds, some small transverse cracks are observed at the beginning of the weld seam near the fusion boundary, which deviate rapidly to propagate in a large longitudinal crack (0.6 mm in average width). However, when welding speeds are lower, there is a succession of fine transverse cracks (less than 30 μm in width) that do not reach the central equiaxed zone. For each welding speed, except the lowest, there is a critical welding current above which cracking is observed. It can be observed also that large longitudinal cracks are only observed for high welding speed (up to 20 mm/s).

Application to the RDG Model

To finish this study based on the combination of experimental and numerical results to better understand interactions between process parameters, microstructure and loading of the solidification zone, and hot cracking sensitivity, the crack initiation model developed by Rappaz, Drezet, Gremaud (RDG criterion) [1] has been used to compare hot cracking sensitivity of samples with or without initial tensile stress. This model is applied as post processing of finite elements calculations presented before. The RDG model is based on the calculation of a depression ΔP_{max} in the residual liquid on dendrite foot (Eq. 1–2) produced by volume changes due to solidification shrinkage, thermal contraction of the solid skeleton, and mechanical strains. The liquid ability to flow between dendrites is calculated using Carman Kozeny permeability model developed for porous media. μ and β correspond respectively to the liquid viscosity and to the solidification shrinkage (μ = 0.001Pa.s, β = 0.061). The λ_2 parameter, corresponding to spaces between the secondary dendrite arms, has been extracted from the micrographs. The evolution of the solid fraction fs(T) is calculated using Scheil-Gulliver relation. The temperature T, the temperature gradients G(T), the strain rate $\dot{\varepsilon}_P(T)$ and the solidification rate v_T were extracted from the results of numerical simulation presented before.

$$\Delta P_{\max} = \frac{180.(1+\beta).\mu}{\lambda_2^2} \int_{T_S}^{T_L} \frac{E(T).f_s(T)^2}{(1-f_s(T))^3.G(T)} dT + \frac{180.v_T.\beta.\mu}{\lambda_2^2} \int_{T_S}^{T_L} \frac{f_s(T)^2}{(1-f_s(T))^2.G(T)} dT \quad (1)$$

$$E(T) = \frac{1}{G(T)} \int_{T_S}^{T} f_s(T).\dot{\varepsilon}_P(T) dT \quad (2)$$

The RDG model is based on the hypothesis of a columnar dendritic structure, so it was calculated in our case only in the columnar zone. The columnar grain direction in our experiment being nearly transverse to the welding direction, we calculated the depression in the solidification zone between the weld pool boundary and the dendrite foot on a line perpendicular to the fusion line (red line on Figs. 13 and 14).

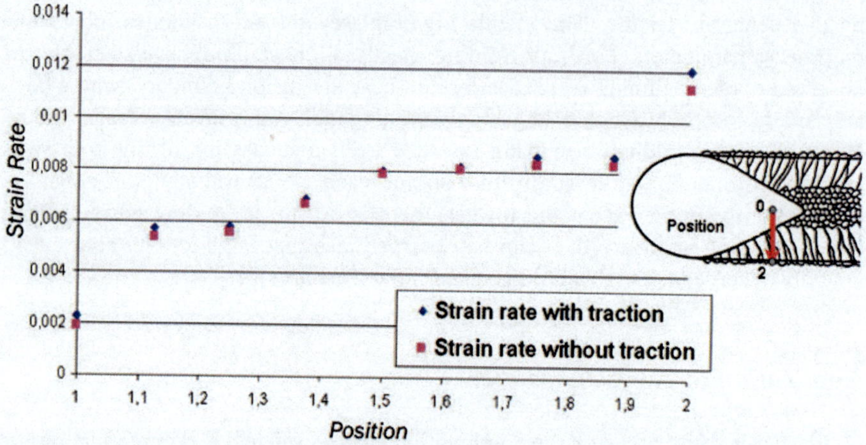

Fig. 13 Evolution of the strain rate in the columnar zone of the weld seam, with and without initial tensile stress

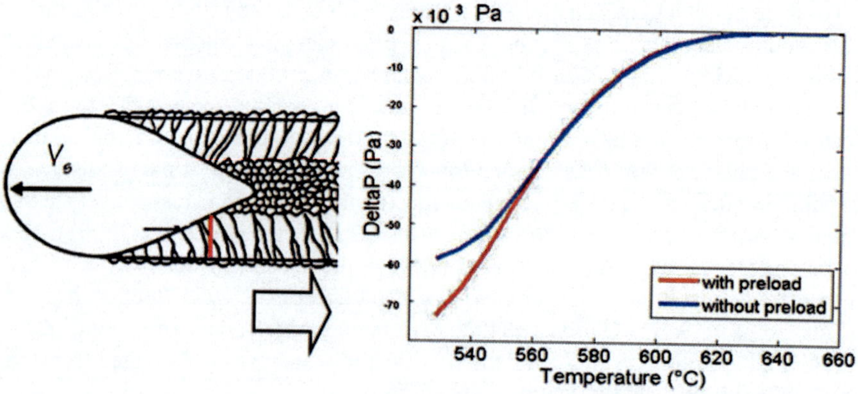

Fig. 14 Influence of the initial tensile stress on the pressure drop between the columnar grains calculated using RDG model

The evolution of the total strain rate (elastic plus plastic) transverse to the columnar grain direction, extracted from the numerical simulation, is presented in Fig. 13. We can see on this figure that with an initial tensile stress, the strain rate becomes higher only at the fusion boundary. This zone corresponds to the columnar dendrite foot, which is supposed to be the more sensitive zone with respect to hot cracking.

The pressure evolution, calculated along columnar dendritic grains, is shown on Fig. 14. The pressure drop increase at the end of solidification, due to the permeability fall when solid fraction tends towards 1. The higher depression drop observed with initial tensile stress given by RDG model confirms all previous observations concerning the detrimental effect of this initial loading with respect to hot cracking sensitivity of samples.

In the future, the RDG model, which allows taking into account some morphological aspects of the solidification zone, should be used in complement to numerical simulation to evaluate the effect of welding parameters with respect to crack initiation.

Conclusions

It has been demonstrated that solidification cracking is a faceted problem, with many influencing parameters. Many studies examined hot tearing but few of them have studied the relationship between mechanical aspect induced by the process and microstructure. A new, original and simple test has been developed to promote hot cracking initiation. Applying an initial tensile stress on the welding direction promotes crack initiation for our welding conditions. Cracks initiated during the tests at the fusion boundary, transversally to the welding direction.

Numerical simulation allowed us to show that fusion boundary suffers the highest strain in the longitudinal direction, which explains the crack initiation in this zone. Simulation results also allowed understanding the effect of an initial tensile stress on the loading of the solidification zone.

Observation using high speed camera helped us to better understand the mechanisms of crack initiation and the bifurcation at the grain scale. A correlation has been made between the grain morphology and the cracking sensitivity, the peripheral columnar grain zone being more sensitive than the central equiaxed grain zone. Cracking phenomenon then results in the combination of sensitive microstructure and critical mechanical loading.

Moreover, a mapping representing the cracking sensitivity according to welding conditions has shown the importance of welding power and welding speed on the crack initiation and on the different failure types.

Finally, the RDG model has been coupled as post processing to numerical simulation results, in order to combine microstructural parameters to the mechanical parameters provided by simulation, to help us to predict hot cracking sensitivity.

The general methodology developed in this study, based on the combination of experimental tests and characterization, and on numerical simulation, will be used in the future to improve welding operations with respect to hot cracking issue. Two ways of improvement will be studied: the first one concerns microstructural change in the solidification zone (grain orientation, grain size, width of columnar zone ...) by acting on the process parameters; the second one concerns the change of mechanical loading in the solidification zone, by acting on the heating distribution on the samples [11] and on the boundary conditions.

References

1. Rappaz M, Drezet JM and Gremaud M (1999) A new hot tearing criterion. Metallurgical and materials transactions A 30A:449.

2. Cross CE (2005) On the origin of weld solidification cracking – Hot Cracking Phenomena in Welds – p3.
3. Chihoski RA (1972) Weld cracking in aluminium alloy. Welding journal.
4. Farrar JCM (2006) Hot cracking tests. Hot Cracking Phenomena in Welds, pp. 291–304.
5. Maisonnette (2010), Influences mécaniques et métallurgiques de procédés haute température sur un alliage d'aluminium 6061-T6 – PHD, INSA Lyon.
6. Gaumann M (1997) Nucleation ahead of the advancing interface in directional solidification. Materials science and engineering A226-228.
7. DebRoy T (1995) Physical processes in fusion welding. Reviews of Modern Physics 67(1).
8. Norman AF (1999) Effect of welding parameters on the solidification microstructure of autogenous TIG welds in an Al-Cu-Mg-Mn alloy. Materials Science and Engineering A259:53–64.
9. Hunziker O (2000) On formation of a centerline grain boundary during fusion welding. Acta Materiala 48:4191–4201.
10. Wang N (2004) Solidification cracking of super alloy single and bi crystals. Acta Materiala 52:3173–3182.
11. Hernandez LE (1984) The influence of external local heating in preventing cracking during welding of aluminium alloy sheet. Welding Journal 63(3):84 s.

Hot Cracking Susceptibility of Wrought 6005 and 6082 Aluminum Alloys

Paul Kah, Jukka Martikainen, Esa Hiltunen, Fisseha Brhane, and Victor Karkhin

Introduction

Outstanding in their unique combination of light weight, high strength, high toughness, extreme temperature capability, excellent corrosion resistance and versatility of extruding and recycling capabilities, aluminum alloys have applications in almost every manufacturing sector such as in transportation, construction and building. A typical application of 6005-T6 and 6082-T6 heat-treatable wrought aluminum alloys is structural and architectural. Aluminum-based alloys can be successfully arc welded without any cracking related problems or with only minor ones. There are three areas that can significantly influence the probability of hot cracking in an aluminum welded structure. These are the susceptible base alloy chemistry, the selection and use of the most appropriate filler alloy and choosing the most appropriate joint design. The 6xxx series alloys are very sensitive to cracking if the base metal composition remains close to the filler metal composition. During arc welding, the cracking tendency of these alloys is adjusted to acceptable levels by the dilution of the base material with excess magnesium (by the use of the 5xxx series Al-Mg filler alloys) or excess silicon (by the use of the 4xxx series Al-Si filler alloys). The most appropriate and successful method used to prevent cracking in the 6xxx series base materials is to ensure that an adequate filler alloy is added during the welding operation [1–10].

Much research has been conducted fully understand the main causes of hot cracking with Al-Mg-Si alloys during welding [4–13]. To avoid or minimize the cracking effect it is recommended to use appropriate filler alloys [5–10]. Filler alloys 5356 with 5% Mg and 4043 and with 5% Si have been used to compare the effect of filler metal composition on hot cracking. It was concluded that a weld with filler 4043 is less prone to hot cracking than that with filler 5356 due to the narrow solidification temperature range and the lower eutectic temperature of the weld metal which

P. Kah (✉)
Laboratory of Welding Technology and Laser Processing, Lappeenranta University of Technology, Lappeenranta, Finland
e-mail: Paul.Kah@lut.fi

J. Lippold et al. (eds.), *Hot Cracking Phenomena in Welds III*,
DOI 10.1007/978-3-642-16864-2_4, © Springer-Verlag Berlin Heidelberg 2011

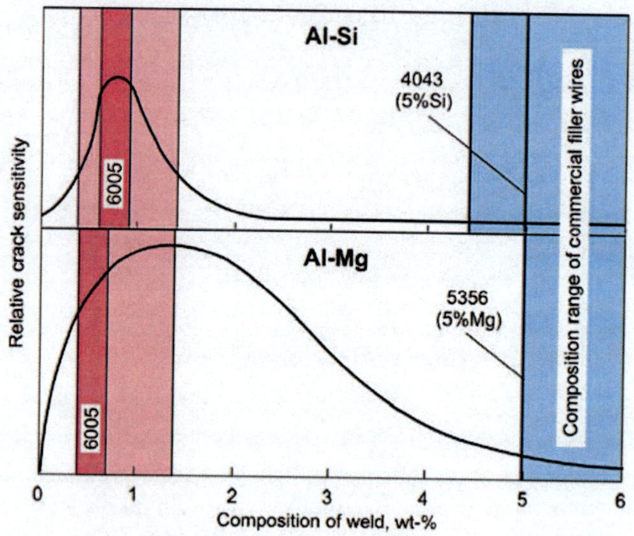

Fig. 1 The effect of composition on crack sensitivity of binary Al-Si and Al-Mg alloys [17]

enable the base metal to solidify first (Fig. 1). The problem with 4043 is that it is easily anodized after welding, producing dark weld metal and a highly visible weld result. This is due to the amount of silicon in the composition. Thus, there is no significant difference in hardness when using either of the two filler alloys. In fact the specimens welded with filler 4043 showed a heat affected zone that was, on average, 2 mm wider than those welded with filler 5356 with both GTAW and GMAW.

GTA and GMA welding were employed to compare different heat inputs, where the heat input in pulsed GTAW was about four times higher than that in pulsed GMAW. Some of the test specimens were also preheated to reduce tensile stresses and the cooling rate so as to decrease the liquation cracking effect. Liquation cracking occurs in the partially melted zone or PMZ of a weld i.e. just next to the fusion zone. Liquation can occur along the grain boundary as well as in the grain interior. Grain boundary liquation makes the PMZ susceptible to liquation cracking [14, 15]. Hot cracking is a high-temperature cracking mechanism and mainly a function of how metal alloy systems solidify.

The degree of restraint also significantly affects liquation cracking. The more restraint the material shows, the higher the probability of liquation cracking. Because aluminum alloys have high thermal contraction and if they are not relaxed to solidify, they develop high tensile stress that tends to tear or open the liquated grains. Selecting a base metal which is fine grained and less susceptible to liquation cracking, such as a material that does not contain low melting impurities or segregates, can also help to reduce this problem if it is feasible [9, 16].

Experimental Procedures

10 mm thick 6005-T6 and 6082-T6 wrought aluminum alloy specimens have been tested for cracking based on different criteria. First, filler alloys were used to understand their effect on cracking. The actual chemical compositions of the base and nominal composition of filler metals are shown in Table 1. The influence of the different welding processes, preheating and the base alloys were examined on hot cracking. Mechanized 90 Hz pulsed GTAW and GMAW with short circuiting droplet transfer were used for all the experiments for good repeatability. The welding parameters used were the following: The ambient temperature was approximately 20°C, and some of the samples were heated to about 120°C before welding to examine the effect of preheating. The welding currents used were 172 A, 210 A and 352 A on both base alloys with and without preheating, and the voltages were kept at 17 V, 22 V and 26 V respectively. The welding and feeding rates ranged from 2.8 to 10 mm/s and 0.34–13.5 m/min. Pure argon (Ar + 0.03% NO) was used as a shielding gas in both cases, and flow rates were 14 l/min and 10 l/min for GMAW and GTAW, respectively. The filler wire electrode diameter was 1.2 and 2.4 mm in GMAW and GTAW, respectively, and stick-out in GMAW was 15 mm long. The bead-on-plate technique was used in all cases.

Sufficient cleaning of the surface was done using a stainless wire brush prior to welding, with the goal of removing all oxides, oils and loose particles from the surface to be welded. This is especially important because of the susceptibility of the aluminium weld to porosity due to hydrogen and the dross due to oxygen. The materials to be welded should be rigid enough in order to prevent them from contracting without restraint during welding. This allowed hot cracking to occur during welding and the crack susceptibility to be evaluated [18–20].

For the GTAW tests, a new Fronius Majicwave 5000 Job G/F GTAW machine was used. The pulsed alternating current from this machine could be adjusted and measured. An ESAB CWF1 wire feeder was used for the wire feeding purpose. In the case of GMA welding, a Pro GMA 530 Kemppi welding machine with pulsed alternating current was used in all the experiments.

The resultant welds were cut, polished and etched with a solution of 8% HF and 12% HCl in water for micro structural examination by optical microscopy. The

Table 1 Chemical composition of the base metals and filler wire (wt%)

Al-alloys	Si	Fe	Cu	Mn	Mg	Cr	Zn	Others	Al
6005	0.6	0.21	0.12	0.15	0.54	0.028	0.01	0.15	Bal.
6082	1.2	0.33	0.08	0.50	0.78	0.14	0.05	0.15	Bal.
Nominal composition of alloying in filler wire (wt%)									
	Si	Fe	Cu	Mn	Mg	Zn	Ti	B	Al
4043	5	0.8	0.3	0.05	0.05	0.10	0.20	0.0008	Bal.
5356	0.25	0.4	0.1	0.05–0.20	5	0.14	0.10	0.0008	Bal.

transverse cross-sectional area of each weld was determined with a digital camera and a computer using commercial software.

Results and Discussion

Macro and micro photos of the weld cross-sections and Vickers hardness tests (3 kg) were performed so as to investigate the cracks and the resultant loss of mechanical properties, i.e. hardness due to welding. Each hardness graph shows the hardness across the weld. The acronyms "NoH" and "PH" on the hardness graphs denote not preheated and preheated, respectively. Besides, the letters "FL" represent the fusion line which means the weld interface.

An experiment carried out without filler metal to see the effect it has on crack sensitivity is shown in Fig. 2. When there is no filler metal, hot cracks are located in the weld and heat affected zone. The material used was 6005 – T6 alloy welded by GTAW, current $I = 300$ A and welding speed $v = 2.5$ mm/s, and no filler metal nor preheating (20°C) was applied.

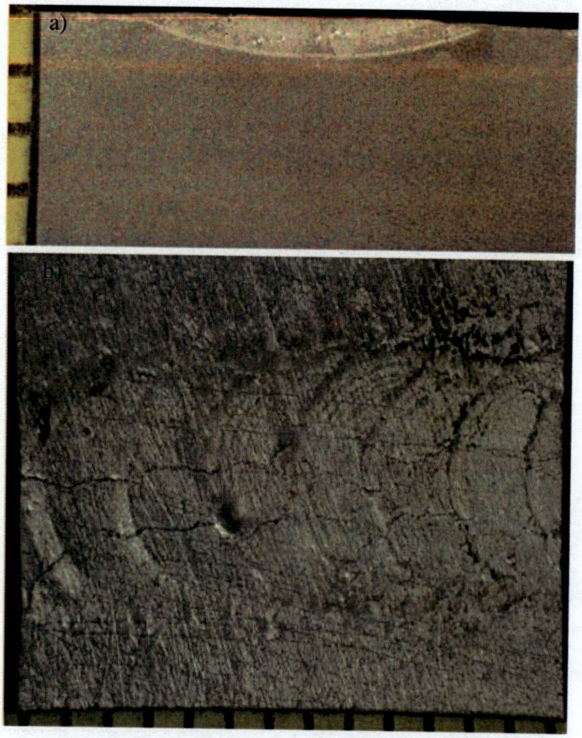

Fig. 2 An example of hot cracking (**a**) in the weld cross-section and (**b**) from top view

Effect of Filler Alloys

Qualitative evaluation was used to present hot cracking test susceptibility; thus, the extent of cracks due to the filler alloys was simply assessed by an observation of the intensity of cracks in the weld cross-section using high magnification of 100 X.

Figure 3a, b show the macrographs of the weld welded with a heat input Q = 440 J/mm and 426 J/mm with alloy 6005 welded with fillers 5356 and 4043, respectively. Open liquation cracks are evident along the outer edge and the root of the weld. The effect of these filler wires in the reduction of cracks can be seen from the macro photos (Fig. 3), and their influence on hardness can be observed from the graphs of hardness tests (Fig. 4).

As can be seen from Fig. 4, on the whole, the hardness distributions are similar in the weld metal, heat-affected zone (HAZ) and base metal. Figure 4a, b present the hardness distributions of 6005 specimens by GTAW and GMAW welded without preheating. The minimum values in both GTAW and GMAW for those welds using filler 4043 are less than 60 HV3. If 6082 metal alloy is welded with the same filler wires, the situation is different. The hardness distribution goes below 60 and slightly above 80 HV3. Thus, there is no significant difference in hardness when using either of the two filler alloys. In fact the specimens welded with filler 4043 showed a HAZ that was, on average, 2 mm wider than those welded with filler 5356 in both GTAW and GMAW.

Fewer and smaller cracks can be observed in the joint that was GMAW welded with 4043 filler wire (Fig. 4a). Hence, it can be concluded that 4043 filler wire is preferable to 5356 wire from the viewpoint of liquation cracking susceptibility in the GTAW and GMAW welding of the 6005 alloy.

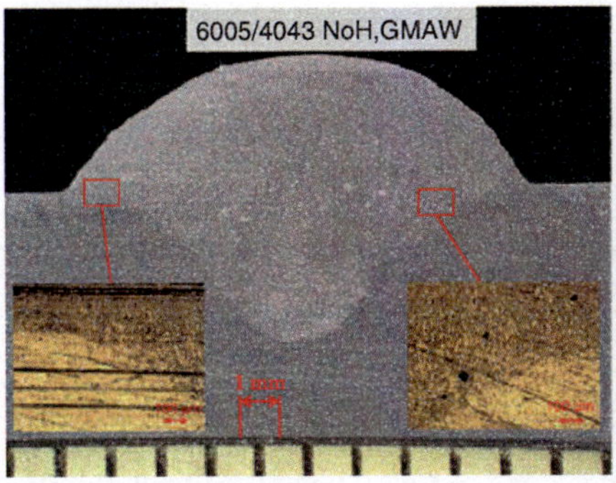

a)

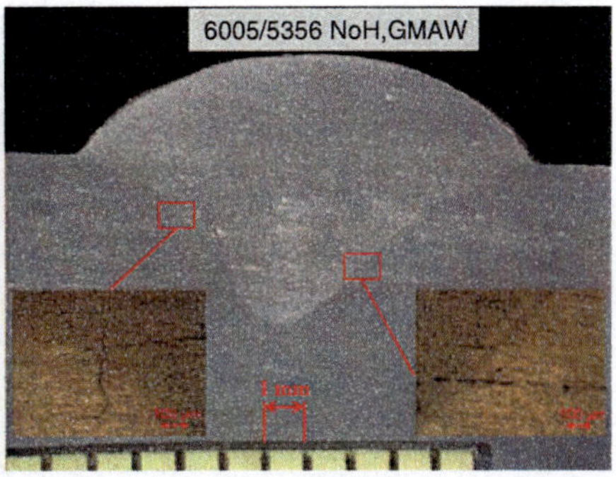

b)

Fig. 3 Macro and micro-sections of the effect of filler metal on liquation cracking

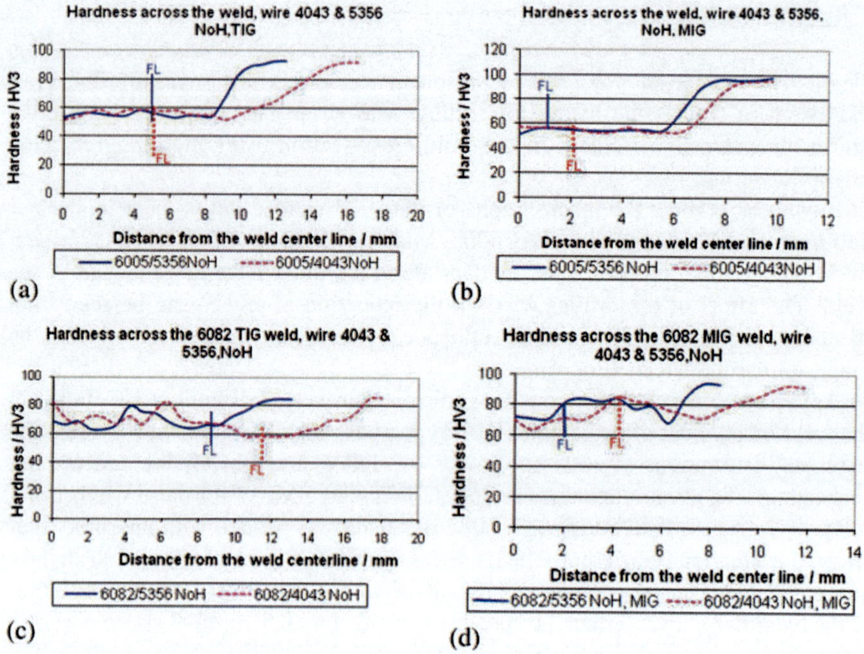

Fig. 4 Hardness profiles of 6005 and 6082 welds with different filler wires

Effect of Welding Process

The dependence of the intensity of cracks on GTAW and GMAW welding processes can be found by comparing Fig. 5a (Q = 440 J/mm) and b (1290 J/mm). There are cracks and few smaller cracks in Fig. 5a, but only one large crack and fewer small cracks in Fig. 5b. But there can be more cracks, including invisible cracks. Accordingly there are more cracks in the GTAW welded specimens than in the corresponding GMAW welded ones. Thus, the GMAW welded samples show less cracks and higher or similar hardness values with a narrower HAZ, than those in GTAW welded ones (Fig. 6).

The difference in the hardness values of the specimens welded with GTAW and GMAW is significant especially with the 6082 alloy; this difference is about 9 HV3 on average (Fig. 6c and d). The smallest difference in hardness values can be observed in the 6005 alloy which is about 1.5 HV3 (Fig. 6a and b). The width of the HAZ in a GTAW welded joint is 4–5 mm greater than that in a GMAW welded joint due to the higher heat input.

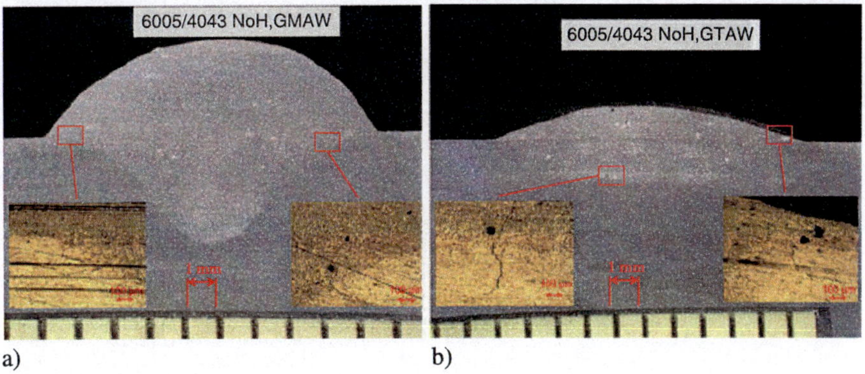

Fig. 5 Macro and micro photos of weld samples welded with 4043 filler metals by GMA & GTA welding processes

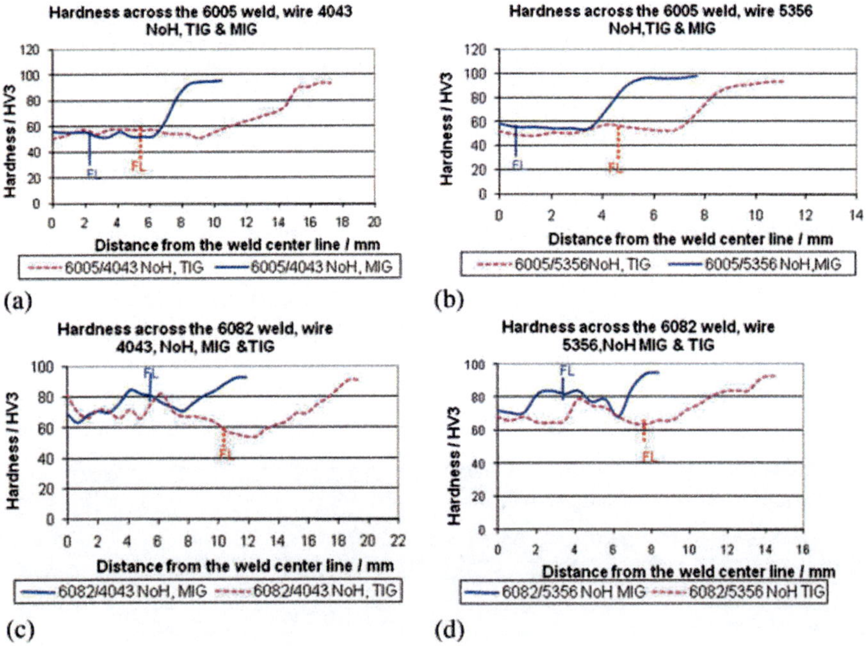

Fig. 6 Hardness graphs in GMAW and GMAW welded joints of alloys 6005 and 6082

Effect of Preheating

Preheating helps in minimizing hot cracking by reduces stresses during cooling. Thus, a lower heat input can also be used. Preheating is heating the whole base metal or part of it up to a desired temperature before welding, if it is recommended. This preheating temperature is mostly found in the welding specification procedure for the alloy at hand, but generally it can range from 110 to 140°C for most commercial aluminum alloys. Preheating can be done in a furnace, if the structure is small, and a bank of heating torches, electrical strip heaters or induction heaters can be required, if the structure is larger. The major benefit of preheating is the lower cooling rate in the weld and the base metal, which leads to better ductility and more resistance to cracking. Especially for highly restrained structures and joints it minimizes the shrinkage stresses in the weld and the adjacent base metal; besides slow cooling enables hydrogen to diffuse before it makes a problem after solidification [21–23]. Aluminum–magnesium–silicon alloys are sensitive to overheating which can form liquation cracking in the HAZ, so proper care has to be taken when preheating [12].

Figure 7 shows macro and micro photos of a weld sample with 4043 filler metal by the GMA welding process with and without preheating. It was noticed that with preheating, the penetration tends to be slightly deeper than without preheating. The reduction of cracks due to preheating is not remarkable, as can be seen when comparing Fig. 7a (Q = 440 J/mm) and b (Q = 440 J/mm). Eventhough in the case of the 6005 preheated GTAW welded with filler 5356 specimen, the size of the cracks decreases somewhat, but there is still approximately the same number of cracks.

The hardness profiles in Fig. 8 demonstrate that there is not more than a 1.5 HV3 difference in the hardness profiles with and without preheating, and there is a 1–2 mm wider HAZ in the preheated specimens. When preheating, the softened area enlarges due to the rise in the temperature.

In all three cases, when we combined both the loss of hardness and number of cracks formed, without considerable hardness differences, the intensity of cracks is less when using the 4043 filler wire than the 5356 wire, as well with the GMAW welded specimen as with the GTAW welded ones. However, the effect of preheating does not show any reduction in the number of cracks.

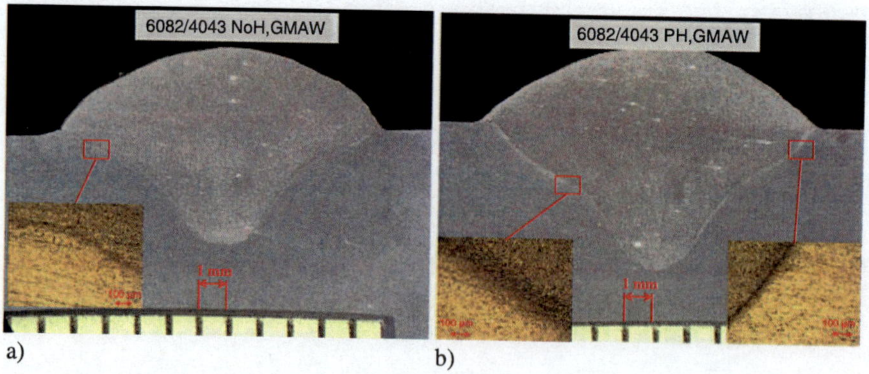

Fig. 7 Macro and micro photos of weld samples welded with 4043 filler metals by GMAW welding process with and without preheating

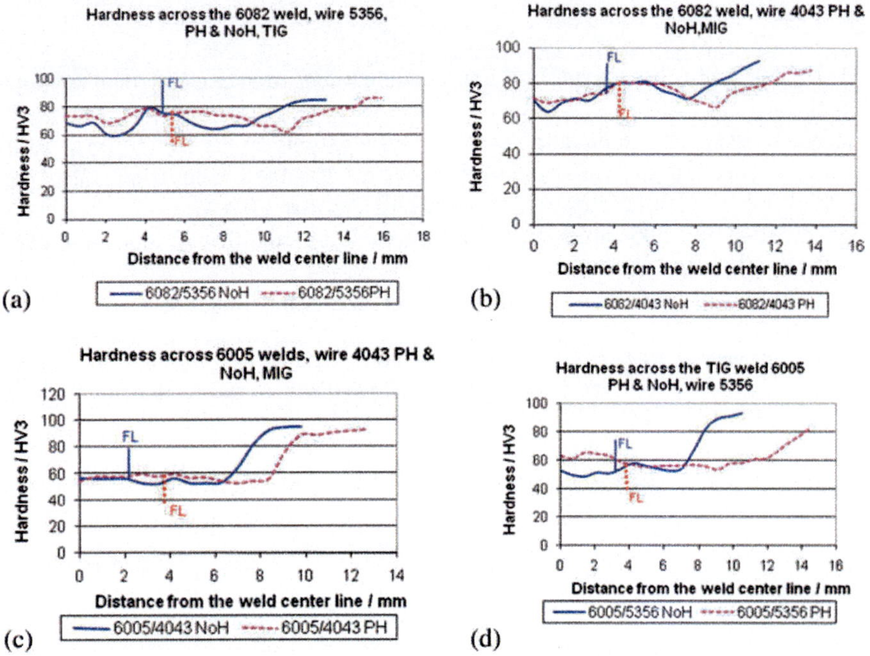

Fig. 8 Hardness graphs of specimens welded with preheating (PH) and without preheating (NoH)

Effect of Base Metal

Thus, it follows that according to this experiment the 6082 alloy is less susceptible to liquation cracking than the 6005 alloy (Fig. 9). The heat input Q = 440 J/mm was equal in both cases.

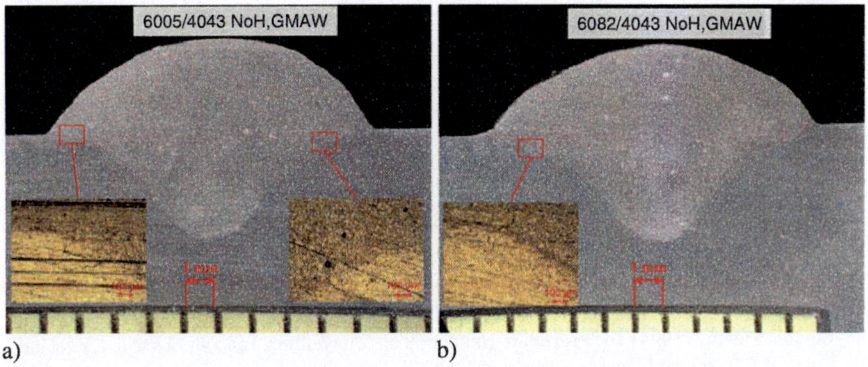

Fig. 9 Macro and micro photos of weld samples welded 6005 and 6082 with 4043 filler metals by the GMAW process without preheating

Conclusions

1. The 6082 alloy base metal is less susceptible to liquation cracking than the 6005 alloy.
2. When assessing the liquation cracking susceptibility in GTA and GMA welding of 6005 and 6082 base alloys, GMAW results in fewer liquation cracks and higher or similar hardness with a narrower HAZ than GTAW.
3. The 4043 and 5356 filler metals are immune to solidification cracking, even though liquation cracking can occur.
4. Preheating has little or no significant effect on avoiding liquation cracks.

Acknowledgements This work was supported by the Finnish Funding Agency for Technology and Innovation (TEKES) under Grant (MODUVA). The authors are grateful to Antti Kähkönen and Antti Heikkinen for providing the test materials.

References

1. Adamowski J and Szkodo M (2007) Friction Stir Welds of Aluminum Alloy AW6082-T6. Journal of Achievements in Materials and Manufacturing Engineering 20:403–406.
2. Cross CE and Böllinghaus T (2006) The effect of restraint on weld solidification cracking in Aluminum. Welding in the world 50:51–54.
3. Mazzolani FM (1995) Aluminium alloy structures, 2nd edition. Chapman & Hall, London.
4. Gittos NF and Scott MH (1981) Heat Affected Zone Cracking of Al-Mg-Si Alloys. Welding Journal 60(6):95–103.
5. Kou S (2003) Solidification and liquation cracking issues in welding. Welding Journal of minerals, metals and materials 55(6):37–41.
6. Huang C and Kou S (2000, May) Partially melted zone in aluminum welds – liquation mechanism and directional solidification. Supl. to Welding Journal 79:113–120.
7. Huang C and Kou S (2002, October) Liquation mechanisms in multicomponent aluminum alloys during welding. Welding Journal:211–222.
8. Huang C and Kou S (2003, July) Liquation Cracking in Partial- Penetration Al-Mg-Si Welds. Welding Journal 82:184–193.
9. Huang C and Kou S (2004, April) Liquation Cracking in Full Penetration Al-Mg-Si Welds. Welding Journal 83:111–121.
10. Cao G and Kou S (2006, January) Predicting and reducing liquation-cracking susceptibility based on Temperature Vs fraction solid. Welding Journal 85:9–18.
11. Hunziker O, Dye D and Reed RC (2000) On formation of a centerline grain boundary during fusion welding. Acta Materialia 48(17):4191–4201.
12. Messler RW Jr (2004) Principle of Welding. Wiley-Vch Verlag GmbH & Co. KGaA, Weinheim.
13. Davis JR Jr (1994) Aluminum and Aluminum alloys. ASM specialty hand book
14. Kou S (2003) Welding Metallurgy, 2nd edition. Wiley, New York, NY
15. Ma T and Den Ouden G (1999) International Journal for the Joining of Materials Denmark 11(3):61–67.
16. Yeomans SR (1990) Successful welding of aluminium and its alloys. Australian Welding Journal, fourth quarter.
17. Grong O (1994) Metallurgical Modeling of Welding, 2nd Edition.
18. Lincoln Electric (1994) The Procedure Handbook of Arc Welding. Lincoln Electric, Cleveland
19. Rao KP, Ramanaiah N and Viswanathan N (2008) Partially melted zone cracking in AA6061 welds. Materials and Design 29:179–186.

20. Schenk T, Richardson IM, Kraska M and Ohnimus S (2009) A study on the influence of clamping on welding distortion. Computational Materials Science 45:999–1005.
21. Akahter R, Ivanchev L and Burger HP (2007) Effect of pre/post T6 heat treatment on the mechanical properties of laser welded SSM cast A356 aluminum alloy. Materials Science and Engineering A 447:192–196.
22. Funderburk S (1998) What is Preheat? Engineering Services the Lincoln Electric Company, Cleveland.
23. Lyndon BJ (2006) Process Specification for the Heat Treatment of aluminum alloys space center. NASA, Houston TX.

Prediction of Liquation Crack Initiation in Al-Mg-Si Alloy Welded Joints

Jukka Martikainen, Esa Hiltunen, Fisseha Brhane, Victor Karkhin, and S. Ivanov

Symbols and Abbreviations

E	elastic modulus, Pa
f_L	liquid fraction
f_S	solid fraction
ΔH	enthalpy, J g^{-1}
L	latent heat, J g^{-1}
T	temperature, K
T_0	initial (preheating) temperature, K
T_C	coherency temperature, K
T_{CB}	coherency temperature of base metal, K
T_E	eutectic temperature, K
T_{EB}	eutectic temperature of base metal, K
T_{EW}	eutectic temperature of bulk weld metal, K
T_L	liquidus temperature, K
T_{LB}	liquidus temperature of base metal, K
T_{LW}	liquidus temperature of bulk weld metal, K
$T_{\max}$	peak temperature, K
T_S	solidus temperature, K
t	time, s
X	parameter of liquation crack initiation
X_{cr}	critical parameter of liquation crack initiation
ε	strain
ε^T	thermal strain
ε_a^p	accumulated plastic strain
$\varepsilon_{a\,cr}^p$	critical accumulated plastic strain
$\varepsilon_{eq\,a}^p$	accumulated equivalent plastic strain
$\varepsilon_{eq\,a\,cr}^p$	critical accumulated equivalent plastic strain

J. Martikainen (✉)
Department of Mechanical Engineering, Lappeenranta University of Technology, Lappeenranta, Finland
e-mail: jukka.martikainen@lut.fi

J. Lippold et al. (eds.), *Hot Cracking Phenomena in Welds III*,
DOI 10.1007/978-3-642-16864-2_5,

$\varepsilon_x^p, \varepsilon_y^p, \varepsilon_z^p$	components of plastic strain
$\sigma_x, \sigma_y, \sigma_z$	stress, Pa
σ_Y	yield stress, Pa
FDZ	Filler-deficient zone
PMZ	Partially melted zone

Introduction

Al-Mg-Si alloys are susceptible to liquation cracking in the partially melted zone (PMZ) [1–10]. This zone is the region immediately outside the fusion zone where liquation occurs during welding due to heating above the eutectic temperature.

Liquation cracking has also been called edge-of-weld cracking, base-metal cracking, heat-affected zone cracking, and hot cracking. These terms refer to a weldability concept. According to standard ISO 581 the latter includes a combination of four variables: material, process, type of construction and purpose. This means that for liquation cracking not only do material properties influence liquation cracking, but also welding conditions and structure rigidity.

Two kinds of hot cracking criteria are available in the literature: non-mechanical and mechanical. The former type deals with vulnerable temperature range, phase diagram, and process parameters [8, 9, 11]. It allows one to select welding consumables, however it does not allow, for example, such important factors as restraint to be taken into account. The latter type of criteria involves critical stress, critical strain or critical strain rate. A comprehensive coverage of mechanical criteria has been presented by L. Katgerman and coworkers [12, 13]. The ultimate liquation cracking criterion needs to combine aspects of thermal history, shrinkage, evolution of the semi-solid microstructure and mechanical behavior. However, none of the existing criteria can give the appropriate answer on whether the liquation crack will appear or not and what will be its position, shape and length. The main reason for this is the lack of understanding of the mechanisms involved in liquation crack formation.

The aim of this study is to develop a technique to predict liquation crack initiation in an aluminum weld joint with consideration for the material properties, welding conditions and welded structure rigidity. The technique is demonstrated with the example of arc welding of a heat-treatable 6005-T6 alloy.

Liquation Cracking Criterion

Liquation cracking is an example of mechanical failure and, therefore, it can be treated adequately using the apparatus of fracture mechanics. When formulating a mechanical criterion the following three scale levels of the model can be used. (i) In a microscopic model a small two phase volume element is treated with the element size being the distance between neighboring columnar dendritic axes and of the length of the mushy zone [14]. It operates with macroscopic mechanical behavior and microscopic phenomena like feeding and porosity formation. The main difficulty of this model is to assess the critical state of a liquid film which strongly depends on, for example, the local composition (microsegregation) and the hydrogen content [15]. (ii) In a macroscopic model, the volume is mushy and its mechanical properties are space-averaged over the volume. Critical values of stress and strain can be found experimentally. (iii) The megascopic model is a big and rigid structure including the weld, base metal, restraints, etc. Such technological probes allow one to compare the effects of base metal, welding conditions, welding consumables, constraints and give answers in the form "crack-no crack", "better-worse", but does not practically allow a look inside the phenomenon to give an answer to the question: will a liquation crack occur in the real welded structure?

In the present study, a macroscopic model is taken. The condition of liquation crack initiation can be written as

$$X = X_{cr}$$

where X is a parameter characterizing a local mechanical state of the vulnerable weld zone, and X_{cr} is the critical value of this parameter characterizing the local mechanical properties of the material in this zone.

Hot cracking is generally believed to result from the tensile fracture of a liquid film at the grain boundary. Such fracture has to be expressed in terms of hydrostatic stress. However so far determination of the local stress and, especially, its critical value (cavitation pressure) faces insuperable difficulties. This problem is more difficult to solve than another well-known problem which remains to be solved: to predict ultimate tensile strength at room temperature using detailed information about microstructure only. Therefore, in this study an accumulated plastic strain ε_a^p is taken as the fracture parameter for the following reasons. Assume that there is no fluid flow from the molten pool to the partially melted zone (PMZ) to compensate shrinkage (this condition can exist as it will be shown later). Our calculations have shown that under solidification the metal at the PMZ undergoes nearly proportional (simple) loading. Consequently the relationship between averaged accumulated strain and tensile hydrostatic stress in a liquid film is single-valued, that is, the liquid state can be evaluated in terms of strain. The advantage of such an approach lies in the possibility to find the critical value of the strain experimentally.

This approach raises the question of what temperature range should be taken for the calculation of strain ε_a^p? To give the answer let us consider the zones of the welded joint when the liquidus temperature of bulk weld metal T_{LW} is lower than that of the base metal T_{LB} (Fig. 1). For instance, the condition $T_{LW} \leq T_{LB}$ corresponds to recommended welding of 6005 aluminum alloy ($T_L = 925$ K) with 4043 filler wire ($T_L = 632$ K) or 5356 filler wire ($T_L = 635$ K). The solidification front is no longer isothermal as in welding without a dissimilar filler metal.

A liquid stagnant layer (other names for it are laminar-flow layer, filler-deficient zone) extends along the molten pool boundary and its thickness may range up to 200 μm [16] or one order of magnitude greater than the subgrain diameter

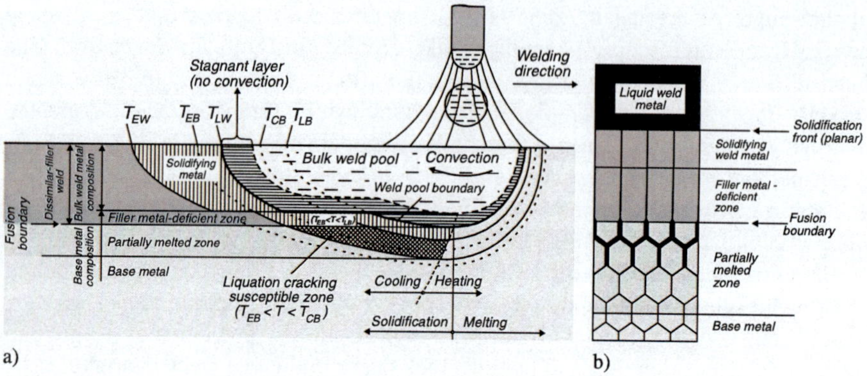

Fig. 1 Schematic of weld zones in longitudinal section (**a**) and microstructure (**b**). The liquidus temperature of bulk weld metal T_{LW} is lower than the liquidus temperature of base metal

(5 ... 20 μm). The stagnant layer can solidify as a filler-deficient beach along the weld interface, often thin and discontinuous because of weld pool convection. The actual thickness of the layer depends much on the strength and direction of weld convection. Solidification mode of the stagnant layer is partially or entirely planar (Fig. 1b) as solidification rate is low and the temperature gradient is high.

The composition of the filler-deficient zone (FDZ) is uniform as the diffusion coefficients in liquid are relatively high and low-velocity laminar flow along the solidification front equalizes additionally the concentration even if microsegregation prior to welding was considerable. The local chemical composition of FDZ can be calculated by solving the diffusion problem with consideration for main factors [17]. Many experiments have demonstrated that FDZ is chemically similar to the base metal [18].

As the FDZ is chemically uniform and solidification mode is planar its metal solidifies at the liquation temperature of the base metal T_{LB}. Just after solidification of FDZ the subgrain boundaries in PMZ are partially liquid because the solutes have not enough time to diffuse and diffusion coefficients for solid are relatively low. Solid film in the FDZ can freely move towards the PMZ when the latter shrinks and is a slurry. Later, at the coherency temperature T_C, when the solid network in the PMZ is coherent, it can sustain and transmit stresses. Therefore, the coherency temperature has to be taken as the temperature onset for the accumulated strains ε_a^p.

It should be emphasized that through time the semi-solid material can accommodate the imposed thermal strain by elastic and plastic deformation without filling of the gaps and pores with the solid. Here time is a parameter of loading (accumulation of strains and damages) and is not a parameter of diffusion process. Temperature can also be such a parameter. If it is granted that the derivative of strain with respect to temperature is constant, $\partial\varepsilon/\partial T = \text{const}$, the accumulated strain is uniquely determined by this derivative. Such a mechanical criterion of hot cracking formation based on comparison of derivative $\partial\varepsilon/\partial \mathrm{T}$ with the ductility curve was suggested by N.N. Prokhorov more than 50 years ago [19].

Materials

The base metal was a solution heat-treated and stabilized 6005-T6 aluminum alloy. The chemical composition of the base and filler metals is presented in Table 1.

In order to find the plastic strain ε_a^p accumulated within the coherency range of the base metal (between the coherency temperature T_C and the eutectic temperature T_E) it is necessary to solve the heat conduction and thermomechanical problems. This requires the temperature-dependent properties including enthalpy, dilatogram (dilatation curve), yield stress, elastic modulus and ductility. These properties are scarcely available and need to be determined experimentally using reliable experimental techniques.

The Gleeble® 3800 System were used in this study. This physical simulator allows for testing of hot cracking susceptibility [20]. The test specimens were 6 mm in diameter and 100 mm long. The water-cooled copper jaws ensured the required cooling rate and uniform temperature distribution across the specimen and over a distance more than 10 mm. The temperature was measured using a Ni-Cr vs Ni-Al thermocouple 0.2 mm in diameter embedded in the hole and located in the middle of the specimen. The temperature was maintained with an accuracy of 1 K during holding. The C-gauge was used to measure an axial displacement at the center of the specimen with an accuracy of 10 μm. The load was fixed accurate to 0.1 N.

The enthalpy and liquid fraction have been calculated by using Polytherm codes developed at the St. Petersburg Polytechnic University [21] (Fig. 2). These codes allow one to calculate phase fractions, enthalpy of the aluminum-based alloys and their phases. It is seen from Fig. 2 that the latent heat L constitutes 40% of the enthalpy at the liquidus temperature T_L. The liquid fraction f_L is assumed to be independent on cooling rate (back diffusion in solid is neglected).

Figure 3a shows that the thermal cycle type does not influence the thermal strain profile $\varepsilon^T(T)$. The curve for the base metal was calculated using the Polytherm codes, the other two were calculated knowing the fraction solid vs. temperature (Fig. 3b) [9]. It is evident from the curves that the solidification shrinkage is much higher than the thermal contraction of the solid and it influences strongly the stress formation at temperature above the eutectic temperature T_E. Above the temperature

Table 1 Composition of base and filler metals

	Element, wt-%				
Alloy	Mg	Si	Cu	Mn	Zn
6005[a]	0.582	0.704	0.108	0.199	0.045
4043[b]	0.05	5.2	0.30	0.05	0.10
5356[b]	5.00	0.25	0.10	0.05	0.10

[a]Measured by using PolySpek-F (ARUN technology)
[b]Huang and Kou [9]

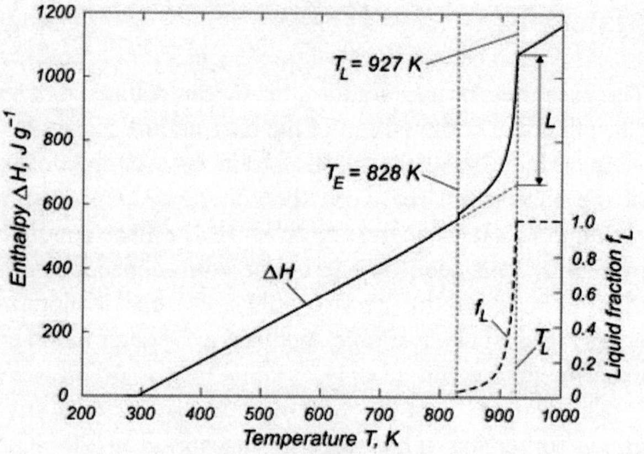

Fig. 2 Enthalpy ΔH and liquid fraction f_L vs. temperature for 6005 alloy

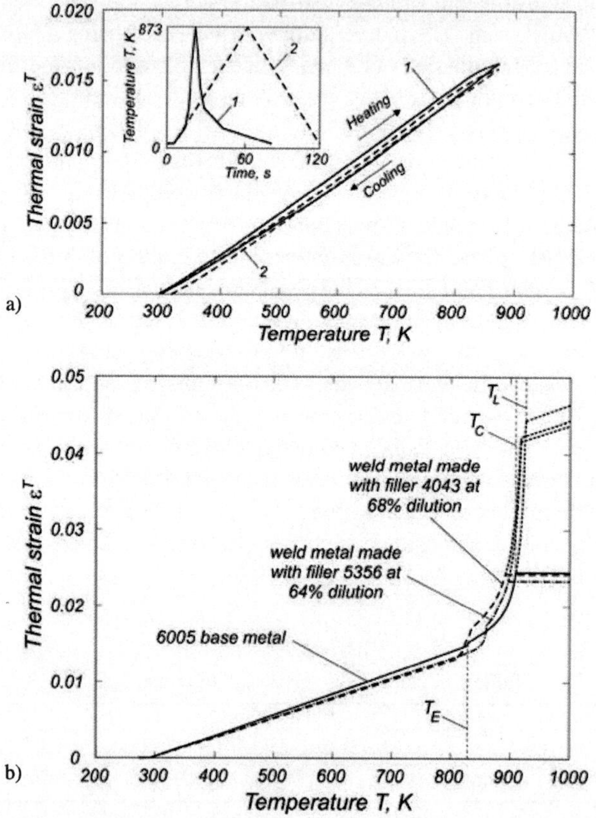

Fig. 3 Dilatograms: **(a)** – experimental curves for 6005 alloy (1 – welding thermal cycle, 2 – thermal cycle at constant heating and cooling rates); **(b)** – curves used in calculation for 6005 alloy and weld metal

T_E the thermal strain can be neglected because the metal is a slurry (flow is free without hot tears).

Heating of 6005-T6 alloy leads to a very sharp reduction in the yield stress σ_Y in the PMZ of the weld (Fig. 4). Residual yield stress decreases dramatically after heating above 600 K. The temperature dependence of the elastic modulus E is more conservative (Fig. 5).

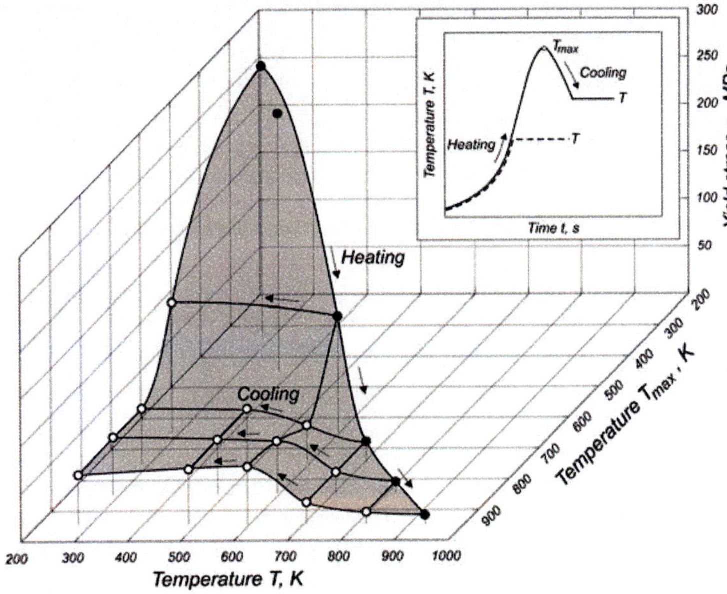

Fig. 4 Yield stress σ_Y vs. temperature T for 6005-T6 alloy tested at temperature T during heating and cooling

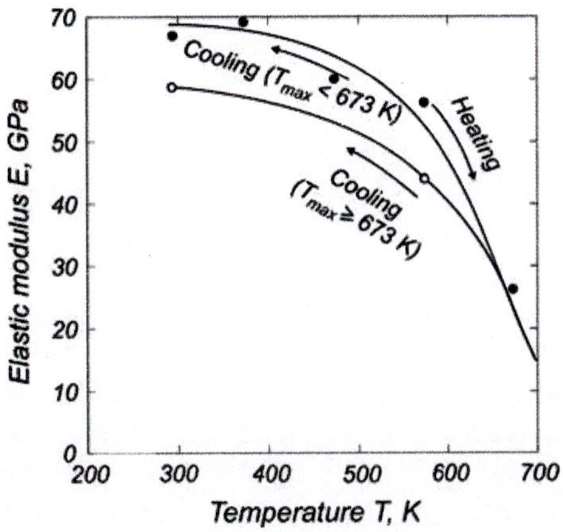

Fig. 5 Temperature dependence of elastic modulus for 6005-T6 alloy

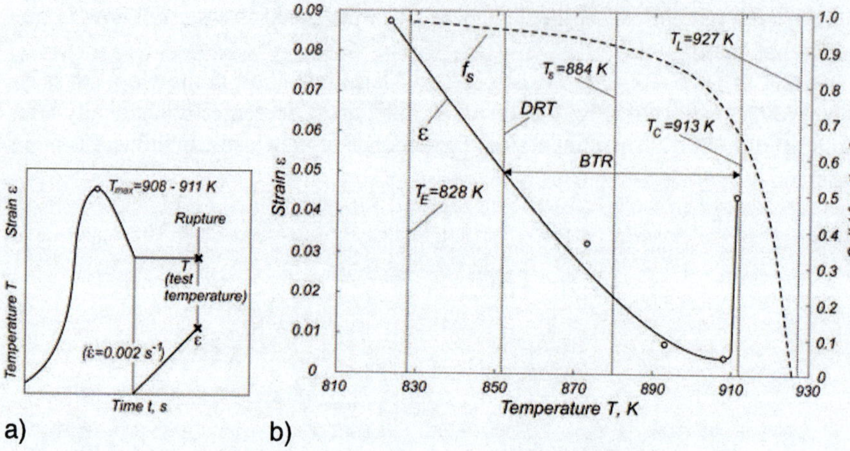

Fig. 6 Schematic view of thermal cycles and tensile loading under constant strain rate $\dot{\varepsilon}$ and constant temperature T (**a**) and ductility curve ε and solid fraction fs (**b**)

Ductility is the ability of a material to be plastically deformed without fracture (Fig. 6). After heating and cooling, the metal was tensile tested under constant temperature T and constant strain rate $\partial\varepsilon/\partial t = 0.002\,\text{s}^{-1}$ which corresponds to the average thermal strain rate in welding (Fig. 6a). The brittle temperature range BTR lies within 852 to 813 K. The ductility recovery temperature DRT relates to 5% strain. The coherency temperature T_C is 913 K which corresponds to the solid fraction $fs = 0.65$, that is, it falls the range between 0.50 and 0.80 [13]. The ductility curve shows a well-defined trough between the solidus temperature T_S and the coherency temperature T_C. The curve shape is similar to that for Al-Mg-Si alloys 6061 (0.88% Mg, 0.66% Si) and 6063 (0.50% Mg, 0.45% Si) [12]. We emphasize that in the experiments the metal is only partially melted and filling phenomenon does not take place in the testing.

The thermophysical properties of the weld metal were taken as those of the base metal. The dilatogram is presented in Fig. 3b.

Welding Conditions

Gas metal arc welding (GMAW) and gas tungsten arc welding (GTAW) with and without filler metal were used (Table 2). Initial (preheating) temperature T_0 was 293 K and 393 K. Pure argon was used as a shielding gas. Tungsten electrode diameter was 3.2 mm. Filler wire electrode diameter was 2.4 mm, and stickout was 14 mm. The weld was made on a beam using bead-on-plate technique (Fig. 7a). The beam was mechanically free.

Table 2 Welding conditions

Test	Process	Filler wire	Preheating temperature (K)	Current (A)	Voltage (V)	Welding speed (mm s^{-1})	Gross heat input (J mm^{-1})
1	GMAW	4043	393	245	27.6	8.3	815
2	GTAW	4043	393	300	13.7	2.5	1,644
3	GTAW	4043	293	300	13.7	2.5	1,644
4	GMAW	5356	393	245	27.3	8.3	806
5	GTAW	–	293	300	15.0	2.5	1,800

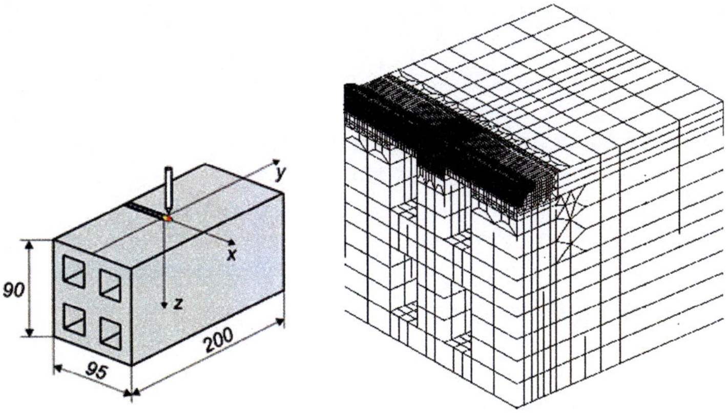

a) b)

Fig. 7 Beam and reference frame (**a**) and finite element mesh (half of the beam, y≥0) (**b**)

Metallography

In the absence of a filler metal, many solidification cracks in the weld and liquation cracks in PMZ were observed (Test 5, Table 2). By using 4043 and 5356 filler wires in GMAW and GTAW the solidification cracking was completely prevented. The dilution ratio (the ratio of the amount of the base metal melted to the total amount of the weld metal) lies from 0.38 to 0.61. The disadvantage of 4043 wire is non-uniformity of color comparing base alloy and weld metal after anodizing.

The experiments have shown that liquation cracking occurs under all welding conditions (Table 2). While some of the liquation cracks are in contact with the molten pool, others seem to emerge at a distance from the fusion boundary. Figure 8 shows liquation cracking in the weld cross-section. The cracks are seen predominantly at the grain boundaries where solute concentration is high and the metal is last to solidify. It is worth noting that it is difficult to judge the entire crack surface shape and dimensions from the cross-section. For instance, if the angle between the crack surface and the weld cross-section is small then the intersection line can be parallel to the plate surface as is seen in Fig. 8.

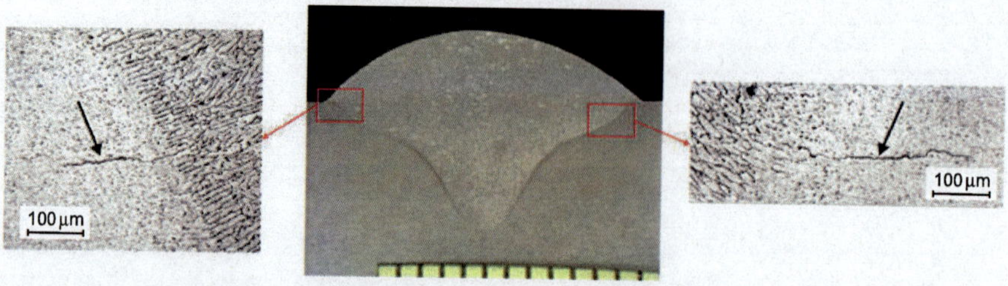

Fig. 8 Liquation cracking in weld cross-section (GMAW, 4043 filler, T_o=393 K)

Thermal and Mechanical Modeling

As stated in the section "Liquation cracking criterion", the liquation cracking sensitivity is linked to the accumulated plastic strains undergone by the PMZ metal during cooling below the coherency temperature T_C. The present section details the finite element model used to assess the temperature distributions and accumulated plastic strains.

Geometry and Mesh

Due to symmetry only one half of the whole domain is meshed (Fig. 7b). The arc travels along x-axis and mesh is refined in the regions that undergo melting and solidification, especially in the PMZ. In these regions the 8-node brick elements are 0.556 mm long and of 0.167 mm × 0.185 mm section. The mesh has 66,810 elements and 69783 nodes. ANSYS codes were used.

Heat Source and Heat Conduction Problem

The parameters of the apparent heat source which simulates the effects of welding arc, deposited liquid metal and convection are obtained by using inverse modeling [22]. It was assumed that the thermal conductivity was 0.172 W mm^{-1} K^{-1}, and the coefficient of heat transfer was 0.00005 W mm^{-2} K^{-1}. The enthalpy is presented in Fig. 2.

A comparison between calculated and experimental molten pool shapes, weld cross-sections and thermal cycles shows good agreement (Figs. 9 and 10).

Mechanical Modeling

We assumed that the heat conduction problem and the thermomechanical problem are not coupled. The mechanical properties of the base metal are shown in Figs. 3–5.

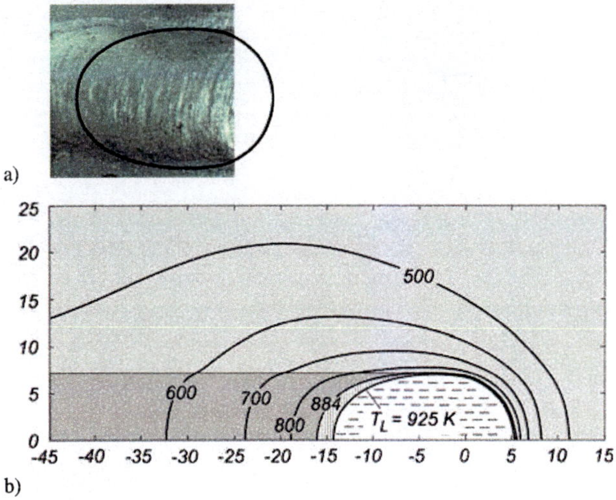

Fig. 9 Top view of the weld (calculated weld pool contour is shown as a *black line*) (**a**) and calculated temperature field around arc on the top of the beam at 5.7 s after start of welding (symmetry half is shown) (**b**) (GMAW, 4043 filler, $T_o = 393$ K)

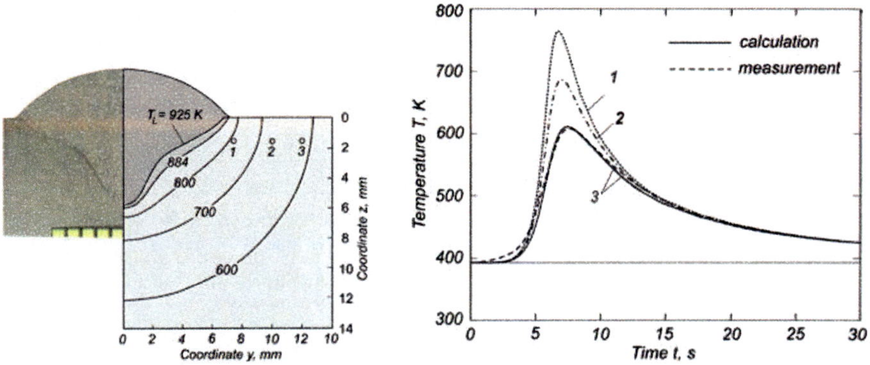

Fig. 10 Weld cross-section and peak temperature field (**a**) and thermal cycles at points 1–3 of heat-affected zone (**b**)

Results and Discussion

Transient Strains

Let us consider the following base case: GMAW, 4043 filler wire, preheating temperature is 393 K (Table 2, Fig. 8). For analysis of transient stresses and strains, two separate characteristic points in the PMZ were considered at the middle cross-section ($x = 0$): point A near the weld toe and point B just below the weld interface (Fig. 11a).

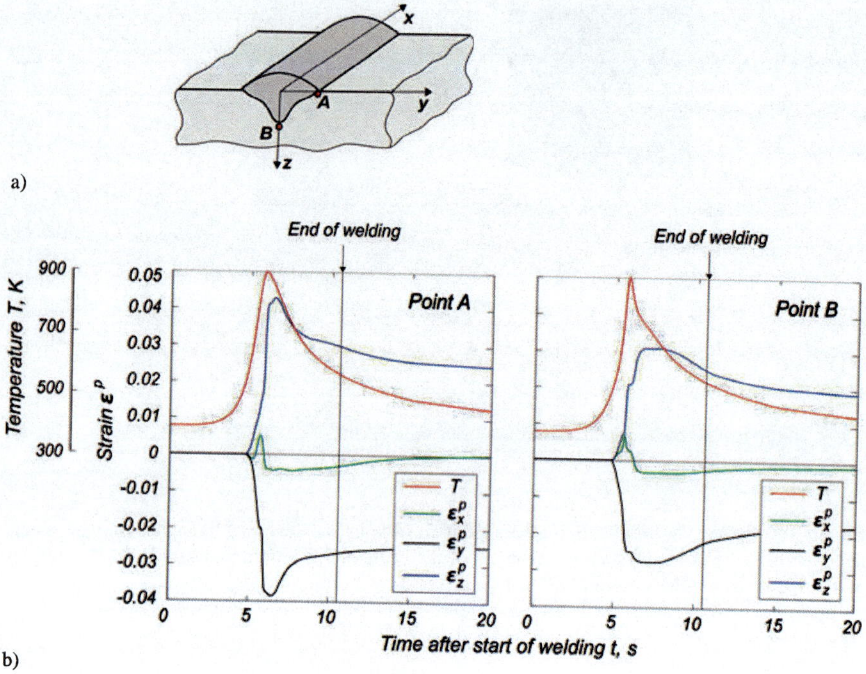

Fig. 11 Coordinate system (**a**), thermal cycles and transient plastic strains (**b**) in the middle cross section of the weld (GMAW, 4043 filler, $T_o = 393$ K)

On heating, the longitudinal plastic strains ε_x^p and transverse plastic strains ε_y^p are negative (Fig. 11b). On cooling, the strain rates are positive, $\partial\varepsilon_x^p/\partial t > 0$ and $\partial\varepsilon_y^p/\partial t > 0$. Pronounced strain rates are observed at the brittle temperature range, from 913 K down to 852 K. Accumulated plastic strains in this range determine sensitivity of the welded joint to liquation cracking.

Prediction of Liquation Crack Initiation

For prediction of liquation cracking initiation, at first the following question should be answered: which plastic strain components should be taken into account in analysis? Our observation suggests that liquation cracks are oriented predominantly along the stretched grains. This means that the critical strain value $\varepsilon_{a\,cr}^p$ depends on grain boundary orientation. For simplicity, take the orientation-independent equivalent plastic strain accumulated at the temperature below the coherency temperature T_C:

$$\varepsilon_{eq\,a}^p(T) = \int_{T_c}^{T} d\varepsilon_{eq}^p(T)$$

where

$$d\varepsilon_{eq}^{p} = \frac{\sqrt{2}}{3}\left[\left(d\varepsilon_{x}^{p} - d\varepsilon_{y}^{p}\right)^{2} + \left(d\varepsilon_{y}^{p} - d\varepsilon_{z}^{p}\right)^{2} + \left(d\varepsilon_{z}^{p} - d\varepsilon_{x}^{p}\right)^{2} + \frac{3}{2}\left(d\gamma_{xy}^{p\,2} + d\gamma_{yz}^{p\,2} + d\gamma_{zx}^{p\,2}\right)\right]^{1/2}$$

In general, material damage is conditioned by accumulated equivalent plastic strains.

Assume that the ductility curve obtained under one-dimensional tension is valid also for a three-dimensional strain state (the hypothesis of a single curve is valid). Then it is possible to predict the location and moment of liquation crack initiation by comparing the calculated curve $\varepsilon_{eq\,a}^{p}(T)$ with the experimental ductility curve (critical value) $\varepsilon_{eq\,a\,cr}^{p}(T)$. Hence the condition for liquation crack initiation is as follows:

$$\varepsilon_{eq\,a}^{p} = \varepsilon_{eq\,a\,cr}^{p}$$

The local state of material $\varepsilon_{eq\,a}^{p}$ depends on structure rigidity, properties of base and weld metals, welding conditions, etc. It can also depend on external loading as, for example, in Varestraint, Transvarestraint, and MVT test [23]. The local mechanical property of the PMZ metal $\varepsilon_{eq\,a\,cr}^{p}$ depends on local composition, fracture solid, grain orientation, etc.

Figure 12 shows that in the case under study, liquation cracks form at relatively high temperature (open circle indicates liquation crack initiation). The weld toe

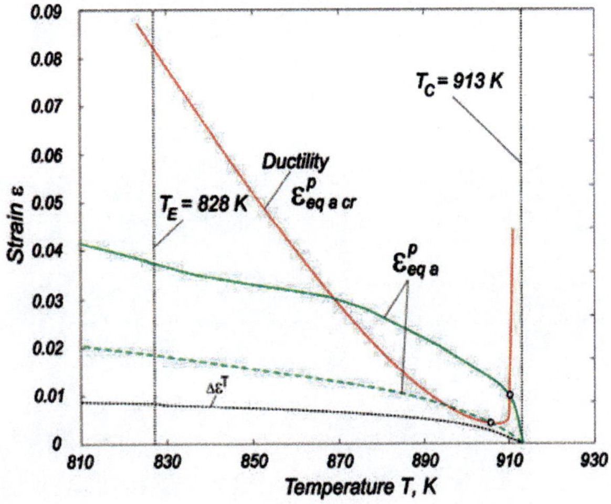

Fig. 12 Comparison of the ductility curve $\varepsilon_{eq\,a\,cr}^{p}$ and calculated accumulated equivalent plastic strains $\varepsilon_{eq\,a}^{p}$ at vulnerable parts of partially melted zone (- point A, - - - - - point B) (GMAW, 4043 filler, $T_o = 393$ K)

(point A) is more susceptible to liquation cracking than the area just under the weld (point B) because it has higher strains. The plastic strains are far above the accumulated thermal strain $\Delta\varepsilon^T(T) = \Delta\varepsilon^T(T_C) - \Delta\varepsilon^T(T)$ due to a strain concentration effect. The thermal strain can be used in liquation cracking analysis with caution (the error is not on the safe side).

The microcracks can propagate according to multi-phase fracture mechanics because the transient and residual stresses in the PMZ are tensile (Fig. 13). Residual stresses are relatively low because the residual yield stress of the unhardened alloy is low.

Hence, the developed technique allows the assessment of liquation crack initiation. To do this requires the base metal properties, including the ductility curve, and solution of the thermomechanical problem. More adequate mathematical models can be developed only on the basis of more accurate physical models. At the

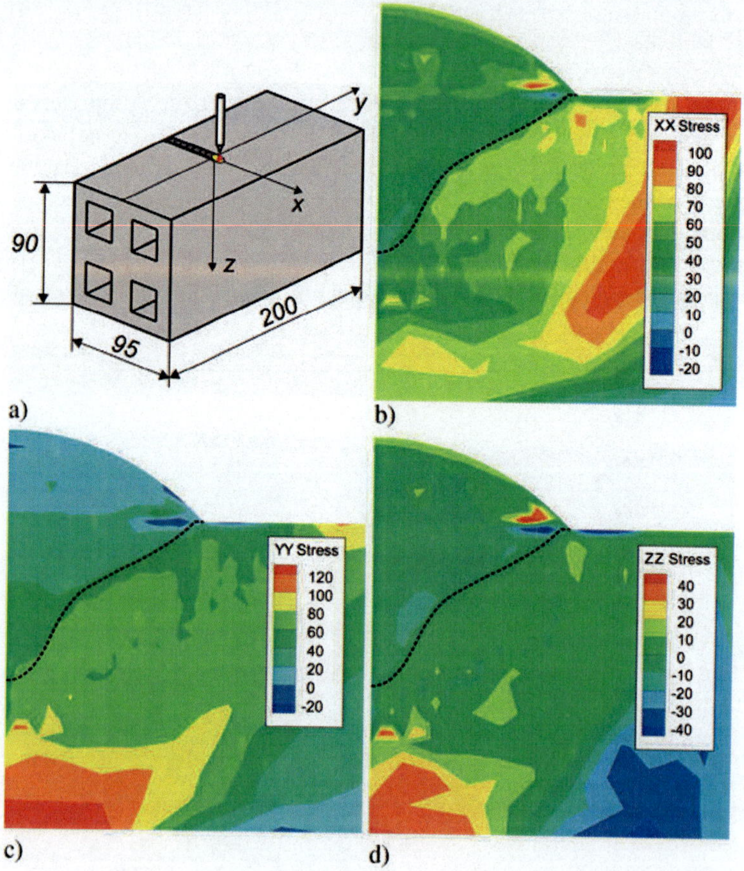

Fig. 13 Reference frame (**a**) and residual stress field σ_x (**b**), σ_y (**c**), and σ_z (**d**) in the middle cross-section of the beam during GMA welding (4043 filler, preheating 393 K)

moment the main impediment to the development of such models is lack of understanding of the mechanisms of liquation crack initiation and propagation. Further study of metallurgical and mechanical behavior of two-phase metal is necessary to understand these mechanisms in order to find more accurate criterion of liquation cracking.

Conclusions

1. The developed technique allows the prediction of liquation crack initiation in a rigid structure during gas metal and tungsten arc welding of aluminum heat-treatable 6005-T6 alloy.
2. Heating of 6005-T6 alloy leads to a sharp reduction in the yield stress in the heat-affected zone
3. The 6005-T6 alloy has a pronounced ductility drop at 850–910 K when solid fraction is over 65%.
4. The semi-solid metal in the partially melted zone of the rigid welded joint is subjected to relatively high plastic strain in the range between the coherency temperature and the eutectic temperature. The accumulated strains exceed the ductility of the base metal resulting in liquation cracking.
5. On cooling, the transient and residual stresses at the partially melted zone are tensile, and they may cause the liquation cracks to propagate
6. Further study of metallurgical and mechanical behavior is necessary to understand the mechanisms of liquation cracking in Al-Mg-Si alloy welded joints in order to find more accurate criterion of cracking.

Acknowledgements The authors thank Professor C.E. Cross for discussion and constructive remarks. Professor V.M. Golod is greatly thanked for providing the Polytherm computations.

References

1. Metzger G (1967) Some mechanical properties of welds in 6061 aluminum alloy sheet. Welding Journal 46:457-s–469-s.
2. Gittos N and Scott M (1981) Heat-affected zone cracking of Al-Mg-Si alloys. Welding Journal 60:95-s–103-s.
3. Kerr H and Katoh M (1987) Investigation of heat-affected zone cracking of GTA welds of Al-Mg-Si alloys using the Varestraint test. Welding Journal 66:251-s–259-s.
4. Katoh M and Kerr H (1987) Investigation of heat-affected zone cracking of GTA welds of Al-Mg-Si alloys using the Varestraint test. Welding Journal 66:360-s–368-s.
5. Miyazaki M, Nishio K, Katoh M, Mukae S and Kerr H (1990) Quantitative investigation of heat-affected zone cracking in aluminum alloy 6061. Welding Journal 69:362-s–371-s.
6. Ellis M, Gittos N and Hadley I (1997) Significance of liquation cracks in thick section Al-Mg-Si alloy plate. The Welding Institute Journal (UK) 6:213–255.
7. Kou S (2003) Welding metallurgy, 2nd edition. Wiley, New York, NY.
8. Huang C, Cao G and Kou S (2004) Liquation cracking in partial penetration aluminium welds: assessing tendencies to liquate, crack and backfill. Science and Technology of Welding and Joining 9:149–157.

9. Huang C and Kou S (2004) Liquation cracking in full-penetration Al-Mg-Si welds. Welding Journal 83:111-s–122-s.
10. Rao K, Ramanaiah N and Viswanathan N (2008) Partially melted zone cracking in AA6061 welds. Materials and Design 29:179–186.
11. Cao G and Kou S (2006) Predicting and reducing liquation-cracking susceptibility based on temperature vs. fraction solid. Welding Journal 85:9-s–18-s.
12. Eskin D, Suyitno and Katgerman L (2004) Mechanical properties in the semi-solid state and hot tearing of aluminium alloy. Progress in Materials Science 49:629–711.
13. Katgerman L and Eskin D (2008) In search of the prediction of hot cracking in aluminium alloys. In: Boellinghaus, T, Herold, H, Cross, C, Lippold, J (Eds) Hot Cracking Phenomena in Welds II. Springer, Berlin, pp. 3–18.
14. Rappaz M, Drezet J and Gremaud M (1999) A new hot-tearing criterion. Metallurgical and Materials Transactions A 30A:449–455.
15. Coniglio N and Cross C (2009) Mechanism of solidification crack initiation and growth in aluminium welding. Metallurgical and Materials Transactions A 40A:2718–2728.
16. Yang Y and Kou S (2007) Fusion boundary macrosegregation in dissimilar-filler metal Al-Cu welds. Welding Journal 86:331-s–339-s.
17. Rajamaki P, Karkhin V and Homich P (2010) Analysis of macrosegregation near fusion boundary in fusion welding. Science and Technology of Welding and Joining 15:31–39.
18. Yang Y and Kou S (2010) Microsegregation in Al-Si welds made with dissimilar filler metals. Science and Technology of Welding and Joining 15:1–14.
19. Prokhorov N (1952) Hot cracking in welding. Mashgiz, Moscow (in Russian).
20. Mandzey S (2005) Testing for susceptibility to hot cracking on Gleeble physical simulator. In: Boellinghaus T, Herold H (Eds) Hot Cracking Phenomena in Welds. Springer, Berlin, pp. 347–376.
21. Savelyev K and Golod V (2010) Program module for determination of temperature-dependent characteristics of aluminum-based alloys in modeling of casting processes. Proceedings of the 8th Conference "Castings Today and Tomorrow", SPbGPU, St. Petersburg, Russia, pp. 373–386 (in Russian)
22. Karkhin V, Plochikhine V, Ilyin A and Bergmann H (2002) Inverse modelling of fusion welding processes. In: Cerjak H, Bhadeshia HKDH (Eds) Mathematical Modelling of Weld Phenomena 6. Maney Publishing, London, pp. 1017–1042.
23. Lippold J (2005) Recent developments in weldability testing. In: Boellinghaus T, Herold H (Eds) Hot Cracking Phenomena in Welds. Springer, Berlin, pp. 271–290.

Hot Tearing Sensitivity of Al-Mg-Si Alloys Evaluated by X-Ray Microtomography After Constrained Solidification at High Cooling Rate

E. Giraud, M. Suéry, J. Adrien, E. Maire, and M. Coret

Introduction

It is well-known that some aluminum alloys are sensitive to hot cracking such as Al-Cu or Al-Mg-Si alloys [1–3]. During welding processes, cracks can form in the molten zone thus limiting their applications. To reduce this phenomenon, many solutions are available among them changing the alloy composition by using a filler wire [1, 2]. In the case of Al-Mg-Si alloys, it is common to use a filler alloy from the Al-Si or Al-Mg series. However, the composition of the molten zone, which is required to reduce hot tearing susceptibility, is still not precisely known since it depends on the base alloy, the alloy used for the filler wire and the welding conditions. Consequently, to determine the appropriate composition, it is necessary to investigate the behavior of various mixtures of the base and filler alloys under conditions close to those encountered during welding.

Many techniques have been developed to assess hot tearing susceptibility [3]. The ring mould testing is frequently used to evaluate the hot tearing tendency of aluminum alloys [3, 4]. The mould consists of a ring and a core. The liquid is poured into the space between the two parts of the mould. During solidification, the core hinders solidification shrinkage and thermal contraction of the solidifying mixture therefore leading to hot tearing. The hot tearing susceptibility is then related to the length of the main crack. The major limitation of this technique is that the solidification conditions are usually different from those encountered during welding. A more appropriate technique is the Varestraint test [5]. It consists in applying a strain of known magnitude to a weld coupon during the welding process. The deformation is made by bending the weld coupon over a mandrel and forcing it to conform to the radius of the mandrel. The crack susceptibility is then determined by measuring the total length of cracks obtained for a particular applied strain. However, another type of tests allows reaching the solidification conditions encountered during welding without performing joints, namely tensile tests during solidification [3, 6–8]. In

E. Giraud (✉)
SIMaP, Universités de Grenoble, CNRS UMR5266, Saint Martin d'Hères Cedex, France
e-mail: eliane.giraud@simap-grenoble-inp.fr

J. Lippold et al. (eds.), *Hot Cracking Phenomena in Welds III*,
DOI 10.1007/978-3-642-16864-2_6, © Springer-Verlag Berlin Heidelberg 2011

this type of test, a sample is mounted on a tensile testing machine and is locally melted. A tensile strain is then applied during solidification of the molten zone by either moving or keeping fixed the crosshead of the tensile machine. In this latter case named constrained solidification, the tensile strain generated in the molten zone is only due to thermal contraction and solidification shrinkage in the specimen. In both cases, the sample is connected to a load cell which allows recording the stress resulting from the deformation of the solid as a function of temperature.

The aim of this paper is to report the results of constrained solidification experiments carried out both on a 6061 alloy and on alloys with various Si or Mg contents. In order to reproduce conditions close to welding, the experiments were carried out at high cooling rate starting from the liquid. After the experiments, the zone of the specimen which was initially liquid has been observed by X-ray microtomography in order to detect possible cracks. The hot cracking sensitivity of the various alloys was then determined based on the volume fraction and the number of open cracks in the molten zone.

Experimental Procedure

Materials

The 6061 alloy was used as the base metal and 4043, 4047 and 5356 alloys were used to change the composition of the base metal since they are employed as filler alloys during welding. The AA6061 alloy was supplied as rolled plates, 50 mm in thickness and in the T6 condition (solution heat treated and then artificially aged). The AA4043 (Al-5Si) and AA5356 (Al-5 Mg) alloys were supplied as wire of 1.6 mm diameter. The AA4047 (Al-12Si) was supplied as wire of 1.4 mm diameter. The compositions of the various alloys are given in Table 1.

Alloys with various compositions were prepared by mixing various amounts of the base metal and the filler wire. The weight percent of Mg or Si in the mixture is then given by:

$$[X]_{Mixture} = \frac{m_{BA}}{M}.[X]_{BA} + \frac{m_{FA}}{M}.[X]_{FA} \tag{1}$$

Table 1 Chemical composition (wt%) of the various alloys

	Si	Mg	Fe	Cr	Mn	Cu	Al
6061	0.61	0.93	0.26	0.20	0.12	0.28	Bal.
4043	4.7	0.004	0.3	–	0.01	0.01	Bal.
4047	11.25	0.001	0.20	–	0.001	0.0009	Bal.
5356	0.03	5.06	0.02	0.12	0.14	–	Bal.

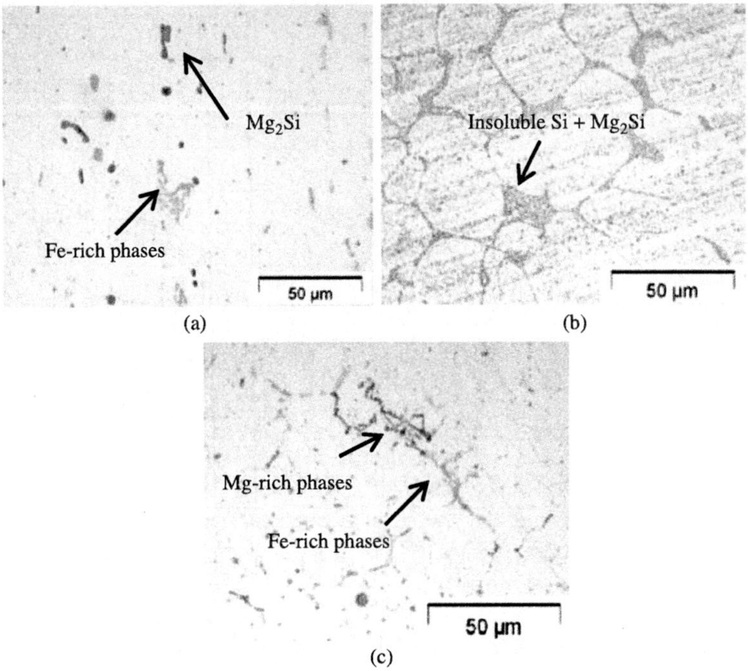

Fig. 1 Microstructure of 6061 (**a**), alloy containing 3 wt% of Si (**b**), alloy containing 3 wt% of Mg (**c**) – No etching

with: [X] the weight percent of the considered element (Si or Mg) in the mixture, in the base alloy (BA) and in the filler alloy (FA); m is the mass of the base or filler alloy; M the total mass.

Nine alloys were studied: the 6061 alloy (as reference); alloys containing 1, 2 and 3 wt% of Si (obtained by mixing 6061 and 4043); alloys containing 4 and 5 wt% of Si (obtained by mixing 6061 and 4047); and alloys containing 2, 3 and 4 wt% of Mg. Figure 1 shows typical microstructures of the alloys. In the base metal, typical Mg_2Si (dark) and iron-rich phases (light) are present. When the Si content increases, Si phases appear besides those encountered in the base alloy. No microstructural differences are found between mixtures obtained by using the 4043 or the 4047 alloy. When increasing the Mg content, iron-rich phases (light) and magnesium-rich phases (dark) are present.

Mechanical Tests Under Constrained Solidification Conditions

Figure 2 shows the experimental device used for the mechanical tests. An Adamel DY34 machine equipped with a 10 kN load cell was employed. The samples were cylinders of 9.5 mm diameter and 120 mm length.

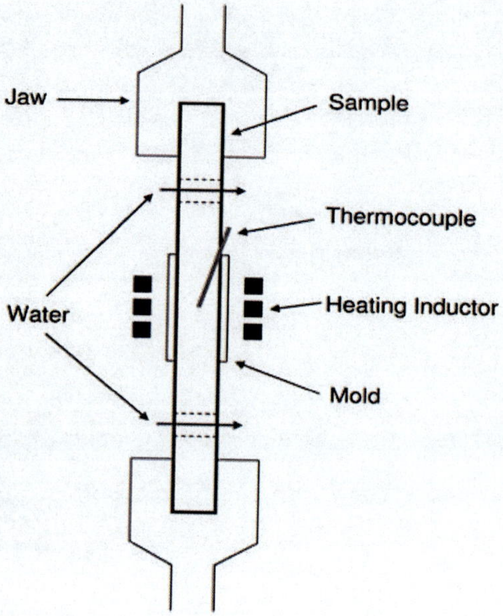

Fig. 2 Sketch of the experimental device used for the constrained solidification

The central part of the sample was heated by induction at 2 K/s until the liquid state. The temperature was measured by a K-type thermocouple of 0.5 mm diameter located in the central part of the sample. The deformations due to the heating were accommodated by moving the crosshead of the machine. An alumina crucible was placed around the heated zone to maintain the liquid in the center of the sample when very high liquid fractions are reached. Once the central part was liquid, the sample was cooled by keeping the crosshead fixed. Thus, the accommodation of thermal contraction and solidification shrinkage was prevented leading to constrained solidification. The cooling rate was about 80 K/s thanks to a water flow on both sides of the molten zone. The evolution of the load generated by constrained solidification was recorded in real time as a function of temperature. The temperature was linked to solid fraction by using the Alcan ProPhase software (based on the diffusion model of Sigli et al. [9]). Table 2 gives the evolution of solid fraction with temperature for the various alloys.

X-Ray Microtomography Procedure

Specimens of 2 mm diameter were machined from the zone of each sample which has been completely melted and solidified under constrained conditions. The specimens were then imaged by X-Ray microtomography carried out in INSA Lyon. During the rotation of the specimen over 180°, 1400 transmission images were recorded. The pixel size was set to 2 μm. After data acquisition, three-dimensional (3-D) images of the specimen were reconstructed using standard reconstruction tools.

Table 2 Temperature (K) and corresponding solid fraction (Fs) for the various alloys investigated

Alloy/Fs	0.6	0.80	0.90	0.95	0.97	0.99	1
6061	910	894	873	861	853	806	779
1% Si	906	888	863	838	821	807	779
2% Si	895	869	827	824	822	810	779
3% Si	883	870	831	824	822	811	749
4% Si	868	836	826	825	823	813	779
5% Si	853	837	829	825	824	816	779
2% Mg	901	880	869	864	845	727	692
3% Mg	893	868	853	809	724	692	692
4% Mg	884	857	821	727	707	699	692

The hot tearing tendency of each alloy was evaluated by determining the quantity of cracks present in the 3-D images. This quantity was evaluated through two parameters: the volume fraction and the number of open cracks. These data were obtained by digital processing of the 3-D images. A conventional threshold (black/white) was used to separate the cracks and the solid, a crack corresponding to black pixels connected with each other. It is then easy to determine the number of cracks per unit volume. In the same way, the volume percent of open cracks is equal to the sum of the volume of each crack divided by the volume of the analyzed zone. The volume of each crack is deduced from the corresponding number of pixels, the pixel size being known.

Experimental Results

Mechanical Tests Under Constrained Solidification Conditions

Figure 3 shows the variation of stress as a function of solid fraction for the various alloys. The influence of the Si content is shown in Fig. 3a and the influence of the Mg content in Fig. 3b. The behavior of the alloys containing only 3 and 5 wt% Si and 3 and 4 wt% Mg are shown for a better clarity of the figures. The other alloys exhibit the same trend as that shown in Fig. 3. Stress is obtained by dividing the measured load by the cross section of the sample.

For the 6061 alloy, stress starts increasing for a solid fraction of 0.6 and continues increasing up to a solid fraction of 1. However, at a solid fraction of 0.97, the curve exhibits a sharp transition. Below 0.97, stress increases slowly, while, at higher solid fractions, stress increases much more rapidly.

Whatever the Si or Mg content, the stress-solid fraction curve exhibits the same shape with the stress starting to increase at a solid fraction of 0.6 and a transition in its slope at a solid fraction which depends on the Mg content only. Indeed the transition is shifted toward lower solid fractions when the Mg content increases.

Fig. 3 Evolution of the measured stress as a function of the solid fraction during the solidification of alloys with different Si contents (**a**) and with different Mg contents (**b**)

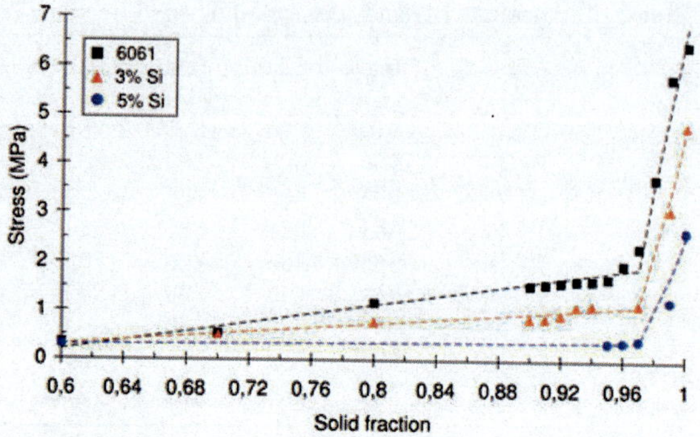

(a)

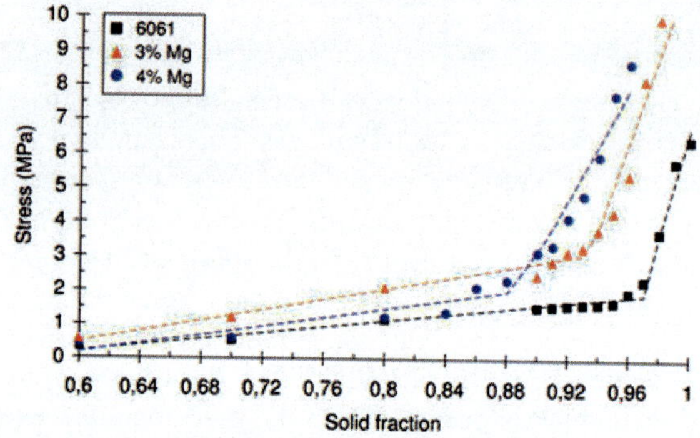

(b)

It should be noted that the stress level developed during solidification depends on the composition of the alloy. When the Si content increases, smaller stresses are developed compared to the 6061 alloy. An inverse trend is observed when the Mg content increases.

X-Ray Microtomography Observations

Figure 4 shows 3-D images obtained by X-ray microtomography of the molten zone of some specimens after solidification. The open cracks are in yellow and the thermal gradient is in the vertical direction.

The crack distribution is not homogeneous along the molten zone but the cracks are concentrated in the center. Since the cooling of the sample is due to water flow on both sides of the solidifying zone, the solidification front moves from the edge of the molten zone to the center of the sample. Thus, the central part of the specimen remains the hot point during the cooling step so that this zone experienced the larger strain.

The figure shows also that the Si and Mg contents significantly influenced the number of cracks: a significant decrease compared to the 6061 alloy is indeed observed for Si and Mg contents equal to 5 and 3 wt%, respectively.

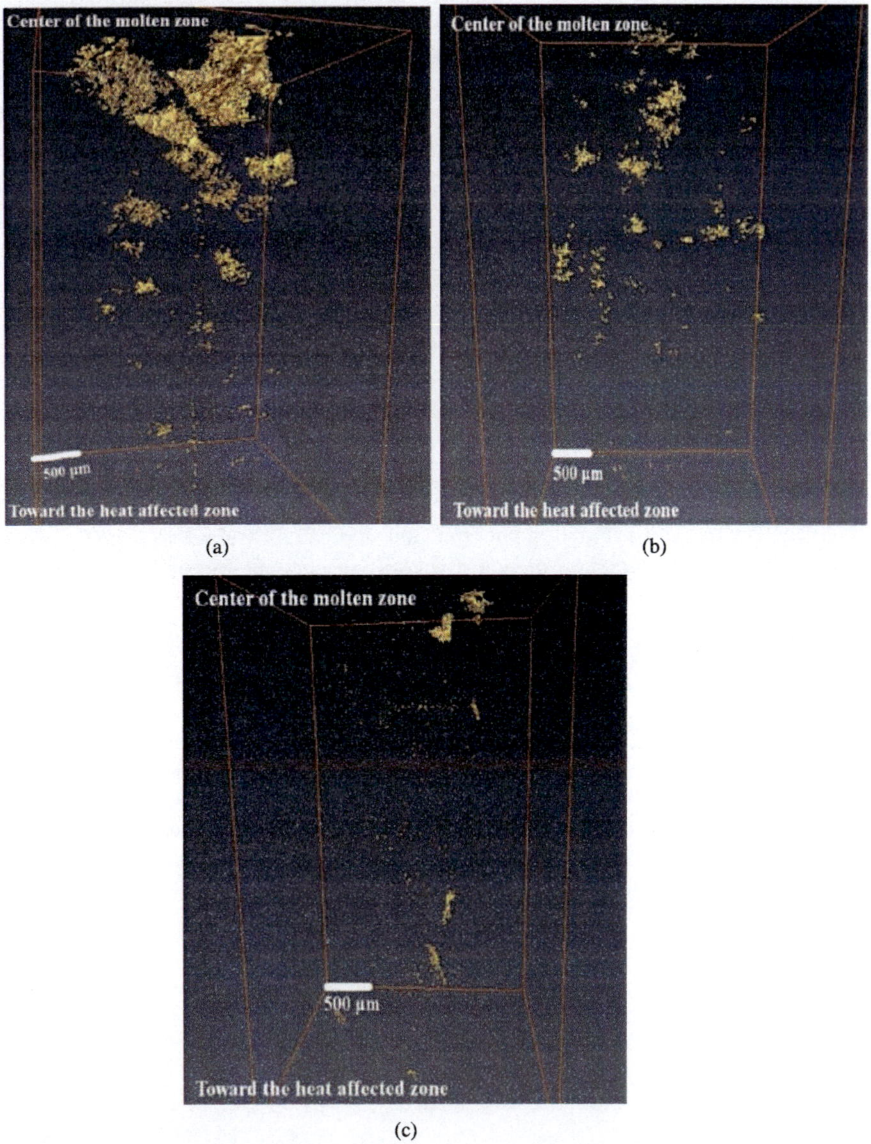

Fig. 4 3-D images of the molten zone after X-ray microtomography observations performed on: 6061 alloy (**a**), alloy with 3 wt% of Mg (**b**), and alloy with 5 wt% of Si (**c**)

Discussion

The agreement between solidification conditions occurring in the mechanical test and those encountered during welding will be discussed first. Then the behavior of the 6061 alloy during constrained solidification in comparison with alloys containing various Si or Mg contents will be discussed. Finally, the sensitivity to hot cracking will be evaluated and compared to the literature.

Comparison Between Conditions Encountered During Constrained Solidification And Welding

The tests are very close to the solidification conditions encountered during welding for three main reasons. Firstly, constrained solidification is encountered during welding owing to the clamping which is usually applied to the parts to be joined [10]. Secondly, a cooling rate of 80 K/s is reached during the solidification of the samples which is of the same order of magnitude as in welding. Thirdly, the microstructure along the sample (Fig. 5) which results from the tests is similar to that found in a welded joint [11]. The molten zone with equiaxed dendrites is observed in the center and more columnar dendrites are found when getting closer to the heat affected zone. The heat affected zone on both sides of the molten zone exhibits a granular structure and a liquation phenomenon (indicated by arrows on the figure).

However, the phenomenon of liquid backfilling encountered in welding is not present here. Therefore the solidification conditions are more severe than in welding.

Behavior of the Alloys Under Constrained Solidification

The variation of the stress observed when the solid fraction increases is linked to the microstructure changes of the molten zone during solidification. This evolution has been detailed in numerous studies concerning the behavior of semi-solid alloys under tensile strains [3, 11–14]. At the beginning of solidification, dendrites are free to move into the liquid. Deformations are accommodated by liquid flow which is easy or by rearrangement of grains. Then dendrites begin to come into contact: the liquid surrounds the dendrites and can circulate between the dendrite arms to accommodate the deformations. This transition is called coherency. In the last stages of solidification, the solid skeleton becomes more and more dense hindering

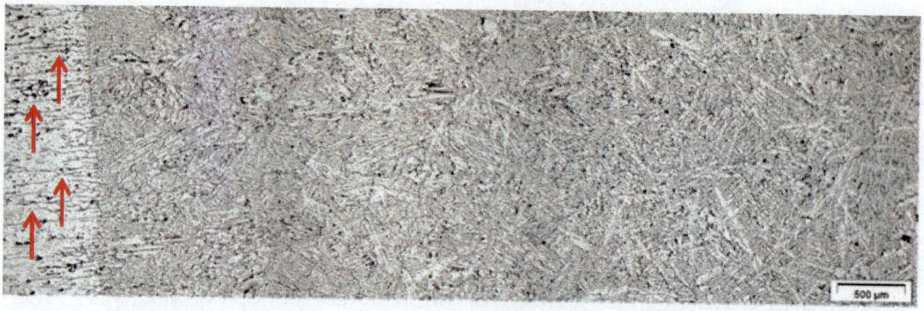

Fig. 5 Microstructure of a 6061 sample tested under constrained solidification (with the heat affect zone on the left and the center of the molten zone on the right) – Keller etching

the liquid flow which always surrounds the dendrites but in the form of thin films. Finally, at the end of solidification, the surface area of solid-solid contacts increases rapidly and liquid is in the form of isolated pockets. This transition is called coalescence. Consequently, during solidification, two important transitions are present in the microstructure of the semi-solid state: the coherency and the coalescence transition. It is obvious that these microstructural changes will have an impact on the mechanical properties as observed in the curves of Fig. 3.

The solid fraction at which the stress starts increasing underlines the coherency solid fraction which is close to 0.6 whatever the composition of the alloy. The transition for which a change in slope is noticed is due to the coalescence: an important increase of solid-solid contacts results in a faster increase of the resistance of the semi-solid state. The coalescence solid fraction is independent of Si content but depends on Mg content. It is shifted toward lower solid fractions when Mg content increases; which means that the solid skeleton develops earlier in the Mg-rich alloys.

Moreover, Si and Mg contents have also an impact on the stress level developed during solidification. For high Si contents, the lower level can be linked to the presence of lower thermal contraction and solidification shrinkage compared to the 6061 alloy. For high Mg contents, the higher level is the result of the combined effect of a smaller coalescence solid fraction and of lower temperatures compared with the 6061 alloy.

In terms of hot cracking, it seems consequently more interesting to modify the Mg content. During their solidification, the aluminum alloys exhibit a transition in their failure mode: ductile/brittle/ductile [3, 11, 14, 15]. Thus a solid fraction range exists for which the behavior of the semi-solid state is brittle and sensitive to hot cracking. In general, it corresponds to solid fractions ranging between 0.90 and 0.97. By increasing the Mg content, it is therefore possible to form more solid bridges which hinder crack propagation in this critical solid fraction range.

Sensitivity to Hot Cracking of the Various Alloys

X-ray microtomography observations have shown that cracks are present in the molten zone (Fig. 4). However, it is necessary to know whether these cracks formed in the semi-solid state or after solidification has been completed. To answer this question, tensile tests have been performed at room temperature on samples previously tested under constrained solidification. The samples have failed in their central part and the fracture surface, shown in Fig. 6, exhibits ductile deformation zones (Fig. 6a) and smooth dendritic zones (Fig. 6b). The presence of smooth dendritic zones suggests that pores and/or cracks are present before the tensile test. Since the as-received alloy does not contain any pores or cracks, it can be concluded that these defects were formed in the semi-solid state during the constrained solidification experiment. Hot cracking has thus occurred during solidification by failure of the liquid films [11, 16] and the cracks observed in the 3-D images are due to this phenomenon.

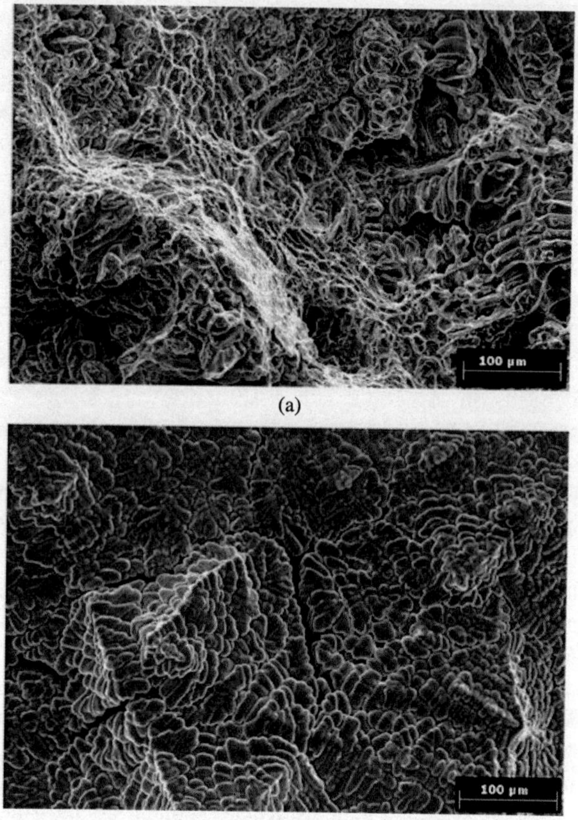

(a)

(b)

Fig. 6 Fracture surfaces obtained after a tensile test at room temperature on the 6061 alloy previously tested under constrained solidification

From the 3-D images obtained by X-Ray tomography, the sensitivity to hot tearing can be determined as a function of the Si or Mg contents. Two parameters have been defined: the volume percent of open cracks (Fig. 7a) and the number of cracks per unit volume (Fig. 7b). For both parameters, a standard lambda curve is found as a function of the Si or Mg contents with a peak susceptibility occurring for Si or Mg contents in the range between 1 and 2 wt%. The alloys become less sensitive to hot tearing than the 6061 alloy when the Si or Mg content exceeds 3 or 2 wt% respectively. The minimum sensitivity is reached for 5 wt% Si or 3 wt% Mg. However, the susceptibility of the alloy containing 4 wt% Mg remains unclear: it seems to be more sensitive than the alloy with 3 wt%. This tendency is more important on the number of cracks per unit volume than on the volume percent of open cracks which suggest the presence of very small pores and/or cracks. This result can be attributed to the threshold procedure and perhaps to the presence of small defects in the initial alloy. It could be also possible that an increase of the Mg content beyond 3 wt% increases the hot tearing tendency. Shimizu [17] has studied the hot tearing susceptibility during electron beam welding of 6061 alloy with 4043, 4047 and 5356 alloys used as

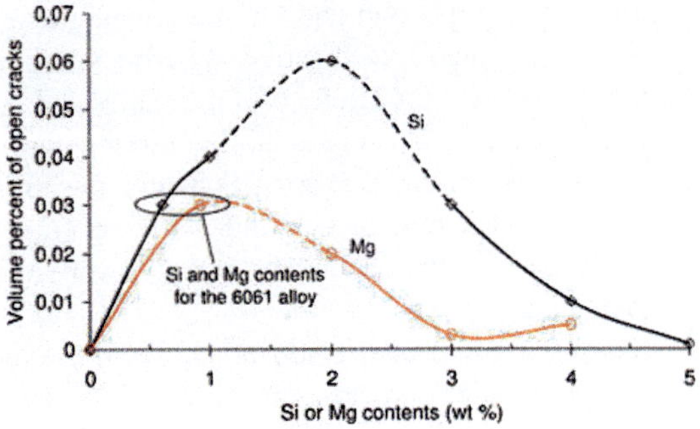

(a)

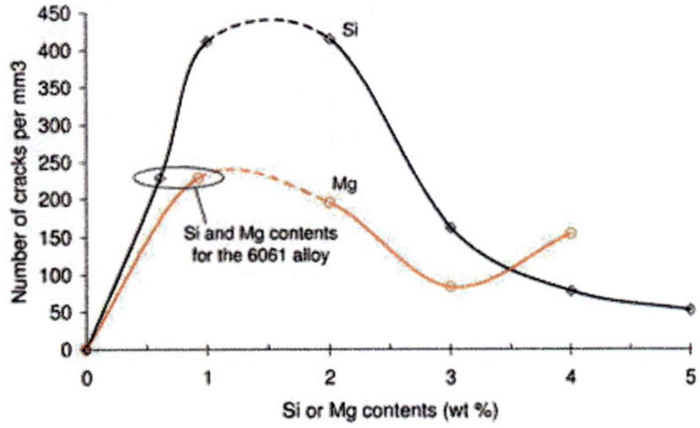

(b)

Fig. 7 Hot cracking sensitivity based on the volume percent of open cracks (**a**) or the number of cracks per unit volume (**b**)

filler wire. In his study, cracks were found to disappear when the Si content exceeds 3.2 wt% or the Mg content exceeds 2.3 wt%. For the behavior of alloys containing different Mg contents, the minimum sensitivity found by Shimizu is in agreement with our study. However, this is not the case when considering alloys with various Si contents. It can be assumed that this difference is due to the absence of liquid feeding during constrained solidification in contrast with the welding process.

Moreover, it can be noticed that the alloys containing various Mg contents exhibit a lower susceptibility to hot tearing than those containing various Si contents. This result is in agreement with the conclusion concerning the effect of Mg on the strength of the alloy. Increasing Mg content indeed allows obtaining a more resistant microstructure (Fig. 3). This higher resistance is due to the fact that coalescence occurs at a lower solid fraction and that the corresponding temperature ise much lower than that for the base alloy (Table 2).

Conclusions

1. The tensile test used in this study allows reproducing the conditions encountered during welding in terms of microstructure (equiaxed dendritic molten zone

and liquation phenomenon in the granular heat affected zone) and solidification (i.e. constrained solidification at high cooling rate). The test allows also measuring in real time the strength of the material as a function of solid fraction.

2. The various alloys prepared by mixing the base 6061 alloy with Si-or Mg rich alloys exhibit the same behavior during solidification. Coherency occurs at a solid fraction of 0.6 and coalescence is observed at a solid fraction of 0.97 whatever the Si content and at lower solid fractions when the Mg content increases.
3. X-ray microtomography observations of the molten zone allow quantifying the sensitivity to hot cracking of the various alloys. A lambda curve is found with a peak susceptibility occurring for Si or Mg contents in the range from 1 to 2 wt% and a minimum susceptibility at 5 wt% Si or 3 wt% Mg.
4. The use of a filler alloy from the Al-Mg series seems to be the most appropriate solution to reduce hot cracking during welding of the 6061 alloy owing to the lower susceptibility to hot cracking of alloys containing Mg compared to those containing Si. Indeed, during constrained solidification of Mg-rich alloys, it was observed that the coalescence of the solid skeleton occurs at lower solid fraction compared to Si-rich alloys. The alloy shows then a higher resistance to crack propagation in the critical solid fraction range.

Acknowledgements One of the authors (EG) is grateful to CNRS (French National Center for Scientific Research) and AREVA for financial support through a scholarship. The authors thank Cédric Gasquères, ALCAN CRV, for providing the ProPhase calculations and Luc Salvo, SIMaP, for his assistance during the processing of the microtomography data.

References

1. Anderson T (2005) How to avoid cracking in aluminum alloys. Welding Journal:25–27.
2. Cicala E, Duffet G, Andrzejewski H, Grevey D and Ignat S (2005) Hot cracking in Al-Mg-Si alloy laser welding – operating parameters and their effects. Materials Science & Engineering 395A:1–9.
3. Eskin DG, Suyitno and Katgerman L (2004) Mechanical properties in the semi solid state and hot tearing of aluminium alloys. Progress in Materials Science 49:629–711.
4. Drezet JM (1999) Validation of a new hot tearing criterion using the ring mould test. Journal of Physics IV France 9:53–62.
5. Cross CE, Olson DL, Liu S (2003) Aluminum welding. In: Totten, GE, MacKenzie, DS (Eds) Handbook of aluminium, Vol. 1. Marcel Dekker, Inc, pp. 481–532.
6. Langlais J and Gruzleski JE (2000) A novel approach to assessing the hot tearing susceptibility of aluminium alloys. Materials Science Forum 331–337:167–172.
7. Nagaumi H, Takateru U (2003) Study of the crack sensitivity of 6xxx and 7xxx aluminium alloys. Materials Science Forum 426–432:465–470.
8. Fabregue D, Deschamps A, Suéry M and Drezet JM (2006) Non-isothermal tensile test during solidification of Al-Mg-Si-Cu alloys: Mechanical properties in relation to the phenomenon of hot tearing. Acta Materialia 54:5209–5220.
9. Sigli C, Dif R, Commet B and Warner T (2003) Computer-aided optimization of aluminium alloys for airframe application. Materials Science Forum 426–432:351–356.
10. Cross CE, Boellinghaus T (2006) The effect of restraint on weld solidification cracking in aluminium. Welding World 50:50–54.

11. Fabregue D (2004) Phd Thesis, INPG, France.
12. Rappaz M, Drezet JM and Gremaud M (1999) A new hot tearing criterion. Metallurgical and Materials Transactions 30A:449–455.
13. Dahle AK and Arnberg L (1996) The rheological properties of solidifying aluminium foundry alloys. JOM 48:34–37.
14. Giraud E, Suéry M and Coret M (2010) Mechanical behaviour of AA6061 aluminum in the semi solid state obtained by partial melting and partial solidification. Metallurgical and Materials Transactions 41A:2257–2268.
15. Magnin B, Maenner L, Katgerman L and Engler S (1996) Ductility and rheology of an Al-4.5%Cu alloy from room temperature to coherency temperature. Materials Science Forum 217–222:1209–1214.
16. Lin S, Aliravci C and Pekguleryuz MO (2007) Hot tear susceptibility of aluminium wrought alloys and the effect of grain refining. Metallurgical and Materials Transactions 38A:1056–1068.
17. Shimizu S (2002) Study on crack prevention in electron beam welding of 6061 aluminum alloy. Journal of Light Metal Welding and Construction 40:23–30.

Part II
Steels and Stainless Steels

Short Term Metallurgy and Hot Cracking During Laser Beam Welding of Austenitic Stainless Steels

Thomas Böllinghaus, A. Gumenyuk, and V. Quiroz

Introduction

Austenitic stainless steels have a wide application range in numerous industrial branches, like the oil and gas industry as well as the pulp and paper industry or the textile industry, predominantly due to their corrosion resistance and mechanical properties at high and low temperatures. Their mechanical properties also have gained more interest in the automotive industry in the recent time for improvement of the crash resistance of the structural components.

However, fabrication welding of austenitic stainless steel products has to be paid special attention, in order not to reduce the corrosion resistance by the loss or precipitation of Cr and respective deterioration of the surface conditions after welding. For such reasons, low heat input welding processes like GTAW or plasma welding are usually preferred in manufacturing Austenitic stainless steel components. As very promising welding techniques providing even less heat input, application of laser and laser hybrid welding gained increasing interest in numerous industrial branches. The low heat input during laser and laser hybrid welding of austenitic stainless steel components also reveals much lower distortions, eliminating the necessity of additional technological steps and costs to bring the product its net shape. As further advantages, the high automation degree and the flexibility of respective laser welding production lines enormously widen the product spectrum to be manufactured with the same. Typical double or higher welding speeds as those during arc welding additionally increase the equipment efficiency by much shorter production cycles. For longitudinally or spiral welding of austenitic stainless steel tubular products CO_2-lasers thus became meanwhile a standard [1, 2].

More recently, high power solid state lasers are dominating the market, because of their larger flexibility by using an optical fibre instead of mirror guide systems, higher energy efficiency and smaller footprints. As a further advantage of these systems, the reduced formation of the plasma has to be addressed, saving costly

T. Böllinghaus (✉)
BAM Federal Institute for Materials Research and Testing, D-12205 Berlin, Germany
e-mail: Thomas.Boellinghaus@bam.de

J. Lippold et al. (eds.), *Hot Cracking Phenomena in Welds III*,
DOI 10.1007/978-3-642-16864-2_7,

He or He/Ar mixtures instead of Ar as a process or shielding gas. Recently, high power laser systems are available on the market and become increasingly attractive for manufacturing of thick walled components, providing a serious competitor for electron beam technology applied in this sector up to now [3].

Economically, the relatively high investment costs of laser welding have to be considered, requiring a specific overall cost calculation necessary for each application. Also, the cost intensive high precision of joint preparation required for defect free welding has to be taken into account. However, this can partially be alleviated by using hybrid welding instead of pure laser welding, which significantly increase the range of permissible tolerances, like gap and mismatch.

However, during fabrication welding of austenitic stainless steels with laser or hybrid processes special attention has to be paid to the avoidance of hot cracking. Especially austenitic stainless steels on the primary ferrite side of the eutectic rim –the three phases (liquid-δ-γ) equilibrium region in the Fe-Cr-Ni constitutional diagram might undergo a primary solidification mode change towards primary austenite due to higher solidification rate during laser and laser hybrid welding causing segregation and undercooling effects [4, 5, 6, 7]. Consequently, a deeper understanding of the metallurgical aspects associated with such innovative laser and hybrid welding is required for a defect free fabrication of welded austenitic stainless steel components.

The present paper represents a brief overview of the research work on the laser weldability of austenitic steels which has been done at the BAM during the past decade [8, 9, 10] and summarizes the results obtained in the framework of different research project carried out at the BAM during that time.

Rapid Solidification Phenomena During Welding of Austenitic Stainless Steels

The theory of constitutional undercooling is applied for prediction of solidification structures and morphologies under consideration of the interdependencies between cooling rates, solidification velocity at the solidification front, temperature gradients, concentration of the constituent elements and the distribution coefficient.

Normally, if the diffusion rate is slower than the solidification front progress, a concentration gradient appears in front of solid-liquid interface. The concentration of the solute can be expressed as a function of the distance from the solid-liquid interface. The temperature of solute correspondent to the temperature of liquidus $T_l(x)$ can be obtained from the phase equilibrium diagram. If the real temperature $T(x)$ is lower than hypothetical temperature $T_l(x)$ of the concentration layer, a region of constitutional undercooling appears ahead of the solidification front. The plane solidification front becomes unstable [11]. Great progress in understanding and modelling of the solidification phenomena was achieved by Mullins and Sekerka in the 1960s by their application of the perturbation theory to the analysis of the solidification phenomena in binary alloys [4]. Under consideration of the surface tension it provides a method, which use sinusoidal perturbations at plane interface and allows prediction about their stability.

The ratio between solidification velocity of the dendrite tip, its curvature and degree of undercooling can be obtained using the KGT (Kurz-Giovanola-Triverdi) model [5]. The dendrite tip radius corresponds to the maximum of the wavelength of the perturbation, at which a stable growth is possible. As a result, the dendrite tip radius can be found from the solution of the equation, describing the degree of supersaturation, which is taken from analytical solution for diffusion field around an ellipsoid of revolution representing a dendrite tip. Quite a number of further models based on the phase field method [12, 13, 14, 15] have been developed in recent time which describe the dependence of the curvature of the solidification front, i. e. of the dendrite complexity, on the respective cooling rate in the melt at the solidification front.

Such models are based on the assumption of a near-equilibrium limit, in which the solidification speed is largely controlled by the rate of transport of latent heat away from the solid-liquid interface. This has widely been accepted for the range of solidification processes taking place during conventional arc welding. In such cases, the crystal growth occurs slowly and the distribution between the concentration in solid and liquid phase on either side of the interface can be described by equilibrium distribution coefficient. Such situation corresponds to the condition of so called local interfacial equilibrium [6], which is valid in case of slow velocity of the solidification front.

However, during rapid solidification, as for instance during laser beam welding with high welding speeds, substantial deviations from the equilibrium at the interface may occur. It was found that although the total free energy of the liquid drops as a consequence of solidification, the chemical potential of the minor component in dilute binary alloys often increases. Consequently, solute concentrations far larger than the equilibrium solid solubility limit might be established. This phenomenon is referred to as "solute trapping" [16]. As the solidification interface is moving at high speed, the selection and distribution of solute occurs less easily, i.e. the distribution coefficient reaches its maximum at infinite high velocity. Aziz [7] proposed the following formulation of solidification velocity dependence for non-equilibrium distribution coefficient k:

$$k = \frac{k_0 + a_0 \times V/D}{1 + a_0 \times V/D} \tag{1}$$

where V- is the solidification velocity, k_0 – is the value of the distribution coefficient corresponding to the equilibrium phase diagram, D- is the value of the diffusion coefficient in the liquid phase and a_0 – is an input parameter related to interface thickness

Such significant deviations from the equilibrium conditions, i.e. solidus and liquidus concentration, are referred to as kinetic undercooling in contrast to the constitutional undercooling described above.

During welding, austenitic stainless steels experience either a primary ferritic (δ) or a primary austenitic (γ) solidification. If the composition of the melt in the Fe-Cr-Ni constitution diagram is near the eutectic rim and the solidification passes the field melt – δ – γ, both mixed crystals, δ and γ, can be formed. At such conditions, a primary ferritic solidification is followed by a secondary transformation in solid solution into an austenitic microstructure with residual delta ferrite (Fig. 1, left).

However, in the case of rapid solidification processes occuring during beam welding, for instance, quite a number of Fe-Cr-Ni alloys located on the ferritic side near the eutectic rim on account of their chemical composition experience primary austenitic solidification instead of a primary ferritic solidification. Such solidification rate dependent alterations of the primary solidification mode are usually described by the mechanisms producing a cellular or a dendritic growth, respectively. Thus, for a better understanding of the solidification morphologies formed during laser welding of austenitic steels near the eutectic rim specific metallurgical short-time kinetics have to be considered [17]. The factors determining these mechanisms are the above described kinetic undercooling and the respective supersaturation of the melt ahead of the dendrite tips. Figure 1 (right) illustrates schematically such a phase selection as modelled by Kurz [8] for the Fe-Ni binary system. It has to be considered that during solidification the metastable phase, as for instance the γ mixed crystal, might compete with the stable phase, represented by the δ mixed crystal.

For a melt having the initial concentration C_0 the thermodynamically induced solidification mode change is not visible in the equilibrium diagram shown on the left hand side in Fig. 1, but is given by a rate-dependent kinetic phase selection as

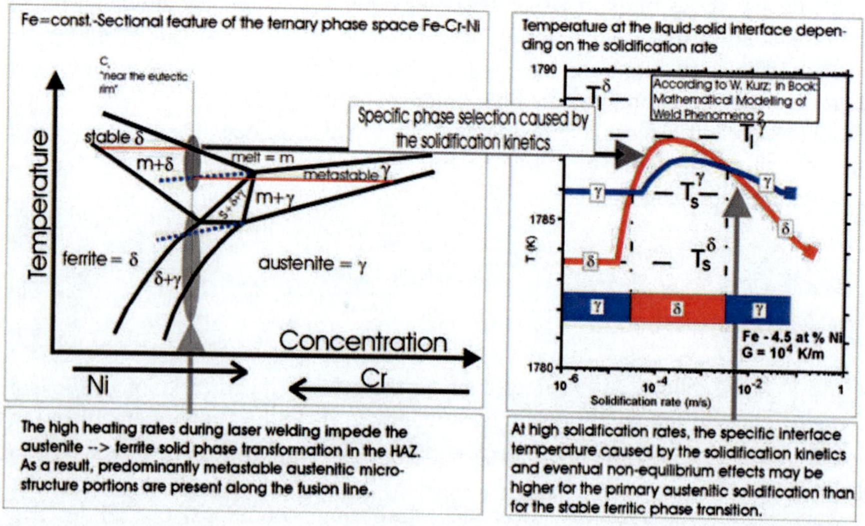

Fig. 1 Illustration of the change of the primary solidification mode in austenitic stainless steels near the eutectic rim [9, 17]

illustrated on the right hand side in Fig. 1. Following the considerations of Kurz [8], that phase will be formed which solidifies at the highest liquid solid interface temperature. This is one explanation for the primary solidification mode change from ferrite to austenite during austenitic stainless steel laser welding. It is based on the presumption that the nucleation for such a process is not prevented and it can be anticipated, however, that a weld pool provides a sufficient number of such nuclei to initiate or change, respectively, the solidification mode.

In this context, it can also be expected that predominantly austenitic crystals are present along the fusion line, due to the incomplete austenite decomposition during rapid heating of the material. Thus, as another explanation, the primary ferritic solidification might be prevented, because the number of ferritic nuclei present in the melt is too small. In particular, epitaxial growth effects due to the incomplete austenite decomposition at the fusion line have to be taken into account.

As a consequence of the crystal growth via the metastable austenitic primary phase, larger liquidus-solidus temperature intervals and larger shrinkages are encountered. This not only provides larger segregation at the liquidus-solidus interface causing additional constitutional undercooling. It also leads to a situation that the residual melt cannot be fed back into the dendrite spaces completely and also that the alloy at the remains longer in the brittle temperature range (BTR) at the solidification front [18, 19]. Consequently, an increasing risk of hot solidification and liquation cracking is associated with such primary solidification mode changes towards austenite [10, 20, 21, 22, 23].

Fe-Cr-Ni Alloys

Primary Solidification Mode Changes in Laser Welds

In order to investigate to what extent a specific selection of austenitic stainless steels (Table 1) solidify in the respective modes under the conditions of short-term metallurgy, remelting experiments were performed by bead–on–plate welding using a 6 kW CO_2 laser as well as the Gas Tungsten Arc Welding (GTAW) process [9]. The welding power and speed were selected under the presumption that the solidification rate is the decisive process–dependent parameter influencing the primary solidification mode. The primary solidification modes were determined from colour etched transverse sections of the welds.

It turned out that the investigated heats of the austenitic stainless steels AISI 304L (1.4306), 309 (1.4828) and 316L (1.4404) tend to change their solidification from the primary ferritic to the primary austenitic mode during laser welding particularly at low heat inputs, i. e. at high solidification rates.

For the material AISI 316 Ti (1.4571) no change was found. In Fig. 2, microstructures with respective primarily solidified austenitic portions are shown in cross sectional areas of the materials AISI 304L and 309.

Table 1 Chemical composition of the test materials [wt-%]

	Nr	1	2	3	4	5	6	7	8	9	10	11	12
	DIN	1.4306	1.4306	1.4306	1.4828	1.4828	1.4828	1.4828	1.4571	1.4571	1.4571	1.4318	1.4371
	AISI	304 L	304 L	304 L	309[a]	309[a]	309[a]	309[a]	316 Ti	316 Ti	316 Ti	301 L	201 L
	UNS	30403	30403	30403	30900	30900	30900	30900	31635	31635	31635	30103	20103
	Thickness, mm	2	3	6	3	3	3	3	5	3	2.5	3.2	3.2
chemical composition wt-%	C	0.01	0.02	0.02	0.05	0.05	0.05	0.05	0.04	0.03	0.04	0.03	0.04
	Si	0.64	0.62	0.36	1.71	1.65	1.92	2.11	0.58	0.61	0.52	0.51	0.40
	Mn	1.13	1.10	1.34	1.39	1.32	0.68	0.67	1.03	1.75	1.24	1.09	6.92
	P	0.02	0.03	0.03	0.03	0.03	0.02	0.02	0.03	0.02	0.02	0.03	0.03
	S	0.003	0.003	0.002	0.003	0.007	0.005	0.003	0.004	0.002	0.001	0.003	0.01
	Cr	18.18	18.14	17.96	19.27	19.29	18.70	19.00	16.38	16.80	16.39	17.52	16.64
	Nl	9.74	9.29	10.12	10.43	10.66	11.02	11.00	10.62	9.90	10.62	7.49	4.91
	Mo	0.22	0.23	0.23	0.12	0.20	0.23	0.29	2.04	2.07	1.95	0.17	0.09
	Tl	–	0.02	0.01	0.01	0.01	0.01	0.02	0.29	0.31	0.49	0.01	0.01
	Nb	–	0.02	–	0.01	0.02	0.01	0.01	–	0.02	0.02	0.02	0.02
	N	0.05	0.05	0.05	0.03	0.02	0.05	0.06	0.02	0.02	0.01	0.11	0.13
	Cu	0.17	0.33	0.30	0.22	0.35	0.10	0.12	0.26	0.21	0.23	0.19	0.32
	Co	–	0.18	–	0.07	0.11	–	–	–	0.13	–	0.10	0.11
	V	–	0.07	–	0.04	0.09	–	0.02	–	0.05	–	0.05	0.04
	W	–	–	–	–	–	–	–	–	–	–	0.02	0.02
	Al	0.002	0.002	–	0.002	0.001	–	0.002	–	0.020	–	0.005	–
Schaeffler	Cr-eq/Ni-eq	1.82	1.84	1.64	1.73	1.71	1.68	1.75	1.58	1.71	1.54	2.07	1.81
WRC1992	Cr-eq/Ni-eq	1.65	1.67	1.53	1.50	1.49	1.37	1.39	1.49	1.68	1.49	1.65	1.88
DeLong	Cr-eq/Ni-eq	1.61	1.63	1.46	1.61	1.62	1.52	1.54	1.50	1.64	1.49	1.51	1.29
Hammar Svenson	Cr-eq/Ni-eq	1.74	1.76	1.58	1.74	1.71	1.67	1.71	1.70	1.87	1.74	1.84	1.78

[a] AISI 309 Grade does not match exactly DIN EN 1.4828, deviating in Cr and Si–contents

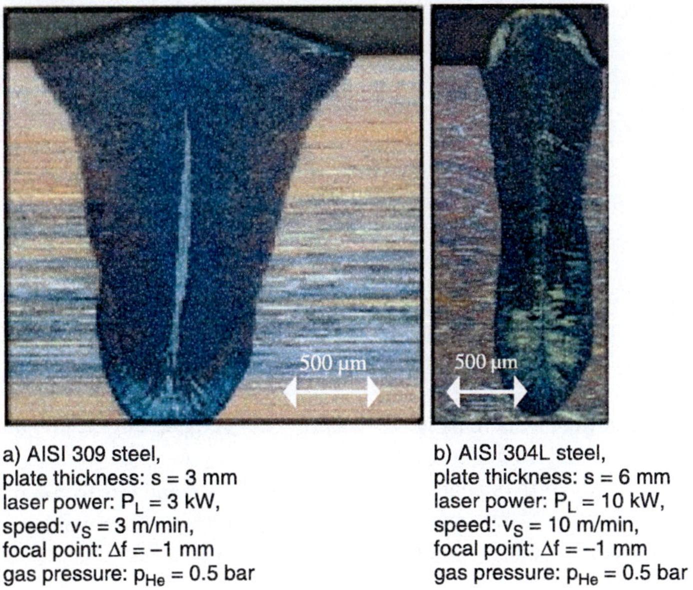

Fig. 2 Cross section of austenitic stainless steel welds showing a change of the primary solidification mode from primary ferritic (*dark*) towards primary austenitic (*light*) structure in the weld metal

The sections of primary austenitic solidification clearly differ from those of ferritic solidification and can be determined quantitatively. It turned out that the size of these portions increases with the welding speed. At high welding speeds and low penetration depths, weld metals with almost complete primary austenitic microstructure are produced.

To the diagrams in Fig. 3, the primary austenitic portions of the entire weld cross section were assigned dependent on the welding parameters and compared with the respective penetration depths. By such diagrams, parameter combinations required

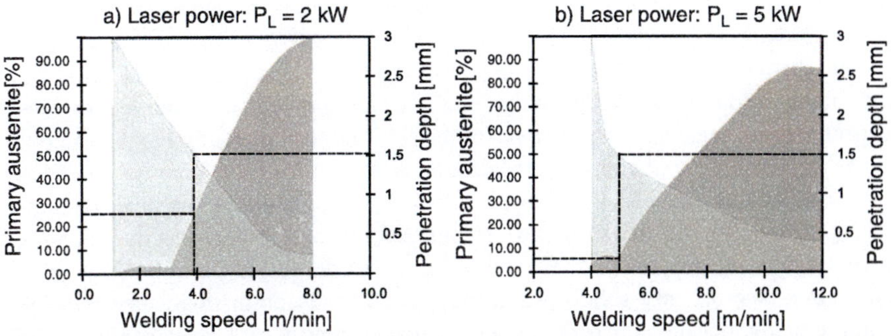

Fig. 3 Portion of primary austenite and penetration depth in the weld metal of the re-melted stainless steel AISI 309 dependent on the welding speed

for sufficient penetration at a minimum of primary austenite targeted at hot cracking avoidance can be determined. If, for instance, a depth of 1.5 mm is required this can be achieved with a welding speed of about 3.8 m/min at a laser power of $P_L = 2$ kW producing a 25% portion of primary austenite in the weld. At a higher laser power of $P_L = 5$ kW, the same depth is required with a reduced amount of 5% primary austenite in the microstructure, in order to avoid hot cracking.

Such diagrams are specific for each particular alloy and in case in the Fig. 3 related to the steel grade AISI 309.

These results demonstrate that for practical application the hot cracking resistance of laser welded austenitic stainless steels near the eutectic rim might be recovered by suppressing the primary solidification mode change selecting the welding parameters carefully and by a respective laser optimized joint design. This in turn means, overwelding of laser welds and considerable alterations of the welding parameters, in particular increasing the welding speed, have still to be regarded as critical with respect to the hot cracking resistance.

In this context, it has to be mentioned that the specific standards for laser welding [24] should include guidance regarding the risk and respective avoidance of hot cracking in austenitic stainless steel laser welds.

Effect of Nitrogen on the Primary Solidification Mode

Nitrogen generally affects the solidification behaviour as an austenite stabilizer [25]. In order to investigate the influence of nitrogen on the solidification behaviour of the studied austenitic steels during rapid laser welding processes, different amount of nitrogen was introduced in the materials. For such experiments, various re-melts of base material by CO_2-laser using one, two or three times by side-by-side passes where nitrogen was added to the process gas and to the shielding gas, respectively, as shown in the Fig. 4. After such nitrogen introduction into the weld metal by welding, the nitrogen content was homogenized by with a GTA torch in argon. Alternatively the re-melting of the base materials by GTA-welding with nitrogen as a shielding gas has been performed. A comparison of the nitrogen contents before and after GTA dressing did not reveal any noticeable losses of nitrogen by such homogenisation process.

The change of the primary solidification mode of these slowly welded GTA- or homogenized laser welds depending on the welding speed and solidification rate, respectively, was taken as an indicator for the shift of the composition to the eutectic rim as a consequence of a respective nitrogen pick-up.

Cross sections of such welds have been subjected metallographic investigations after Lichtenegger and Bloech [26] colour etching and phase-selective etching, respectively, with respect to their solidification mode, phase portions as well as hot cracks. The primary solidification modes can easily be distinguished by such colour etching, because the residual delta ferrite of a primary ferritic solidification appears clearly in the austenitic transformation products at room temperature (ferrite dark, austenite bright). The chemical composition of the material was determined

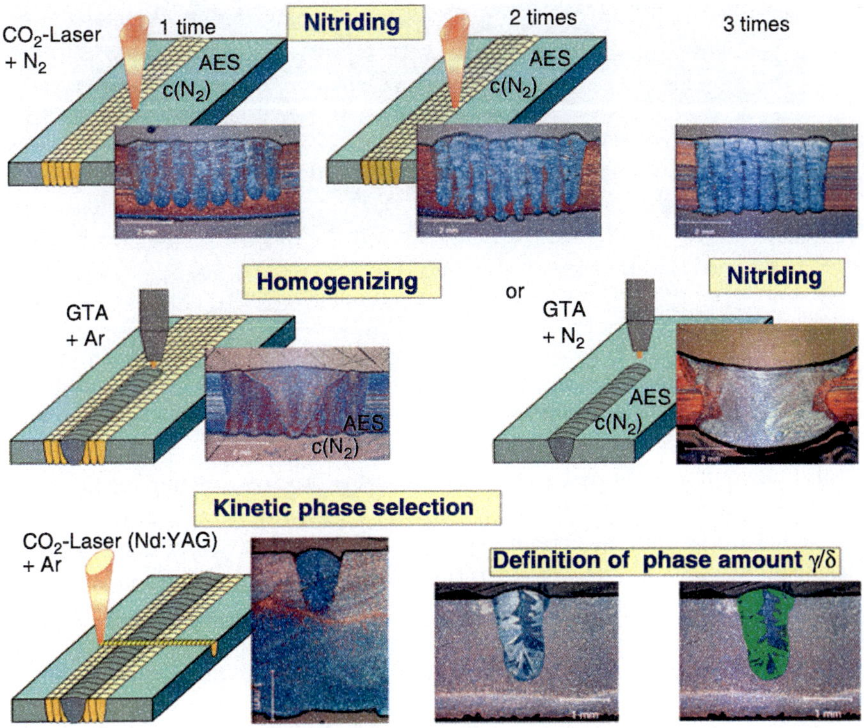

Fig. 4 Scheme of experimental procedure for nitrogen introduction

before and after welding using spark atomic emission spectrometry (AES) at a good reproducibility. In particular, the deviations of the nitrogen content ranged below 2%. Comparative measurements were carried out using ion-induced nuclear resonance analysis as well as glow discharge optical emission spectroscopy (GDOES). Segregation and element distribution measurements were performed using electron beam microanalysis plus energy-dispersive X-ray analysis (EDX).

For the following investigations, the solubility of molecular nitrogen in steel melts as opposed to that of atomic nitrogen has been considered as to be insignificantly small. Accordingly, it has been assumed that the nitrogen is present in the plasma in ionized or thermally dissociated form, respectively, during the re-melting experiments with nitrogen as process gas during CO_2-laser re-melting or as an addition to the shielding gas during the arc welding. It can also be taken as evident that only the nitrogen content in the melt during liquid/solid transition is relevant for primary solidification. As a further assumption, precipitation of nitrides in the melt has not been considered. Phases formed at lower temperatures in the solid state do not influence the primary solidification mode change. This means that nitrogen has been presumed to be present in the crystal lattice during phase transition predominantly in interstitial solid solution.

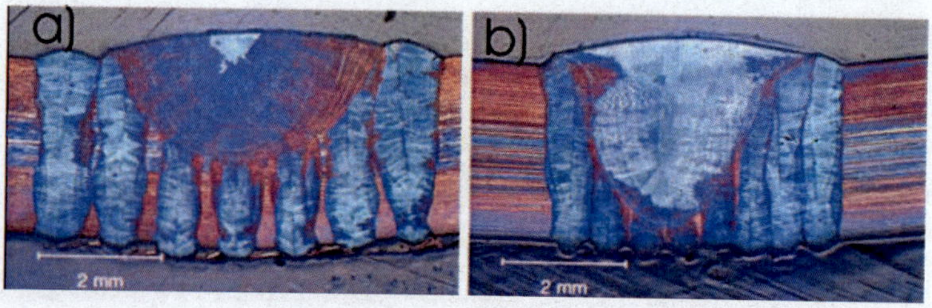

Fig. 5 GTA-re-melts into twofold (**a**) and threefold nitrided CO_2 laser remelts

Examples of the obtained weld metals are shown in Fig. 5. In Fig. 5a, it is illustrated that the twice laser re-melted region solidify nearly completely in primary austenitic mode (bright areas in the narrow laser welds). After slow GTA-overwelding, by which the laser welded area is homogenized, a primary ferritic remelt product is existent again, indicating that the primary austenite of the laser welds was related to the kinetic processes described above. But, after three times laser welding (Fig. 3b), the GTA-re-melt also experiences austenitic solidification. This means that the weld metal composition has been shifted across the eutectic rim to the austenite area, due to the addition of somewhat more nitrogen during laser welding three times.

In Fig. 6, it is shown for the material AISI 309(6) that with one, two and three times CO_2-laser welding the nitrogen pick-up increases with the number of passes

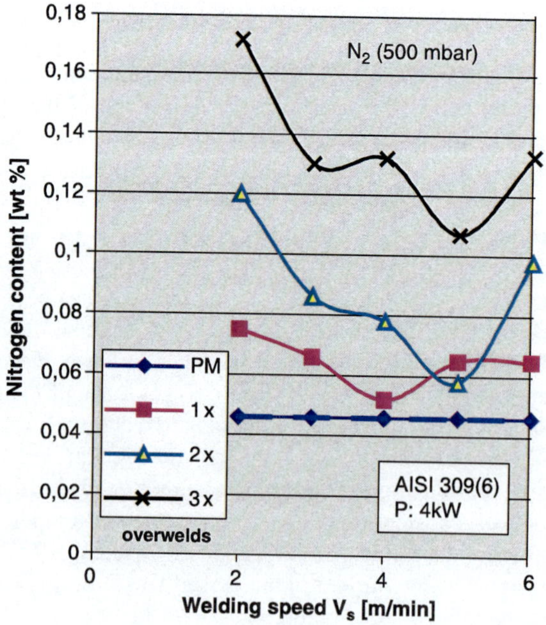

Fig. 6 Nitrogen pick-up during CO_2-laser welding depending on the welding speed

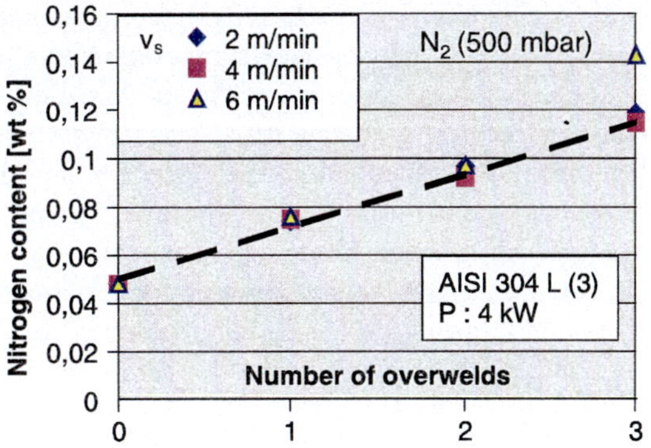

Fig. 7 Nitrogen pick-up during CO_2-laser welding depending on the number of passes

without any indication of saturation. It drops at first with increasing welding speed and then rises again at higher speeds. This reversal can be explained by the decreasing melt volume at higher speeds and constant laser power. For the material AISI 304L(3) (Fig. 7), the nitrogen pick-up increases also with the number of laser passes, but it is nearly independent of the welding speed, i.e. the two reverse tendencies depending of the welding speed obviously cancel each other out.

For comparison, investigations of nitrogen pick-up were carried out in the same way for multiple Nd:YAG-laser welding. The same tendencies have been found, but at a 3–10 times lower nitrogen pick-up as compared to CO_2-laser welding. As one reason the fact has to be addressed that in Nd:YAG-laser welding only a thermal dissociation of nitrogen takes place while during CO_2 laser welding an ionisation takes place.

Kinetic phase selection. In order to study the impact of nitrogen contents on the specific short-term metallurgical primary solidification change, the remelts described above exhibiting various nitrogen contents were again subjected to laser welding at different welding speeds. During these operations, argon was used as process as well as shielding gas. After welding, the portions of the primary solidification modes were determined. During such investigations, it turned out that the second laser welding is principally suitable for such investigations, but more reliable results could be achieved, if the GTAW process was applied by adding nitrogen to the process gas. By addition of various volume fractions of nitrogen to the shielding gas the GTAW process allows to introduce different levels of nitrogen in the remelted la ser welds.

For the material AISI 309(5), the diagram in Fig. 8 shows the influence of nitrogen on the solidification rate necessary for the change of the solidification mode. The base material which was not nitrogen treated had a nitrogen content of 0.024 wt% and changes the solidification mode from a primary ferritic to an austenitic one at a laser power of 1.5 kW above a welding speed of 0.5 m/min. Introducing higher nitrogen contents at the same welding speeds appreciably greater primary austenitic portions appear in the weld metal, which even at an extrapolation to $Vs = 0$, i.e. at slow equilibrium solidification, still differ from zero. Hence, with a nitrogen content of approximately 0.05 wt%, this material has eutectic composition which allows both solidification modes coexisting even in equilibrium without undercooling.

The principal dependence of the solidification change and an associated dependence of the hot cracking resistance can be drawn from a comparison of the diagrams in Figs. 8 and 9. For instance, the non-nitrogen treated alloy AISI 309(7) with a basic content of 0.057 wt-% shows still no sign of any solidification change even at welding speeds of up to 5 m/min.

Only at nitrogen contents of above approximately 0.08 wt% increasing austenite portions appeared. These are, however, minor fractions as compared to those of the material AISI 309(5). From this it can be deduced that the distance from the

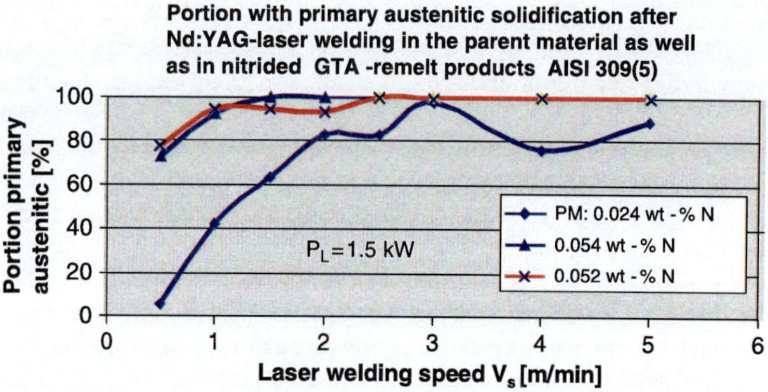

Fig. 8 Influence of the nitrogen content on primary solidification of AISI 309(5)

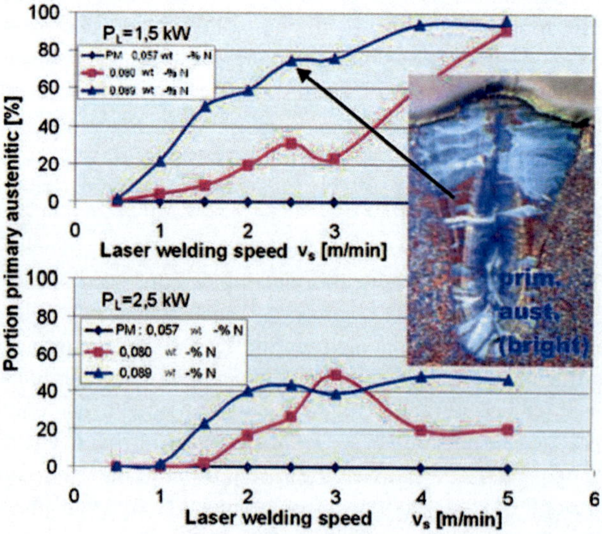

Fig. 9 Influence of the nitrogen content on primary solidification of AISI 309(7)

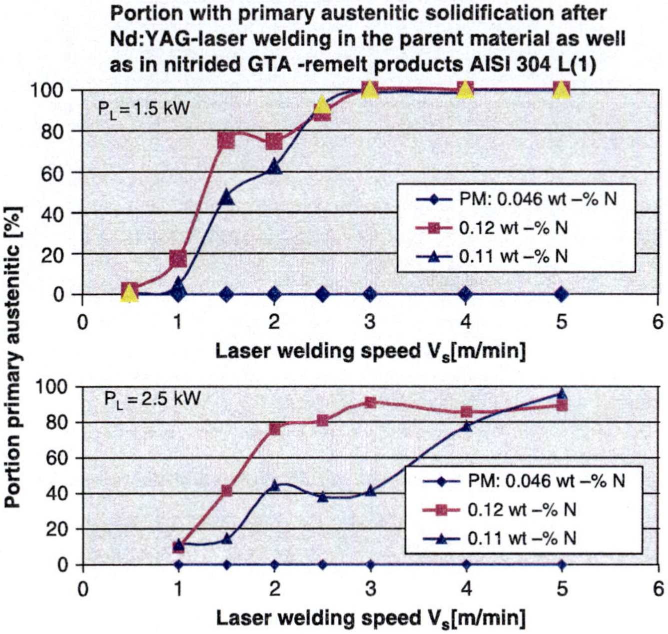

Fig. 10 Influence of the nitrogen content on primary solidification of AISI 304L (1)

eutectic rim and the influence that is exerted on it by nitrogen additionally depends strongly on the further alloying elements. Figure 10 shows that also for the material AISI 304L(1) an increase of the nitrogen content to about 0.1 wt% causes a solidification mode change already at low welding speeds.

From the viewpoint of practical application there is a risk of undesirable nitrogen pick-up particularly during laser welding, if the process gas is coaxially led to the beam through a welding nozzle. In the case of unfavourable configuration it may occur that air is sucked in from the beam guidance system. This effect is illustrated in the comparison made in Fig. 11. Here, CO_2-laser welding was in one case performed with a conventional coaxial nozzle configuration and in the other case with a modification, in which process gas overpressure was generated via a bypass at the beam entrance in front of the nozzle to make sure that less air is brought in from the beam guidance system. As can be seen from Fig. 11, by such technical modification the beginning of the unfavourable solidification mode change is shifted towards higher welding speeds.

Metastable States and Epitaxial Effects at the Fusion Line

In the context of a solidification change in laser welding it has to be considered that in the parent material areas adjacent to the fusion line the diffusion-controlled austenite-ferrite transformation in the solid state is largely incomplete as a result

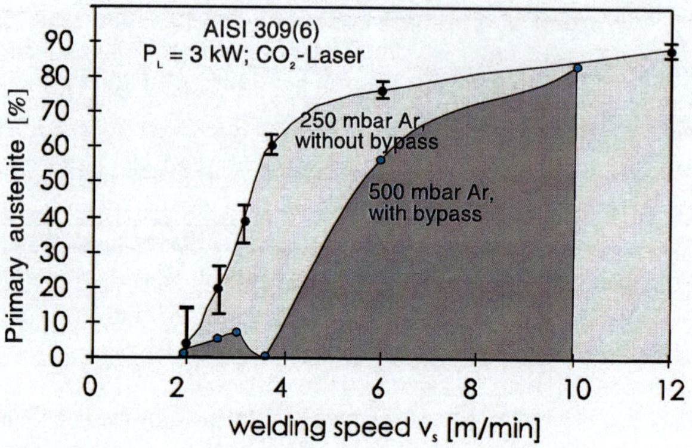

Fig. 11 Effect of air introduced via the welding nozzle

of the rapid heating processes. Thus, for epitaxial crystallization of the melt, predominantly austenitic nuclei are present at the fusion line. Due to the lack of ferritic nuclei for ferritic equilibrium solidification, the melt may, also in these areas of very low solidification rates, be undercooled and retain this state up to the metastable austenitic solidification area. It can be seen from Fig. 12 that primary austenitic borders are formed at the fusion line of material AISI 316Ti(10), for which no kinetic-specific change of the primary solidification mode was found even at high welding speeds, and which is hence relatively far away from the eutectic rim on the ferritic side. But this austenitic solidification comes to an end after a relatively short time because it is superseded by the ferritic equilibrium solidification enabled by elevated temperatures.

In laser welds of materials which, by contrast, exhibit a kinetic-specific change of the primary solidification mode, cellular austenitic morphologies may, in case, adjoin an austenitic border at the fusion line and extend over larger areas. Figure 13 illustrates a laser weld for steel AISI 309(7) with otherwise completely primary ferritic solidification, however with marked primary austenitic zones starting directly at the fusion line. This effect must be considered as critical, particularly in conjunction with smaller isolated hot cracks which cannot be excluded in such areas. If the material to be welded is austenitized by any arising nitrogen pick-up, the epitaxy- and kinetic-specific effects may combine and form pronounced primary austenitic morphologies involving the risk of hot cracking. The appearance of metastable

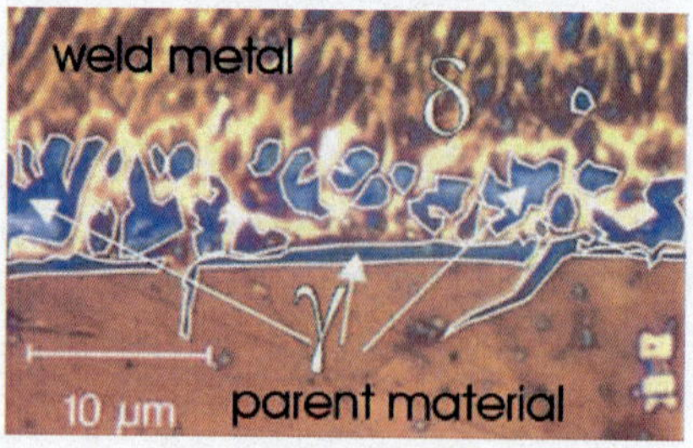

Fig. 12 Primary austenitic border (γ) at the fusion line; AISI 316Ti(10); v_S: 10 m/min, P_L: 3 kW

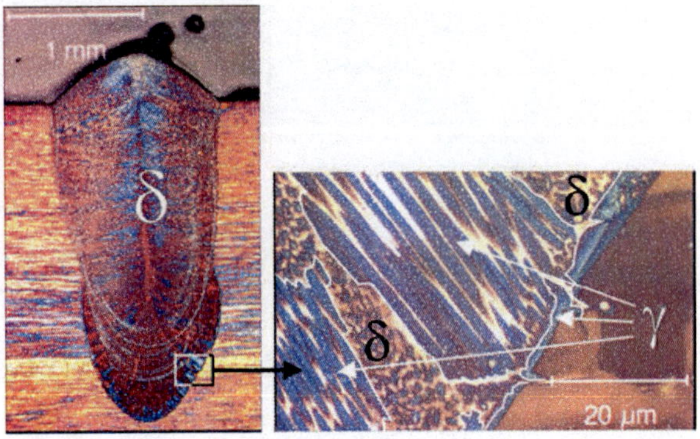

Fig. 13 Epitaxial primary austenitic solidification (γ) at the border AISI 309(7); v_s: 4 m/min; P_L: 3 kW

ferritic solidification of an austenitic melt has been proved in quenching experiments and containerless solidification experiments, respectively [27, 28]. The fact that the nucleation energy for ferritic crystallization is lower than that for austenitic solidification turned out to be propitious to the crystallization of metastable ferritic states from undercooled melts.

The question of whether such undercooling is reached also in beam welding processes has thus been investigated. Accordingly, gradual nitriding and the associated austenitization of a starting composition with originally ferritic solidification has deliberately been used to create austenitic melt compositions near the eutectic rim to elucidate their solidification modes during rapid solidification with the corresponding kinetic undercooling. The respective investigations were performed with the material AISI 309(4). Twofold CO_2-laser welding with nitrogen as process gas and subsequent homogenizing GTA-overwelding still engenders completely primary ferritic solidified weld metal. In Fig. 14a, this area is marked by the symbol **δ**. The material nitrided in this way has subsequently been provided with a further rapidly executed laser weld feeding additional nitrogen to it. This further nitriding entails nearly complete primary austenitic solidification which is marked in Fig. 12a and b by the symbol **γ**. Only at the end of the weld pool, i.e. in the weld centre (Fig. 14b), where the solidification rate reach its maximum large numbers of equiaxed dendritic crystals are formed. These clearly had to be identified as primary ferritic solidification products owing to the irregularly enclosed residual ferrites (Fig. 14c). Undercooling and supersaturation, respectively, must inevitably be concluded from the equiaxed dendritic morphology.

This correlates very well with the solidification site, i.e. with the high solidification rates and with the kinetic undercooling at the end of the weld pool. This undercooling is so pronounced as to allow equiaxial dendritic crystallization to take place requiring that the solidification heat is carried away to the outside and led into the undercooled melt. In this connection, metastable *ferritic* solidification favoured by energetics during nucleation is obviously predominant at this point. Hence, for an austenitic initial composition, there is thermodynamically no stable area for the existence of ferrite at temperatures below the solidification interval in solid state, and particularly no area of coexistence in equilibrium, i.e. transformation area, for ferrite and austenite, so that a ferrite, formed in metastable state, with properly

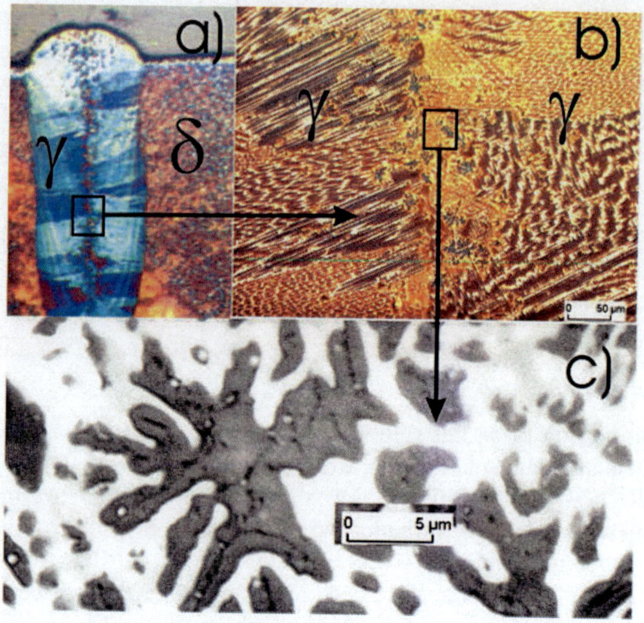

Fig. 14 Metastable ferritic crystallization of equiaxed dendritic morphologies in an austenitic melt. AISI 316 Ti(9, nitrided); v_s: 10 m/min; P_L: 5 kW

speaking austenitic composition is immediately forced in a non-equilibrium reaction to transform into austenite and to freeze in most unstable state, respectively. This is confirmed by the very thin black inclusions (unstable residual ferrite) in the equiaxed dendrites. Post-heating of the specimen containing the solidification products shown in Fig. 14 in a furnace to about 800°C and subsequent cooling at a slow rate prompts the black ferrites in the equiaxed dendrites in the weld centre to nearly totally disappear (Fig. 15a), i.e. to completely undergo austenitization. The co-annealed stable residual delta ferrites (δ-area in Fig. 14a), however,

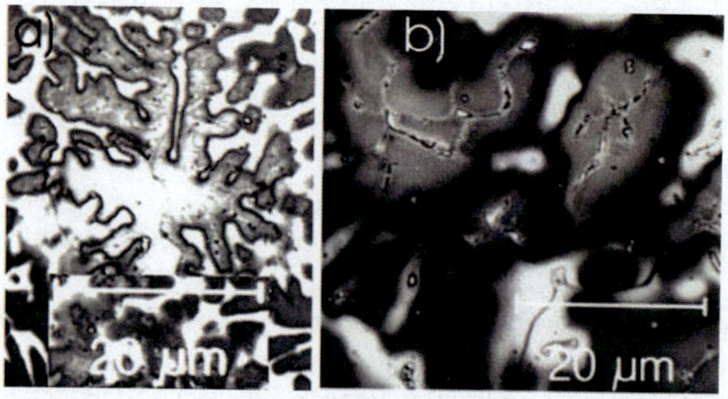

Fig. 15 Metastable (*left hand side*) and stable ferritic solidification products in AISI 316 Ti (9) after annealing up to 800°C

clearly subsist, as can be seen from Fig. 15b. This fact, too, confirms the metastable primary ferritic solidification mode of the morphologies in the weld centre (Fig. 14c). The repetition of this procedure for a specimen with a higher nitrogen content, which exhibits stable austenitic solidification already in the GTAW-homogenized area, yielded similar results.

Manganese Alloyed Austenitic Stainless Steels

Austenitic and austenitic-ferritic (duplex) Cr-Mn-Ni- steels represent an economically interesting alternative to austenitic Cr-Ni- steels, due to the high costs for Nickel. In Cr-Mn-Ni steels, nickel is partially substituted by the less expensive austenite former Manganese and small amounts of Nitrogen. These steel grades are known for about 70 years now, but, in contrast to Europe, they have received widespread industrial distribution predominantly in those regions having less access to Nickel, such as in South-Eastern Asia. Their market share has been doubled between 2001 and 2005 from 5 to 10% [27, 28]. One of the reasons preventing the wide industrial application of Cr-Mn-Ni steels in the European market is the difficulty of forming during manufacturing and processing. Another reason is the lack of information and correspondent research in the last decades, in particular related to their corrosion resistance and mechanical properties. Also, wel-ding of Cr-Mn-Ni steels has not been thoroughly investigated up to now.

For such reasons, laser welding studies for comparison of a Cr-Ni (AISI 301 L) to a Cr-Mn-Ni (AISI 201 L) steel grade have been carried out, with two laser sources, the CO_2- and fibre laser. The chemical composition of both steels and the calculated Cr_{eq}/Ni_{eq} are listed in Table 1. From Fig. 16, it can be drawn that both steels range in the primary ferritic solidification mode region of the WRC 1992 diagram, even though the given Ni-equivalent does not take into account the austenite stabilizing influence of Mn. The content of sulphur and phosphorus does not give any serious reason for concern about hot crack susceptibility of both materials.

First trials were done as bead on plate welds with the fibre laser on plates with dimensions 100 mm × 150 mm and thickness of 3.2 mm. Seve-ral weld runs were made on the same plate simultaneously at the same welding parameters for both materials. Additionally overlap welds were produced on the 3.2 mm sheets with the dimensions 40 mm × 150 mm with the CO_2-laser.

For bead on plate welds, the laser power was kept at 4 kW for all wel-ding tests (except for the welds made at the lowest welding velocity of 2 m/min where it was set at 2.5 kW) whereas the welding velocity was changed between 2 and 8 m/min. In all experiments, Argon was used as shielding gas at the flow rate of approximately 15 l/min by using two different configurations of the shielding gas nozzle, a coaxial and a trail gas nozzle. In every case the plates were welded along the rolling direction.

The lap weld joints were produced with 5 kW laser power at the focal point position of –2 mm, whereas the welding velocity was changed from 1.8 to 3 m/min.

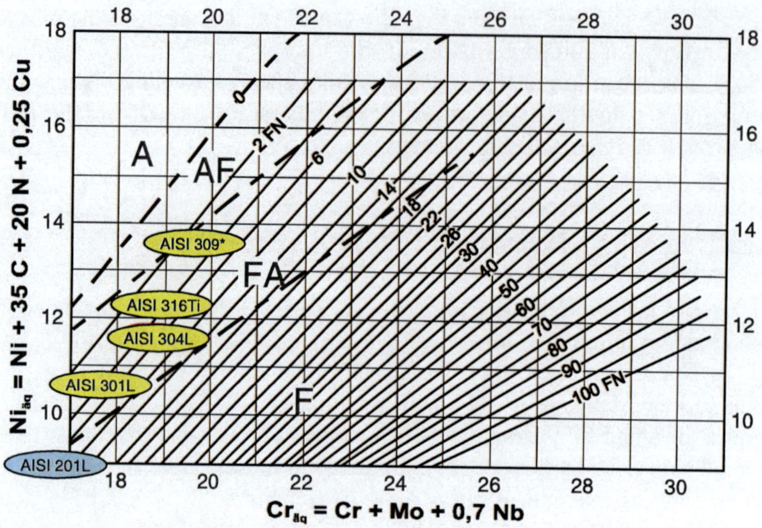

Fig. 16 Location of the used austenitic steel grades in the WRC 1992 diagram

A 50 vol-% Ar/He mixture was used as shielding gas at a flow rate of ca. 10 l/min, which was applied coaxially.

The visual inspection of the bead on plate welds revealed that the Mn-alloyed material is prone to spatter formation during welding whereas in case of AISI 301L a spatter free welding process could be observed (Fig. 17). This can be attributed to the effect of selective evaporation of Manganese, which is supported by the measurement of chemical composition of weld metal. The results of NDT made

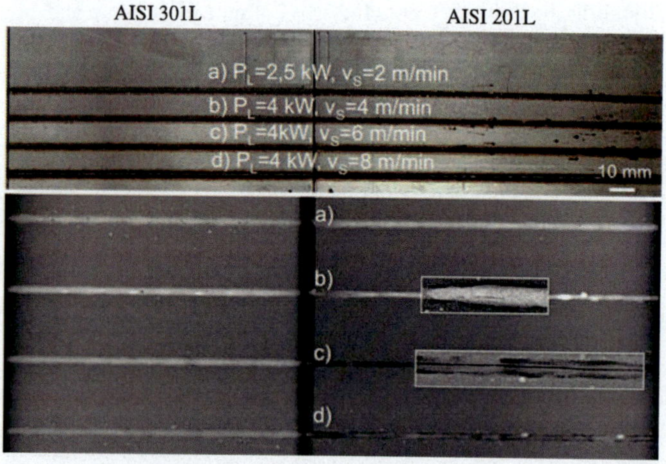

Fig. 17 Weld appearance and radiographic pictures of bead on plate laser welds of AISI 301L and AISI 201 L stainless steel sheets (*crack* are shown under magnification in the *rectangular areas*)

by the radiographic method showed that the welds of the material AISI 201L produced at 6 m/min contain a longitudinal crack almost over the whole welding distance (Fig. 17c). These could be observed for both variations of the shielding gas supply configuration. For a welding speed of 4 m/min (Fig. 17b), the cracks were locally distributed and the total length comprised only a few millimetres. In contrast, the specimens welded at extremely high (8 m/min) and extremely low (2 m/min) welding speed did not reveal any cracks. Since no cracks were detected in the AISI 301 L steel welds for any welding parameters, it can tentatively be concluded that the Cr-Mn-Ni steel is more prone to solidification cracking during laser welding.

The weld cross-sections confirmed the formation of solidification cracks at the weld centreline (Fig. 18). Since crack occurrence in these material correlates with a topside undercut, it might be assumed that cracking is related to the weld metal shrinkage process.

The analysis of the microstructure of solidified welds in Fig. 19 reveals an existence of residual delta ferrite in both investigated materials and gives the evidence of primary ferritic solidification mode in case of AISI 201 L and ferritic-austenitic in case of AISI 301 L.

In the overlap welds produced with the CO_2-laser, cracks appeared only in the Manganese alloyed material at the highest penetration depths resul-ting at the lowest investigated welding speeds ($v_s = 1.8$ mm/min and v_s=2 mm/min). The cracks are located in the lower weld part (Fig. 20). The crack appearance is similar to that found on bead on plate welds. As a specific feature, it has been observed that the contours of the crack flanges fit together (Fig. 21) and that the cracks are located in a region where the growing dendrites encounter each other. SEM examinations of the crack on the weld cross section by rotating the sample 20° revealed a

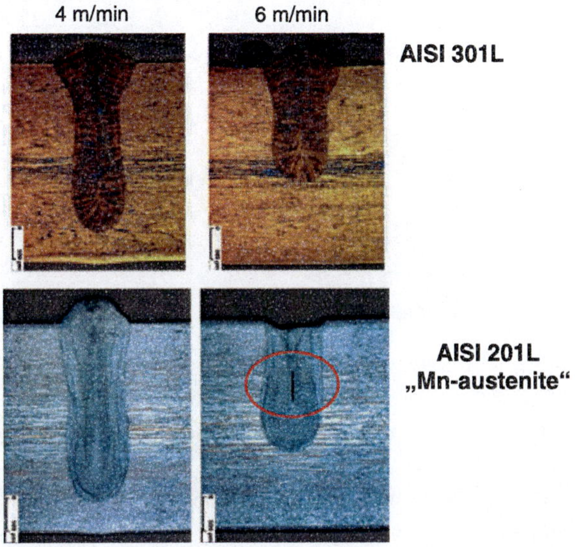

Fig. 18 Hot cracks in the AISI 201L weld material during laser bead on plate welding

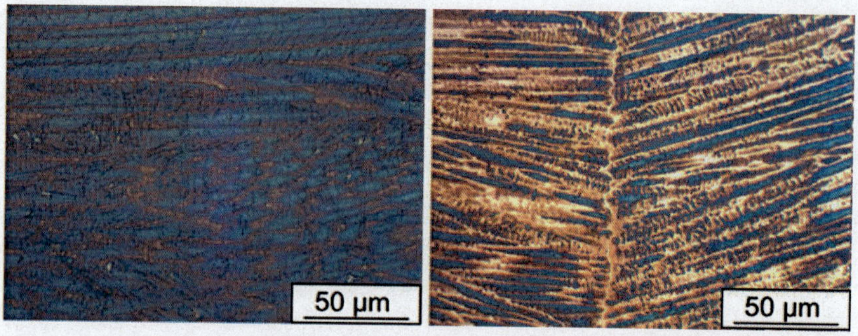

Fig. 19 Micro structure of solidified metal in AISI 201 L (*left*) and AISI 301 L (*right*) welds

Fig. 20 Overlap weld cross sections of both investigated materials welded at v_s = 2 m/min and P_L = 5 kW

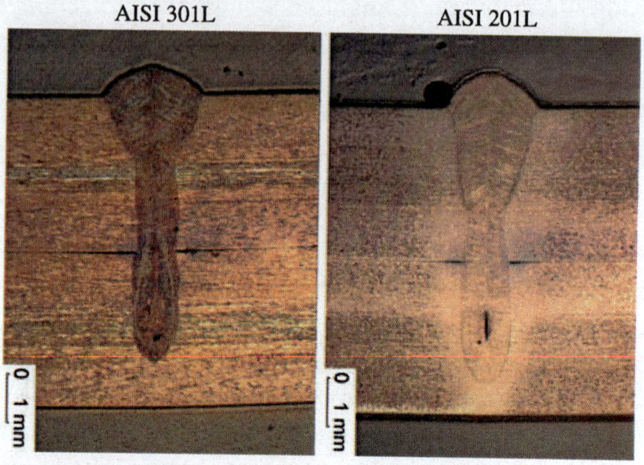

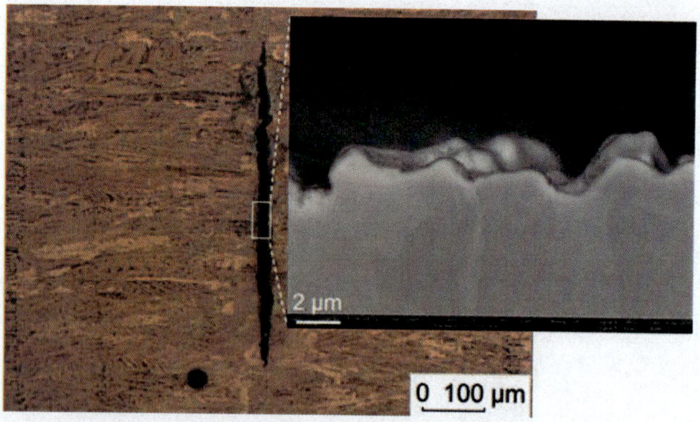

Fig. 21 SEM examination of the formed crack

dendritic structure of the crack surface (Fig. 19), which is typical for solidification cracks.

At the present time, about the reasons for the hot crack formation in the Cr-Mn-Ni steel during laser welding can only be speculated. The following facts might be associated with the formation of the cracks:

1. The as-delivered Cr-Mn-Ni steels has up to 25% higher yield strength than the investigated Cr-Ni steel might provide other deformation interactions between the weld metal and the vicinity at high temperatures, causing some different cracking behaviour.
2. Another possible reason, which could be responsible for the observed volume change is an intensive evaporation of Mn during welding promoting the spatter formation. The measured concentration profiles by means of EPMA confirmed this hypothesis. The local concentration of Mn in the welded metal decreased from initially 7.13% to the 6.61%.
3. The primary solidification mode does not explain the hot crackong behaviour in those materials and can not be attributed to the appearance of the hot cracks in contrast to the Cr-Ni austenitic stainless steels.

Comparison of Hot Cracking Tests

In order to investigate the hot cracking resistance of the Cr-Ni Austenitic stainless steels in the first section during laser beam welding, a special test facility was designed (Fig. 22). This test set up enables particularly the variation of key parameters, i.e. strain and strain rate that determine weld metal hot cracking during the solidification process up to slightly below the solidus temperature. Furthermore, it allows a better handling under the relatively immobile CO_2-lasers as compared to the larger test set ups required for Varestraint Tests and other procedures [30].

During the hot cracking test a sheet or plate specimen is feeded between two transportation rolls and wrapped around a mandrel (Fig. 22). The laser runs perpendicular to this arrangement and produces a quasi–diagonal remelt weld across the specimen. The drive is provided by a round axle coupled with the control unit of the laser system. The strain rate can be varied by different diameters of the bending mandrel providing various respective angular velocities. This ensures, within satisfactory limits, free selection and direct programming of the welding speed, of the strain rate as well as of the heat input via the laser power. The critical strain rate

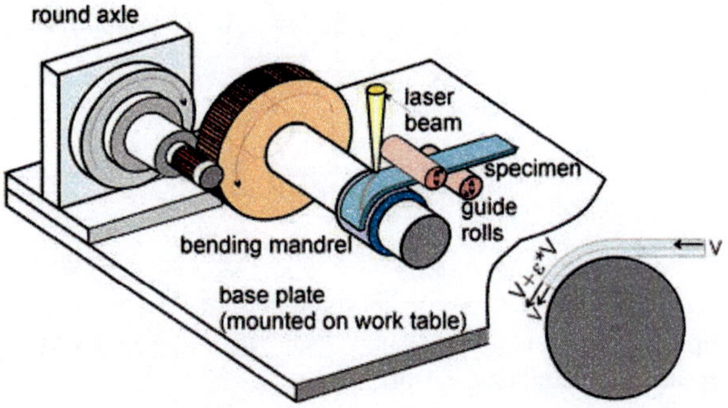

Fig. 22 Test set up for investigation of the hot cracking resistance of materials during laser beam welding

at which a macro–crack appears at the specimen surface is taken as one criterion for the hot cracking susceptibility. Except with marginal devaitions of the collateral heat distribution, the test conditions over the weld length are constant and allow reliable comparisons of the hot cracking resistance during laser welding of different materials. The test, however, can only be regarded as a ranking procedure and has only little transferability of the results to real components, if the real strain rate of the component weld is not known.

However, such ranking tests revealed with very proper consistency that materials, as for instance an AISI 316Ti stainless steel, which do not undergo a primary solidification mode change during laser beam welding are not subjected to hot cracking, even at high strains and respective strain rates. But, the stainless steels with the AISI classification 304L, 309 and 316L which change their primary solidification mode from Ferrite to Austenite during high speed laser welding showed cracking at higher strain rates without exception. These were evidently identified as hot cracks by respective light microscopy and SEM investigations

From the detailed evaluation of sections from welds produced in the testing facility using different loading conditions and welding parameters, diagrams can be drawn as shown in Fig. 23. By plotting the occurrence of cracking versus the welding and loading conditions, materials can be classified with respect to their hot cracking resistance at short term metallurgical effects occurring during high speed laser welding. As demonstrated by Fig. 23, in the AISI 304 L stainless steel welds hot cracks are only formed at elevated strain rates and thus, this material is less susceptible to hot cracking than the steel AISI 309 at the same conditions.

Recently, an improved new test facility has been developed, the Controlled Tensile Weldability (CTW) Test [31]. This test procedure is based on a controlled tensile load or displacement rate on the specimen during welding (Fig. 24). Since the

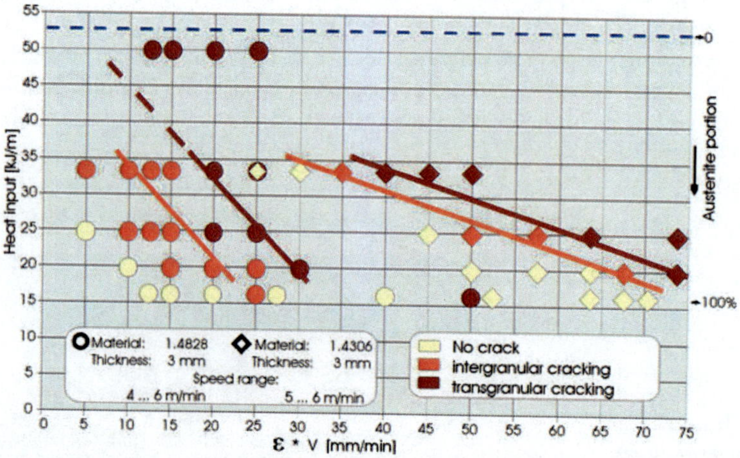

Fig. 23 Hot cracking in stainless steels 304L and 309 at 4% strain of the specimen surface dependent on heat input and total strain rate

Fig. 24 CTW-Test set up for studying hot cracking behaviour of materials during laser beam welding

test set up is easily movable, it allows also studies of the hot crack susceptibility of different materials for different welding processes. The advantage of using this test for laser welding applications is the load change within in a very short time, typical for high speed welds. As a particular advantage in contrast to the above described procedure, the CTW Test allows studying weld cracking phenomena also on thicker materials. As a further benefit compared to the classical Varestraint and other tests, thicker welds can be tested perpendicular to the welding direction at high deformation rates, simulating shrinkage restraint processes in the transversal direction. This also allows a better transfer of the real component welding conditions to the test set up and vice versa.

To clarify the hot cracking origin in the Cr-Mn-Ni steel laser welds, preliminary experiments were carried out with the CTW-Test on the two previously considered steel grades. All specimens were welded with a power of 6 kW with a CO_2-laser at a speed of 1.5 m/min at full penetration by setting the focal point position to – 1 mm and using a 50 vol-% Ar/He mixture. The 150 mm long and 30 mm wide samples with a thickness of 3.2 mm were arc welded to clamping devices with the dimensions 150 mm × 250 mm × 8 mm, as shown in Fig. 24.

The cross head velocity affecting the local strain rates in the weld was changed in the range between 25 and 50 mm/min.

Laser beam welding was started at 10 mm from the plate edge and strain was applied at 32.5 mm after the weld start as schematically shown in Fig. 25. The cross head travel, determining the total sample elongation was set to 1 mm, so that the duration of strain application varied according to the cross-head velocity.

Hot crack formation was observed to occur near the middle of the weld. The position of crack initiation relative to the weld start was shifted to higher values with decreasing cross-head velocity (Fig. 25).

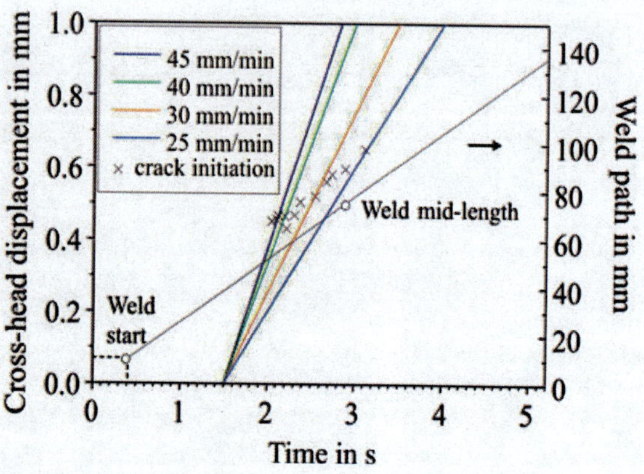

Fig. 25 Diagram showing the CTW-Test procedure and position of crack initiation in the weld

Hot cracking occurred in the weld centreline in the lower part of the welds (Fig. 26), as observed in previous welding experiments. The detection of cracks was carried out by radiographic inspections (Fig. 27).

The critical rate for crack formation was determined to be below 25 mm/min for the steel grade AISI 201 L, as all samples tested at the different investigated cross-head speeds contained cracks. The comparison between both materials shows for example, that the cross head velocity of 40 mm/min does not produce any cracks in the AISI 301 L welds, whereas in the steel grade AISI 201 L a crack of ca. 20 mm length is formed (Figs. 25 and 27). This is indicative of the better hot cracking resistance of the steel grade AISI 301 L.

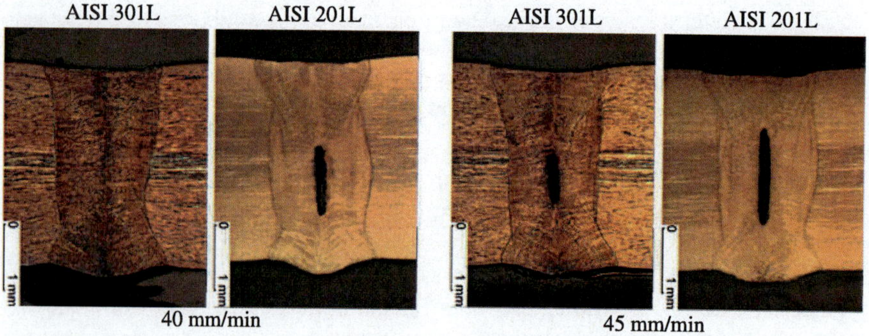

Fig. 26 Hot cracking in stainless steels 301L and 201L, tested under tensile load, at 40 mm/min cross head speed (*left*) and at 45 mm/min (*right*)

Fig. 27 Radiographic film showing a *centreline* hot crack in the laser w

By increasing the cross head speed to 45 mm/min, cracks in both materials appear, but with significant difference regarding the crack proportions of both materials (Fig. 26, right). The crack length is about four times higher for the Manganese alloyed steel AISI 201 L, which also accounts for its higher hot crack susceptibility.

Conclusions

1. Regarding Cr-Ni austenic stainless steels, with increasing welding speed and thus, solidification rate, the primary solidification mode of the materials AISI 304 L, 309 and 316 L steel laser welds changes from the primary ferrite to the primary austenite, while the AISI 304 material remains unaffected. Such solidification mode changes generally entail a restricted hot cracking resistance.
2. The primary solidification behaviour is strongly influenced by the effect of nitrogen dissolved in the melt by using nitrogen in the shielding gas during laser welding of austenitic stainless steels.
3. During laser welding of steels with fully austenitic behaviour, undercooling and supersaturation occur at high solidification rates which may entail metastable ferritic crystallisation. In comparison to non-gravity vacuum experiments [28], this demonstrates that also during welding of Fe-Cr-Ni alloys, a solidification mode change can take place in both directions.
4. Preliminary tests with new Fe-Cr-Mn-Ni austenitic stainless steel grades reveal a larger tendency to the formation of centreline hot cracks at the same welding parameters than observed for Cr-Ni alloys. The critical strain rates for solidification cracking appearance are significantly lower than those for a standard austenitic stainless steel. The reason for appearance of hot crack in this case is not directly linked with the solidification mode and must be further investigated.
5. A new developed procedure, the Controlled Thermal Weldability (CTW) Test enables more flexible testing with respect to hot cracking, also matching better the particular conditions of laser welding. The new test procedure is particularly applicable for the evaluation of critical strain rates as the most relevant parameter for hot cracking. Together with testing perpendicular to the welding direction and also testing of thicker sections, this procedure provides transferability to real components and vice versa than any other hot cracking test.

Acknowledgements The presented results have been supported by different funding organisation in frame of research projects running in the different time at the BAM:

BMBF-Project 13N68717 Laser 2000/Relations Between Solidification Morphology and Hot Crack Initiation During Laser Welding of Predominant Austenitic Materials

DFG-Project BO 1720/4-1 „Short time metallurgical phenomena during rapid melting processes in Fe-Cr-Ni system near eutectic rim “.

AiF-Project 16208 N, “Laser beam welding of austenitic and austenitic-ferritic CrMnNi-steels”.

The authors greatly appreciate the contributions of the former and current colleagues: Helmut Schobbert, Martin Wolf, Karen Stelling, Carl Cross, Nicolas Coniglio, Andreas Hannemann, Marco Lammers, Karin Schlechter and Marina Marten to this paper.

References

1. Processing Duplex/Superduplex, Butting, Informational prospect (source www.butting.de)
2. Tubular News N.2 (2007) Outokumpu Stainless Tubular Products, p 6 (source www.outocumpu.com)
3. Jones LP, Aubert P, Avilov V, Coste F, Daenner W, Jokinen T, Nightingale KR and Wykes M (2003) Towards advanced welding methods for the ITER vacuum vessel sectors, Fusion Engineering and Design, v. 69, Nr. 1-4, 215–200.
4. Mullins WW and Sekerka RF (1964) Stability of a planar interface during solidification of a dilute binary alloy. Journal of Applied Physics 35:444–451.
5. Kurz W, Giovanola B and Trivedi R (1986) Theory of microstructural development during rapid solidification. Acta Metallurgica 34:823–830.
6. Kurz W and Fisher DJ (1998) Fundamentals of Solidification, 4th revised edition. Trans Tech Publications, Switzerland.
7. Aziz M (1982) Model for solute redistribution during rapid solidification. Journal of Applied Physics 53:1158–1168.
8. Böllinghaus Th and Schobbert H. Laser Weld Metallurgy of Austenitic Stainless Steels. In Conference Proceedings Stainless Steel World 2003, Maastricht, pp. 243–257.
9. Schobbert H. Relations Between Solidification Morphology and Hot Crack Initiation During Laser Welding of Predominant Austenitic Materials, BMBF – Final Report No. 13N68717, in German
10. Schobbert H, Böllinghaus Th and Wolf M. Hot Cracking Resistance of Laser and Hybrid Welded Austenitic Stainless Steels, Trends in Welding Research, Proceedings of the 6th International Conference 15.-19. April 2002, S. 76–81, ASM International, USA 2003.
11. Tiller W (1970) Solidification. In:Cahn RW (Ed) Physical Metallurgy, 2nd edition. North-Holland, Amsterdam, p. 403.
12. Caginalpg G (1986) An analysis of a phase field model of a free boundary. Archive for Rational Mechanics and Analysis 92:205–245.
13. Kobayashi R (1992) Simulations of three dimensional dendrites. In: Kai S (Ed) Pattern Formation in Complex Dissipative Systems. World Science, Singapore and Philadelphia, pp. 121–128.
14. Wheeler A, Murray BT and Schaefer RJ (1993) Computation of dendrites using a phase field model. Physica D 66:243–262.
15. Warren JA and Boettinger WJ (1995) Prediction of Dendritic Growth and Microsegregation Patterns in a Binary Alloy Using the Phase-Field Method. Acta Metallurgica et Materialia 43:689–703.
16. Baker JC and Cahn JW (1971) Thermodynamics of Solidification. In: Solidification. ASM, Metals Park, OH, p. 23–58.
17. Kurz W (1995) Dendrite Growth in Welding. Mathematical Modelling of Weld Phenomena 2. In: Cerjak H (Ed) The Institute of Materials. London, pp. 40–53.
18. Rappaz M, Drezet JM and Gremaud M (1999) A New Hot-Tearing Criterion. Metallurgical and Materials Transactions A 30A:449–455
19. Prokhorov N (1962) The technological strength of metals while crystallizing during welding. Welding Production 9:1–8.
20. Elmer JW, Allen SM and Eagar TW (1989) Microstructural development during solidification of stainless steel alloys. Metallurgical and Materials Transactions A 20A:2117–2131.

21. David SA, Vitek JM and Hebble TL (1987) Effect of rapid solidification on stainless steel weld metal microstructures and its implications on the Schaeffler diagram. Welding Journal 66:289s–300s.
22. Lippold JC (1994) Solidification behaviour and cracking susceptibility of pulsed laser welds in austenitic stainless steels. Welding Journal 73:129s–139-s.
23. Lienert TJ and Lippold JC (2003) Improved weldability diagram for pulsed laser welded austenitic stainless steels. Science and Technology of Welding and Joining 8:1–9
24. EN ISO 15614-11 Specification and qualification of welding procedures for metallic materials – Welding procedure test – Part 11: Electron and laser beam welding
25. Lippold JC and Kotecki D (2005) Welding Metallurgy and Weldability of Stainless Steel. Wiley-Interscience, New York, NY.
26. Lichtenegger P and Blöch R (1975) Colour etching of high alloy steels. Practical Metallography 12:567–573.
27. Koseki T and Flemings MC (1996) Solidification of Undercooled Fe-Cr-Ni Alloys. Part II. Microstructural Evolution. Metallurgical and Materials Transactions A 27A:3226–3240.
28. Volkmann TW and Löser DM (1997) Herlach Nucleation and Phase Selection in Undercooled Fe-Cr-Ni Melts: Part II. Containerless Solidification Experiments. Metallurgical and Materials Transactions A 28A:461–469.
29. Singhal LK (2006) Stainless Steels Recent Developments and Outlook: Demand Capacity and Product Development. DSTI/SU/SC (2006) 15, presented on "Joint India/OECD/IISI Workshop" New Delhi (India) 16–17 Mai 2006, p. 21.
30. ISO 17641-3 Destructive tests on welds in metallic materials Hot cracking tests for weldments - Arc welding processes Part 3:
31. Kromm A, Kannengieser Th (2007) Influence of Local Weld Deformation on the Solidification Cracking Susceptibility of a Fully Austenitic Stainless Steel. Hot Cracking Phenomena in Welds II. Springer, Berlin, pp. 127–145.

Laser Weldability of Stainless Steel

Carl E. Cross, N. Coniglio, E.M. Westin, and A. Gumenyuk

Introduction

The formation of centerline solidification cracks can sometimes prove troublesome when laser welding certain crack sensitive grades of stainless steel. It would be beneficial, for purposes of alloy selection and process development, if a reliable testing method could be used to rank different stainless grades as to their relative laser weldability. Also of interest is to quantify critical cracking conditions for purposes of mechanistic modeling. The use of standard weldability test methods and crack evaluations for arc welding are not always suitable for laser welding. The varestraint test, for example, poses a problem with regard to the rapid travel speeds sometimes encountered in laser welding (e.g. travel speeds 1–100 times faster than arc welding), relative to the rate of test strain application.

Questions regarding weldability normally arise when developing laser welds for new grades, not knowing how the new material will behave in comparison with more thoroughly characterized grades. An example of this is the laser welding of new high-manganese, low-nickel stainless steel alloys (e.g. 201 and LDX® 2101). From a metallurgical standpoint, alloy rankings based upon arc welds do not necessarily reflect accurately upon laser weldability. One case in point is the shift in solidification mode due to rapid solidification conditions (favoring primary austenite) when laser welding, thereby lowering weldability. Keyhole instability in laser welding is another factor that may influence cracking susceptibility.

With the availability of new high power (20 kW) fiber lasers, there is a trend toward the use of high travel speeds (20–50 mm/s). This results in high solidification rates and temperature gradients that produce unique differences in grain structure, solidification mode, segregation, and weldability. Similarly, the local strain distribution around rapidly moving laser welds is expected to react differently to global

C.E. Cross (✉)
Federal Institute for Materials Research and Testing, Berlin, Germany
e-mail: cecross@bajabb.com

J. Lippold et al. (eds.), *Hot Cracking Phenomena in Welds III*,
DOI 10.1007/978-3-642-16864-2_8, © Springer-Verlag Berlin Heidelberg 2011

restraining forces. It follows, therefore, that a weldability test is needed that is compatible with the laser welding process conditions. Only a limited number of laser weldability studies have been made to date [1–5].

In this work the Controlled Tensile Weldability (CTW) test, with procedures initially developed for gas-tungsten arc welding, has been suitably adapted for the laser welding of stainless steels. In particular, a test procedure was specifically developed for CO_2 laser, full-penetration keyhole welds made on 3 mm thick plate for a variety of austenitic and duplex stainless steel grades. Details of this test procedure and its limitations will be examined and results will be presented. A ranking of laser weldability (i.e. susceptibility to solidification cracking) for different stainless grades is made and compared with expected behavior from available literature.

Background

Weldability of Stainless Steel

The weldability of stainless steel, examined in terms of susceptibility to solidification cracking in the weld metal, has been shown to be predominantly controlled by alloy composition and impurity content [6]. Impurities sulfur and phosphorous, in particular, are known to be associated with cracking, extending the solidification range by forming low melting eutectic liquid [7]. Boron and copper may also play a similar role. Impurity-rich liquid segregates to grain boundaries where liquid films are exposed to local tensile strains resulting from thermal contraction and solidification shrinkage.

The primary solidification mode is known to have a predominant effect on weldability. Alloys that solidify as primary austenite (γ) and remain fully austenitic (A-solidification), corresponding to Hammar-Svensson chromium-nickel equivalent ratios (Cr_{eq}/Ni_{eq}) less than 1.48 for arc welds [8, 9], are most susceptible to cracking. The well-defined, straight and continuous nature of the grain boundaries associated with this mode of solidification is believed to be at least partially responsible for its high cracking susceptibility [10].

For alloys with Cr_{eq}/Ni_{eq} ratios higher than 1.48, where primary solidification occurs as ferrite (δ) and then austenite forms subsequently as a peritectic-eutectic (FA-solidification), the resistance to cracking is exceptional. For these alloys, weld metal grain boundaries are irregular and ill-defined making crack propagation more difficult [10]. Also, it has been argued that impurity liquid films do not wet austenite/austenite boundaries as well as ferrite/austenite boundaries [11]. For alloys still higher in Cr_{eq}/Ni_{eq} ratio such as duplex grades, solidification occurs completely as ferrite (F-solidification) and susceptibility to cracking again increases, but not to the level experienced by fully austenitic grades. The nature of grain boundaries reverts back to being straight and continuous, and wettability improves in the absence of austenite allowing for more continuous liquid films.

Examples of the above behavior regarding weldability and Cr_{eq}/Ni_{eq} ratio have been observed in varestraint tests comparing total crack lengths for alloys exhibiting different primary solidification modes in gas-tungsten arc welds [12, 13]. Nelson et al. [12] showed that the weldability of duplex grades 2205 and Ferralium™ 255 are similar and that they rank between two 304L variants, one having primary-γ (A-solidification, ferrite number 1) and one having primary-δ (FA-solidification, ferrite number 4.5). Varol et al. [13] reported a similar trend, thus indicating that austenitic grades that solidify as primary-δ (FA solidification) have improved weldability over duplex grades.

In the case of laser or electron beam welds, there is a complicating factor when trying to apply weldability ratings established for arc welds. Under solidification conditions characteristic of beam welding, e.g. high temperature gradient and (in some cases) high growth rate, there is higher undercooling experienced at the solidification front favoring austenite [14–16]. This has the general effect of shifting the upper Cr_{eq}/Ni_{eq} ratio limit for primary-γ solidification to higher values [1]. Also, there is observed the possibility to have both modes of solidification (primary-γ + primary-δ) occur intermixed within the weld metal.

Critical Strain Rate

Solidification cracking has for a long time been related to the inability of a particular alloy to accommodate strain accumulated over a brittle temperature range in the weld mushy zone [17–19]. As a result of this ductility approach to cracking, the rate of strain accumulation must be considered important in determining whether a tolerable level of strain is exceeded within a fixed solidification time. Consequently, a variety of different weldability tests have been specifically developed to measure critical strain rates needed for cracking [20–24]. More recent developments have suggested that strain rate may play a more direct role in controlling crack initiation and growth mechanisms [25, 26]. Critical strain rates in and of themselves may be taken as a measure of weldability, where a high critical rate represents greater resistance to cracking [27].

Laser Weldability Testing

Laser weldability testing to date has been largely limited to evaluating bead-on-plate welds made under controlled conditions. This has proven useful in providing comparisons of alloys based simply upon crack-no crack observations. Using such an approach, Lippold [1] and Leinert and Lippold [2] have demonstrated an extension in primary austenite solidification (and associated cracking susceptibility) to higher Cr_{eq}/Ni_{eq} ratios for stainless steel laser welds. However, such comparisons only apply to the specific welding conditions (e.g. heat input, travel speed, restraint intensity) under which these tests were made.

A few researchers have either developed or adapted weldability tests specifically for laser welding. Schobbert et al. [3] introduced a special test based upon the varestraint test concept [28]. Like the varestraint test, a weld coupon is wrapped around a curved mandrel during welding, imposing a fixed strain behind the weld pool at the weld surface. However, unlike the varestraint test, strain in this test is applied at an oblique angle to the welding direction. This test proved useful in comparing two different austenitic stainless steel alloys (Grades 309 and 304), establishing critical deformation rates needed to form cracks. Grade 309 was found to solidify as primary-γ at high travel speeds. Consequently, the critical deformation rate to form cracking was found to be less than for Grade 304, which consistently solidified as primary-δ. This test has numerous drawbacks, including 1) the oblique angle of strain application- not normally experienced in welding, 2) the inability to calculate or measure local strain or strain rate, 3) the non-uniformity of strain and strain rate through the specimen thickness, and 4) the hinging of material along the weld (i.e. non-conformity to die).

Weise [29] applied the hot deformation rate (HDR) test [23] to laser welds on a structural steel. By testing at different applied deformation rates, he was able to identify the critical condition for cracking, expressed in terms of strain over temperature drop (2.7×10^{-5}/K) following the approach of Prokhorov [18]. It is conceivable that a variable-speed trans-varestraint test could be utilized in a similar fashion to the HDR. However, due to the mechanics of bending encountered in all such tests, it is not possible to generate a uniform local strain rate anywhere in the weld, thus making it difficult to relate the applied global deformation rate to anything meaningful on a local scale.

Nishimoto and Mori [4] used the preloading tensile strain (PLTS) cracking test to investigate the weldability of CO_2 laser welded 304 stainless steel with variable nitrogen content. Similar in concept to the Sigmajig test [30], the PLTS test fixture preloads the weld specimen (12–18 kg/mm^2) resulting in enhanced strain during welding and allowing crack-no crack comparisons. It was convincingly shown that the upper Cr_{eq}/Ni_{eq} ratio boundary for primary-γ solidification increases with dendrite growth velocity (i.e. weld travel speed). However, the relation between pre-load and local strain rate was not established.

Experimental

Material

A variety of different types of austenitic and duplex stainless steels have been investigated in this study as listed in Table 1 with corresponding compositions. Alloys are grouped in this table as super-austenitic (310, 904L, 254 SMO®), austenitic (304L, 301LN, 201, 316L, 321), and duplex (2205, 2304, LDX® 2101). Alloys are shown in order of increasing chromium-nickel equivalent ratio (see Cr_{eq}/Ni_{eq} in Table 2). Based upon.

Table 1 Measured compositions for investigated stainless steel grades (in wt%)

Grade	EN	C	Si	Mn	P	S	Cr	Ni	Mo	N	Cu	Ti
904L	1.4539	0.015	0.35	1.61	0.019	0.001	19.99	23.93	4.40	0.049	1.42	–
310	1.4845	0.046	0.56	0.99	0.012	0.001	25.06	19.36	0.13	0.038	0.10	0.001
254 SMO®	1.4547	0.014	0.39	0.55	0.014	0.001	20.04	17.75	6.06	0.194	0.76	–
316L	1.4404	0.022	0.49	1.70	0.027	0.002	16.82	10.49	2.52	0.043	0.33	–
304L	1.4307	0.015	0.35	1.74	0.031	0.001	18.15	8.44	0.41	0.069	0.25	0.001
301LN	1.4318	0.023	0.50	1.02	0.024	0.002	17.17	7.18	–	0.122	–	–
321	1.4541	0.039	0.50	1.42	0.025	0.001	17.32	9.08	0.48	0.012	0.37	0.47
201	1.4372	0.025	0.39	6.70	0.023	0.004	16.45	4.21	–	0.140	–	–
2205	1.4462	0.018	0.38	1.40	0.019	0.002	22.46	5.80	3.22	0.188	0.23	0.003
LDX 2101®	1.4162	0.025	0.65	5.13	0.019	0.001	21.57	1.56	0.28	0.229	0.30	–
2304	1.4362	0.023	0.39	1.46	0.025	0.002	23.20	4.83	0.45	0.116	0.22	0.001

Table 2 Comparison of Chromium-Nickel equivalent ratios for stainless steel grades based upon compositions in Table 1

Grade	Cr_{eq}/Ni_{eq}- WRC 1992	Cr_{eq}/Ni_{eq}- H&S
904L	0.95	0.99
310	1.16	1.22
254 SMO®	1.17	1.33
316L	1.59	1.69
304L	1.78	1.83
301LN	1.65	1.84
321	1.65	1.84
201	2.08	1.93
2205	2.51	2.88
LDX 2101®	3.08	3.16
2304	2.95	3.68

WRC 1992 [32]:
$Cr_{eq} = (\%Cr)+(\%Mo)+0.7(\%Nb)$
$Ni_{eq} = (\%Ni)+35(\%C)+20(\%N)+0.25(\%Cu)$

Hammar & Svensson [33]:
$Cr_{eq} = (\%Cr)+1.37(\%Mo)+1.5(\%Si)+2(\%Nb)+3(\%Ti)$
$Ni_{eq} = (\%Ni)+22(\%C)+1.42(\%N)+0.31(\%Mn)+(\%Cu)$

Cr_{eq}/Ni_{eq} ratio and behavior established for arc welds, all of the super-austenitic grades would be expected to be fully austenitic (A-solidification), the austenitic grades as primary-δ (FA-solidification), and the duplex grades as primary-δ (F-solidification).

The super-austentics are highly alloyed for exceptional corrosion resistance, but traditionally exhibit poor weldability due to their solidification as primary austenite. Alloy 304L is a low carbon version of the standard 304 (18–10 type) stainless steel, used to avoid sensitization during welding. Alloy 301LN is a slightly lower alloy (lower cost) replacement for 304L, with reduced corrosion resistance, and alloy 201 is a low cost (low nickel, high manganese) replacement for 301LN. Alloys 316L and 321 are different variations on the 304L grade, where 316L contains molybdenum for improved pitting resistance, and 321 is stabilized against sensitization with titanium. Alloy 2205 is the standard grade of duplex, and 2304 is a low molybdenum version of 2205. LDX® 2101 is a lean duplex, containing high manganese and nitrogen as a cost-effective replacement for nickel.

Weld Coupons

Each test coupon (150 mm × 30 mm × 3 mm thick) was arc welded along its 150 mm length to two 304 stainless steel extension plates (150 mm × 250 mm × 8 mm thick) as shown in Fig. 1. These extension plates were suitable for clamping

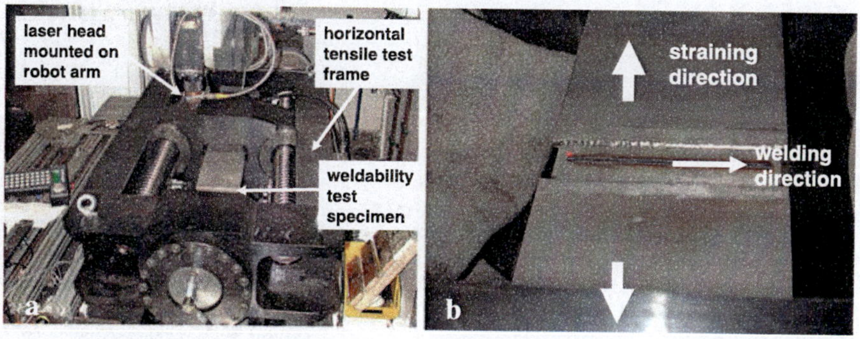

Fig. 1 Controlled Tensile Weldability (CTW) test fixture showing overview of (**a**) test frame and (**b**) test specimen

to the CTW test frame. For weldability testing, a single laser weld was made in the center of each coupon along its 150 mm length. Following welding, the weld coupon was cut out for analysis and the large extension plates reused.

Welding Parameters

All welds in this study were made with a 6 kW CO_2 laser, run at full power at a travel speed of 25 mm/s. The laser beam was focused 1 mm below the plate surface, with a 200 mm focal distance. Welds were made bead-on-plate, full-penetration, in a keyhole mode. A mixture of argon and helium (50/50) was used for shielding gas. Coupons were degreased with acetone just prior to welding. It is believed that these parameters provided somewhat higher heat input than the minimum needed to achieve full penetration thus producing a wider bead and a higher propensity for cracking.

CTW Test

The Controlled Tensile Weldability (CTW) test involved the application of a controlled tensile displacement during welding, applied transverse to the welding direction at a controlled rate. This was accomplished by affixing the test coupon to a horizontal 500 kN screw-type tensile test machine (see Fig. 1). The concept of this test is to control the strain rate across the mushy zone by systematically controlling the globally applied displacement rate via a fixed cross-head speed. By looking for crack-no crack conditions, one can identify the critical applied displacement rate needed for crack formation. A series of tests were run for a given alloy, each at a progressively higher cross-head speed, until a critical speed was found that produced a continuous centerline crack near the weld mid-length. Critical cross-head speeds were observed over a range between 0 and 1 mm/s for the alloys examined. The CTW test sequence consisted of the following steps:

1. *clamp test plate into CTW test frame*
2. *apply 2 kN pre-load*
3. *start weld 5 mm from edge of plate*
4. *initiate cross-head travel 32.5 mm after weld start*
5. *terminate weld at 120 mm weld length*
6. *stop cross-head travel at weld termination (or 1 mm maximum cross-head displacement)*
7. *remove test plate, cut-out weld coupon for radiography*

Crack Detection

Whenever the critical displacement rate was exceeded for a given alloy, a continuous crack was found to occur along the weld centerline, contained within the weld metal (see Fig. 2). An exception to this was found with the super-austenitics, where the centerline crack exited both crown and root surfaces. Thus, radiography was routinely used to detect and characterize crack formation for each test. Cracks were found to form near the weld mid-length, roughly 20–50 mm beyond the point where displacement was first initiated, and then grow continuously to different lengths. Cracks initiated sooner for those tests run at higher displacement rate.

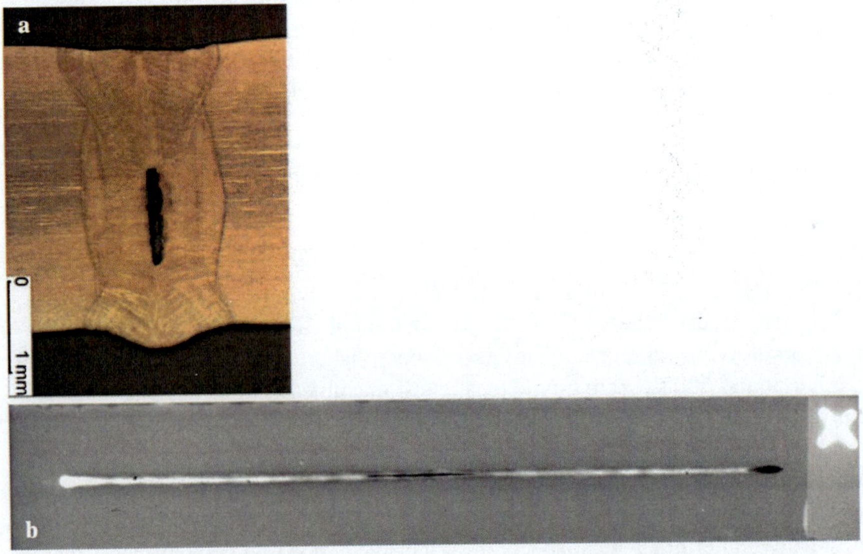

Fig. 2 CTW test specimen showing (**a**) weld cross-section with embedded solidification crack, and (**b**) radiograph of weldment with centerline solidification crack

Results and Discussion

Development of CTW Test Procedure

The general concept and procedure for operating this test were based upon experience gained from a previous study examining the solidification cracking in Gas-Tungsten Arc (GTA) welded aluminum [31]. However, there were notable differences in behavior that required changes in procedure for laser welding as outlined here. The most notable difference was that cracks did not form instantaneously with strain application, as was the case with aluminum GTA welds.

With higher welding speeds used for laser welding (laser: 25 mm/s vs. GTA: 4 mm/s), the corresponding time for testing is much shorter (laser: 6 s vs. GTA: 25 s). Also, higher cross-head speeds were required to achieve cracking (laser: 0–1 mm/s vs. GTA: 0–0.1 mm/s). This means that with laser CTW testing of stainless steel, higher maximum strains are encountered and, due to a higher elastic modulus, higher loads are generated (up to 180 kN). This resulted in problems with failure in the auxiliary welds used to join the coupon to the extension plates when testing at the high end of cross-head speed (i.e. near 1 mm/s).

To avoid auxiliary weld failure, the maximum cross-head displacement was limited to 1 mm, thus limiting the maximum load to below 180 kN. This means that for the high end of cross-head speed, strain could only be applied for 1 s (i.e. 25 mm weld travel). It is conceivable that for highly weldable alloys requiring even higher cross-head speeds in order to crack, this limitation in the application of strain over distance may prove unworkable.

Critical Displacement Rates

A bar graph comparing the critical cross-head speeds needed to generate cracking for different grades is given in Fig. 3. In particular, the location of crack initiation is indicated relative to the weld mid-length. It is observed that there is a consistent trend for each alloy, where the crack forms closer to the point of strain application with increasing cross-head speed. This suggests that a uniform strain rate across the specimen is not established instantaneously, but requires a certain amount of displacement over which to develop. At higher applied cross-head speeds, it could be that the critical local displacement rate is established sooner. However, it may also be that a certain time is required for crack nucleation depending upon the local conditions.

The height of bars for each alloy in Fig. 3 indicates approximately the cross-head speed for crack initiation at the weld mid-length. Using this as a criteria to compare different alloys, it is found that cracks for the super-austenitic stainless steels formed at the lowest cross-head speeds (0.30–0.40 mm/s), accurately reflecting upon their poor weldability. The austenitic stainless steels showed better weldability, with critical displacement rates of 0.50–0.65 mm/s. The duplex alloys showed the best

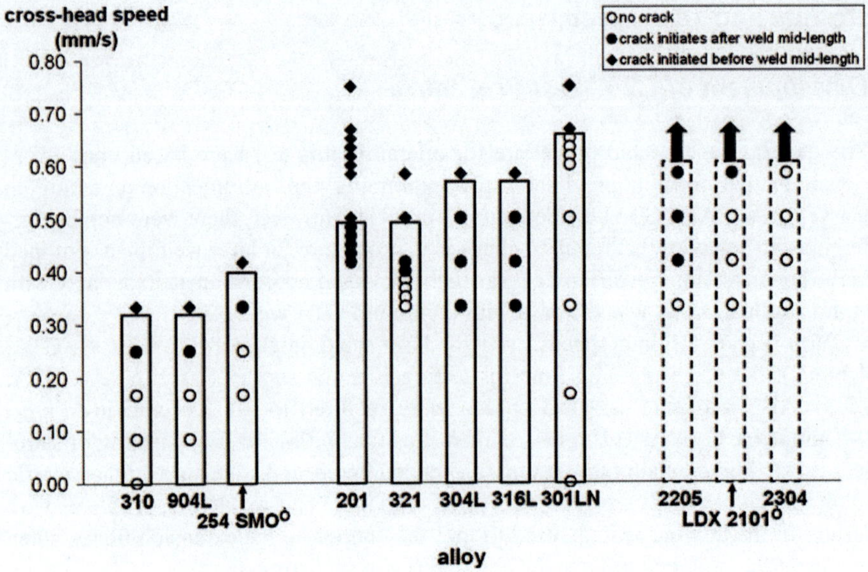

Fig. 3 Comparison of CTW test results for super-austenitic, austenitic, and duplex stainless steels showing observed solidification cracking behavior varying with applied cross-weld displacement rate (i.e. cross-head speed)

weldability with critical displacement rates all above 0.6 mm/s, although these limits were not firmly established within the frame of experiments performed. In particular, Grade 2304 showed the greatest resistance to cracking as demonstrated in the relative weldability comparison of in Fig. 4. It is of interest to note that the high manganese grades (201 and LDX® 2101) demonstrated good performance comparable to their respective grades.

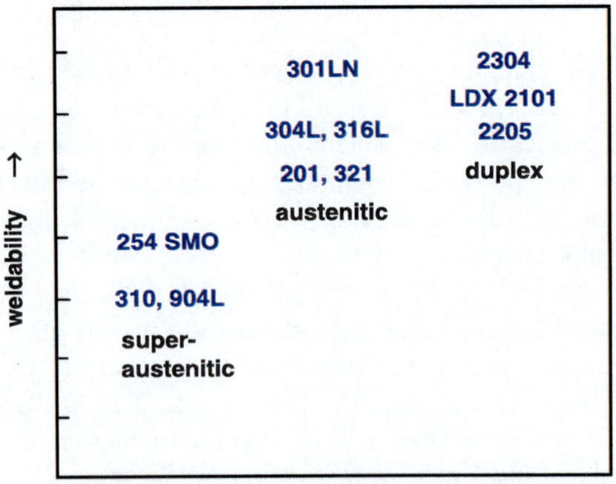

Fig. 4 Comparison of relative weldability of stainless steel grades based upon critical displacement rates from Fig. 3

Cr/Ni Equivalent and Weldability

A comparison of Cr_{eq}/Ni_{eq} ratios for the alloys examined in this study is given in Table 2. Two different formulations for chromium and nickel equivalents have been used: one based upon WRC-1992 [32] and one based upon Hammar and Svensson [33]. The former does not account for the alloying element manganese, which probably does not represent as well the high manganese grades 201 and 2101. However, the general trend between the three groups remains the same with the super-austenitics having the lowest ratios (primary-γ, A-solidification) and the duplex alloys having the highest ratios (primary-δ, F-solidification).

With regard to the comparisons made in Fig. 4, the exceptional laser weldability observed for duplex grades relative to the austenitic grades is normally not experienced in arc welding. This may be explained by a shifting of primary ferrite solidification to higher Cr_{eq}/Ni_{eq} ratios at high cooling rates as demonstrated in Fig. 5. Grades with Cr_{eq}/Ni_{eq} around 1.7 (e.g. 304L) will experience reduced laser weldability in comparison with arc weldability, because there may be some primary-γ formed. Whereas, at the high Cr_{eq}/Ni_{eq} ratio of duplex grades, there is no effect of cooling rate on weldability and no difference between laser and arc weldability.

It must be emphasized that any weldability test provides only a relative comparison of alloys and does not necessarily represent what will actually be experienced in practice. Additional factors including welding parameters, keyhole stability, weld geometry, and degree of restraint can provide significant influences on cracking behavior. Also, weldability in general is comprised of many other factors besides resistance to solidification cracking, including resistance to corrosion and liquation cracking. Thus, any conclusions regarding the weldability of the alloys in this study must be carefully considered with this in mind.

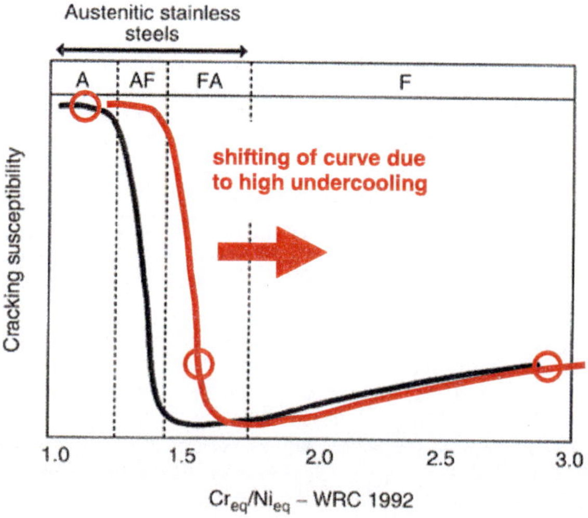

Fig. 5 Effect of rapid solidification rate of laser welding shifting primary solidification mode to higher Cr_{eq}/Ni_{eq} ratios (original curve from Ref. [6]) reducing the weldability of primary-δ, FA solidification austenitic grades

Summary

The CTW test has been applied to evaluate the laser weldability (i.e. resistance to solidification cracking) of several different grades of stainless steel, including super-austenitic, austenitic, and duplex. This test, which involves the application of transverse strain during welding, required displacement rates in the range 0–1 mm/s to generate centerline solidification cracking. Unlike most laser weldability testing of the crack-no crack type, the CTW test allows for the possibility to quantify critical conditions needed for cracking.

When compared in terms of the minimum displacement rate need to generate cracking at specimen mid-length, the duplex grades showed the highest weldability (>0.6 mm/s), whereas the super-austenitics showed the lowest weldability (0.30–0.40 mm/s). The apparent exceptional laser weldability of duplex grades may be a relative effect, caused by the reduced weldability of primary-δ, FA solidification austenitic grades due to rapid cooling.

Some difficulties were encountered in CTW testing related to the rapid speed of laser welding. The location of crack initiation varied with the applied displacement rate, moving to shorter distances at higher cross-head speeds. The reason for this behavior remains unclear at this point. Problems were also encountered with high stresses generated at the upper end of displacement rate, causing failure in auxiliary welds. The relationship between applied displacement rate and local strain rate in the mushy zone must still be established.

Acknowledgements The authors wish to acknowledge colleagues at BAM, in particular K. Stelling, A. Hannemann, K. Schlechter, K. Scheideck, and M. Richter for help with CTW testing, M. Marten for metallography, and H.-J. Malitte for radiography. Also, material donated by Outokumpu Stainless was greatly appreciated.

References

1. Lippold JC (1994) Solidification behavior and cracking susceptibility of pulsed-laser welds in austenitic stainless steels. Welding Journal 73:129s–139s.
2. Lienert TJ and Lippold JC (2003) Improved weldability diagram for pulsed laser welded austenitic stainless steels. Science and Technology of Welding and Joining 8:1–9.
3. Schobbert H, Böllinghaus Th and Wolf M (2003) Hot cracking resistance of laser and hybrid welded austenitic stainless steels. In: Trends in Welding Research, ASM International, pp. 76–81.
4. Nishimoto K and Mori H (2004) Hot cracking susceptibility in laser weld metal of high nitrogen stainless steels. Science and Technology of Advanced Materials 5:231–240.
5. Juhl TW and Flemming OO (2003) Assessment of hot crack properties of laser welded stainless steel alloys: In: SPIE Proceedings of the 1st International Symposium High-Power Laser Macroprocessing, pp. 224–229.
6. Lippold JC and Kotecki DJ (2005) Welding metallurgy and weldability of stainless steels, Wiley (ISBN-13: 978-0471473794).
7. Li L and Messler, RW (1999) The effects of phosphorus and sulfur on susceptibility to weld hot cracking in austenitic stainless steels. Welding Journal 78:387s–396s.

8. Kujanpää V, Suutala N, Takalo T and Moiso T (1979) Correlation between solidification cracking and mirostructure in austenitic and austenitic-ferritic stainless steel welds. Welding Research International 9:55–76.
9. Suutala N, Takalo T and Moisio T (1979) Relationship between solidification and microstructure in austenitic and austenitic-ferritic stainless steel welds. Metallurgical Transactions 10A:512–514.
10. Brooks JA, Thompson AW and Williams JC (1984) A fundamental study of the beneficial effects of delta ferrite in reducing weld cracking. Welding Journal 63:71s–83s.
11. Hull FC (1967) Effect of delta ferrite on the hot cracking of stainless steel. Welding Journal 46:399s–409s.
12. Nelson DE, Baeslack III WA and Lippold JC (1987) An investigation of weld hot cracking in duplex stainless steels. Welding Journal 66:241s–250s.
13. Varol I, Baeslack III WA and Lippold JC (1997) Characterization of weld solidification cracking in duplex stainless steel. Materials Characterization 39:555–573.
14. David SA, Vitek JM and Hebble TL (1987) Effect of rapid solidification on stainless steel weld metal mirostructure and its implications on the Schaeffler diagram. Welding Journal 66:289s–300s.
15. Elmer JW, Allen SM and Eagar TW (1990) The influence of cooling rate on the ferrite content of stainless steel alloys. In: Recent Trends in Welding Science and Technology, ASM International, pp. 165–170.
16. Kurz W and Trivedi R (1995) Modern solidification theory applied to welding. In: Trends in Welding Research, ASM International, pp. 115–120.
17. Pellini WS (1952) Strain theory for hot tearing. Foundry 80:125–199.
18. Prokhorov NN (1956) The problem of the strength of metals while solidifying during welding. Svarochnoe Proizvodstvo 6:5–11.
19. Senda T, Matsuda F, Takano G, Watanabe K, Kobayashi T and Masuzaka T (1971) Experimental investigations on solidification crack susceptibility for weld metals with trans-varestraint test. Transactions JWS 2:141–162.
20. Tamura H, Kato N, Ochiai S and Katagiri Y (1977) Cracking study of aluminum alloys by the variable tensile strain hot cracking test. Transactions JWS 8:16–22.
21. Arata Y, Matsuda F, Nakata K and Shinozaki K (1977) Solidification crack susceptibility of aluminum alloy weld metals (Report II)- Effect of straining rate on cracking threshold in weld metal during solidification. Transactions JWRI 6:91–104.
22. Matsuda F, Nakagawa H, Nakata K and Okada H (1979) The VDR cracking test for solidification crack susceptibility on weld metals and its application to aluminum alloys. Transactions JWRI 8:85–95.
23. Wilken K and Kleistner H (1985) Remarks on the classification and evaluation of hot cracking tests. IIW-Doc IX-1379-85, International Institute of Welding, Paris.
24. Herold H, Streitenberger M and Pchennikov A (2001) Modelling of the PVR test to examine the origin of different hot cracking types. In: Mathematical Modelling of Weld Phenomena 5, Institute of Materials, pp. 783–792.
25. Rappaz M, Drezet J-M and Gremaud M (1999) A new hot-tearing criterion. Metallurgical and Materials Transactions 30A:449–455.
26. Braccini M, Martin CL, Suéry M and Bréchet Y (2000) Relation between the mushy zone rheology and hot tearing phenomena in Al-Cu alloys. In: Modelling of Casting, Welding, and Advanced Solidification Processes IX. Shaker Verlag, pp. 18–24
27. Cross CE and Coniglio N (2008) Weld solidification cracking- critical conditions for crack initiation and growth. In: Hot Cracking Phenomena in Welds II. Springer, pp. 39–58.
28. Savage WF and Lundin CD (1965) The varestraint test. Welding Journal 44:433s–442s.
29. Weise S (1998) Heißrißbildung beim laserstrahlschweißen von baustählen. Thesis Dissertation, University of Bremen.

30. Goodwin GM (1987) Development of a new hot-cracking test- the sigmajig. Welding Journal 66:33s–38s.
31. Coniglio N, Cross CE, Michael Th and Lammers M (2008) Defining a critical weld dilution to avoid solidification cracking in aluminum. Welding Journal 87:237s–247s.
32. Kotecki D and Siewert TA (1992) WRC-1992 constitution diagram for stainless steel weld metals- a modification to the WRC-1988 diagram. Welding Journal 71:171s–178s.
33. Hammar Ö and Svensson U (1979) Influence of steel composition on segregation and microstructure during solidification of austenitic stainless steels. In: Solidification and Casting of Metals. The Metals Society, London, pp. 401–410.

Weld Solidification Cracking in 304 to 304L Stainless Steel

P.W. Hochanadel, T.J. Lienert, J.N. Martinez, R.J. Martinez, and M.Q. Johnson

Introduction

The fact that austenitic stainless steels may be susceptible to weld solidification cracking is not new. In fact, correlations between cracking susceptibility and both the primary solidification mode (PSM) and the impurity content of these alloys have been made for a number of years [1–15]. These investigations have shown that with sufficient specific impurities and primary solidification of austenite, the weld metal is susceptible to weld solidification cracking.

In addition, constitution diagrams, such as the Schaeffler Diagram and the Welding Research Council (WRC) diagrams have been used to select filler metals and/or predict resulting weld metal constituents in stainless steel welds [16–19]. These constitution diagrams and weld cracking susceptibility diagrams are specific to welding conditions, such as cooling rate. Many investigators have shown that the primary solidification mode changes with changes in solidification rate [20–23]. It is still convenient, however, to refer to weldability diagrams as the arc welding or pulsed laser beam welding (LBW) weldability diagram, without considering potential variations in solidification within each category. Little work, however, has been done in documenting weld cracking susceptibility in "similar alloy", dissimilar heat austenitic stainless steel welds [23–24]. Hochanadel, et al., [24] have shown that weldability diagrams may be used to analyze the weldability of two "similar" dissimilar materials by noting the composition end-points and utilizing a tie-line between the two points. In changing an existing process from pulsed LBW to pulsed GTAW, the weldability is typically expected to be improved in that the "no cracking" region of the weldability diagram is expanded in the GTAW case. This work was done to investigate the weldability response of welding two dissimilar austenitic stainless steel heats together and comparing the response when welding with both a pulsed LBW and pulsed GTAW.

P.W. Hochanadel (✉)
Los Alamos National Laboratory, Los Alamos, NM, USA
e-mail: phoch@lanl.gov

J. Lippold et al. (eds.), *Hot Cracking Phenomena in Welds III*,
DOI 10.1007/978-3-642-16864-2_9,

Procedure

Several annulus welds were made between 304L (outer tube) and 304 (inner tube) materials with both pulsed LBW and pulsed GTAW. Figure 1 shows the top view of tubes prior to welding. The 304L outer tube was machined from bar stock and the 304 inner tube was seamless tube material. Table 1 lists the nominal chemical composition of the materials investigated.

A Lumonics JK706 Nd:YAG pulsed laser welder was used to make the laser welds. The average welding power ranged from 35 to 110 W (verified by a "hockey-puck" or "lollipop" calorimeter), and the pulse widths investigated were either 3 ms

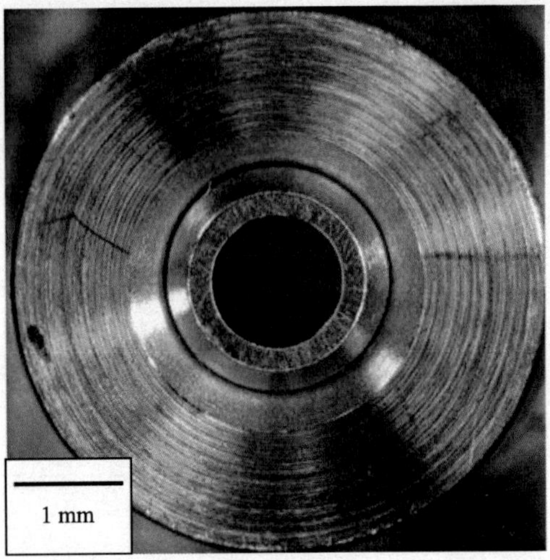

Fig. 1 Top view of the tubes prior to welding

Table 1 Nominal alloy compositions (in weight percent)

	Outer tube (304L)	Inner tube (304)
Cr	18.5	19.9
Ni	11.6	10.0
Mn	1.6	1.8
Si	0.36	0.46
Co	0.025	N/A
Mo	0.044	N/A
C	0.027	0.051
S	0.002	0.016
P	0.010	0.020
N	0.037	N/A
Hammar and Svensson Cr_{eq}/Ni_{eq}	1.44	1.72
S+P	0.012	0.036

or 5 ms, giving a duty cycle of 3 or 5%, since the pulse frequency was held constant at 10 Hz. The travel speed was held constant at 0.5 mm/s, and the beam diameter was approximately 1 mm. All laser welds were made at sharp visual focus.

The pulsed GTA welds were made using an Arc Machines, Incorporated (AMI) 307 power supply/controller using a special weld head capable of performing a tube-to-tube annulus weld (see Fig. 2). The weld was made with a 4 level weld and Table 2 lists the pulsing parameters used. The arc gap was at approximately 0.4 mm using a 1 mm diameter electrode with an 18° grind angle and the tip diameter was 0.25 mm. The travel speed was approximately 4.5 mm/s.

Leak checking of the welds was performed with an Alcatel ASM 180t leak detector. Subsequent microstructural observation was performed after sectioning and polishing using standard metallographic procedures. Electrolytic etching was performed with a 10 pct. oxalic acid solution.

Fig. 2 Specialized annulus tube weld head for use with AMI 307 welding power supply

Table 2 Pulsed gas tungsten arc welding parameters used in this investigation

Level	Level time (sec.)	Primary current (A) note: pulse duration is 0.08 s	Background current (A) note: pulse duration is 0.03 s
1	1.8	28.0	8.4
2	1.8	27.7	8.3
3	1.8	27.3	8.2
4	1.8	27.0	8.1

Note: Primary pulse duration is 0.08 s and the background pulse duration is 0.03 s.

Results and Discussion

Figure 3 shows the top views of the pulsed LBW (Fig. 3a) and the pulsed GTAW (Fig. 3b), while Fig. 4 shows a cross-section of a typical pulsed LBW processed weld with the original tube configurations shown for comparison. Solidification cracking was observed in the cross-sections of many of the arc and laser welds.

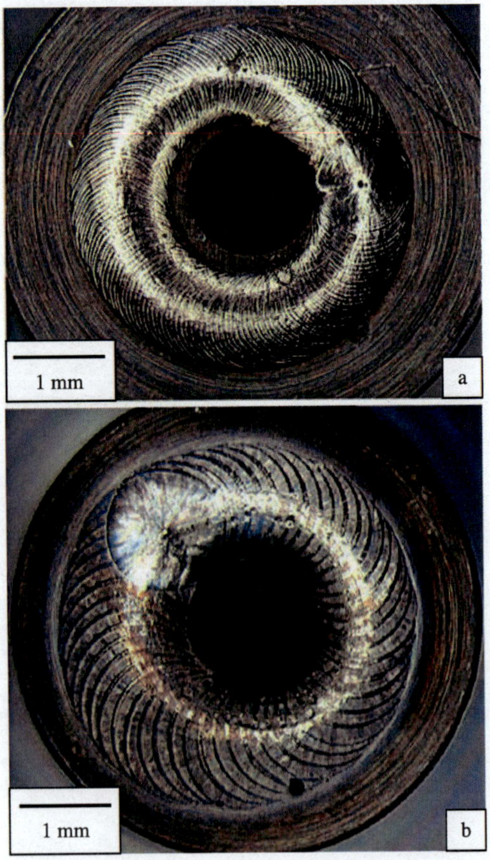

Fig. 3 Top view of the tubes after welding. The pulsed laser beam welded tubes are in 3a and the pulsed gas tungsten arc welded tubes are in 3b

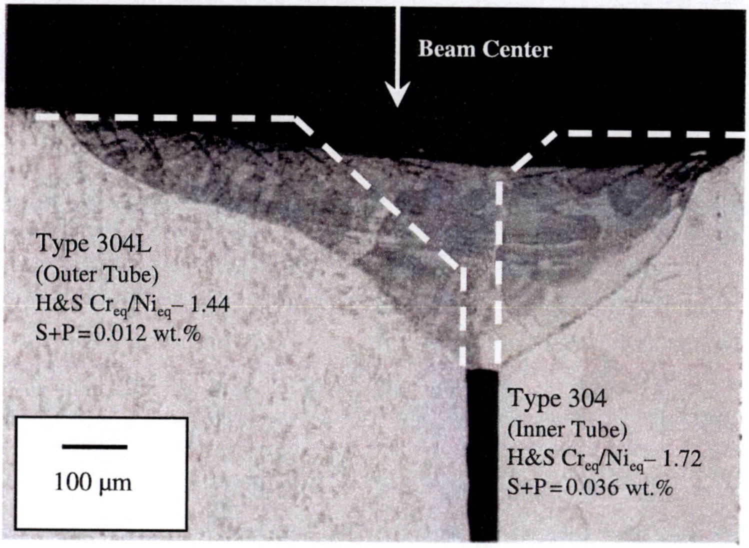

Fig. 4 Cross-section of weldment and joint schematic (*dashed lines*) of the 304–304L stainless steel combination. Average power of the weld was 45 W and the pulse width was 5 ms

Laser Welds

The higher power pulsed laser welds (at approximately 90 W average power or higher) were found to be susceptible to weld solidification cracking, as shown in Fig. 5. Although many of the cracks were found to be in the final weld pulse, cracks were found throughout the weld.

Normal welding conditions dictate that the top both tubes are flush. A cross-section of this condition is seen in Fig. 6. In one of the off-normal conditions, the tube-to-tube offset was moved from zero (*i.e.*, flush tubes, as seen in Fig. 6) to 0.25 mm (in Fig. 7). Figure 7 shows weld solidification cracking, and the average weld metal composition was estimated to be at approximately 30% outer tube/70% inner tube based upon area fraction analysis.

In the previous investigation [24], the resulting lower power welds showed a dual solidification mode (solidification as both primary austenite on one side and primary ferrite on the other, as seen in the micrograph) in the 304–304L weld joint, as shown in Fig. 8. Figure 8 shows how epitaxial growth of primary austenite occurred initially upon solidification for a short distance, until the solidification mode shifted to primary ferrite (both in the root in Fig. 8a and along the fusion boundary in 8b). This phenomena was also observed by Lippold when pulsed laser welding austenitic stainless steels [8], and he suggested that the diffusion controlled transformation of austenite to ferrite in the heat affected zone was suppressed, allowing for epitaxial growth of austenite. Figure 8 also shows how the solidification mode

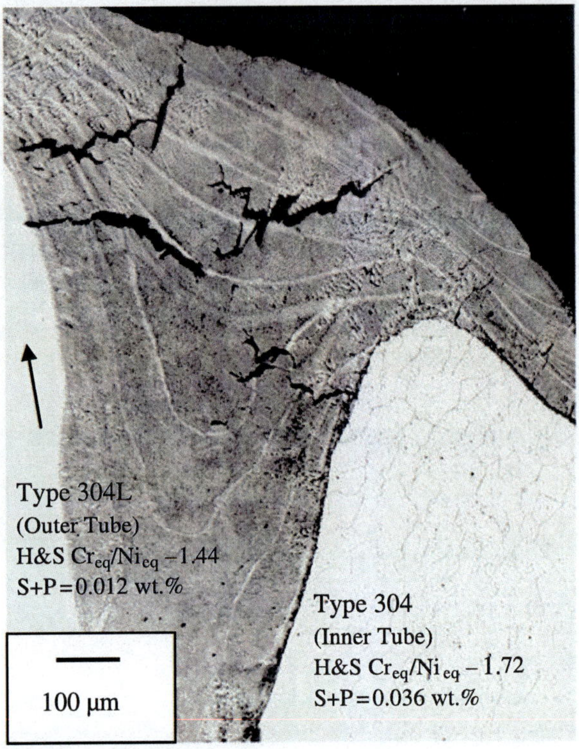

Fig. 5 Weld solidification cracking along solidification grain boundaries in a higher power laser weld (95 W average power and 3 ms pulse width)

transitioned back to primary austenite, as depicted by the "A" on the right-hand side of the micrographs. This could be caused by either a change in solidification growth rate, a change in composition (as expected), or a combination of both (pulse boundary).

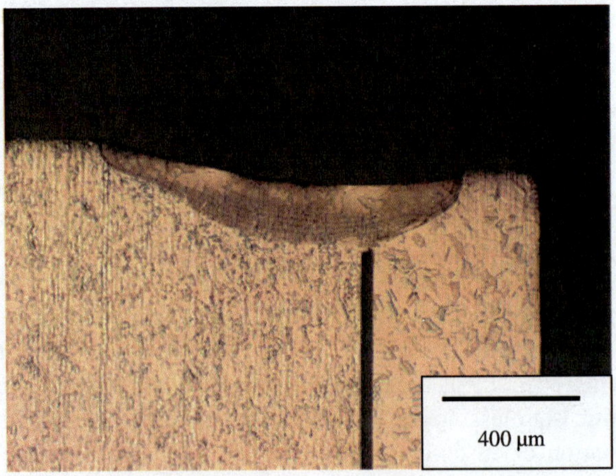

Fig. 6 Cross-section of the shallow penetration pulsed laser beam weld at nominal conditions (60 W average power, 5 ms pulse width, 10 Hz)

Fig. 7 Cross-section of the shallow penetration pulsed laser beam weld with a 0.25-mm tube offset (60 W, 5 ms pulse width, 10 Hz)

Figure 9 shows the fusion zone in the higher power weld. It appears that this higher power weld had only a primary solidification mode of austenite or austenite/eutectic ferrite (no evidence for primary ferrite solidification was observed). In the higher power welds (greater than 75 W average power with 5 ms pulse), the welds were found to be keyhole mode welds, while at powers lower than 75 W, conduction mode welds were observed. Possible causes for the different solidification mode observations were either the difference in solidification rates between the two weld modes (conduction mode versus keyhole mode), the difference in mixing, or a combination of both.

Arc Welds

A series of arc welds were also made for comparison. Figure 10 shows a cross-section of an arc weld made at nominal weld parameters and with the tubes flush to one another. Weld solidification cracking was observed in many of the arc welded materials. Note in Fig. 10a that much of the weld metal appears to have a PSM of A, even though the cross-section is a low magnification. In addition, a dual primary solidification mode was observed, which was similar to some of the observations of the laser welded materials. This is shown in Fig. 10b. The FA region shown in is in the region of the inner tube material – the material with the Hammar and Svensson $Cr_{eq}/Ni/_{eq}$ of 1.72.

Discussion

By plotting the compositional window of the 304 or 304L stainless steels on the WRC 1992 constitution diagram, it is easily seen that several resulting solidification microstructures may be realized depending upon the material compositions and

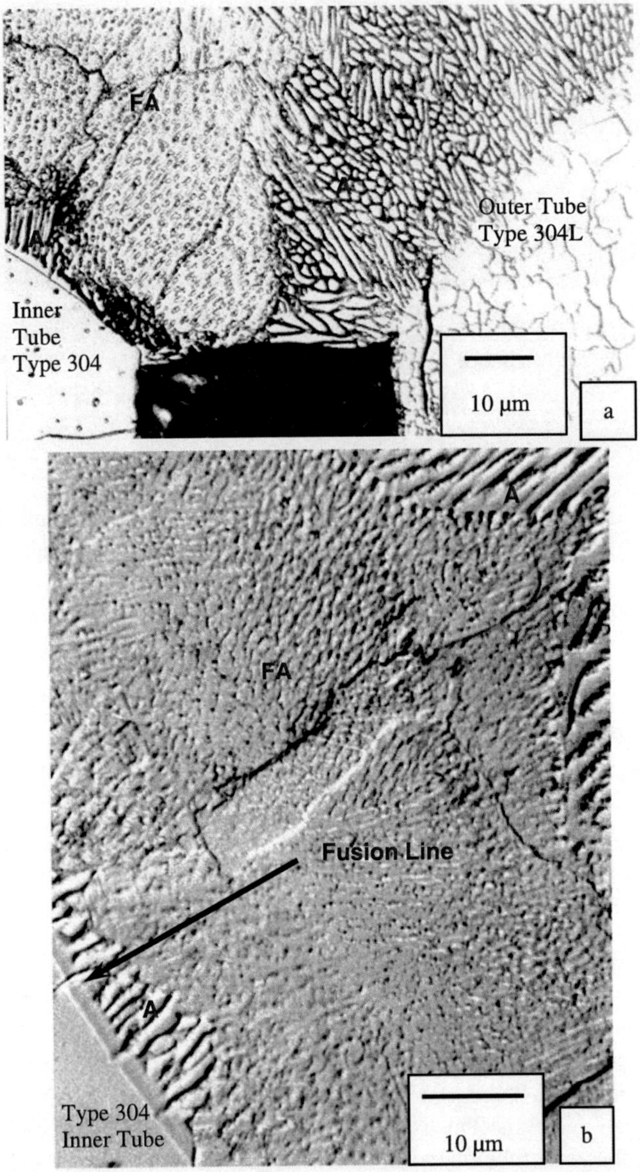

Fig. 8 Results from previous investigation showing dual primary solidification mode in the conduction mode pulsed laser beam weld. Micrograph depicting the variation in solidification microstructure at the root of the weld (8a). Fusion boundary region depicting the heat affected zone, unmixed zone and transition regions (8b). Adjacent regions of primary austenite (A) and primary ferrite (F) solidification are seen in a lower power weld (45 W average power and 5 ms pulse width)

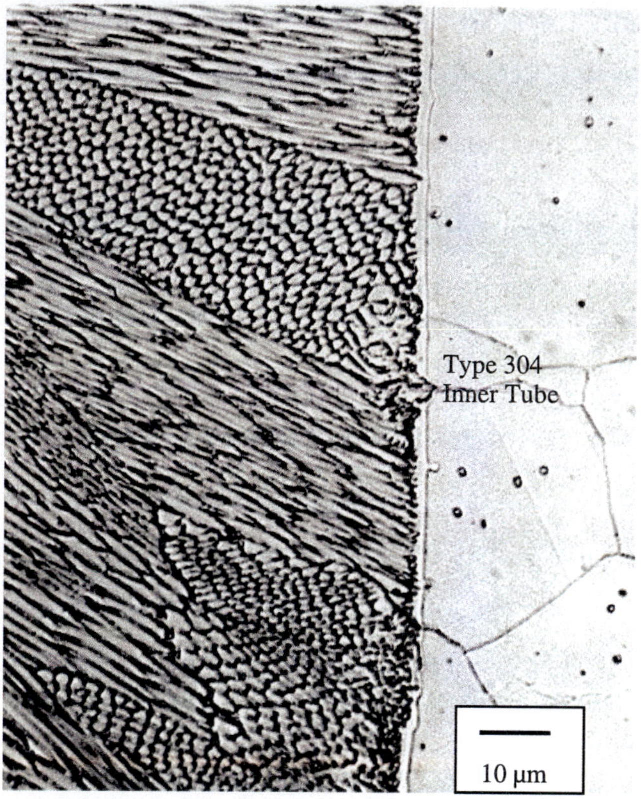

Fig. 9 Fusion zone in a higher power laser weld (95 W average power and 3 ms pulse width) showing epitaxial growth and primary austenite solidification along the high Cr_{eq}/Ni_{eq} boundary of the inner tube

the extent of mixing/dilution. This is seen in Fig. 11. This figure also shows the compositions of interest plotted for comparison. The WRC constitution diagram predicts that the combination of these two materials should solidify as delta ferrite and would not be expected to experience solidification cracking under "normal" solidification conditions in arc welding.

The weldability diagram modified by Lienert and Lippold to display cracking tendencies in pulsed laser welded materials is shown in Fig. 12 [15]. The two compositions investigated in the current study are plotted as the Hammar and Svensson Cr_{eq}/Ni_{eq} versus the total weight % (S + P). This diagram shows clearly that both materials are readily weldable if they are laser welded autogenously. However, by placing a tie line between the two compositions, a region is found where the combination of these materials can yield a crack susceptible composition. In fact, when looking at the original weld joint configuration, an average composition of the weld metal may be calculated based upon area fraction analysis compared to the original joint configuration. It should be noted that the average weld metal composition does

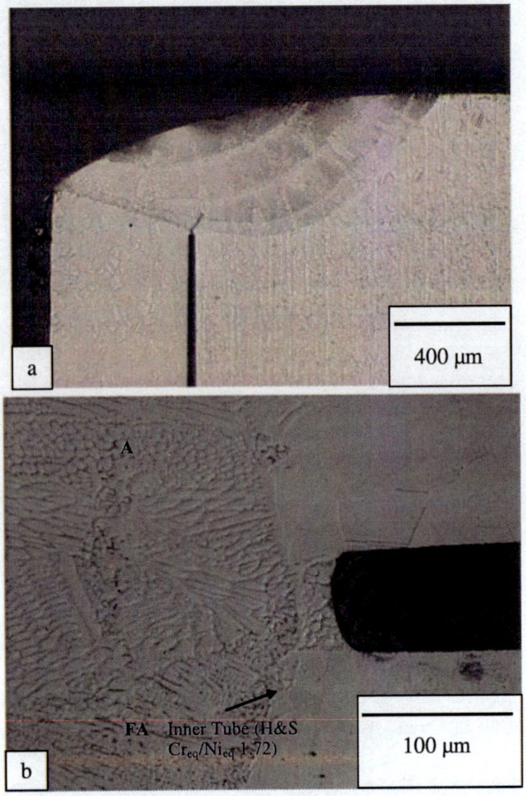

Fig. 10 Cross-section of GTAW processed materials showing weld solidification cracking in (**a**) and variable PSM in (**b**)

not necessarily represent local conditions were compositional variations are known to occur. In the case of the shallow penetration pulsed laser weld, found in Fig. 6, it is seen that the average weld metal composition was found to be approximately 70% outer tube/30% inner tube. This average composition is shown as a red star on Fig. 12. It is seen with no tube-to-tube offset, the average weld metal composition is in a region that borders the crack susceptibility region. Similarly, the example of tube offset of 0.25 mm (Fig. 7) shows that the average composition in a region in which solidification cracking is possible (and apparent, in this case). Recall that solidification cracking is both metallurgical and mechanical – the tube offset allows sufficient mechanical driver for solidification cracking was present which may cause cracking.

It was believed that changing the welding process from pulsed LBW to pulsed GTAW would be beneficial, since the demarcation line in the weldability diagram moves from a Hammar and Svensson Cr_{eq}/Ni_{eq} of nearly 1.6 ("Solidification cracking likely") or 1.7 ("Solidification cracking possible") in the pulsed LBW weldability diagram seen in Fig. 12 [15] to approximately 1.5 in the original GTAW weldability diagram [2], which is shown in Fig. 13. Based upon the original joint,

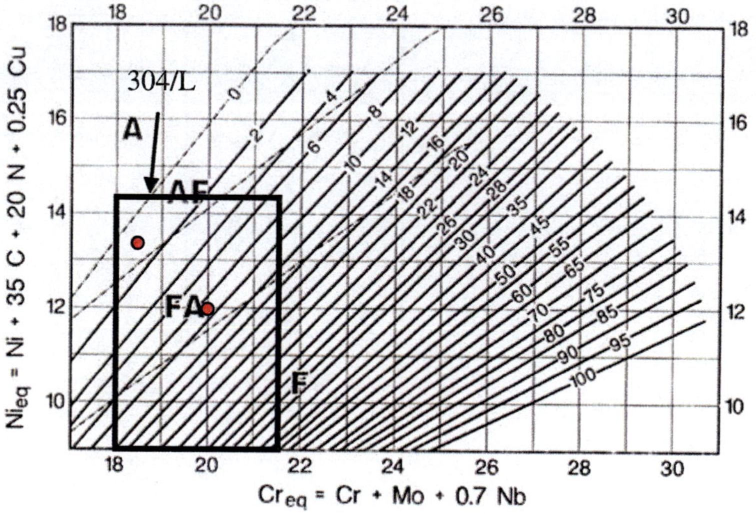

Fig. 11 WRC-1992 diagram [19] showing the range for 304/L stainless steel composition. The two points are shown for the composition of materials investigated in the current study

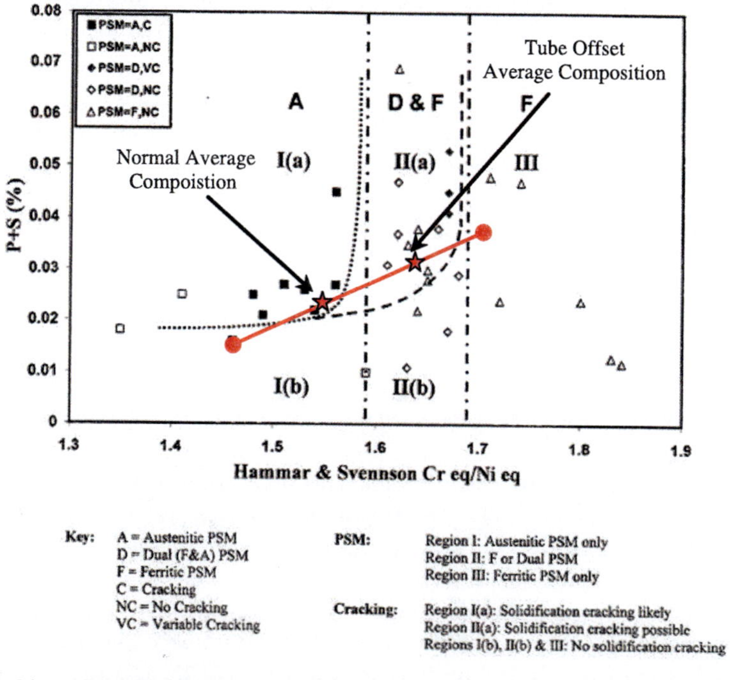

Fig. 12 Lienert/Lippold-modified weldability diagram [15] for pulsed laser beam welds. The end-points of the tie-lines indicate the material compositions, while the stars indicate the average composition based upon area fractions

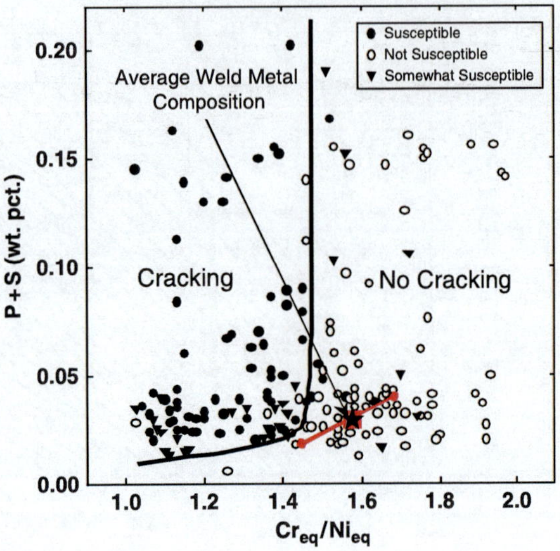

Fig. 13 Original weldability diagram [2] based upon GTAW with tie-line shown for current compositions

the average weld metal composition was estimated to be nearly 50% outer tube/50% inner tube (area fraction analysis *versus* the original joint). The composition end points and tie-line, along with the 50/50 average composition point is shown in Fig. 13. Based upon this diagram, the combination of the materials should be weldable over the entire range of compositions possible. However, as seen in Fig. 10, weld solidification cracking was possible.

A first order approximation of the cooling rate during solidificationwas shown first by Katayama and Matsunawa [25] and then used by Elmer, et al. [21] to be predicted by the following

$$\lambda_1 = 80(\varepsilon)^{-0.33} \tag{1}$$

where λ_1 is the cell spacing (in μm) and ε is the cooling rate (in K/s). The cooling rate was not found to vary much between the lower power and higher power pulsed laser beam welds (*i.e.* conduction mode versus keyhole mode), all of which were estimated to be between 10^5 and 10^6 K/s. This observation is in good agreement with the predicted cooling rates reported in literature for pulsed laser welds [25, 26]. Thus, it was determined that the cooling rate did not influence the observed change in solidification mode. It is important to note that the primary solidification mode is controlled by the solidification rate. High solidification rates lead to high dendrite tip undercooling which drives the shift in the primary solidification mode [8]. Based upon the results of a previous investigation [24], the mixing was believed to have the greatest influence on the solidification mode *locally*, since the composition variations in the conduction mode weld allowed for a sizable unmixed zone and nonuniform mixing. The keyhole mode is a potent mode for vigorous mixing in the weld metal, as shown by the results of the previous investigation [24], and as such, allowed for a more uniform weld metal composition, which resulted in a more uniform weld metal microstructure. However, caution should be used in simply looking just at the composition and general cooling rate data. Zacharia, et al. [26] have shown that local variations in cooling rates are experienced at the different solidification fronts during pulsed laser welding, and different solidification modes and structures may result from these local variations.

Utilizing the approximation of cooling rate from the cell spacing for the pulsed GTA welded specimens, the cooling rate was estimated to be between 10^3 and 10^4 K/s.

Lippold [8] Elmer, et al. [21] and David et al. [20] have shown that a material that is predicted to have a PSM of ferrite-austenite may solidify as primary austenite at higher solidification rates. In fact, this was Lippold's premise when developing an understanding of why the cracking curve is shifted from the left to the right in the weldability diagram when employing high energy density welding processes [8].

Both Lippold [8] and Elmer, et al., [21] have developed "solidification maps", which show expected primary solidification mode as a function of composition and some variable, such as solidification rate or welding travel speed. The current work utilizes Elmer's map, which shows PSM as a function of composition and EB scan speed, since the processes are both pulsed processes. Further, the EB scan speed related to cooling rate during solidification through Equation 1. Based upon this, the cooling rates were plotted as a range Elmer's diagram as an "Equivalent EB scan speed" and composition. Figure 14 shows Elmer's diagram with the pulsed LBW and pulsed GTAW ranges along with the composition ranges. It is easily seen from this figure that the PSM is predicted to be either fully A (pulsed LBW) or A/AF (pulsed GTAW). This map helps to explain why GTAW processing the dissimilar combination results in a weld metal susceptible to solidification cracking. This map does not fully explain the dual mode of primary solidification in the pulsed laser welded combination, and it was derived from 60% iron ternary alloys. Finally, even though the original weldability diagram (in Fig. 12) was developed on GTAW, this work continues to emphasize that not all GTAW processes are equivalent – the solidification rates and resulting PSMs could be dramatically different depending upon the composition, the material dimensions and the welding parameters. Because of this, some caution should be used when employing the weldability diagrams available.

Finally, it should be noted that weld solidification cracking needs not only a crack-susceptible microstructure, but also sufficient tensile stress to drive a crack open. Certainly, a difference in stress between the conduction mode and keyhole mode exists. The resultant shape of the keyhole mode weld lends itself to higher resultant stress. In addition, both the pulsed LBW and pulsed GTAW processed materials showed both convex and concave weld profiles, which will certainly result in varied solidification stresses. The difference in stresses, however, was not quantified nor studied in this investigation.

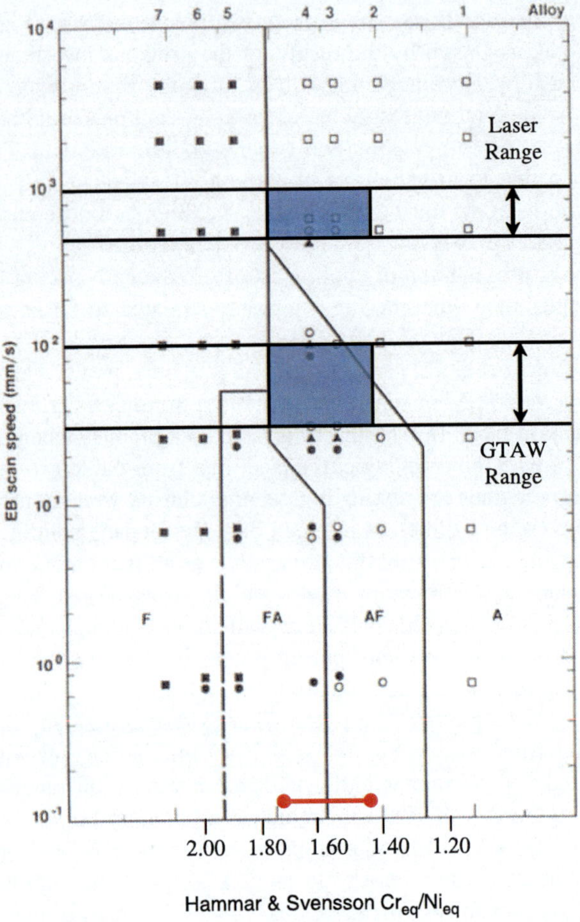

Fig. 14 Solidification mode map adapted from Elmer showing possible composition ranges of materials welded [21]

Conclusions

1. Welding dissimilar materials that are "weldable" when welded independently may result in a weld metal that is susceptible to weld solidification cracking.
2. A modified weldability diagram with a tie-line between two compositions may be useful in predicting the susceptibility of the resulting weld metal to weld solidification cracking. This presumes that the weldability diagram used correlates to the applied welding conditions.
3. When utilizing low power and high travel speed GTAW, a similar shift in the primary solidification mode is observed as in pulsed LBW.
4. When utilizing the weldability diagrams, the user should verify that the diagram is applicable. For instance, even though the original weldability diagram was

developed for GTAW processes, it is not applicable when the GTAW process is drastically different. This study showed that the original weldability diagram was not applicable for the pulsed GTA welds made for this investigation.

Acknowledgements This work was funded under the auspices of the United States Department of Energy, Los Alamos National Laboratory, which is operated by the Los Alamos National Security, LLC for the National Nuclear Security Administration of the U.S. Department of Energy under contract DE-AC52-06NA25396. The authors would like to thank Dr. Paul Burgardt and Mr. John Milewski for many useful discussions. The authors would also like to thank Dr. Mark Paffett for his timely and insightful review.

References

1. Arata RJ, Matsuda F and Katayama S (1976) Fundamental Investigation on Solidification Behavior of Fully Austenitic and Duplex Structures and Effect of Ferrite on Microsegregation. Transactions of JWRI:35–51.
2. Kujanpää V, Suutala N, Takalo T and Moisio T (1979) Correlation Between Solidification Cracking and Microstructure in Austenitic and Austenitic-Ferritic Stainless Steel. Welding Research International 9(2):55–76.
3. Takalo T, Suutala N and Moisio T (1979, August) Austenitic Solidification Mode in Austenitic Stainless Steel Welds. Metallurgical Transactions A:1173–1181.
4. Suutala N, Takalo T and Moisio T (1979, April) The Relationship Between Solidification and Microstructure in Austenitic and Austenitic-Ferritic Stainless Steel Welds. Metallurgical Transactions A:512–514.
5. Suutala N, Takalo T and Moisio T (1980, May) Ferritic-Austenitic Solidification Mode in Austenitic Stainless Steel Welds. Metallurgical Transactions A:717–725.
6. Kujanpää V, Suutala N, Takalo T and Moisio T (1980, June) Solidification Cracking – Estimation of the Susceptibility of Austenitic and Austenitic-Ferritic Stainless Steel Welds. Metal Construction:282–285.
7. Suutala N (1983, February) Effect of Solidification Conditions on the Solidification Mode in Austenitic Stainless Steels. Metallurgical Transactions A:191–197.
8. Lippold JC (1994, June) Solidification Behavior and Cracking Susceptibilty of Pulsed-Laser Welds in Austenitic Stainless Steel. Welding Journal:129-s–139-s.
9. Hull FC (1967, September) Effect of Delta Ferrite on Hot Cracking of Stainless Steel. Welding Journal:399-s–409-s.
10. Masumoto I, Tamaki K and Kutsuna M (1972) Hot Cracking of Austenitic Stainless Steel Weld Metal. Transactions of the Japan Welding Society:1306–1314.
11. David SA, Goodwin GM and Braski DN (1979, November) Solidification Behavior of Austenitic Stainless Steel Filler Metals. Welding Journal:330-s–336-s.
12. Lippold JC (1982) Weld Cracking Mechanism in Austenitic Stainless Steels, Proceedings from the Trends in Welding Research in the U.S., ASM Conference Proceedings, Materials Park, Ohio, 1982, pp. 209–247.
13. Brooks JA and Thompson AW (1991) Micorstructural Development and Solidification Cracking Susceptibility of Austenitic Stainless Steel Welds. International Materials Reviews:16–44.
14. Brooks JA, Thompson AW and Williams JC (1984, March) A Fundamental Study of the Beneficial Effects of Delta Ferrite in Reducing Weld Cracking. Welding Journal:71-s–83-s.
15. Lienert TJ and Lippold JC (2003, January) Improved Weldability Diagram for Pulsed Laser Welded Austenitic Stainless Steels. Science and Technology of Welding and Joining:1–9.
16. Schaeffler AL (1949) Constitution Diagram for Stainless Steel Weld Metal. Metal Progress:680–680B.

17. DeLong WT, Ostrom GA and Szumachowski ER (1956, November) Measurement and Calculation of Ferrite in Stainless Steel Weld Metal. Welding Journal:526-s–33-s.
18. Siewert TA, McCowan CN and Olson DL (1988, December) Ferrite Number Prediction to 100 FN in Stainless Steel Weld Metal. Welding Journal:289-s–298-s.
19. Kotecki D and Siewert TA (1992, May) WRC-1992 Constitution Diagram for Stainless Steel Weld Metals. Welding Journal:171-s–178-s.
20. David SA, Vitek JM and Hebble TL (1987, October) Effect of Rapid Solidification on Stainless Steel Weld Metal Microstructures and Its Implications on the Schaeffler Diagram. Welding Journal:289-s–300-s
21. Elmer JW, Allen SM and Eager TW (1989, October) Microstructural Development During Solidification of Stainless Steel Alloys. Metallurgical Transasctions A:2117–2131.
22. Viteck JM, Dasgupta A and David SA (1983, September) Microstructural Modification of Austenitic Stainless Steels by Rapid Solidification. Metallurgical Transactions A:1833–1841.
23. Lippold JC (1985, May) Centerline Cracking in Deep Penetration Electron Beam Welds in Type 304L Stainless Steel. Welding Journal:127-s–136-s.
24. Hochanadel PW, Cola MJ, Kelly AM and Papin PA (2002) Pulsed Laser Beam Welding of 304 to 304L Stainless Steel: Effects of Welding Parameters on Cracking and Phase Transformations, Proceedings of Joining of Advanced and Specialty Materials IV, Indianapolis, Indiana, 2001, published by ASM International, pp. 16–21.
25. Katayama S and Matsunawa A (1985) Solidification Microstructure of Laser Welded Stainless Steels, Proceedings of the Materials Processing Symposium -ICALEO '84, LIA-Laser Institute of America, Toledo, OH, pp. 60–67
26. Zacharia T, David SA, Vitek JM and Debroy T (1989, May) Heat Transfer During Nd:YAG Pulsed Laser Welding and Its Effect On Solidification Structure of Austenitic Stainless Steels. Metallurgical Transasctions A:957–967.

Solidification Cracking Studies in Multi Pass Laser Hybrid Welding of Thick Section Austenitic Stainless Steel

Miikka Karhu and Veli Kujanpää

Introduction

VTT- Technical Research Centre of Finland has carried out several R&D subtasks during the international FUSION technology program in which a suitable welding method has been considered for the manufacturing of vacuum vessel for International Thermonuclear Experimental Reactor (ITER). The tasks have been involved hybrid Nd:YAG laser and electron beam welding. Because the walls of vacuum vessel will be made of 60 mm thick stainless steel, above mentioned high energy density beam processes have shown great potential of being high efficient joining method. In work of R&D tasks, the laser welding with filler wire and the hybrid welding were used in welding of thick section austenitic stainless steel for a very narrow gap and using a multi pass technique. The results have shown that by using the narrow groove with one pass per layer procedure the number of passes can be decreased significantly and the efficiency of the welding can be increased. In addition, the total heat input stays smaller than in conventional arc welding processes, which leads to lesser welding distortions.

Despite the successful welding process development it has been realized that metallurgical issues like occurrence of weld solidification cracking could cause troubles if it is not taken care of with certain precautions which means e.g. carefully selected weld chemistry. In the tests of basic AISI 304L and AISI 316LN laser and hybrid laser multi pass welding experiments, clear indication of hot cracking was seen in many test welds [1]. Therefore it may be well judged that used vessel material AISI 316L(N)-IG ITER steel grade could be susceptible to hot cracking, if other conditions favoring hot cracking, e.g. increased section thickness and rigidity of structure to be welded, primary austenitic weld solidification, etc. are present [2–5]. Stem from above mentioned observations, a hot cracking study was conducted. The purpose of the study was to develop a test set-up for studying hot cracking in multi pass laser hybrid welding of thick section austenitic stainless

M. Karhu (✉)
VTT-Technical Research Centre of Finland, FI-53851 Lappeenranta, Finland
e-mail: mikka.karhu@vtt.fi

J. Lippold et al. (eds.), *Hot Cracking Phenomena in Welds III*,
DOI 10.1007/978-3-642-16864-2_10,

steel and then use this set-up for evaluation of hot cracking susceptibility of welds produced into the special customized heat AISI 316L(N)-IG ITER-grade by using certain reference filler wire material (THERMANIT 19/15). The concept of the test set-up was to use a straight forward approach in designing: It was built a very rigid clamping table together with fastening system and a test piece-design which is very rigid itself and produces self restraint.

Experimental Procedures

Materials

Parent material used in the experiments was a specially customized heat AISI 316L(N)-IG ITER-grade austenitic stainless steel. Concerning chemical composition of the ITER-grade stainless steel, it is specified to be inside certain limits and it is known as a nominal chemical composition, Table 1. Filler wire used in the experiments was an austenitic stainless steel wire THERMANIT 19/15 with a diameter of 1.2 mm. Regarding the test materials used in this experiments both parent and filler material is from the known heat and known chemical composition according to EN 10204-3.1. The exact chemical compositions of the materials are shown in Table 2.

Initial Assessment of Weldability for Parent and Filler Material

Various constitutional diagrams have been developed in order to evaluate a weldability of stainless steels [7–11]. For example, the use of those diagrams could help

Table 1 Nominal chemical composition (wt-%) of AISI 316L(N)-IG stainless steel. [6]

Element	Min	Max
C	0.015	0.030
Mn	1.6	2.0
Si	–	0.50
P	–	0.025
S	0.005	0.01
Cr	17.0	18.0
Ni	12.0	12.5
Mo	2.30	2.70
Ti[a]	–	0.15
Ta	–	0.01
Nb	–	0.01
Cu	–	0.3
B	–	0.0020
Co	–	0.05
N	0.060	0.080

[a]Nb+Ta+Ti < 0.15 wt%

Table 2 Exact chemical composition (wt-%) of the parent material AISI 316L(N)-IG and filler wire material THERMANIT 19/15 according to EN 10204-3.1

Element	AISI 316L(N)-IG	THERMANIT 19/15
C	0.025	0.015
Mn	1.7	7.32
Si	0.36	0.46
P	0.021	0.013
S	0.0001	0.005
Cr	17.6	20.18
Ni	12.2	15.08
Mo	2.41	2.79
Ti	0.005	–
Ta	0.003	–
Nb	0.005	–
Cu	0.1	0.095
B	0.0004	–
Co	0.03	0.016
N	0.067	0.187

to predict as-welded microstructure and avoid unfavorable parent material or filler material compositions which could lead e.g. to solidification mode or microstructure favoring hot cracking. The diagrams are based on chromium (Cr_{eq}) and nickel (Ni_{eq}) equivalents and their ratio (Cr_{eq}/Ni_{eq}) which can be calculated when a chemical composition of parent material and filler material is available. In the case of conventional arc welding where solidification rate is slow, it is found that if the ratio of chromium and nickel equivalent (Cr_{eq}/Ni_{eq}) in weld metal is below a certain level, e.g. 1.5, the susceptibility for solidification cracking is much higher than in welds where chromium and nickel equivalent ratio is above 1.5. The critical ratio is depending on the solidification rate such that at the higher solidification rates the critical ratio is slightly increased (Cr_{eq}/Ni_{eq} ~1.7). This is observed to be case e.g. in laser or electron beam welding. The impurities, especially sulphur and phosphorus also play a role in the susceptibility of hot cracking. If the impurity content is very low, below 0.01%, the susceptibility is much decreased [12–18].

The hot cracking susceptibility is also dependent on the level of strains affected by the rigidity of the structure. If the structure is thick and very rigid, stresses induced by welding heat in-put, are less likely become relieved by deformation of the structure. If deformations are hindered, strains could become stronger and more concentrated in the welding zone. In above case the strains causing hot cracking develop much easier [2, 3]. Furthermore, it is commonly recognized that in applications, where fully austenitic stainless steel welds are in certain specific reason required (e.g. anti-magnetic requirements etc.), hot cracking in weld metals could cause remarkable problems.

For the above reasons the ordinary austenitic stainless steel composition is often balanced such that usually ordinary AISI 316L or AISI 316LN composition is on the safe composition range, i.e. the ratio of chromium and nickel equivalents is over 1.5, which should ensure that solidification is primary ferritic resulting in approx. 5–15% delta ferrite in the weld metal at room temperature. This has been observed to prohibit hot cracking. Another import factor for hot cracking prevention is reduction of impurity content (S+P); for safety reason it is usually limited to 0.03 ... 0.04%.

Initial requirement governed by the ITER-application was a fully austenitic microstructure for the base and weld metal. Therefore it was anticipated that there could be a risk for weld solidification cracking. Consequently, initial assessments of weldability were made using chromium and nickel equivalents. The ratio of chromium and nickel equivalents in used vacuum vessel material ITER grade austenitic stainless steel AISI 316L(N)-IG was calculated according to nominal chemical composition and for five different chromium and nickel equivalent equations, Tables 1 and 3.

Calculations revealed that depending on the variation of the element contents, Cr_{eq}/Ni_{eq} -ratio is in the range of 1.15–1.68. This indicates potential risk of hot cracking, especially when Cr_{eq}/Ni_{eq} –ratio remains lower than 1.5. The resulting regions defined by points of calculations are plotted in xy-coordinate which is featured in Fig. 1. Depending on obtained heat, marked areas in Fig. 1 gives an idea, how much the range of Cr_{eq} – and Ni_{eq} –ratios can vary in AISI 316L(N)-ITER grade within the permissible composition range (according the Table 1).

Furthermore, chromium and nickel equivalents according to exact chemical compositions (Table 2) for both parent material and filler material were calculated. In that case, only Hammar & Svensson diagram was decided to use, because it will best correlate weld solidification conditions [19].

It can be seen from the diagram in Fig. 2, that produced weld metal will follow the segment line starting from point 1.5 and moving along the segment line towards to point 1.2 depending on dilution ratio. Consequently, produced weld metal will have even smaller Cr_{eq}/Ni_{eq}-ratio than 1.5, because of used filler material. For example, if dilution rate of 40% is assumed, weld metal will have a Cr_{eq}/Ni_{eq}-ratio of 1.3, which means possible risk of hot cracking tendency.

Table 3 Chromium and nickel equivalents provided by different authors [7–11]

Author / name of diagram	Chromium equivalent (Cr_{eq})	Nickel equivalent (Ni_{eq})
Schaeffler	$Cr_{eq} = Cr + Mo + 1.5 \cdot Si + 0.5 \cdot Nb$	$Ni_{eq} = Ni + 0.5 \cdot Mn + 30 \cdot C$
DeLong	$Cr_{eq} = Cr + Mo + 1.5 \cdot Si + 0.5 \cdot Nb$	$Ni_{eq} = Ni + 0{,}5 \cdot Mn + 30 \cdot C + 30 \cdot N$
Hull	$Cr_{eq} = Cr + 1.21 \cdot Mo + 0.48 \cdot Si + 0.14 \cdot Nb + 2.20 \cdot Ti + 0.72 \cdot W + 0.21 \cdot Ta + 2.27 \cdot V + 2.48 \cdot Al$	$Ni_{eq} = Ni + [0.11 \cdot Mn - (0.0086 \cdot Mn^2)] + 24.5 \cdot C + 18.4 \cdot N + 0.44 \cdot Cu + 0.41 \cdot Co$
Hammar & Svensson	$Cr_{eq} = Cr + 1.37 \cdot Mo + 1.5 \cdot Si + 2 \cdot Nb + 3 \cdot Ti$	$Ni_{eq} = Ni + 0.31 \cdot Mn + 22 \cdot C + 14.2 \cdot N + Cu$
WRC-1992 (by Kotecki & Sievert)	$Cr_{eq} = Cr + Mo + 0.7 \cdot Nb$	$Ni_{eq} = Ni + 35 \cdot C + 20 \cdot N + 0.25 \cdot Cu$

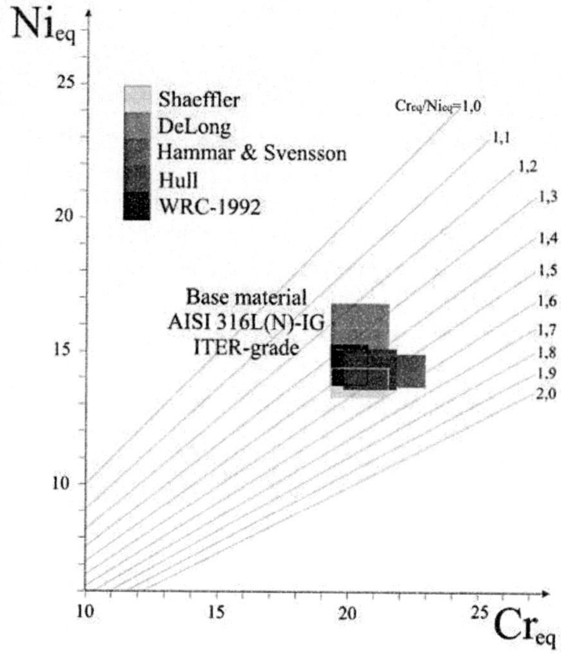

Fig. 1 Chromium (Cr_{eq}) and nickel (Ni_{eq}) equivalents of ITER grade austenitic stainless steel AISI 316L(N)-IG was calculated according to nominal compositions and five different equations. Equivalent regions are coordinated as a chart

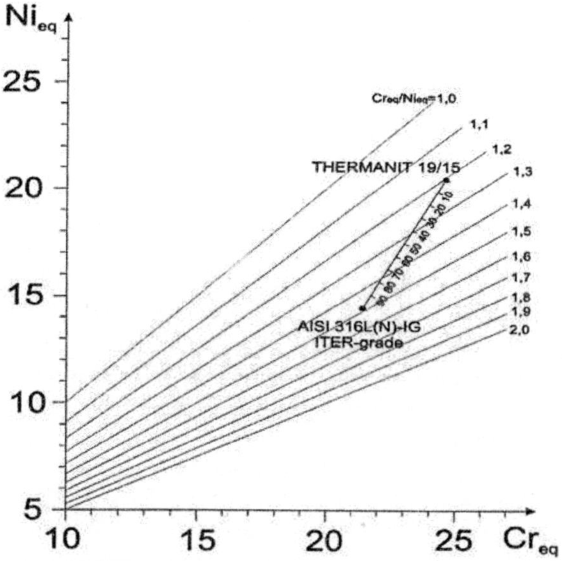

Fig. 2 Cr_{eq} and Ni_{eq} values of parent (AISI 316LN-IG) and filler material (Thermanit 19/15) placed in Hammar & Svensson diagram

Hot Cracking Test Set-Up: Rigid Clamping Table and Test Piece Design

Test set-up was decided to be designed and built in order to evaluate hot cracking susceptibility of ITER-grade heat AISI 316 L(N)-IG austenitic stainless steel when it is being narrow gap multi pass laser hybrid welded under very rigid conditions.

A very rigid clamping system was designed and built in order to emulate rigid welding conditions and strains which can occur in massive components assembly welding. The test system consists of rigid 170 mm thick table and clamping system for that. In clamping system, the total of 6 pieces of Ø 30 mm diameter high strength bolts together with 40 mm thick holder blocks was used to minimize angular distortions caused by welding heat. Each bolt was tightened into the moment of 1,500 Nm with using dial torque wrench. That equals approx. a 300 kN compression per bolt.

Test Pieces

The test piece used was planned to be rigid as it self and to simulate the rigidity of assembly welds. The tested material was AISI 316 L(N)-IG ITER Grade austenitic stainless steel with original thickness of 60 mm. Test piece consisted of two 400 mm × 150 mm plates with thickness of 60 mm. Those 60 mm thick halves were machined in order to posses following features when combined together (Fig. 3): Both ends have close square preparation at the length of 60 mm. The joint thickness, which is intended to be filled using multi pass welds, is 20 mm. The length of the welded joint is 250 mm and the joint has a root gap of 1.2 mm.

As can be seen in section A-A and C-C in Fig. 3, the 20 mm thick joint to be welded is captured near to half way at the whole 60 mm thickness of the test piece. A rectangular cavity for a root of the weld is prepared to be 30 mm × 10 mm. Below the root cavity is a 20 mm thick close square preparation along the whole length of the test piece (400 mm).

After machining of necessary groove geometries, two halves were put together and sealed with using electron beam welding at the sections where close square preparations were prepared, Fig. 4. The aim of above mentioned procedure was to prohibit transverse shrinkage and angular distortions and in that way to produce strains which are correspondent to those which can occur in massive components assembly welding.

Two kinds of 20 mm thick groove geometries were used in hot cracking tests. The groove geometries are shown in Fig. 5. The basic idea was to use wider (Fig. 5a) and narrower (Fig. 5b) groove configurations, in order to get two different multi pass weld shapes together with different depth to width ratios of produced multi pass welds.

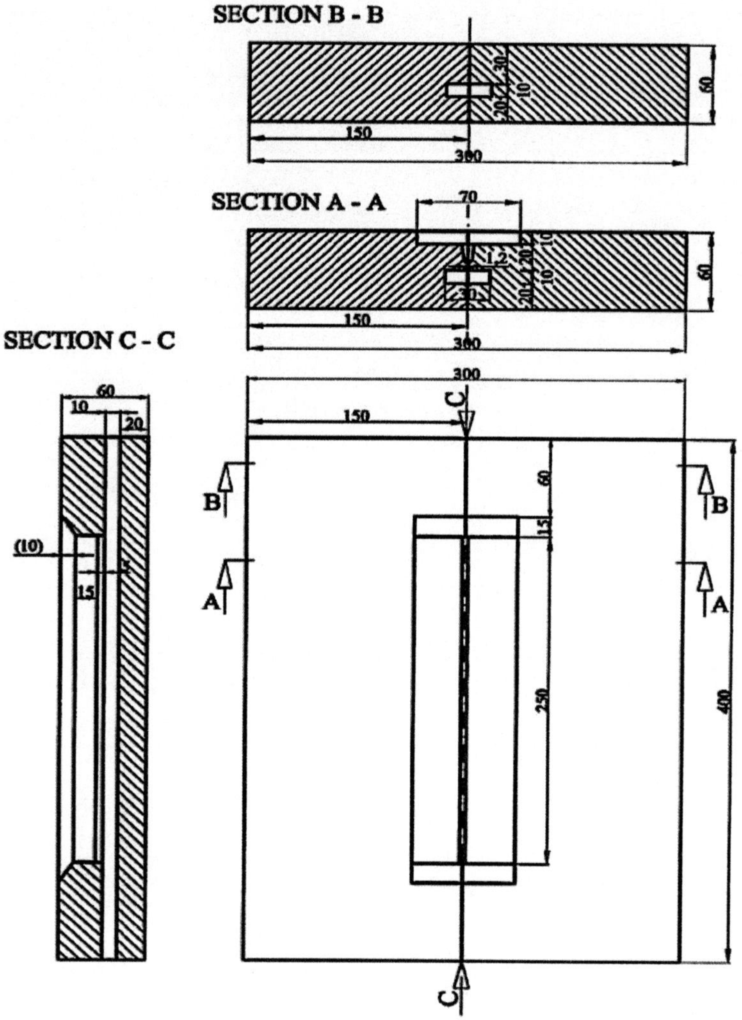

Fig. 3 Test piece used in hot cracking tests

Welding Equipment

Welding experiments were performed by using a combination of Nd:YAG-laser process and MIG-process. The laser used in hybrid process was HAAS-LASER GmbH model HL 3006 D with Ø 0.6 mm optical fiber beam delivery system, Fig. 6. The laser has a maximum output power of 3 kW at the surface of a work piece. The full power of 3 kW was used in all experiments. Used lens focusing optic gives a 200 mm focal length, 6.12 degree focusing angle and can provide a spot diameter of 0.6 mm

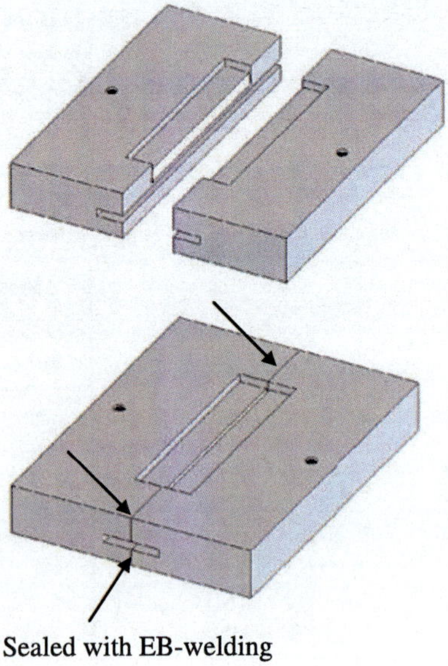

Fig. 4 Test piece used in hot cracking tests. Two machined halves are put together and sealed using electron beam welding into the one test piece

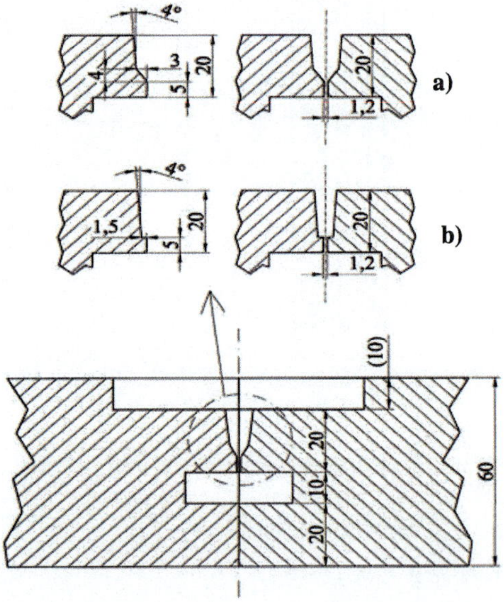

Fig. 5 Two different groove geometries were used in hot cracking tests

in the focal point. The laser system enables a beam parameter product of 25 mmm-rad. GMAW machine used in the hybrid welding was KEMPPI Pro 5200 equipped with ProMIG 501 unit, Fig. 7. KUKA KR 15 robot was used in order to execute required welding movements. Self tailored hybrid welding head was mounted in the wrist flange of KUKA-robot, Fig. 7.

Fig. 6 3 kW Nd:YAG-laser used in the experiments

Fig. 7 Welding equipment used in the experiments. *Left*: Kemppi Pro MIG-welding machine, *right*: Hybrid welding head mounted in KUKA KR 15 robot

Hot Cracking Test Program

As described earlier, two different groove configurations were decided to use such a way that three identical test pieces for both groove geometries were prepared. Consequently, we had two test series with three identical hot cracking welding test repetitions in each series. Test pieces with wider groove geometry, Fig. 5a, belonged to the series A and test pieces with narrow groove geometry, Fig. 5b, belonged to the series B. With using preliminary welding tests proper welding parameters were searched for both groove geometries. Constant welding parameters used in both series are presented in Table 4. As the same parameters in welding of root passes could be used in both series, there were different welding parameters regarding to the welding of filling passes between series A and B, Table 5. That is because filler wire feeding rate, welding speed and focal position needs to be adjusted according to different groove volumes used in test series A and B. Inside the test series, the same parameters were used for welding of each three identical test pieces.

The lay-out of the experimental set-up is shown in Fig. 8.

Before welding experiments, test piece was clamped to the rigid table as described earlier in this study. Welding experiments were executed in such a way that a keyhole mode hybrid welding was first used in welding of root pass (Fig. 9a) and after that a required amount of filling passes (4. . .6 passes) were produced with conduction limited hybrid welding. In conduction limited hybrid method, a power density of an Nd:YAG- laser beam spot was purposely dispersed by using strong defocusing (Fig. 9b). Using the above procedure, the welding process can be brought

Table 4 Constant welding parameters

The constant parameters	
Laser power	3 kW
Focal length	200 mm
Horizontal distance between the laser focal point and filler wire feeding point	2 mm
Diameter of filler wire	1.2 mm
Stick out (electrode extension)	17 mm
Orientation and the angle of arc torch	leading, 55°
Shielding gas (MISON He30= Ar+30%He+0,03NO) and flow rate:	
– Via arc torch's nozzle	20 l/min
– Via extra nozzle	20 l/min

Table 5 Welding parameters of root and filling passes for series A and B

Welding parameters for the root pass		
Filler wire feeding rate	9.5 m/min (28.8 V / 232 A)	
Focal point position	±0 mm (spot sizeØ0.6 mm)	
Welding speed	1.3 m/min	
Welding parameters for the filling passes		
	Wider groove geometry, Fig. 6a Series A	Narrow groove geometry, Fig. 6b Series B
Filler wire feeding rate	9.0. . .9.5 m/min (29. . .30 V / 205. . .260 A)	9.0. . .11 m/min (28. . .31 V / 220. . .295 A)
Focal point position	+50 mm (spot size Ø 5.9 mm)	+30. . .+40 mm (spot size Ø 3.8. . .4.9 mm)

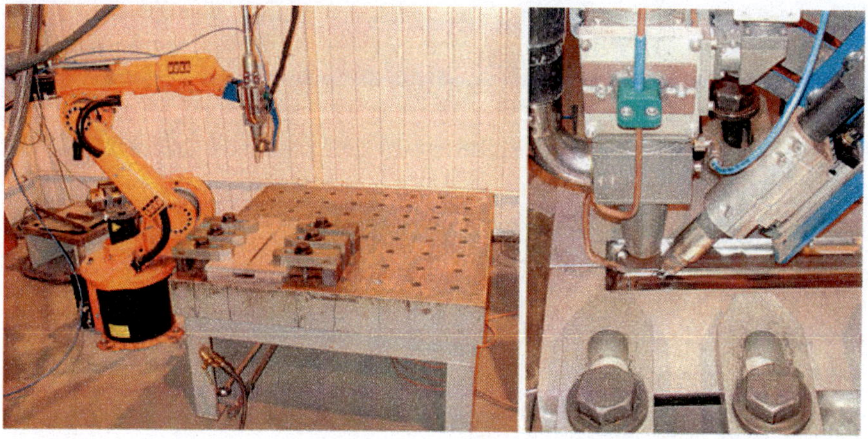

Fig. 8 *Left*: The lay-out of the experimental set-up. *Right*: Close-up from the hybrid welding head configuration

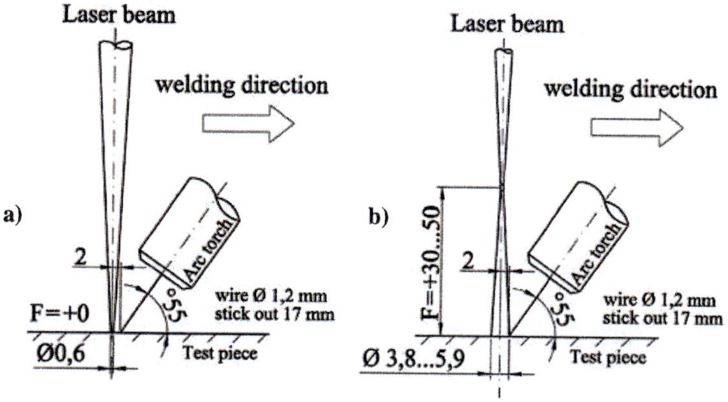

Fig. 9 Set-up used in hybrid welding: (**a**) key-hole mode in welding of root pass and (**b**) conduction limited mode in welding of filling passes

inside the conduction limited regime. In addition, a molten filler metal was synergistically added into the laser induced melt pool via a metal inert arc (MIG, Ar + He shielding) welding process. By doing so, a considerable wide melt zones can be produced, which gave a possibility to fill groove gaps which are impossible to fill with using keyhole hybrid welding alone.

After welding, test-pieces were allowed to cool room temperature before clamping was released. Next, test welds were examined by X-ray radiography. After that, test welds were sectioned for macro- and micro graphical preparation. Evaluations of the welds were done by the visual inspection from welded test pieces and from the macro- and micro graphical cross-sections of the welds and from the X-ray radiographs as well.

In order to get reference data from the decreasing groove cross-section and volume, measuring tasks were carefully performed between welding of each pass. Amount of occurred transverse shrinkage was determined measuring the width of welded test piece perpendicular to weld joint using a calibre rule. In addition, the closure of gap opening was separately measured at the surface of the test piece using a calibre rule.

Finally, measurements and macrographs from the test welds together with chemical analysis of base and filler material were combined as the rate of dilution was determined by calculations. In order to get some reference to dilution calculations, electron probe micro analysis (spot size 500 μm and sampling time 30 s) were selectively performed from the weld cross-sections as well.

Results and Discussion

The scope of the study did not include strain measurement during testing. However, results of welding tests showed that the developed test set-up can produce critical conditions to promote hot cracking in produced test welds.

Observations of Hot Cracking Tendency

Multi pass technique was used in hot cracking welding tests. In test welds of series A (wider groove), a total of 5 passes were needed to fill the groove, whereas in test series B (narrow groove), 7 passes were needed. During the welding experiments, great attention was paid to the visual inspection of the surface of the intermediate passes. The surface of each individual pass was visually inspected before the welding of consequent pass started. The observations from the visual inspections clearly showed that hot cracking, which opened to the surface of the weld pass, did occur. Hot cracking observations concerning individual passes in both series are summarized in Tables 6 and 7.

In test series A, hot cracking occurred only in one test piece of all three: in welding of the first filling pass of multi pass weld H3. In test series B, hot cracking occurred in every test piece. In those test pieces, hot cracking was occurred in

Table 6 Hot cracking observations after completion of individual passes: test series A

	Occurrence of hot cracking		
Test series A	Weld H1	Weld H2	Weld H3
Root pass	No	No	No
1st filling pass	No	No	*Yes*
2nd filling pass	No	No	No
3rd filling pass	No	No	No
4th filling pass	No	No	No

Table 7 Hot cracking observations after completion of individual passes: test series B

	Occurrence of hot cracking		
Test series B	Weld H4	Weld H5	Weld H6
Root pass	No	No	No
1st filling pass	No	*Yes*	*Yes*
2nd filling pass	*Yes*	*Yes*	*Yes*
3rd filling pass	*Yes*	*Yes*	No
4th filling pass	No	No	No
5th filling pass	No	No	No
6th filling pass	No	No	No

the first, second and third filling passes, whereas in root passes no cracking was observed to occur. It was also found that all cracking occurred in welds which were situated 2/3 of plate thickness from the bottom.

It turned out when completed weld cross-sections were observed that cracks found earlier from the surface of the individual passes seemed to be "healed" by

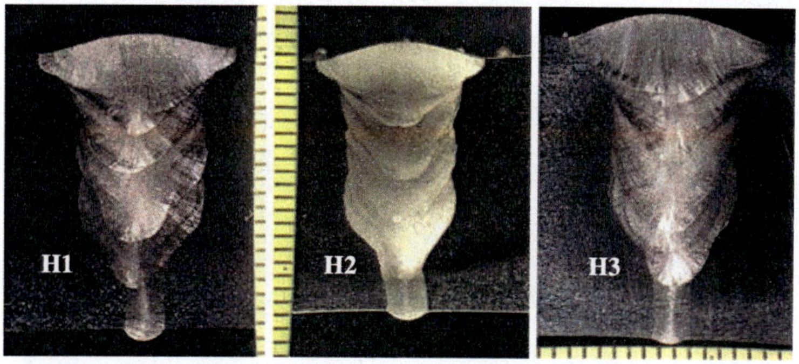

Fig. 10 Welds from the test series A: Weld H1, Weld H2 and Weld H3

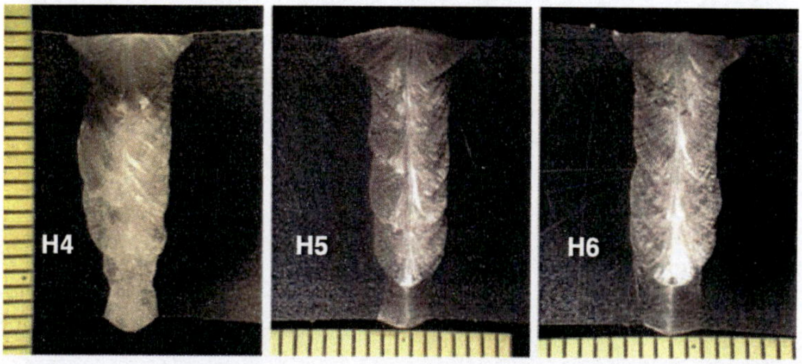

Fig. 11 Welds from the test series B: Weld H4, Weld H5 and Weld H6

re-melting of following overlapping passes. This can be noticed from the macrographs of the weld cross-sections of test series A (weld H1, H2 and H3) and test series B (weld H4, H5 and H6) which are presented in Figs. 10 and 11, respectively.

Effect of Weld Bead Geometry

In both test series A and B, cross-sectional weld bead geometries were measured from the macrographs, especially depth and width of each individual pass (Tables 8 and 9). In addition, depth to width ratios was calculated and those ratio-values were compared to hot cracking tendency.

According to results, narrow groove configuration used in series B tended to cause more hot cracking than wider groove configuration used in series A. However, in root passes hot cracking was not emerged, although they had a large depth to width ratios (D/W~1.29...3.04). In general, depth to width ratio (D/W) of 1 seems to be often a near threshold value concerning the occurrence of hot cracking in filling passes. In multi pass welds H1 and H2, where depth to width ratios of filling passes stayed between 0.5... 0.84, no cracking was observed.

Table 8 Cross-sectional weld bead geometry (depth/width) in test series A

Test series A			
Weld H1	D (mm)	W (mm)	D/W
Root pass	6.4	2.7	2.37
1st filling pass	7.3	8.8	0.83
2nd filling pass	7.6	9.1	0.84
3rd filling pass	6.5	10.5	0.62
4th filling pass	7.0	10.5	0.67
Weld H2	D (mm)	W (mm)	D/W
Root pass	5.8	3.2	1.81
1st filling pass	7.0	8.5	0.82
2nd filling pass	6.3	8.5	0.74
3rd filling pass	6.5	9.5	0.68
4th filling pass	6.5	13.0	0.50
Weld H3	D (mm)	W (mm)	D/W
Root pass	7.0	3.3	3.04
1st filling pass	8.0	8.3	0.96
2nd filling pass	7.5	8.0	0.94
3rd filling pass	6.9	9.5	0.73
4th filling pass	5.8	12.9	0.45

Note: grey color donates occurred hot cracking

Table 9 Cross-sectional weld bead geometry (D = Depth/W = Width) in test series B

Test series B			
Weld H4	D (mm)	W (mm)	D/W
Root pass	5.0	3.6	1.39
1st filling pass	5.4	5.6	0.96
2nd filling pass	4.5	6.3	0.71
3rd filling pass	4.8	6.6	0.73
4th filling pass	4.5	6.3	0.71
5th filling pass	4.6	5.9	0.78
6th filling pass	5.6	8.8	0.64
Weld H5	D (mm)	W (mm)	D/W
Root pass	5.5	3.2	1.72
1st filling pass	6.0	5.1	1.18
2nd filling pass	5.5	5.1	1.08
3rd filling pass	5.8	5.5	1.05
4th filling pass	5.9	5.5	1.07
5th filling pass	5.5	5.9	0.93
6th filling pass	5.5	9.1	0.60
Weld H6	D (mm)	W (mm)	D/W
Root pass	5.7	4.0	1.43
1st filling pass	5.1	5.6	0.91
2nd filling pass	5.8	5.4	1.07
3rd filling pass	5.5	5.1	1.08
4th filling pass	4.3	5.4	0.80
5th filling pass	3.8	5.4	0.70
6th filling pass	4.2	8.3	0.51

Note: grey color denotes occurred hot cracking

Weld Microstructure and Appearance of Hot Cracking

As mentioned earlier, visual evaluation after welding of each filling pass revealed that hot cracks occurred centered at along the weld length and opened to the surface of the weld. In Fig. 12a, it is shown an example from the test series B, where the part of the joint was purposely left without upper filling passes in order to authenticate the presence of hot cracking in that case occurring in the first filling pass. In Fig. 12b, it is shown the top view from the Fig. 12a, which reveals hot cracking opened to the surface of the first filling pass.

From the micro graphs presented in Fig. 13, it can be seen magnifications (450×) taken from the same cross-section as shown in Fig. 13a. Micro graphs show the upper and lower part of the hot crack propagated in the fully austenitic microstructure.

It seems that during the multi pass welding of this study, the following filling pass has been overlapped into the previous pass such much that molten metal has "healed" the hot crack underneath. Evaluations from the X-ray radiography and the

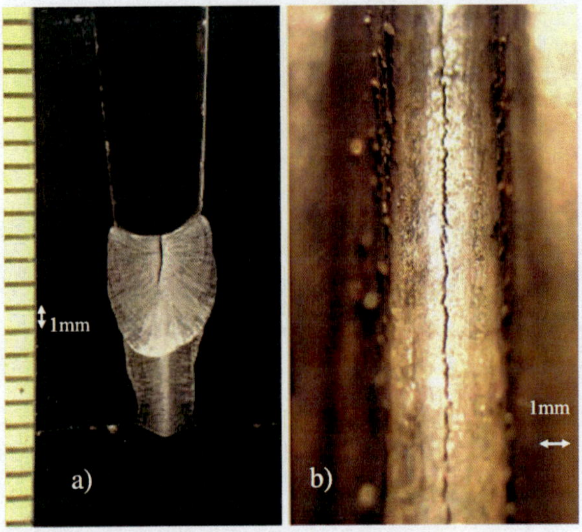

Fig. 12 (**a**) Macro cross-section showing hot cracking propagated into the surface of the first filling pass (magnification 4×). (**b**) The top view from the Fig. 12a showing hot cracking along the weld length (magnification 5,3×)

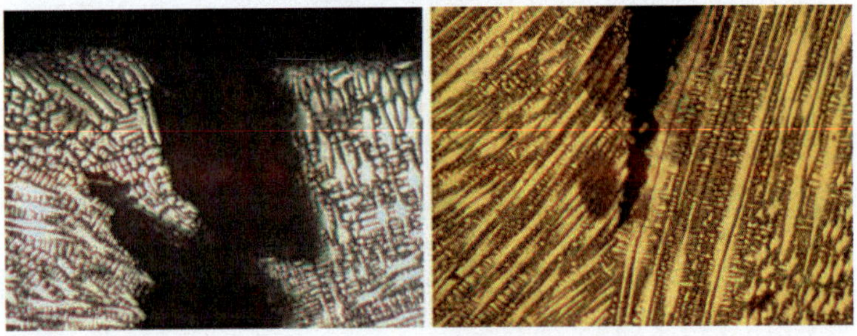

Fig. 13 Micro graphs taken from the same cross-section as shown in Fig. 12a. Micro graphs show the upper (*graph on the left*) and lower part (*graph on the right*) of the hot crack in the fully austenitic microstructure. Magnification 450×

cross-sectional macro graphs (Figs. 10 and 11) of the welds support "healing" observation. Anyhow, in general this healing- effect will not be always 100% certain, which must have taken into considerations in production welds of real applications.

Weld which is a product of primary austenite solidification mode has very distinctive microstructural morphology when viewed metallo-graphically. Above mentioned are e.g. cited in references 19–21. Based on metallographic examination primary austenite solidification can be defined in all test welds of series A (welds: H1, H2 and H3) and B (welds: H4, H5 and H6). There is couple of micro graph example shown from the both series A and B. In Fig. 14a, b it is shown weld metal structures from the intermediate passes of test weld H3 and H4, respectively. Both weld metals have solidified in primary austenite solidification mode. Fully austenitic microstructure of weld metal is presented in Fig. 14a, b.

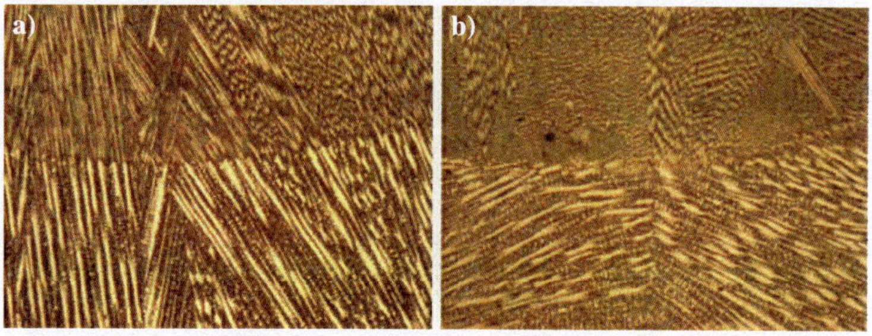

Fig. 14 Micro graphs showing fully austenitic microstructure of weld metal in (**a**) test weld H3 and (**b**) H4. Both weld metals have solidified in primary austenite solidification mode. Magnification 230×

Assessment of Rate of Dilution

From the both test series one test weld was selected for the assessment of rate of dilution: Multi pass weld H1 from the test series A and multi pass weld H5 from the test series B. In order to assess the rate of dilution, following procedure was used: Cross-sectional macrographs of weld H1 and H5 are digitally exported to CAD-software and further converted into CAD-image. The borderlines of groove geometries with known reduction adjustments which are due to transverse shrinkage were super positioned with the weld cross-section into the same image. By using CAD-software's image processing capabilities, individual proportions of each weld pass, melted base- and filler material in terms of cross-sectional area were determined. Finally, rate of dilution in each pass was calculated with using proportions of determined cross-sectional area together with the chemical element's proportions taken from the specifications which show the exact chemical compositions of the base and filler material (Table 2). Since dilution rates were known, chromium and nickel equivalent values were additionally calculated using equations presented in Table 3. Summary of both the rate of dilution and chromium and nickel equivalent calculations (Hammar & Svensson) for the multi pass welds H5 and H1 are presented in Tables 10 and 11.

Dilution in root pass of weld H1 and H5 was determined by calculations to be 63% and 60%, respectively. What comes to filling passes, dilution rates concerning base material are in weld H1 between 6 and 31% (average value 16%) and in weld H5 between 8 and 17.5% (average value 13.25%). Results reveal that each filling pass was some extent diluted by previous passes which naturally contain proportion of elements from the base material. Larger dilutions in root pass results from the narrow square preparation (width ~1 mm) in root face, as groove volume remains very low and much of filler material is not needed.

In order to get some reference to dilution calculations, electron probe micro analysis (EPMA) were performed from the weld cross-sections of welds H1 and H5. EPMA measurements give information about desired element content in weight-%. Electron probe micro analyses were carried out with using a spot size of 500 μm

Table 10 The results of dilution- and Cr/Ni-equivalent calculations concerning multi pass weld H5

Multi pass weld: H5 Pass no.	Proportions of dilution (%)			Hammar & Svensson equivalents		
	Base mat.	Filler mat.	Previous pass	Cr-eq	Ni-eq	Cr/Ni-eq
Root pass[a]	60	40	–	22.48	16.77	1.34
1st filling pass[b]	16.5	44.5	39[a]	23.30	17.99	1.30
2nd filling pass[c]	13.5	51	35.5[b]	23.78	18.75	1.27
3rd filling pass[d]	8	41	51[c]	23.97	19.08	1.26
4th filling pass[e]	11	44	37[d] + 8[c]	23.99	19.13	1.25
5th filling pass[f]	17.5	54.3	28.2[e]	23.92	18.99	1.26
6th filling pass	13	47.9	39.1[f]	23.97	19.06	1.26

For example: 4th filling pass = base material 6% + Filler material 47% + 3rd filling pass 45% + 2nd filling pass 2% = 100%

Table 11 The results of dilution- and Cr/Ni-equivalent calculations concerning multi pass weld H1

Multi pass weld: H1 Pass no.	Proportions of dilution (%)			Hammar & Svensson equivalents		
	Base mat.	Filler mat.	Previous pass	Cr-eq	Ni-eq	Cr/Ni-eq
Root pass[a]	63	37	–	22.41	16.58	1.35
1st filling pass[b]	31	49	20[a]	23.27	17.77	1.31
2nd filling pass[c]	12	46	41[b] +1[a]	23.68	18.58	1.28
3rd filling pass[d]	15	47	38[c]	23.83	18.81	1.27
4th filling pass	6	47	45[d] + 2[c]	24.07	19.29	1.25

and sampling time of 30 s. Each pass in both H1 and H5 test weld was analyzed and further two measurements were made for the base material as well. The points of locations of analyses are shown in Figs. 15 and 16.

When justifying the confidence of EPMA measurements, following comparison was made: The chemical analysis data of the base material AISI 316L(N)-IG

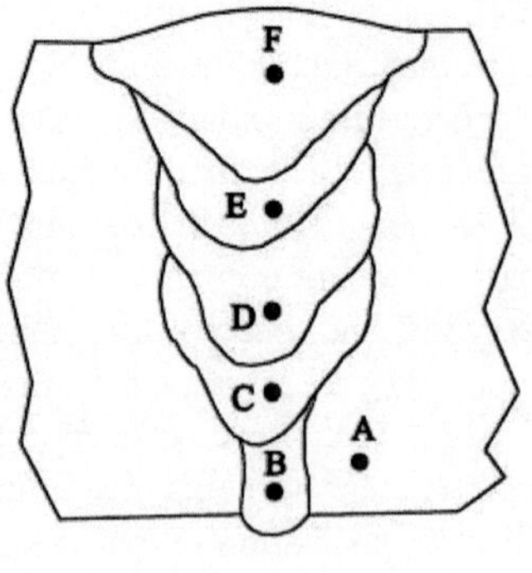

Fig. 15 The points (A…F) showing locations where electron probe micro analyses (EPMA) were taken in multi pass weld H1

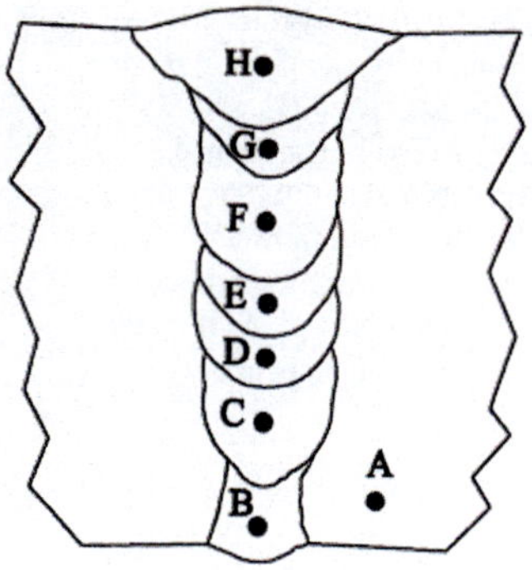

Fig. 16 The points (A...H) showing locations where electron probe micro analyses (EPMA) were taken in multi pass weld H5

according to EN 10204-3.1 (Table 2) was considered as a reference data and that data was compared to data collected from the base material with separate electron probe micro analyses (EPMA). EPMA data differed from the given reference data (EN 10204-3.1) on the range of approx. 2.5–11% depending on the reviewed element. That could be considered as a reasonable matching accuracy between those two independent measurements.

On the other hand, also the results of electron probe micro analysis seemed to be quite consistent when compared to ones which were achieved by using dilution calculations. This confirmed that used procedure for the assessment of rate of dilution can be successfully used. The results including cracking observations and dilution calculations are in good agreement with the initial data, as it is known that both parent material and filler wire used, are fully austenitic. The fact that in test weld H1 and H2 filling passes did not cracked although chromium/nickel-equivalent ratios were at the same level as H4-H6 could have been related to weld cross-sectional geometry. Filling passes in weld H1 and H2 are much wider than the filling passes which were cracked in welds H4-H6 (Tables 8 and 9). When considering test weld H3, it had cracking in the first filling pass, which had the highest depth to width ratio (0.96) of all H3 filling passes, Table 8. Narrower filling passes with large depth to width ratio tended to be more susceptible to hot cracking than the wider ones.

Effect of Weld Metal Composition

In order to speculate what kind of filler material selection could help to prevent weld solidification cracking in the studied case, Cr/Ni-equivalent diagram can be used as a suggestive tool, Fig. 17. In the Fig. 17, the data from the results of dilution- and Cr/Ni-equivalent calculations concerning multi pass weld H1 and H5 are converted to Hammar & Svensson diagram.

As it can be seen from the Fig. 17, in the case of filling passes dilution corresponds to Cr_{eq}/Ni_{eq}-ratio values of 1.25...1.31 in weld metal, whereas in root pass

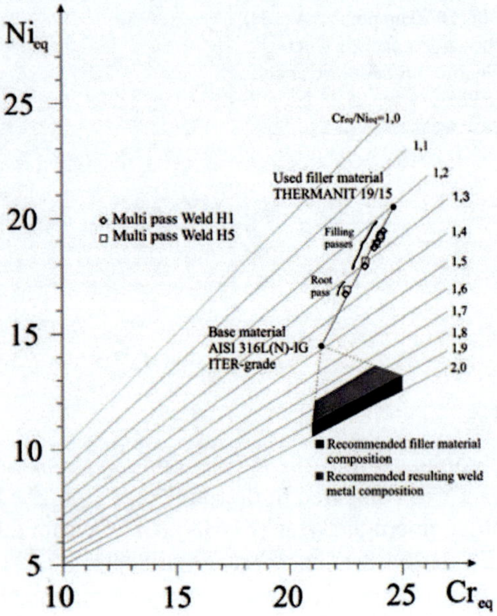

Fig. 17 Cr/Ni-values for filling passes and root pass of multi pass weld H1 and H5 are marked in Hammar & Svensson diagram. *Grey* and *darkened* area shows boundaries for recommended filler material and weld metal composition, respectively

dilution was higher leading to value of 1.35. That is way below compared to risk value of 1.5. In that respect, hot cracks in test welds approved that risk according to Cr_{eq}/Ni_{eq}-calculations is for real.

It should be noticed that the used parent material had quite low impurity content. According to the certificate (see Table 2) phosphorous (P) + sulphurous (S) content was 0.021 + 0.0001 = 0.0211 wt-% whereas the nominal composition spec (see Table 1) allows the total of P + S content of 0.035 wt-% as a maximum. As the results showed that cracking could occur at quite low impurity content (P+S=0,0211 wt%), therefore it could be danger that hot cracking susceptibility strongly enhances if the used ITER-grade heat will hold the maximum amount of impurity content allowed by the nominal composition spec, Table 1.

Selection of Filler Metal to Decrease Cracking Risk

One possibility to overcome cracking risk is a correct filler material choice. In the case of this study, that is to use more ferritic filler wire material, which will have e.g. a value of Cr_{eq}/Ni_{eq}-ratio about 1.9. . .2.0 (see Fig. 17). Using this kind of filler wire and assuming dilution shown in Tables 10 and 11, Cr_{eq}/Ni_{eq}-ratio about 1.75–1.90 and primary ferrite solidification could be achieved in weld metal. However, the welded application itself does not necessarily allow 5–15% room temperature delta ferrite content in weld metal, because of certain requirements or restrictions.

In the case of this study initial requirement was that microstructure of weld metal in joints must be fully austenitic, containing as less as possible delta ferrite at room temperature. It was not fully clear by end-user's side, is ferrite content allowed in assembly welds of ITER-applications and furthermore if ferrite is allowed, what is the maximum allowed content in percent? Thus, it is important to clearly specify the maximum allowed weld ferrite content. If primary ferrite solidification could be achieved in produced weld, weld hot cracking susceptibility would be remarkably reduced compared to situation where a weld is solidified as primary austenitic mode.

Conclusions

In this study the main objective was to find a method for studying hot cracking susceptibility when a thick section austenitic stainless steel is welded using laser hybrid welding process (3 kW Nd:YAG-laser + GMAW) and multi pass technique. The tested parent material was a specially customized heat: AISI 316L(N)-IG ITER-grade austenitic stainless steel. During this study the test system was first developed and tested. It consisted of a very rigid clamping system and specially designed 60 mm thick test piece which was planned to be rigid as itself.

Although the test set-up did not include strain measurement during testing, results of welding tests showed that the developed test set-up can produce critical conditions to promote hot cracking in produced AISI 316L(N)-IG ITER grade test welds in multi pass laser-arc hybrid welding. Results showed that in this case the used combination of the parent/filler material was led to primary austenitic solidification and fully austenitic weld metal microstructure. It was advisable to carry on welding tests with using filler wire composition which may give primary ferrite solidification on weld metal and thus leading to diminishing hot cracking susceptibility. Results also confirmed that hot cracking susceptibility of welded base and filler material combination can be suggestively pre-estimated by applying known weld metal composition and chromium (Cr_{eq}) and nickel (Ni_{eq}) equivalents.

Within the test series, weld passes did not always have same cracking susceptibility although they possessed the same range of Cr/Ni-equivalent values. This can be associated to cross-sectional geometry of individual weld pass. Narrower filling passes with large depth to width ratio tended to be more susceptible to hot cracking than the wider ones. Above could be found in test series B, where weld passes were more susceptible to hot cracking than the passes in series A.

As the scope of this study did not include strain measurement during testing, it has been considered to execute further studies with set-up which include strain measuring at the near weld zone during the welding.

Acknowledgements The authors wish to thank VTT- the Technical Research Centre of Finland, the National Technology Agency of Finland and the European Fusion Development Agreement for funding the project within the experiments of this paper was executed. We like to thank also Mr. Mikko Pesari for his assistance in welding experiments and illustration work for figures.

References

1. Karhu M, Jokinen T and Kujanpää V (2006) Conduction Limited Laser-Arc Hybrid Welding of Austenitic Stainless Steel with Different Joint Geometries, VTT Research report No. TUO25-056299, Espoo, Finland, 25p.
2. Folkhard E (1988) Welding Metallurgy of Stainless Steels. Springer, 279p.
3. Papeleux P, Flipot AJ and Lafontaine I (1973) Welding Austenitic Steel Clads for Fast Reactor Fuel Pins, International Atomic Energy Agency, Document AIEA/SM-173/VII-47, pp. 1–17.
4. Borland JC and Younger RN (1960, January) Some Aspects of Cracking in Welded Cr-Ni Steels. British Welding Journal:22–59.
5. Sorokin LI (2004) Evaluation of crack formation resistance in welding and heat treatment of creep resisting nickel alloys. Welding International 18(5):379–385.
6. Barabash V Material Specification for the supply of 316L(N)-IG austenitic stainless steel plates, G 74 SP 12 03-07-09 W 0.1, Rev.2- 08.07.2003.
7. Schaeffler AL (1949) Constitution Diagram for Stainless Steel Weld Metal. Met. Progr. 56: 680 and 680B.
8. DeLong WT, Ostrom GA and Szumachowski ER (1956) Measurement and Calculation of Ferrite in Stainless Steel Weld Metal. Welding Journal 35:521s–528s.
9. Hull FC (1973) Delta Ferrite and Martensite Formation in Stainless Steels. Welding Journal 52:193s–203s.
10. Hammar Ö and Svensson U (1979) Solidification and casting of metals. The Metals Society, London, pp. 401–410.
11. Kotecki DJ and Siewert DTA (1992) WRC-92 constitution diagram for stainless steel weld metals: a modification of the WRC-1988 diagram. Welding Journal 71(5):171s–178s.
12. Kujanpää V, Suutala N, Takalo T and Moisio T (1979) Correlation between solidification cracking and microstructure in austenitic and austenitic-ferritic stainless steel weld. Welding Research International 9(2):55–76.
13. Lippold JC (1985) Centerline Cracking in Deep Penetration Electron Beam Welds in Type 304L Stainless Steel. Welding Journal 64(5):127s–136s.
14. David SA, Vitek JM and Hebble TL (1987) Effect of Rapid Solidification on Stainless Steel Weld Metal Microstructures and Its Implications on the Schaeffler Diagram. Welding Journal 66(10):289s–300s.
15. Lippold JC (1994) Solidification behaviour and cracking susceptibility of pulsed-laser welds in austenitic stainless steels. Welding Journal 73(6):129s–139s.
16. Lippold JC (1995) Microstructure Evolution in Austenitic Stainless Steel Laser Welds, Proceedings of the Second International Conference on Beam Processing of Advanced Materials, Cleveland, Ohio, USA, pp. 167–178.
17. Lienert TJ (1998) A Combined PSM/Weldability Diagram for Laser Welded Austenitic Stainless Steel, Trends in Welding Research, Proceedings of the 5th International Conference, Pine Mountain, Georgia, USA, pp. 724–728.
18. Lienert TJ and Lippold JC (2003) Improved weldability diagram for pulsed laser welded austenitic stainless steels. Science and Technology of Welding and Joining 8(1):1–9.
19. Suutala N (1982) Solidification Studies on Austenitic Stainless Steels, Doctoral thesis, University of Oulu, Finland.
20. Elmer JW, Allen SM and Eagar TW (1989, October) Microstructural Development during Solidification of Stainless Steel Alloys. Metallurgical Transactions A 20A:2117–2131.
21. Lippold JC and Kotecki DJ (2005) Welding Metallurgy and Weldability of Stainless Steels. Wiley, 357p.

Influence of Minor and Impurity Elements on Hot Cracking Susceptibility of Extra High-Purity Type 310 Stainless Steels

K. Nishimoto, K. Saida, K. Kiuchi, and J. Nakayama

Introduction

Spent nuclear fuel reprocessing facilities require structural materials with high performance and reliability, and in particular a high corrosion resistance against nitric acid. Therefore, ultra-low carbon (ULC) grade nitric acid resistant stainless steels, such as type 304ULC and 316ULC, are currently used for the main components of reprocessing plants [1]. However, it has been found that intergranular corrosion occurs in devices in the reprocessing line after preliminary operations. In order to improve intergranular corrosion resistance, there is an urgent demand that these materials be replaced with ones more resistant to this environment. Very recently, highly pure corrosion-resistant stainless steels containing a large amount of Ni have been developed as nitric acid resistant alloys to replace existing type 304ULC and 316ULC stainless steels [2]. For example, extra high-purity 25 wt%Cr-20 wt%Ni (type 310 stainless steel) and 25 wt%Cr-35 wt%Ni steels, which are manufactured by cold crucible induction melting (CCIM) combined with electron beam cold hearth refining (EB-CHR) methods [3], have been proposed as candidates for corrosion-resistant alloys in the next generation of reprocessing facilities. In these extra high-purity stainless steels, the total sum content of minor and impurity elements, such as C, O, N, Mn, P and S, can be reduced to less than 100 ppm. In order to practically apply such extra high-purity stainless steels, they should also have superior and reliable weldability for fusion welding.

It is well known that austenitic stainless steels such as type 304, 308, 316, 321 and 347 stainless steels are potentially susceptible to weld cracking (especially hot cracking), and that their hot cracking susceptibility is dominantly governed by the Cr_{eq}/Ni_{eq} ratio (*i.e.*, solidification mode, amount of δ-ferrite, etc.) and (P+S) content in a steel [4–9]. In particular, type 310 stainless steel is highly susceptible to solidification cracking due to its fully austenitic structure, and impurity elements such as P and S greatly influence the solidification cracking susceptibility [10–16].

K. Nishimoto (✉)
Osaka University, Suita, Osaka, Japan
e-mail: nishimoto@mapse.eng.osaka-u.ac.jp

J. Lippold et al. (eds.), *Hot Cracking Phenomena in Welds III*,
DOI 10.1007/978-3-642-16864-2_11,

For example, the (P+S) content should be reduced to less than 0.01 wt% in order to obtain excellent crack resistance in the practical use of type 310S stainless steel [10]. In addition, it has been also suggested that Mn improves the hot cracking susceptibility, and that it should be controlled in several percent to inhibit hot cracking in type 310S stainless steel [10, 11]. On the other hand, several researchers reported that the quantitative influence of P on hot cracking was comparable to or larger than (1.1–2.1 times) that of S in type 304 and 310S stainless steels [10, 11, 16], while other researchers conversely suggested that S was more harmful in hot cracking compared with P in austenitic stainless steels (type 308) [17, 18]. These previous investigations covered the hot cracking susceptibility of commercial purity stainless steels containing some amounts of C, P, S as well as Mn. Therefore, the essential influence and quantitative contribution of minor and impurity elements to the hot cracking behaviour in austenitic stainless steels are still unclear.

The objective of this study is to clarify the hot cracking behaviour of extra high-purity type 310 stainless steels. A mechanism to account for the improvement in the hot cracking susceptibility via steel refinement is discussed on the basis of a numerical analysis of solidification segregation in the weld metal. Furthermore, the essential influence of minor and impurity elements such as C, P, S and Mn on the solidification cracking susceptibility is clarified using lab-melted extra high-purity type 310 stainless steels with different amounts of minor and impurity elements. The cause of the increased ductility-dip cracking susceptibility with steel refinement is suggested based on a molecular orbital analysis of the binding strength of the grain boundary.

Materials and Experimental Procedures

Materials

Four types of 310 stainless steels with varing amounts of minor and impurity elements, such as C, P, S and Mn, were used for the base metal. Table 1(a)shows the chemical compositions of the base metals used in this study. The purity of stainless steels used was enhanced in the order of Type 310<Type 310S<Type 310ULC<Type 310EHP steels. Type 310EHP steel is an extra high-purity stainless steel with reduced impurity elements as well as minor elements. In particular, the contents of C, P and S (as well as Mn) in Type 310EHP steel are greatly reduced compared to other steels (to reduce the impurity elements (deoxidation, denitrification), a very small amount of Ti is present). In order to evaluate the quantitative contribution of minor and impurity elements on the hot cracking susceptibility, ten types of lab-melted type 310 stainless steels with different amounts of minor and impurity elements, such as C, P, S and Mn were used for the base metal. Table 1(b) shows the chemical compositions of the lab-melted type 310 stainless steels used in

Table 1 Chemical compositions of steels used (wt%, *: ppm)

(**a**) Type 310, 310S, 310ULC and 310EHP steels

Steel	C	Si	Mn	P	S	Ni	Cr	O	N	Ti	Fe
Type 310	0.12	1.00	1.48	0.020	0.011	20.59	25.12	–	–	–	Bal.
Type 310S	0.05	0.80	1.23	0.016	0.001	19.17	24.94	–	–	–	Bal.
Type 310ULC	0.0057	0.18	1.30	0.004	0.0011	20.39	25.90	0.0079	0.0008	<0.01	Bal.
Type 310EHP	0.0002	0.0025	0.0001	0.0011	0.0008	21.58	23.13	0.0029	0.0022	0.18	Bal.

(**b**) Lab-melted type 310 stainless steels

Steel	C*	Si*	Mn	P*	S	Ni	Cr	O*	N*	B*	Ti	Fe
E1	26	13	0.0001	6	2	22.31	24.64	15	12	0.03	0.16	Bal.
E2	79	75	0.0001	8	1	22.70	25.25	42	42	1.2	0.06	Bal.
E3	110	60	0.0015	7	1	22.54	25.09	47	46	0.59	0.07	Bal.
E4	47	100	0.0001	74	1	21.89	25.89	34	55	0.52	0.12	Bal.
E5	56	100	0.0001	263	1	22.23	25.84	35	62	0.43	0.12	Bal.
E6	53	100	0.0001	7	87	22.51	25.66	40	67	0.45	0.10	Bal.
E7	60	100	0.0001	7	282	22.29	25.77	48	92	0.48	0.10	Bal.
E8	3	40	0.0004	10	26	20.54	24.77	35	9	0.08	0.18	Bal.
E9	5	34	0.15	8	15	20.62	24.72	41	9	0.09	0.19	Bal.
E10	6	22	0.47	6	19	20.83	25.91	11	11	0.11	0.19	Bal.

this study. The basic chemical compositions of the lab-melted steels are comparable to the extra high-purity grade, and vary in the amount of the minor and impurity elements (C, P, S and Mn) individually. The B content of steels is a level of one or sub-ppm (*i.e.*, B-free steels). A very small amount of Ti is also present in these steels.

Experimental Procedures

The hot cracking susceptibility of the steels was evaluated using the transverse-Varestraint test as shown in Fig. 1. The augmented strain was varied between 0.15 and 0.83%. The transverse-welding (GTAW) conditions were as follows: arc current, 150 A; arc voltage, 14 V and welding speed, 1.67 mm/s. The specimens were 5 mm thick. The temperature distribution behind the molten pool in the weld bead was measured by a thermocouple during welding in order to obtain the cracking temperature range. The crack surface occurring in the Varestraint test was observed using a scanning electron microscope (SEM). The hot ductility (reduction of area) of the weld metal was evaluated by the Gleeble test at 1,600–1,685 K applying a crosshead speed of 0.33 mm/s. The dimensions of a specimen for the Geeble hot ductility test are shown in Fig. 2.

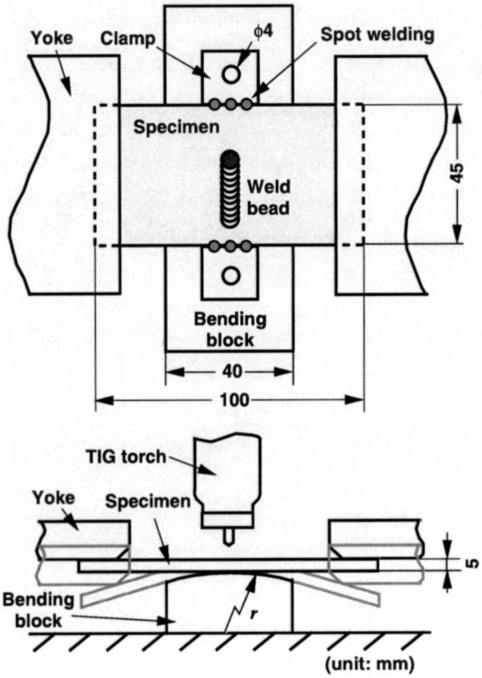

Fig. 1 Schematic illustration of transverse-Varestraint test

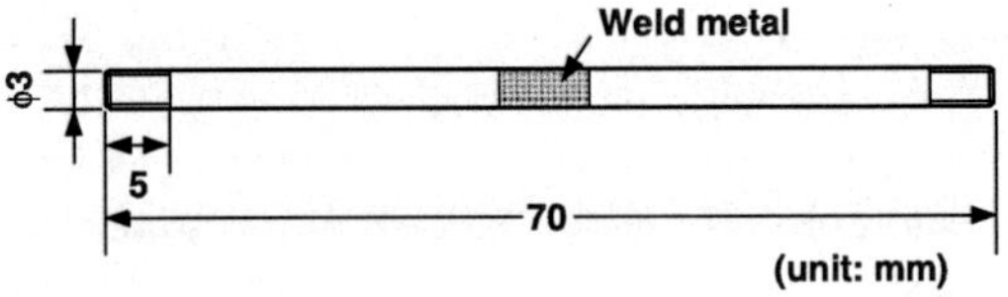

Fig. 2 Dimensions of specimen for Gleeble test

Hot Cracking Behaviour in the Transverse-Varestraint Test

Hot Cracking Susceptibilities of Type 310 Stainless Steels

The hot cracking susceptibilities of Type 310, Type 310S, Type 310ULC and Type 310EHP steels were evaluated by the number of cracks, the total crack length and the maximum crack length after the transverse-Varestraint test. Figure 3a–c show the number of cracks, the total crack length and the maximum crack length (including both solidification and ductility-dip cracks), respectively. The hot cracking susceptibility evaluated by all these parameters drastically decreased in the order of Type 310>Type 310S>Type 310ULC>Type 310EHP steels. Although very few hot cracks occurred in Type 310EHP steel, the hot cracking susceptibility increased as the augmented strain in the Varestraint testing increased.

Fig. 3 Hot cracking susceptibilities of type 310 stainless steels. (**a**) Number of crackes, (**b**) Total crack length, (**c**) Maximum crack length

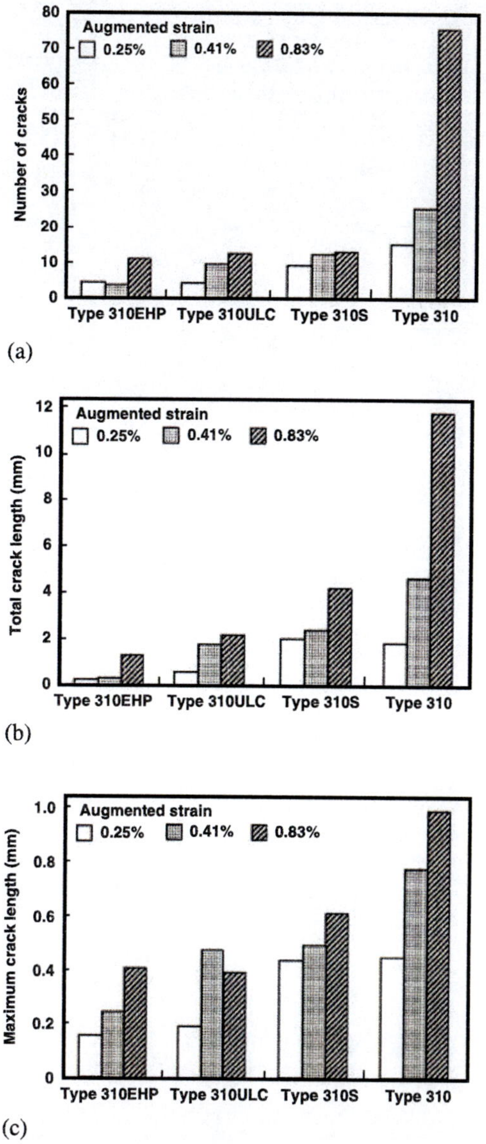

Characterisation of Hot Cracks

Typical examples of the crack surface occurring in the transverse-Varestraint test for Type 310 and Type 310EHP steels are shown in Figs. 4 and 5, respectively. Two types of fracture morphologies could be observed in the crack surface. The crack surface occurring on the higher temperature side adjacent to the fusion line

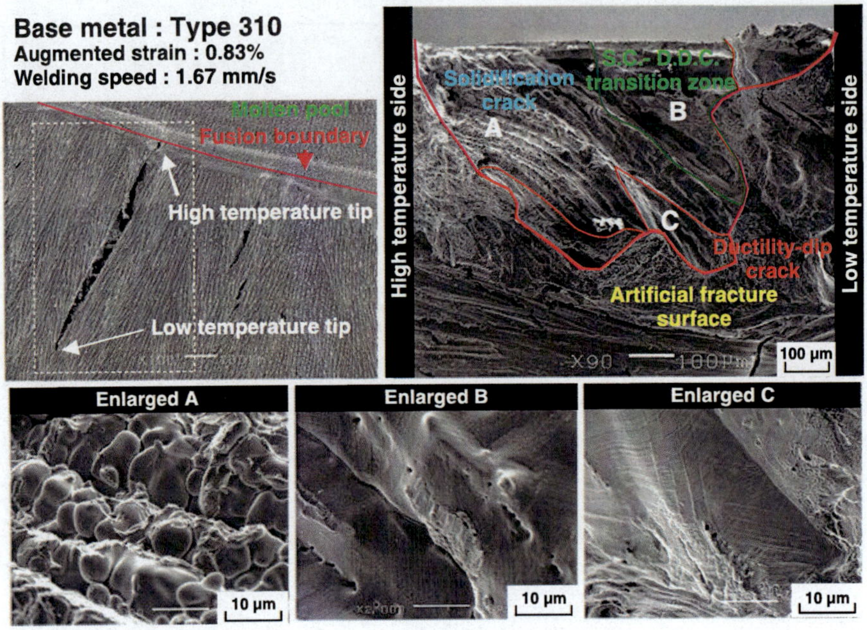

Fig. 4 Microstructures of crack surface occurring in Type 310 steel

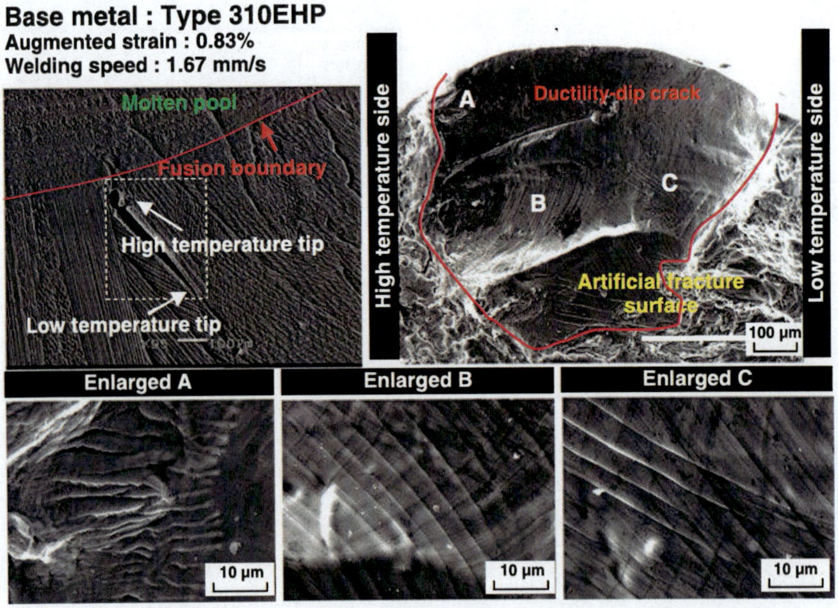

Fig. 5 Microstructures of crack surface occuring in Type 310EHP steel

(solid-liquid interface at time of bending in the transverse-Varestraint test) indicated a dendritic structure with a trace of melting, while that occurring on the lower temperature side furthest from the fusion line (instantaneous solid-liquid interface) indicated an intergranular fracture without melting. Microstructural observations of the crack surface revealed that the former crack could be regarded as solidification cracking, while the latter is due to ductility-dip cracking. In addition, as shown in Fig. 4, some regions containing both morphological features existed on one crack surface (called the "transient zone"). Similar morphologies could be observed in crack surfaces of Type 310S and Type 310ULC steels, but, ductility-dip cracks were not found.

Changes in Solidification and Ductility-Dip Cracking Susceptibilities upon Refining

Based on the cracking location and fracture morphology, every crack within the transverse-Varestraint test was distinguished as either a solidification crack or a ductility-dip crack. Thus, the maximum crack length data shown in Fig. 3c was sorted into three categories according to the type of crack: solidification, ductility-dip or transient zone crack. Figure 6 shows the solidification and ductility-dip cracking susceptibilities in Type 310, Type 310S, Type 310ULC and Type 310EHP steels. The solidification cracking susceptibility drastically decreased as the amounts of minor and impurity elements were reduced (refined), and very few solidification cracks occurred in Type 310EHP steel. Furthermore, as the purity was improved, the ductility-dip cracking susceptibility also decreased, but very few ductility-dip cracks occurred in Type 310S and Type 310ULC steels. However, the ductility-dip cracking susceptibility contrarily increased in Type 310EHP steel.

Figure 7 shows the hot cracking temperature ranges of Type 310, Type 310S, Type 310ULC and Type 310EHP steels obtained from the transverse-Varestraint

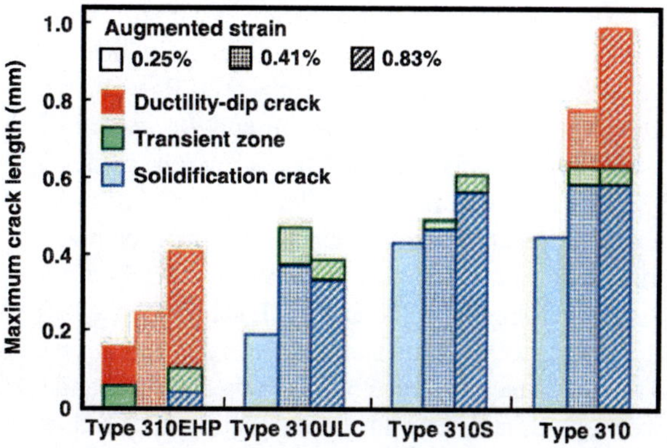

Fig. 6 Hot cracking susceptibilities of type 310 stainless steels

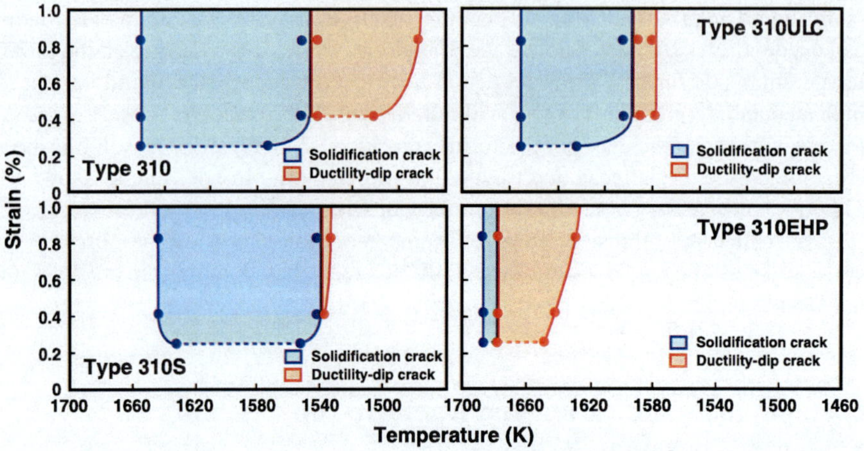

Fig. 7 Hot cracking temperature ranges of type 310 stainless steels

test results. In this figure, the transient zone was approximately divided in half into solidification and ductility-dip cracks. The solidification and ductility-dip cracking temperature ranges enlarged with an increase in the augmented strain at lower strains; however, they were saturated at the augmented strain of around 0.8%. The solidification cracking temperature range monotonically reduced in the order of Type 310>Type 310S>Type 310ULC>Type 310EHP steels. The ductility-dip cracking temperature range of Type 310 steel was relatively wide, while those of Type 310S and Type 310ULC steels were less so. However, the ductility-dip cracking temperature range enlarged again in Type 310EHP steel.

Comparison of Hot Cracking Susceptibility Between Type 310 Stainless Steels and Commercial Alloys

The hot cracking susceptibility of type 310 stainless steels was compared to that of various alloys. The solidification and ductility-dip cracking susceptibilities were respectively evaluated as the solidification brittle temperature range (BTR) and the ductility-dip temperature range (DTR), which were obtained from the cracking temperature ranges. Figures 8 and 9 compare the BTR and DTR between type 310 stainless steels and various commercial alloys, respectively. The BTR of Type 310EHP steel was much smaller than those of type 304 and 316L stainless steels, which are regarded as resistant to solidification cracking because of solidification as primary ferrite (FA mode). However, the BTRs of the other type 310 stainless steels were larger than type 304 and 316L stainless steels. In particular, Type 310 and Type 310S steels were highly susceptible to solidification cracking. On the other hand, the DTRs of Type 310, Type 310S, Type 310ULC and Type 310EHP steels were relatively small, and were far below that of the La-modified Inconel 690 alloy, which

Fig. 8 Comparison of solidification brittle temperature range BTR cracking among various alloys

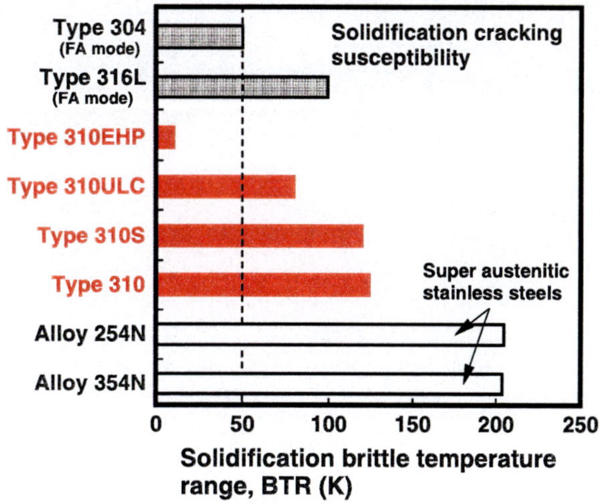

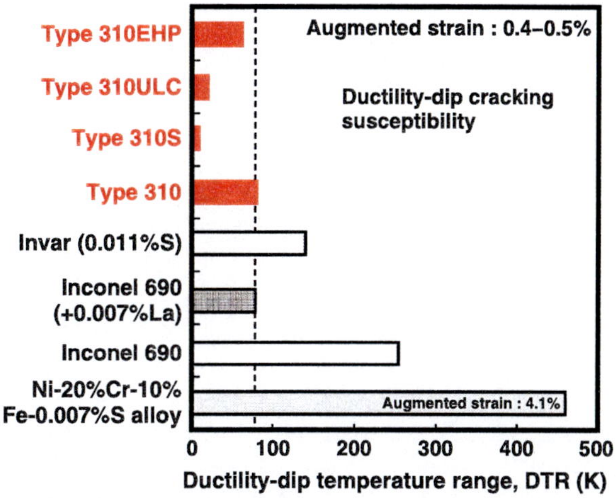

Fig. 9 Comparison of ductility-dip temperature range DTR among various alloys

has a negligible susceptibility to ductility-dip cracking [19]. From these results, it is concluded that the extra high-purity stainless steel (Type 310EHP steel) has a negligible susceptibility to hot cracking (both solidification and ductility-dip cracking) despite being fully austenitic stainless steel. Consequently, refining to the extra high-purity grade would be one of the most innovative countermeasures for solving the hot cracking issue of fully austenitic alloys such as stainless steel and Ni-base alloys.

Improvement Mechanism of Solidification Cracking Susceptibility

Theoretical Analysis of Solidification Segregation

Analysis Model

To clarify the effects of minor and impurity elements on the solidification behaviour of the weld metal, the solidification segregation during welding was numerically analysed. The calculation procedure for the solidification segregation is similar to that previously reported [20, 21]. However, to simplify the computation procedure, the segregation behaviours of C, P and S to the solidification boundary were calculated for pseudo-binary systems of (Fe-25 wt%Cr-20 wt%Ni)-C, P, S using the finite differential method (FDM) scheme. Namely, a binary diffusion model was adopted in the present analysis, and cosegregation of C, P and S was not considered in the present model. Below is a brief description of the theoretical models used in this study.

The distribution of the solute element during the solidification process was determined by solving the diffusion equation. Figure 10 shows a one-dimensional diffusion model in a regular triangle, which assumes that the dendrite morphology is basically a hexagonal prism. Symmetrical boundary conditions were applied to both end segments. The final segment of the liquid phase was defined as the solidification boundary, the cell boundary (C.B.). The diffusive flux J_i from a segment i to a segment i+1 is given by Fick's 1st law;

$$J_i = D\frac{C_{i+1} - C_i}{\Delta x} \quad (1)$$

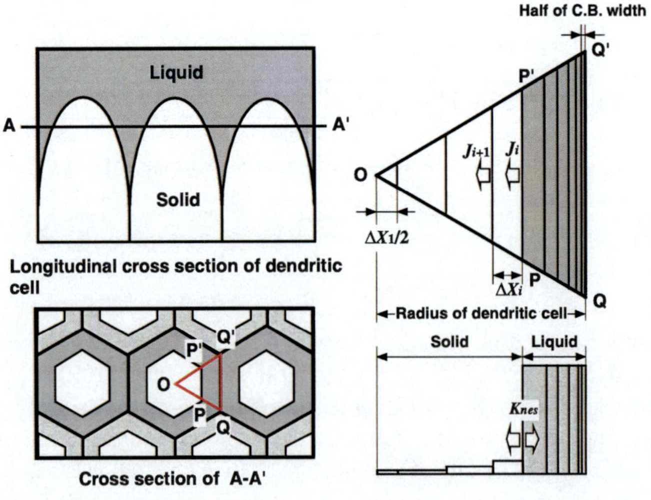

Fig. 10 Schematic illustration of analysed model of segregation during solidification

where D is the diffusion coefficient of the solute, C_i and C_{i+1} are the concentrations in segments i and i+1, respectively and Δx is a segment width. The change in solute concentration ΔC_i during a infinitesimal time interval Δt at a segment i is expressed by;

$$\frac{\Delta C_i}{\Delta t}\frac{S_i + S_{i-1}}{2} \cdot \Delta x_i = J_i S_i - J_{i-1} S_{i-1} \tag{2}$$

where S_i and S_{i-1} are the sectional areas of segments i and i–1, respectively. In the regular triangle model as shown in Fig. 10, S_i is expressed by;

$$S_i = \frac{2}{\sqrt{3}} \sum_{k=1}^{i} \Delta x_k \tag{3}$$

From eqs.(2) and (3), ΔC_i can be given by;

$$\Delta C_i = \frac{2D\Delta t}{\Delta x_i (S_i + S_{i-1})} \cdot \left\{ S_i \frac{C^B_{i+1} - C^B_i}{\Delta x_i} - S_{i-1} \frac{C^B_i - C^B_{i-1}}{\Delta x_{i-1}} \right\} \tag{4}$$

where C^B_{i+1}, C^B_i and C^B_{i-1} are solute concentrations in segments i+1, i and i–1 at the previous Δt, respectively. Assuming the mass conservation law of solute in a dendrite holds, then eq.(5) becomes;

$$\begin{aligned}\left(\sum_{k=1}^{N} \Delta x_k\right)^2 \cdot C_0 = \sum_{i=1}^{N} \left\{ \left(\sum_{k=1}^{i} \Delta x_k + \sum_{k=1}^{i-1} \Delta x_k \right) \cdot \Delta x_i \cdot C^S_i \right\} + \\ \sum_{i=j+1}^{N} \left\{ \left(\sum_{k=1}^{i} \Delta x_k + \sum_{k=1}^{i-1} \Delta x_k \right) \cdot \Delta x_i C^L_i \right\}\end{aligned} \tag{5}$$

where N is the total number of segments, C_0 is the initial solute concentration, C^S_i is the solute concentration in solid phase segment i, C^L_i is the solute concentration in the liquid phase (assuming the solute concentration in the liquid phase is uniform) and j is the segment number of the solid phase at the solid/liquid interface.

The solute concentration at the solid/liquid interface is given by the local equilibrium concept;

$$C^S_j = K_{nes} C^L_{j+1} \tag{6}$$

where K_{nes} is the non-equilibrium distribution coefficient between solid and liquid phases, C^S_j and C^L_{j+1} are solute concentrations of solid and liquid phases at the solid/liquid interface, respectively. The non-equilibrium distribution coefficient can be represented by [22];

$$K_{nes} = \frac{K_{es} + \beta}{1 + \beta}, \ \beta = R_v \frac{\lambda}{2D^L} \tag{7}$$

where K_{es} is the equilibrium distribution coefficient, β is the Peclet number, R_v is the growth velocity of the dendrite (solidification rate), λ is the interatomic spacing and D^L is the diffusion coefficient of the solute in a liquid. The solid/liquid interface migrates according to the solidification velocity determined by the cooling rate during solidification, in other words, the neighbour segment changes from the liquid to the solid phase.

Based on the aforementioned theoretical basis, the solidification segregation behaviours of C, P and S were computed during the welding process. Table 2 summarises the material constants used in the numerical simulation [23, 24]. The cooling rate during solidification was determined to be 200 K/s from the measured thermal cycle in a molten pool. In the present simulation, the width of the final solidification boundary (*i.e.*, cell boundary) was assumed to be 0.4 μm based on the tin-quenched solidification microstructure observed. The final solidification boundary of 0.4 μm thick corresponds to a solid phase fraction of 96% (the fraction of the liquid phase is 4%) in a regular triangle model of solidification segregation as shown in Fig. 10. The final segregated amount of the solute was defined by the solute concentration of the final solidification boundary.

Analysis Results

The calculated concentrations of C, P and S in a liquid phase with progress toward solidification are shown in Fig. 11. Regardless of the type of steel, the C, P and S contents in the liquid phase increased as the solid phase fraction increased (with progress toward solidification). The final segregated concentrations of C, P and S, which remained in the liquid phase, were significantly heightened in the case of Type 310 steel, whereas those in the cases of Type 310ULC and Type 310EHP steels were less due to their lower initial contents. These results suggest that the solidification segregation of minor and impurity elements is suppressed in the extra

Table 2 Material constants used for numerical simulation of solidification segregation

Parameter	Carbon	Sulphur	Phosphorus
Element content (wt%)	Type 310 : 0.12 Type 310S : 0.05 Type 310ULC : 0.057 Type 310EHP : 0.0006	Type 310 : 0.011 Type 310S : 0.001 Type 310ULC : 0.0011 Type 310EHP : 0.0008	Type 310 : 0.020 Type 310S : 0.016 Type 310ULC : 0.004 Type 310EHP : 0.0011
Diffusion coefficient, D (m^2/s) $D = D_g\exp(-Q/RT)$	D_0 : 3.85 × 10^{-4} (m^2/s) Q : 300 (kJ/mol)	D_0 : 7.52 × 10^{-4} (m^2/s) Q : 238.4 (kJ/mol)	D_0 : 2.67 × 10^{-3} (m^2/s) Q : 258 (kJ/mol)
Distribution coefficient, K_{nes}	0.31	0.035	0.13
Half of dendrite radius (m)	1.0 × 10^{-5}	1.0 × 10^{-5}	1.0 × 10^{-5}
Solidification boundary width (m)	4.0 × 10^{-7}	4.0 × 10^{-7}	4.0 × 10^{-7}

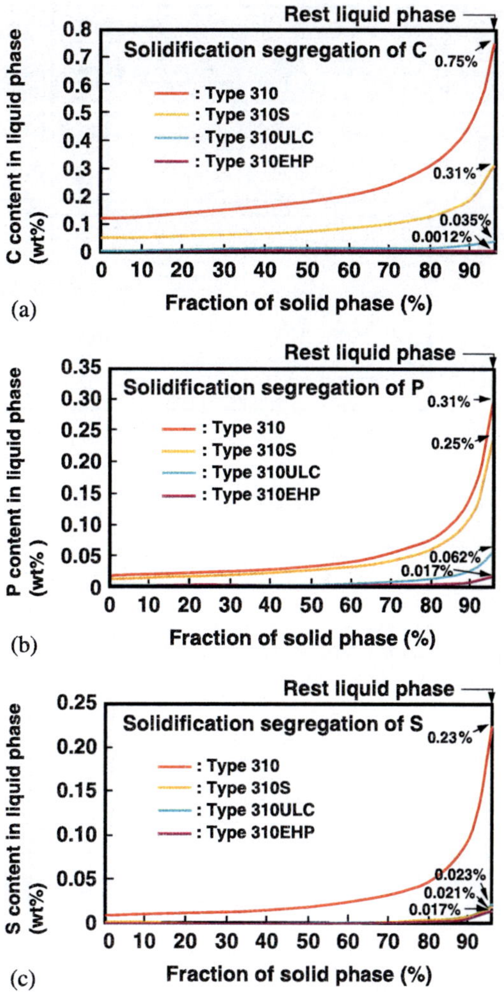

Fig. 11 Solidification segregation behaviours of elements in type 310 stainless steel weld metals. (**a**) Carbon, (**b**) Phosphorus, (**c**) Sulphur

high-purity steel. Figure 12 shows changes in the volume fraction of the liquid phase with temperature for Type 310, Type 310S, Type 310ULC and Type 310EHP steels. In addition, the calculated results using the Thermo-Calc^{TM} equilibrium computation software (database used: TCFe6) are shown, *i.e.*, calculated without considering solidification segregation. For all steels, the fractions of the liquid phase decreased monotonically from the liquidus temperature T_L as the temperature decreased. However, the temperature where the volume fraction of the liquid phase attained 4% (*i.e.*, the final segregated cell boundary in Fig. 10) dipped below the nominal solidus temperature T_S which is determined by equilibrium calculation. We call this final solidification temperature of the remaining liquid phase the "true solidus T_C". The nominal solidus temperature T_S and the true solidus temperature T_C rose in the order of Type 310<Type 310S<Type 310ULC<Type 310EHP

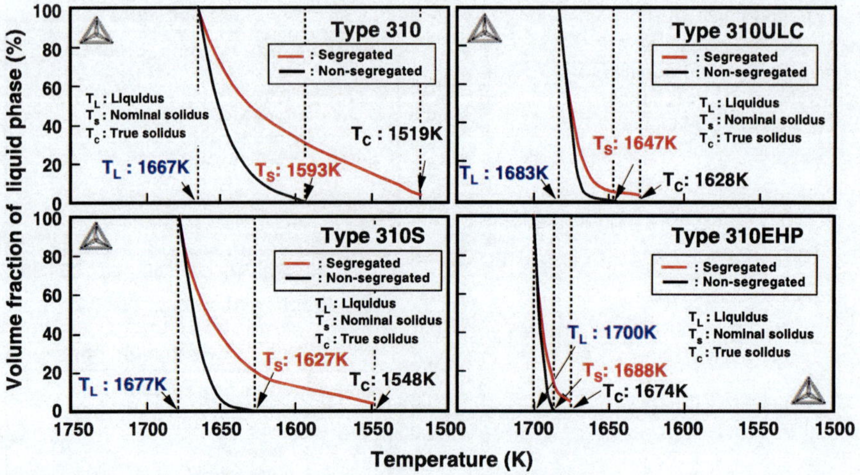

Fig. 12 Change in volume fraction of liquid phase with temperature

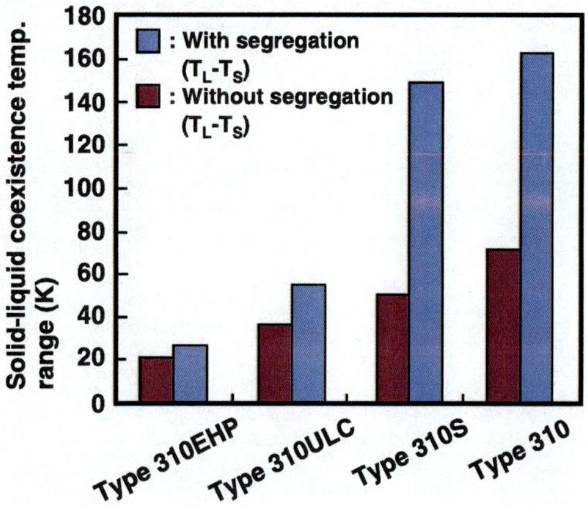

Fig. 13 Temperature difference between liquidus (T_L) and nominal solidus (T_S) or true solidus (T_C)

steels. Figure 13 summarises the temperature differences between the liquidus (T_L) and nominal solidus (T_S), and true solidus (T_C). For all steels, the temperature difference between T_L and T_C (considering solidification segregation) was larger than that between T_L and T_S (corresponding to the equilibrium solidification). In particular, $T_L - T_C$ was much larger than $T_L - T_S$ in Type 310 and Type 310S steels. Namely, the solid-liquid coexistence temperature range was drastically enlarged due to the solidification segregation. Consequently, the solid-liquid coexistence temperature range of Type 310EHP steel with consideration to the solidification segregation was greatly reduced compared with those of Type 310, Type 310S and Type 310ULC steels.

Effect of Solidification Segregation on Solidification Cracking Susceptibility

The relation between the calculated solid-liquid coexistence temperature range and the solidification cracking susceptibility was investigated. Figure 14 shows the relationship between the calculated temperature difference between T_L and T_C ($T_L - T_C$) and the BTR, which is depicted in Fig. 8. Regardless of the type of steel, there was a good positive correlation between them. Therefore, it can be qualitatively deduced that the solidification cracking susceptibility of the extra high-purity steel would be improved by reducing the solid-liquid coexistence temperature range attributed to the depression of the solidification segregation with purification. However, the measured BTRs differed slightly from the calculated values of ($T_L - T_C$), implying that there are differences in the extent of the effect of the remaining liquid phase to the solidification cracking susceptibility (*e.g.* morphology and physical property of the remaining liquid film), even if the solid-liquid coexistence temperature range with the solidification segregation is identical.

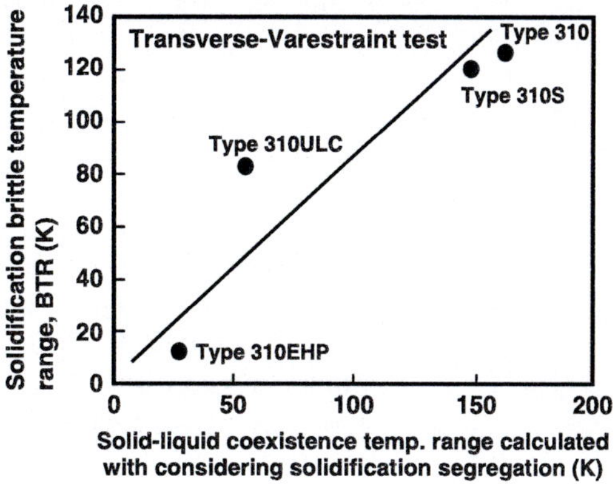

Fig. 14 Relation between calculated solid-liquid coexistence temperature range and solidification brittle temperature range BTR

Hot Ductility of Weld Metal Evaluated by the Gleeble Test

In order to clarify the ductility-dip cracking mechanism in Type 310EHP, the hot ductility of the weld metals was evaluated by the Gleeble test. Figure 15 shows the hot ductility data of the weld metals of Type 310, Type 310S, Type 310ULC and Type 310EHP. In the present study, the hot ductility was evaluated from the reduction of area fractured after the Gleeble test. The testing temperature used as the abscissa was normalised by the lower limit of the BTR (T^*) of each steel, because the cracking temperature range varied by the steel type. The reduction of area of any weld metal fell drastically at higher temperatures where solidification cracks occurred. The reduction of area of Type 310S and Type 310ULC weld metals remained high until the BTR (even in the BTR). In contrast, the reduction of area

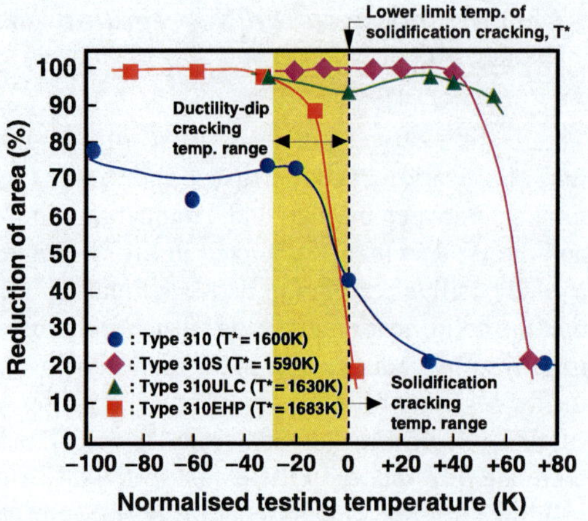

Fig. 15 Hot ductility of type 310 stainless steel weld metals

of Type 310EHP and Type 310 weld metals has been deteriorated in the temperature range below the BTR, where is a possible ductility-dip cracking temperature range. SEM observation of fracture surfaces of Type 310 and 310EHP weld metals after Gleeble test revealed that specimens with the deteriorated ductility were fractured at the grain boundary (GB), and that their fracture morphologies were similar to ductility-dip cracking. These facts suggest that ductility-dip cracking is related to the hot ductility just below T^*. Figure 16 shows the average hot ductility of weld metals in the ductility-dip temperature range (*i.e.*, between 0 and 30 K below T^* indicated by a shadow in Fig. 15). The average hot ductility of Type 310S and Type 310ULC weld metals were sufficiently high at the ductility-dip temperature range. However, the average hot ductility of Type 310EHP and Type 310 weld metals was obviously lowered compared with Type 310S and Type 310ULC weld metals. Therefore, it is

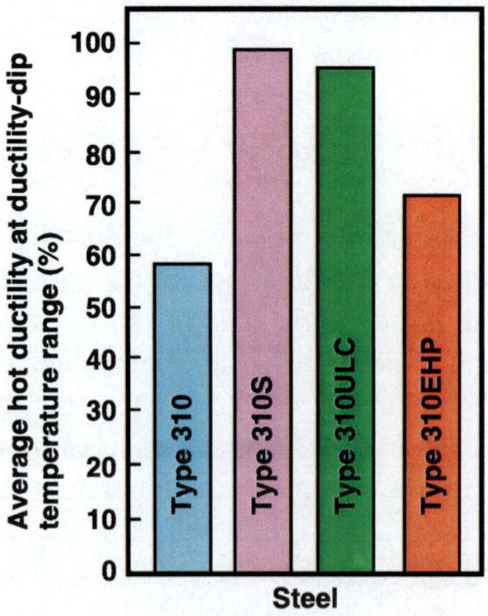

Fig. 16 Average hot ductility at ductility-dip temperature range

suggested that the extra high-purity steel would be slightly susceptible to ductility-dip cracking due to the decreased hot ductility in the ductility-dip temperature range.

Essential Influence of Minor and Impurity Elements on Hot Cracking Susceptibility

Hot Cracking Susceptibility of Lab-Melted Extra High-Purity Stainless Steels

To evaluate the quantitative contribution of C, P, S and Mn to the hot cracking susceptibility, the solidification and ductility-dip cracking susceptibilities were evaluated by the transverse-Varestraint test using lab-melted type 310 stainless steels E1-E10 whose C, P, S and Mn contents were individually varied from the extra high-purity grade. Based on the cracking location and fracture morphology, cracks within the transverse-Varestraint test were distinguished as solidification or ductility-dip crack. The maximum crack length of hot cracks occurred in lab-melted steels E1-E10 is shown in Fig. 17. Only solidification cracking occurred in steels E4-E7 which contained higher amounts of P and S, while ductility-dip crack as well as solidification crack occurred in steels E1-E3 and E8-E10 with lower P and S contents. The solidification cracking susceptibility decreased with a decrease in the P, S and C contents in the steels. In contrast, the ductility-dip cracking susceptibility slightly increased with a decrease in the C content. On the other hand, Mn negligibly affected the solidification and ductility-dip cracking susceptibilities. Figure 18 shows the effects of C, P, S and Mn contents in steel on the BTR of steels E1-E10. The BTR monotonically increased with an increase in the P, S and C contents, while the BTR changed little with variations in the amount of Mn. The role of C in ductility-dip cracking of extra high-purity steels was investigated. Figure 19 shows the effect of the C content on the DTR of steels E1-E3. The DTR obviously increased with a decrease in the C content in the steels. From these results, it is clear that the refining

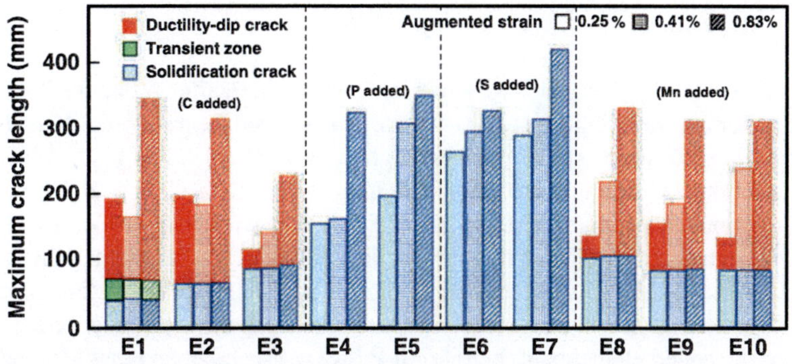

Fig. 17 Hot cracking susceptibilities of lab-melted steels E1-E10

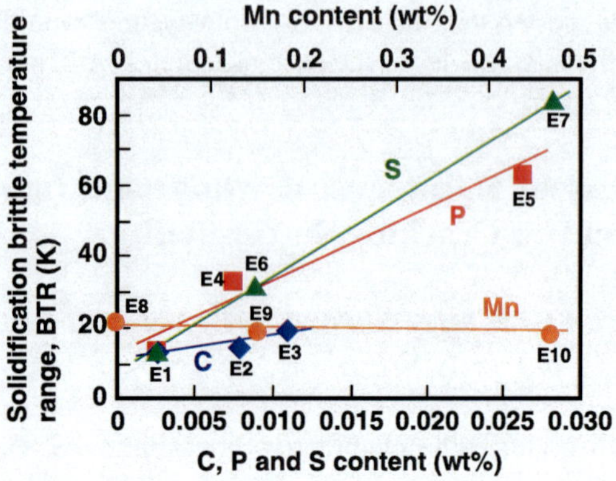

Fig. 18 Effects of C, P, S and Mn contents in steel on solidification brittle temperature range BTR

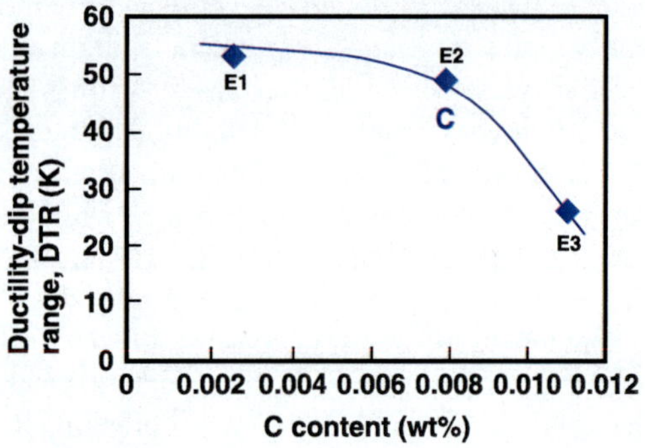

Fig. 19 Effect of C content in steel on ductility-dip temperature range DTR

of P and S would improve solidification cracking, and that of C would also have a beneficial effect on solidification cracking as well as a harmful effect on ductility-dip cracking at the lower P and S steels.

Quantitative Contribution of Minor and Impurity Elements to Solidification Cracking Susceptibility

The change in the BTR per unit wt% of each element (hereafter called the "potency atomic factor for enlarging the BTR") was calculated from the transverse-Varestraint test results, *i.e.*, from the slopes of the straight lines in Fig. 18. Figure 20 shows the calculated potency atomic factors for enlarging the BTR of C, P, S and Mn. The calculated potency atomic factors of C, P, S and Mn were 952, 1,984, 2,571 and –8.5 K/wt%, respectively. The potency atomic factor of S was larger than those of P and C, and that of Mn was negative and quite small. The values of potency atomic factor suggest that the solidification cracking susceptibility is influenced by the nature of the added elements in the ratio of P:S:C = 1:1.3:0.5, and that Mn negligibly influences the solidification cracking susceptibility. The good linear relationship

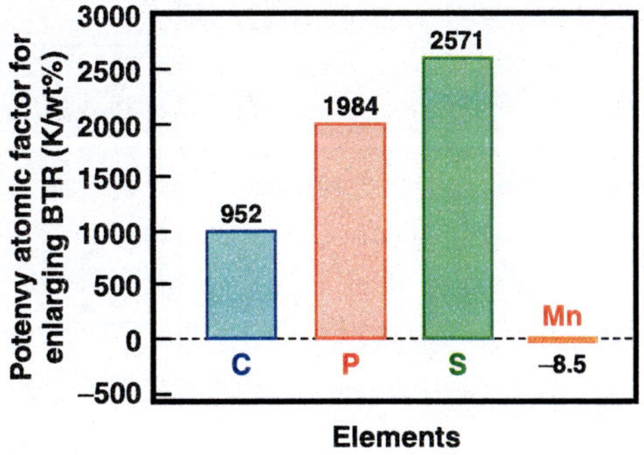

Fig. 20 Potency atomic factors of C, P, S and Mn for enlarging solidification brittle temperature range BTR

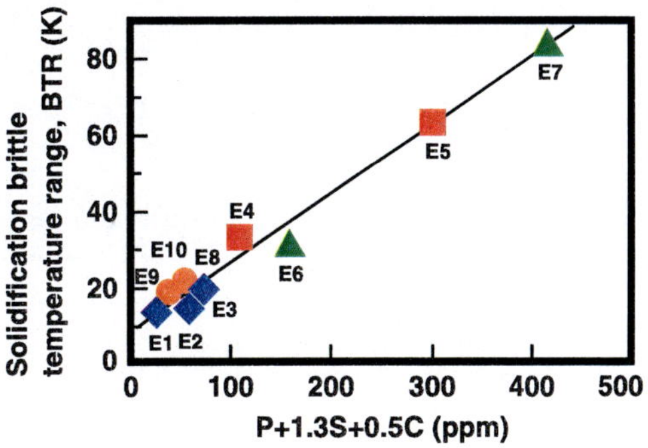

Fig. 21 Relation between compositional parameter P + 1.3S + 0.5C and solidification brittle temperature range BTR

between the compositional parameter of P + 1.3S + 0.5C and the BTR of steels E1-E10 is apparent in Fig. 21, making this a useful parameter for estimating the BTRs of extra high-purity type 310 stainless steels. Namely, the essential influence of S to an increase in the solidification cracking susceptibility is approx. 1.3 times as large as that of P, and that the influence of C is half of P.

On the other hand, there are several reports (described in the Introduction) that the quantitative contribution of P to solidification cracking is comparable to or larger than that of S in commercial austenitic stainless steels. However, the present investigation indicates that the potency atomic factor of S for enlarging the BTR is obviously larger than that of P (and C). In order to examine these competing claims, the microstructure of the Mn-containing weld metal was investigated. Figure 22 shows the microstructure and element profiles analysed by EPMA in Type 310 weld metal. Granular inclusions were observed at the cell boundaries, and both Mn and S were detected at inclusions, as indicated by arrows. This fact suggests that MnS is formed in the Mn-containing weld metal. Considering that commercial-purity austenitic stainless steels generally contain several percent of Mn, the contradiction in findings between our work and that on commercial steels could be explained by the formation of MnS during solidification of the weld metal in the commercial steels. Namely, Mn renders S harmless by stabilising it as MnS in the remaining liquid phase, and accordingly the apparent contribution of S to the solidification

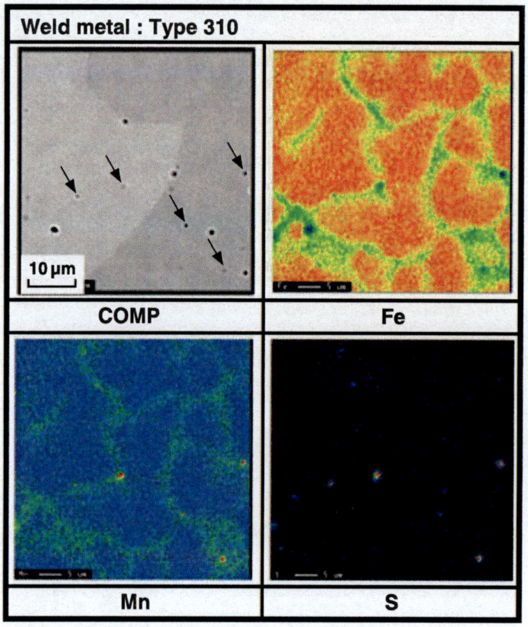

Fig. 22 Element profiles at inclusions in Type 310 weld metal analysed by EPMA

cracking susceptibility would be less than that of P. Thus, it is a logical conclusion that solidification cracking of commercial type 310 stainless steels is governed by not only the S, P, C contents but also the Mn content in the steel.

Role of C on Ductility-Dip Cracking Susceptibility

Molecular Orbital Analysis of Binding Strength of Grain Boundary

A ductility-dip crack occurred in Type 310EHP steel as well as the steels containing lower amounts of P and S (E1-E3, E8-E10), and furthermore the ductility-dip cracking susceptibility increased with a decrease in the C content in the steel. It is generally thought that ductility-dip cracking can be attributed to the GB embrittlement due to the GB segregation of impurity elements such as P and S (see Fig. 23). This conjectures that the ductility-dip cracking susceptibility of the extra high-purity steel (Type 310EHP steel) should be quite low. In order to elucidate the mechanism of ductility-dip cracking in the extra high-purity steel, the binding strength of the GBs was numerically analysed by a molecular orbital analysis (DV-Xα method [25]). Figure 24 schematically illustrates the cluster model used in the present computation, the $\sum 15$ coincidence boundary of γ-Fe consisting of 67 atoms. One or two Fe atoms on the GB were replaced by P, S or C atoms to simulate the GB segregation. The binding strength of the GBs was evaluated from the bond order between intergranular and intragranular Fe atoms in the near neighbourhood of P, S and C

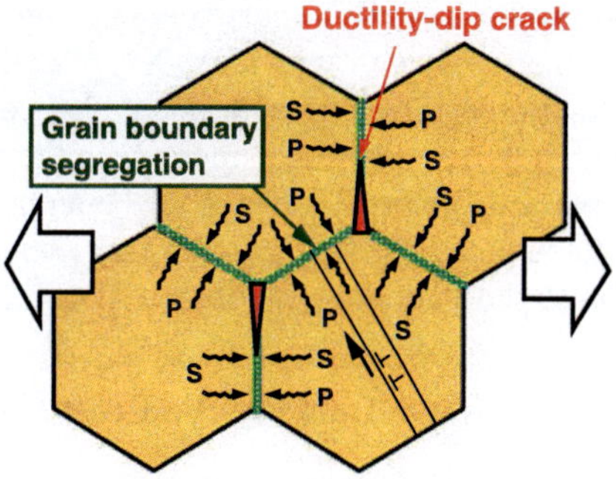

Fig. 23 Schematic illustration of ductility-dip cracking mechanism

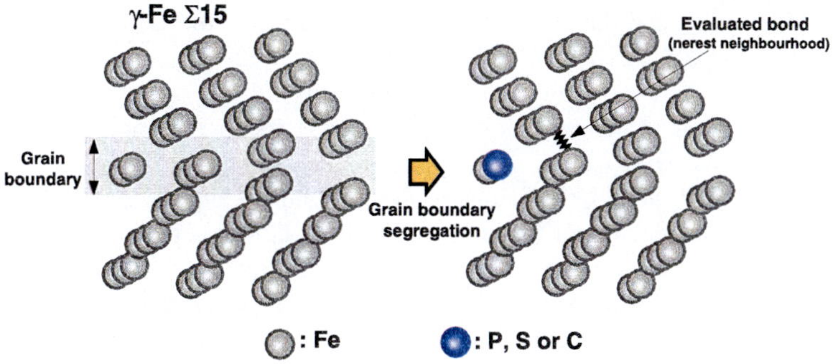

Fig. 24 Schematic illustration of cluster model used (molecular orbital analysis by DV-Xα method)

atoms (it is known that the bond order is linearly correlated with the interatomic bond strength). The calculated bond orders (with no unit) of Fe atom at the GB for different segregation situations are shown in Fig. 25. In this figure, Fe_{67} indicates the non-segregated situation, $Fe_{66}X_1$ and $Fe_{65}X_2$ ($X = P, S, C$) indicate the segregated situations where one and two intergranular Fe atoms were replaced by an X

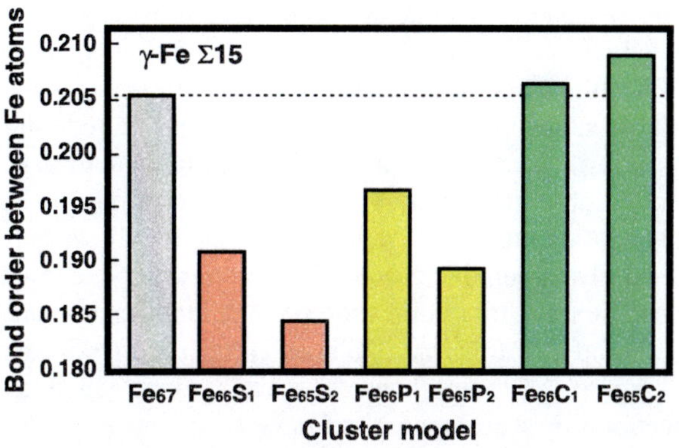

Fig. 25 Calculated bond order between Fe atoms at different grain boundary segregation

atom, respectively. Thus, "$Fe_{65}X_2$" means that the GB segregation has proceeded further compared with "$Fe_{66}X_1$". When P and S were segregated at the GBs, the bond order of Fe atom was reduced compared with the non-segregated situation, and it decreased with an increase in the number of segregated P and S atoms. To the contrary, the bond order of an Fe atom increased compared to the non-segregated situation in the case that C was segregated at the GB, and the bond order of Fe atom increased with progress of the GB segregation of C.

Mechanism of Ductility-Dip Cracking in Extra High-Purity Steels

We also investigated the mechanism by which ductility-dip cracks are produced in the extra high-purity steel (Type 310EHP steel). A molecular orbital analysis suggested that C enhanced the bond strength of Fe atoms at the GB, while P and S reduced it. In Type 310EHP steel used in the present study, not only P, S (and Mn) but also C have been extremely refined. Therefore, the GB embrittlement should be greatly inhibited at the extra low amounts of P and S present, but, the binding strength of the GBs would, on the other hand, be degraded by the extremely low amount of C. Although the details of the mechanism and the quantitative contribution of minor and impurity elements on ductility-dip cracking in the extra high-purity steel have not been clarified yet, this work raises the possibility that, in the ductility-dip cracking mechanism of Type 310EHP steel, the binding strength of the GBs at high temperature (just below the true solidus) would be lowered by extreme levels of refining of GB strengthening elements such as C. We feel that the experimental result that the hot ductility of Type 310EHP weld metal was comparatively deduced (see Fig. 16) supports this conjecture.

Change in Hot Cracking Mechanism upon Extra-Refining

The hot cracking mechanism in conventional purity and extra high-purity austenitic stainless steels are schematically illustrated in Fig. 26. The dominant hot cracking in conventional purity type 310 stainless steels is solidification cracking. The solidification cracking susceptibility is greatly improved in extra high-purity type 310 stainless steels, while the ductility-dip cracking susceptibility in them is slightly increased. Solidification cracks are initiated at the dendritic boundaries (GBs) in the solid-liquid coexistence temperature range by shrinkage strain during cooling. The solidification cracking susceptibility (BTR) is related to the temperature difference between the liquidus and true solidus, and the BTR is significantly enlarged by the solidification segregation during the welding process. It follows that conventional purity steels are highly susceptible to solidification cracking due to the reduction of the solidus temperature by the solidification segregation of minor and impurity elements such as P, S and C. Furthermore, S is essentially more effective in enlarging the BTR than P (approx. 1.3 times), however, the addition of Mn to steel may reduce the harmful effect of S by scavenging it as MnS.

On the other hand, ductility-dip cracks occur at the dendritic boundaries (GBs) and/or the crystallographic migrated GBs in the temperature range below the true

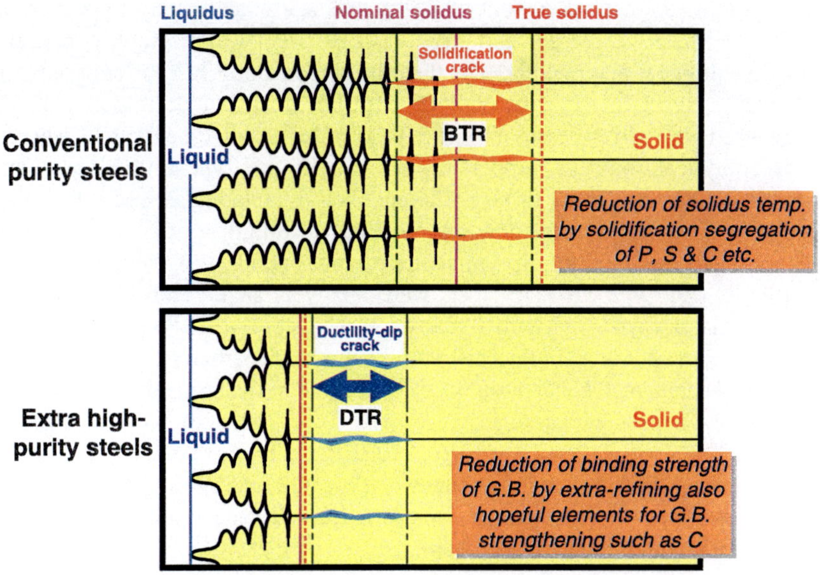

Fig. 26 Schematic illustrations of hot cracking mechanism in conventional purity and extra high-purity steels

solidus (in the solid-state). The ductility-dip cracking susceptibility (DTR) is governed by the imbalance between intergranular and intragranular strength at the elevated temperature. The intergranular strength depends on the binding strength of the GBs. Consequently, extra high-purity steels are slightly susceptible to ductility-dip cracking due to the reduction of the binding strength of the GBs by extra-refining not just the deleterious elements such as P and S, but also the possible beneficial elements such as C for GB strengthening. Furthermore, it is known that ductility-dip cracking is also influenced by the GB pinning and GB tortuosity [26–28]. SEM observation revealed that the microstructure of extra high-purity stainless steel weld metals indicated flat and smooth grain boundary, and accordingly that neither GB pinning nor GB tortuosity was remarkably confirmed because of few inclusions. These facts would also bring increasing ductility-dip cracking susceptibility. However, further investigations are necessary to clarify the effect of the GB morphology and migration on ductility-dip cracking in extra high-purity stainless steels.

Conclusions

In the present study, the hot cracking behaviour of extra high-purity type 310 stainless steel was investigated, and the essential influence of minor and impurity elements on the hot cracking susceptibility was clarified. The results obtained are summarised as follows;

(1) The solidification cracking susceptibility drastically decreased as the amounts of C, P and S decreased, and Type 310EHP steel was negligibly susceptible to solidification cracking. The ductility-dip cracking susceptibility slightly increased again in the case of Type 310EHP steel, but decreased with the enhanced purity in the case of Type 310, Type 310S and Type 310ULC steels.
(2) The solidification brittle temperature range (BTR) of Type 310EHP steel was much smaller than those of type 304 and 316L stainless steels with FA mode solidification, which have a low susceptibility to solidification cracking. On the other hand, the ductility-dip temperature ranges (DTRs) of Type 310S, Type 310ULC and Type 310EHP steels were relatively small. Hence, it was concluded that the extra high-purity stainless steel (Type 310EHP steel) was barely susceptible to hot cracking despite its fully austenitic structure.
(3) The final segregated concentrations of C, P and S in the remaining liquid phase were significantly increased in the case of Type 310 steel, whereas those in the case of Type 310ULC or Type 310EHP steel were reduced. The solid-liquid coexistence temperature ranges of Type 310 and Type 310S steels were enlarged compared with those of Type 310ULC and Type 310EHP steels due to the solidification segregation.
(4) The positive correlation between the calculated solid-liquid coexistence temperature range, which considered the solidification segregation, and the BTR strongly suggested the solidification cracking susceptibility was greatly improved in the extra high-purity steel because the solidification segregation was suppressed during the welding process.
(5) Based on the evaluated hot cracking susceptibility using lab-melted type 310 stainless steels with C, P, S and Mn contents that varied from the extra high-purity materials, the essential influence on solidification cracking was the ratio of P:S:C=1:1.3:0.5; Mn negligibly ameliorated solidification cracking in the extra high-purity steel.
(6) A molecular orbital analysis to estimate the binding strength of the grain boundary suggested that C enhanced the bond strength of Fe atoms at the grain boundary, while P and S reduced it. The proposed mechanism of ductility-dip cracking in the extra high-purity steel (Type 310EHP) raised the possibility that the binding strength of the grain boundaries would be adversely reduced by refining possible beneficial grain boundary strengthening elements such as C to extremely low levels.

Acknowledgements The present study includes the results of work performed under the grant "Research and Development of Nitric Acid Resistant Material Technology Applicable to the Next Generation of Reprocessing Equipment" entrusted to Kobe Steel, Ltd. by the Ministry of Education, Culture, Sports, Science and Technology of Japan (MEXT).

References

1. Inazumi T, Hyakubu T, Takagi Y, Okimoto S, Ehira Y and Ohashi M (1993) "Development of Highly Corrosion Resistant R-SUS304ULC for Nuclear Fuel Reprocessing Plant", NKK Technical Report, No.144, p. 20–25 (in Japanese).

2. Kiuchi K (2006) "Development of Equipment Materials for Purex Reprocessing Process". Journal of Atomic Energy Society Japan 48(11):871–876 (in Japanese).
3. Kunii K, Kusamichi T, Nakayama J and Kiuchi K (2007) JP Patent, No.2007-154214 (in Japanese).
4. Shankar V, Gill TPS, Mannan SL and Sundaresan S (1988) "Fusion Zone and Heat Affected Zone Cracking Susceptibility of Stabilised Austenitic Stainless Steels". Science and Technology Welding and Joining 3(1):17–24.
5. Radhakrishnan VM (2000) "Hot Cracking in Austenitic Stainless Steel Weld Metals". Science and Technology Welding and Joining 5(1):40–44.
6. Lienert TJ and Lippold JC (2003) "Improved Weldability Diagram for Pulsed Laser Welded Austenitic Stainless Steels". Science and Technology Welding & Joining 8(1):1–9.
7. Lundin CD, Chou CPD and Sullivan CJ (1980) "Hot Cracking Resistance of Austenitic Stainless Steel Weld Metals. Welding Journal 59(8):226s–232s.
8. Lippold JC and Savage WF (1982) "Solidification of Austenitic Stainless Steel Weldments: Part III, -The Effect of Solidification Behavior on Hot Cracking Susceptibility". Welding Journal 61(12):388s–396s.
9. Brooks JA and Lambert FJ Jr. (1978) "The Effect of Phosphorus, Sulfur and Ferrite Content on Weld Cracking of Type 309 Stainless Steel". Welding Journal 57(5):139s–143s.
10. Matsuda F, Nakagawa H, Katayama S and Arata Y (1982) "Weld Metal Cracking and Improvement of 25%Cr-20%Ni (AISI310S) Fully Austenitic Stainless Steel", Trans. J.W.S. 13(2):115–132.
11. Ogawa T and Tsunetomi E (1982) "Hot Cracking Susceptibility of Austenitic Stainless Steels". Welding Journal 61(3):82s–93s.
12. Kujanpää VP, David SA and White CL (1986) "Formation of Hot Cracks in Austenitic Stainless Steel Welds – Solidification Cracking". Welding Journal 65(8):203s–212s.
13. Kujanpää VP (1985) "Effect of Steel Type and Impurities in Solidification Cracking of Austenitic Stainless Steel Welds". Metal Construction 17(1):40R–46R.
14. Gooch TG and Honeycombe J (1980) "Welding Variables and Microfissuring in Austenitic Stainless Steel Weld Metal". Welding Journal 59(8):233s–241s.
15. Rabensteiner G, Tosch J and Schabereiter H (1983) "Hot Cracking Problems in Different Fully Austenitic Weld Metals". Welding Journal 62(1):21s–27s.
16. Katayama S (1981) "Improvement in Solidification Cracking of Fully Austenitic Stainless Steels SUS310S", Doctoral Thesis, Osaka University (in Japanese).
17. Li L and Messler RW Jr. (1999) "The Effect of Phosphorus and Sulfur on Susceptibility to Weld Hot Cracking in Austenitic Stainless Steels". Welding Journal 78(12):387s–396s.
18. Li L and Messler RW Jr. (2002) "Segregation of Phosphorus and Sulfur in Heat-Affected Zone Hot Cracking of Type 308 Stainless Steel". Welding Journal 81(5):78s–84s.
19. Nishimoto K, Saida K, Okauchi H and Ohta K (2006) "Prevention of Microcracking in Reheated Weld Metal by La Addition to Filler Metal, – Microcracking in Multipass Weld Metal of Alloy 690 (Part 3) –". Science and Technology of Welding and Joining 11(4):471–479.
20. Nishimoto K, Mori H and Hirata H "Effect of Grain Boundary Segregation of Sulphur on Reheat Cracking Susceptibility in Multipass Weld Metal of Fe-36%Ni Alloy". Proceedings of the 7th International Symposium. J.W.S., Kobe (2001), pp. 827–838.
21. Nishimoto K, Mori H, Esaki K, Hongoh S and Shirai M (1999) "Effect of Sulphur and Thermal Cycles on Reheat Cracking Susceptibility in Multipass Weld Metal of Fe-36%Ni Alloy", IIW Doc. IX-1934-99.
22. Aziz MJ (1982) "Model for Solute Redistribution during Rapid Solidification". Journal of Applied Physics 53:1158–1168.
23. Morita Z and Tanaka T (1988) "Thermodynamics on the Equilibrium Distribution Coefficients of Solute Elements between Solid and Liquid Phases in Iron Alloys". Tetsu-to-Hagane 74(7):1210–1218 (in Japanese).
24. The Japan Institute of Metals (2004) Kinzoku Data Book, Maruzen (in Japanese).

25. Adachi H, Owada H, Tanaka I, Nakamatsu H and Mizuno M (1998) Introduction to Molecular Orbital Calculation DV-Xα method, Sankyo (in Japanese).
26. Noecker II FF and DuPont JN (2009) "Metallurgical Investigation into Ductility Dip Cracking in Ni-Based Alloys: Part II". Welding Journal 88(3):62s–77s.
27. Lippold JC and Nissley NE (2007) "Further Investigations of Ductility-Dip Cracking in High Chromium, Ni-Base Filler Metals". Welding in the World 51(9/10):24–30.
28. Young GA, Capobianco TE, Penik MA, Morris BW and McGee JJ (2008) "The Mechanism of Ductility Dip Cracking in Nickel-Chromium Alloys". Welding Journal 87(2):31s–43s.

Investigations on Hot Cracking of Novel High Manganese TWIP-Steels

D. Keil, M. Zinke, and H. Pries

Introduction

The development of lightweight structures is imperative for the entire automotive industry in order to accommodate environmental requirements and improving safety and comfort without increasing the weight. The development of multi-phase steels with a high degree of ductility and strength enables the production of complex components with lower weights which also meet the requirements of a lightweight steel construction. The latest developments consist of high manganese content steels with FCC-lattices characterised by an exceptional strength and a very high elongation to fracture as a result of the TWIP-effect (Twinning Induced Plasticity). Figure 1 shows the mechanical and technological properties of these novel Fe-Mn-steels compared to other steel concepts. They clearly illustrate the exceptional balance of strength and plasticity of these materials. Elongations to fracture of over 50% at a tensile strength of approx. 1,000 MPa can be achieved.

National and international steel manufacturers are currently about to introduce these steels to the market [1–3]. However, there are as yet no specifics regarding their weldability.

There are essentially two different alloy concepts for high manganese steels: an Fe-Mn-C-alloy with a high C-content of up to 0.7 wt-%, and a reduced density Fe-Mn-Al-Si-alloy with a considerably lower carbon concentration. Both concepts have manganese as their main alloying element, with concentrations between 15 and 30 wt-%. The Fe-Mn-Al-Si-alloys also vary in their Al- and Si-concentrations, which each contribute from 2 to 3 wt-% to the lightweight construction potential of these materials [4].

If, due to the austenite lattice structure which remains stable down to room temperature, parallels are drawn with the welding of FCC Fe-Cr-Ni-steels, it is necessary to take into account both a strong tendency towards carbide precipitation

D. Keil (✉)
Institute of Materials and Joining Technology, University of Magdeburg, Magdeburg, Germany
e-mail: daniel.keil@ovgu.de

J. Lippold et al. (eds.), *Hot Cracking Phenomena in Welds III*,
DOI 10.1007/978-3-642-16864-2_12, © Springer-Verlag Berlin Heidelberg 2011

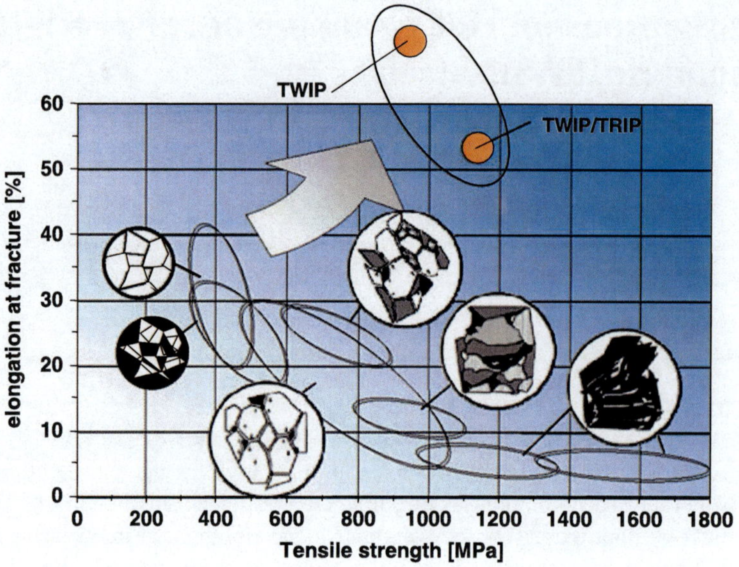

Fig. 1 Development of high strength steel grades for automotive applications [4, 5]

due to the high carbon content, and a very high hot cracking susceptibility (solidification and liquation cracks) due to the lattice structure, particularly when fully austenitic primary solidification occurs. The literature shows that the alloying element manganese can have a positive influence on the hot cracking resistance of Fe-Cr-Ni-steels [6]. It should also be noted that austenite, as compared to ferrite, is characterised by a lower thermal conductivity, a higher thermal expansion coefficient, a greater specific electrical resistance and a greater tendency towards coarse grain formation in the HAZ.

Experimental Procedures

Materials

The test materials consisted of four non-galvanised Fe-Mn-steels with various alloy compositions. All test materials had a high manganese content between 15 and 25 wt-% (Table 1). The manganese in the FeMn-1 to FeMn-3 alloys was replaced with the more cost-effective element carbon. The manganese content therefore ranges from 25 wt-% in FeMn-1 to 15 wt-% in FeMn-3. Differences in the chemical composition occur mainly in terms of the other alloy elements aluminium and silicon. FeMn-1 to FeMn-3 contain between 2.5 and 3 wt-% of Al and Si. However, FeMn-4 contains only 1.33 wt-% aluminium and 0.22 wt-% silicon.

Table 1 Chemical composition (wt-%) of the investigated base metals

Fe-Mn-Steel	Name	C	Mn	Al	Si	P	S
X5MnAlSi25-3-3	FeMn-1	0.08	24.6	2.95	2.71	<0.001	0.0100
X35MnAlSi25-3-3	FeMn-2	0.30	24.7	3.20	2.95	<0.001	0.0038
X70MnAlSi15-3-3	FeMn-3	0.68	15.6	2.51	2.54	<0.001	0.0042
X60MnAl18-1.5	FeMn-4	0.59	17.8	1.33	0.22	0.036	0.0008

Table 2 Thermo-physical properties of the investigated base metals compared to 1.4301 (AISI 304)

Fe-Mn-Steel	T_{Sol}(°C)	T_{Liq}(°C)	ΔT(K)	c at 20°C (J/kgK)	λ at 20°C (W/m·K)	α at RT-500°C (10^{-6} K^{-1})	Primary solidification
X5MnAlSi25-3-3	1,320	1,398	78	551	10.78	20.44	A
X35MnAlSi25-3-3	1,329	1,376	47	552	10.52	20.41	A
X70MnAlSi15-3-3	1,297	1,387	90	538	11.15	21.82	A
X60MnAl18-1.5	1,352	1,416	64	527	13.30	21.66	A
X5CrNi18-10 (1.4301)	1,437	1,467	30	500	15.00	18.00	F

All four Fe-Mn-steels have an austenitic structure. The FeMn-1 and FeMn-2 alloys also exhibit a small δ-ferrite content. The magneto-inductively determined ferrite content of FeMn-1 is 0.53 FN and that of FeMn-2 is 0.05 FN.

The experimental determination of the solidification temperature range by means of differential thermal analysis produced a very high solidification temperature range ΔT of 90 K for FeMn-3 in particular. This value is three times greater than that of the metastable austenitic Cr-Ni-steel 1.4301 (AISI 304). The specific heat capacity c and thermal expansion coefficient α of the Fe-Mn-steels are also considerably greater than those of the Cr-Ni-steel, while the thermal conductivity λ is less (Table 2).

Table 3 Chemical composition (wt-%) of the reference materials

Steel	Mat.-no.	C	Si	Mn	P	S
NiCr23Mo16Al	2.4605	0.003	0.02	0.15	0.002	0.002
X15CrNiSi25-20	1.4841	≤0.20	1.50–2.50	≤2.00	≤0.045	≤0.030
X8CrMnNi19-6-3	1.4376	0.042	0.38	6.80	0.024	0.004
X5CrNi18-10	1.4301	0.040	0.35	1.33	0.030	0.004
		Cr	Ni	N	Mo	Primary solidification
NiCr23Mo16Al	2.4605	22.08	60.50	0.034	15.50	A
X15CrNiSi25-20	1.4841	24.00–26.00	19.00–22.00	–	–	A
X8CrMnNi19-6-3	1.4376	17.90	4.29	0.200	0.15	F
X5CrNi18-10	1.4301	18.15	8.05	0.057	–	F

In order to classify the results of the PVR-test, other materials (Table 3) whose hot cracking resistance is known, were tested in addition to the four Fe-Mn-steels. The Cr-Ni-steels are metastable austenitic materials, which solidify in a primary ferritic solidification mode when fused. For this reason, they are considered to be more resistant to hot cracking than alloys which solidify as primary austenite, and therefore without further transformation [7, 8]. At the opposite end of the scale were the Ni-based alloy NiCr23Mo16Al (2.4605) [9] and the heat-resistant fully austenitic steel X15CrNiSi25-21 (1.4841).

Flat Tensile Test

The programmable deformation cracking ("Programmierter-Verformungs-Riss"/PVR-) test as per DIN EN ISO 17641-3 [10] was used to investigate the hot cracking resistance of the four Fe-Mn-steels (Fig. 2). Due to the sheet thicknesses of the test materials, the PVR specimens had a uniform thickness of 1.5 mm (including reference materials). This deviates from the standard PVR-specimen measurements specified in the aforementioned standard (thickness: 10 mm). In the PVR-test, a flat tensile test is carried out simultaneously with a welding process at a linearly increasing speed with a maximum tension speed of v_{max}. For the test the specimen is clamped in the device and not pre-stressed. The point at which the first hot crack appears corresponds to the critical tension speed v_{cr}. This correlates directly with the critical deformation rate [%/°C] mentioned in the hot cracking theory of Prokhorov [11] and can therefore be used as a criterion for assessing hot cracking resistance.

The critical tension speed is calculated by the following equation:

$$v_{cr} = \frac{a^*L_{HC}}{v_w} \tag{1}$$

where: v_{cr} = critical tension speed (cross-head speed)

a = acceleration of cross-head
L_{HC} = position of the first hot crack
v_w = welding speed

The PVR-test may be used to test specimens of the base material, the pure weld metal or weldments (joint welds as well as deposition welds). The hot cracking test allows a variation of welding processes (TIG, GMA, MMA), the filler metal, the welding consumables (shielding gas, sheathing, filler materials) and the welding parameters. The welds are produced as beads on plate welds or as deposits runs constant, controlled welding conditions. The advantages of the inspection process include the low test complexity, the excellent reproducibility of the test results and the good differentiation between the three types of hot cracking: solidification cracks (SC), liquation cracks (LC) and ductility dip cracks (DDC).

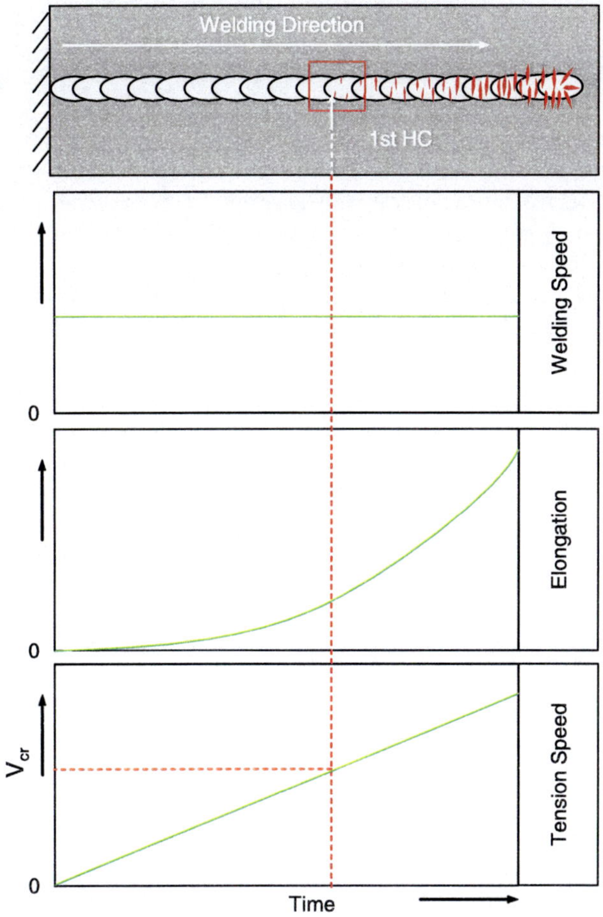

Fig. 2 Flat tensile test – schematic illustration

The fully mechanised TIG welding process was used on the Fe-Mn-steels in the PVR-test. The strength of the current was used as the test parameter. All other parameters such as welding speed (20 cm/min), voltage (8 V) and max. tension rate (55 mm/min) remained constant (Table 4).

Table 4 Welding and test-parameters for PVR-test

Variant (No.)	I (A)	U (V)	v_w (cm/min)	E (kJ/cm)	max. v_{tr} (mm/min)
1	50	8	20	1.2	55
2	55	8	20	1.3	55

Self-Restrained Testing Procedure – Lap Joint Test

In addition to the PVR-tests, a hot cracking test on a self-restrained specimen for arc welding processes tailored to a real weldment geometry was developed.

The self-restrained hot cracking test consisted of the GMA-welding of a lap joint on two clamped test sheets, for which it was possible to vary both the energy input per unit length as well as the overlap (distances to the free edge) between the upper and lower sheet. Similar self-restrained tests for the laser welding of Al-Si-alloys have already been described in [12].

Hot cracking susceptibility is either measured in terms of the critical distances to the free edge over the same energy input per unit length or, where required, the critical energy input per unit length over the same distances to the free edge. They depend on the chemical and thermophysical properties of the filler and base materials, the residual welding stresses, and the strain or strain rate of the specimen. Different overlap lengths of 6, 8 and 10 mm were created by using lower sheets of different widths (Fig. 3). For the purpose of the tests it is assumed that as the overlap decreases and the duration of the welding process increases the accumulation of heat and therefore the elongation of the material increases at the boundary area. In this context the risk of hot cracks appearing on the underside of the lower sheet increases too. Both FeMn-4 and 1.4376 were used as test materials and were welded with the cold metal transfer/pulsed arc process (pulse-mix) and an Fe-Mn-metal-cored wire.

To evaluate the impact of the filler metal on the hot cracking susceptibility an austenitic welding wire was also investigated. The G 18 8 Mn (ER307) is a crack resistant wire for joining dissimilar steels as well as manganese alloyed steels. The chemical composition for both filler metals is provided in Table 5.

All specimens were evaluated using dye penetrant testing (PT) to check for the occurrence of macro-cracks. A further investigation was carried out in which cross-sections of the weld seams of the X60MnAl18-1.5 were obtained at specific points (start, centre and end of seam) and metallographically evaluated.

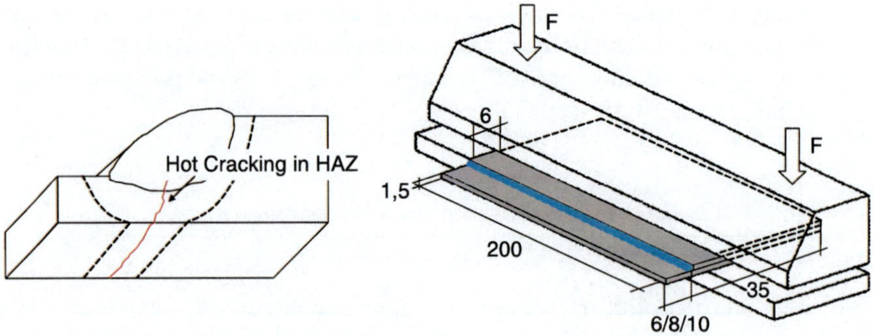

Fig. 3 Lap joint test – schematic illustration

Table 5 Chemical composition (wt-%) of the filler metals

Filler material	mm	C	Si	Mn	Cr	Ni	Mo	V	Al	Structure
DZ01[a]	1.2	0.29	0.12	16.80	0.12	–	–	–	–	fcc
G 18 8 Mn (ER307)[b]	1.0	0.08	0.80	7.00	19.00	9.00	–	–	–	fcc

[a]Test certificate.
[b]Standard analysis/manufacturer's data.

Results and Discussion

Flat Tensile Test Results

The results of the hot crack investigations in the PVR-test are shown in Fig. 4, which show the type of initial hot crack in addition to the critical tension speed. Compared to the Cr-Ni-steels, which are considered to be uncritical in terms of their hot crack susceptibility when processed correctly, all the Fe-Mn-steels subjected to the PVR-test exhibited a considerably higher tendency towards hot cracking. The Ni-based alloy 2.4605 demonstrates a slightly better behaviour and the heat-resistant fully austenitic steel 1.4841 behaves similar to the Fe-Mn-alloys. Among the Fe-Mn-steels, the X70MnAlSi15-3-3 (FeMn-3) is characterised by the highest tendency towards hot cracks. This produces a considerably smaller difference in the position

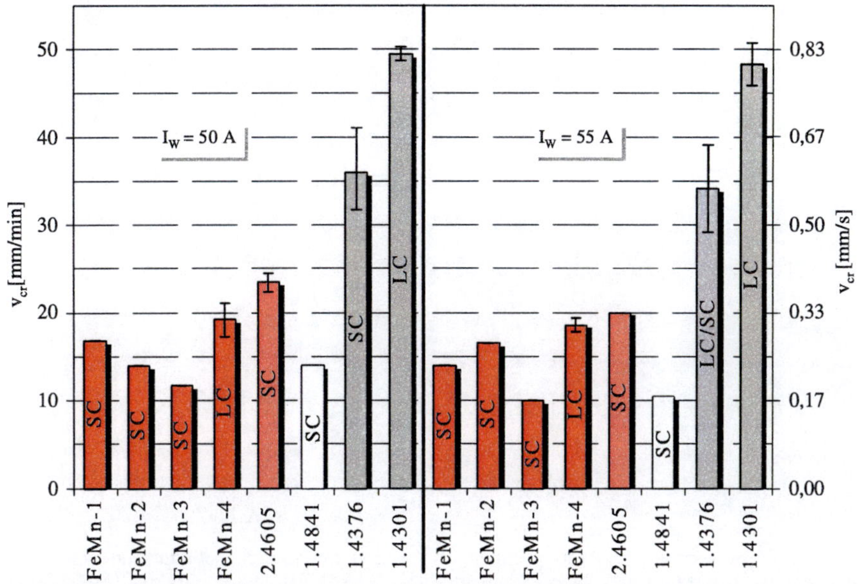

Fig. 4 Results of the PVR-test

of the first hot crack than in the number/formation of the hot cracks. Increasing the current and therefore the energy input per unit length led to a fall in the critical tension speed and therefore to a greater tendency towards hot cracking for nearly all alloys. The ranking towards hot cracking among the materials remained the same.

The first hot cracks experienced by the X5MnAlSi25-3-3 (FeMn-1), X35MnAlSi25-3-3 (FeMn-2) and X70MnAlSi15-3-3 (FeMn-3) alloys were solidification cracks. With regard to X60MnAl18-1.5 (FeMn-4) the specimens tested were found to have liquation cracks. The examination of surface sections confirmed the results of the tests conducted on the stereo microscope with respect to the type and position of the initial hot cracks (Fig. 5).

Since according to [6] the remaining delta ferrite content may also, in addition to the primary crystallisation, represent a factor influencing the hot cracking susceptibility, the δ-ferrite content of the weld metal of the TIG bead on plate welds of the PVR-specimens was determined magneto-inductively. Only the X5MnAlSi25-3-3 and X35MnAlSi25-3-3 Fe-Mn-alloys exhibited an accumulation of δ-ferrite (see also reference [13]). However, it was not possible to ascertain a direct effect on v_{cr} due to the primary austenitic crystallisation.

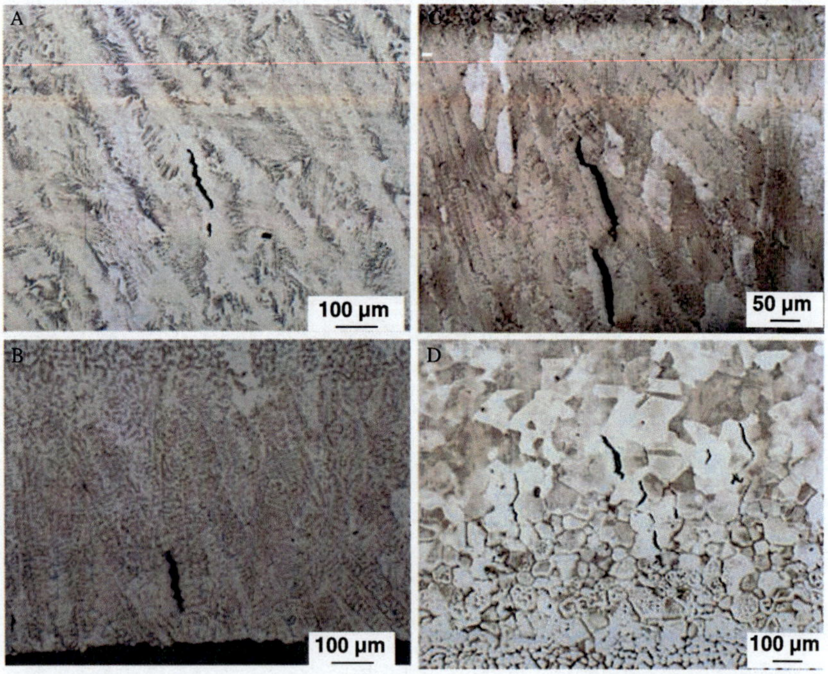

Fig. 5 Surface sections of PVR-specimen of the first hot cracks, solidification cracks in (**a**) FeMn-1, (**b**) FeMn-2, (**C**) FeMn-3, liquation cracks in (**D**) FeMn-4. Note the difference in magnification

Self-Restrained Testing Procedure – Lap Joint Test

For identical conditions, the evaluation of the self-restrained specimens using the dye penetrant test produced a considerably earlier formation of hot cracks in the Fe-Mn-steel than the 1.4376 (Table 6). At distances to the free edge of 6 mm and an effective energy input per unit length of 3.0 kJ/cm, macroscopic longitudinal cracks were detected on the underside of the lower sheet of three of the X60MnAl18-1.5 (FeMn-4) specimens (Fig. 6). No cracks were detected on the 1.4376 under these conditions.

By raising the distance to the free edge to 8 mm, macroscopic cracks were prevented, however, the cross-section exhibited micro-cracks in the form of liquation cracks in the high temperature area of the lower sheet (Fig. 7). The intercrystalline behaviour of these liquation cracks was considerably assisted by the presence of coarse grains in this zone. The grain size in this area was 10 times higher than that of the unaffected base metal.

As the heat input increased (3.5 kJ/cm), the cracks then spread until they reached as far as the underside of the sheet and into the weld metal. No cracks were detected in either penetrant tests or metallographic tests with distances to the free edge of 10 mm and 3.5 kJ/cm energy input per unit length.

Table 6 Results of the PT-tested self-restrained specimens; filler metal: DZ01

Distance to the free edge (mm)	6	8	8	10
X60MnAl18-1,5				
E_{eff} (kJ/cm)	3.0	3.0	3.5	3.5
Number of welds with cracks/number of welds	3/3	0/3	3/3	0/3
1.4376				
E_{eff} (kJ/cm)	3.0	3.0	3.5	3.5
Number of welds with cracks/number of welds	0/3	0/3	0/3	0/3

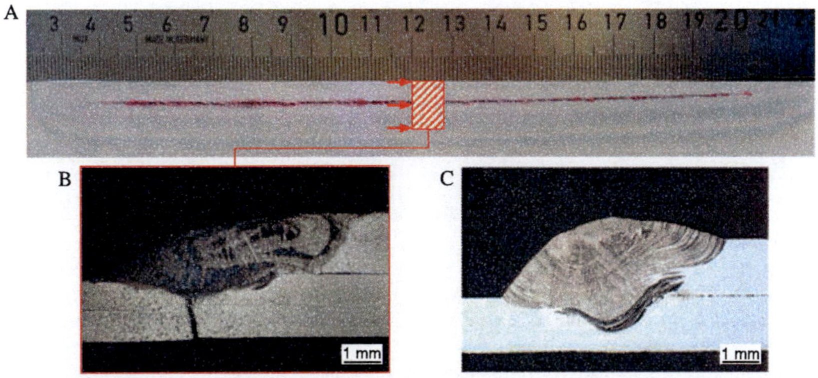

Fig. 6 (**a**) PT-tested bottom of the sheet with 6 mm critical distance from the edge (**b**) cross section of FeMn-4 with LC (**c**) crack free 1.4376

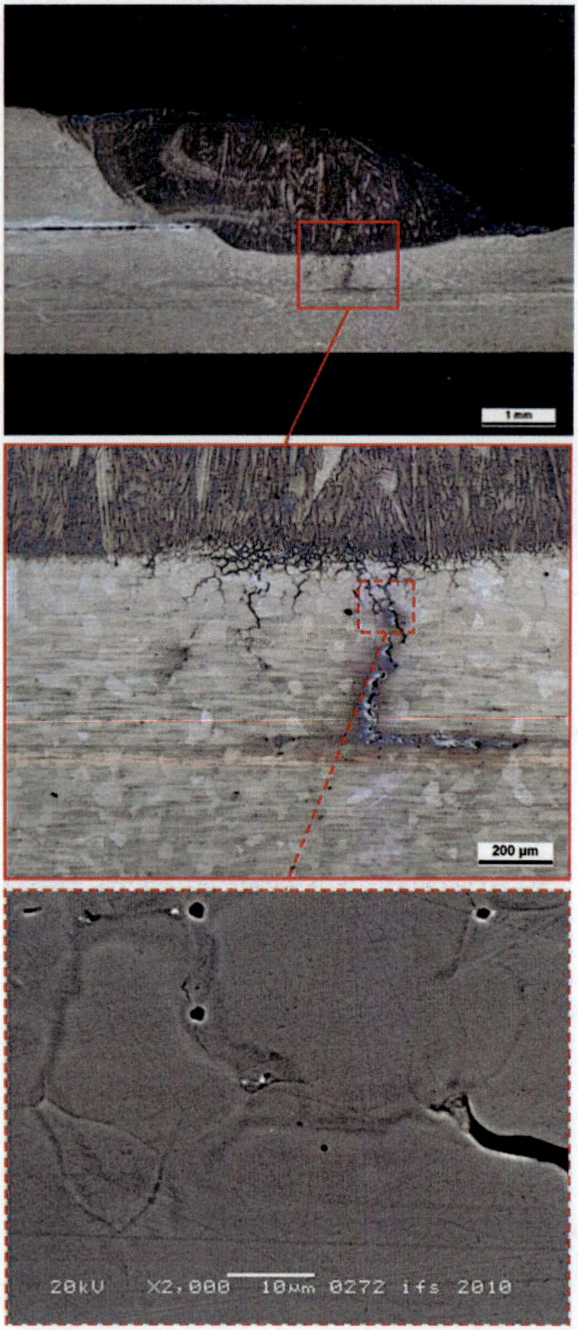

Fig. 7 Liquation crack in the HAZ of the FeMn-4; filler metal: DZ01

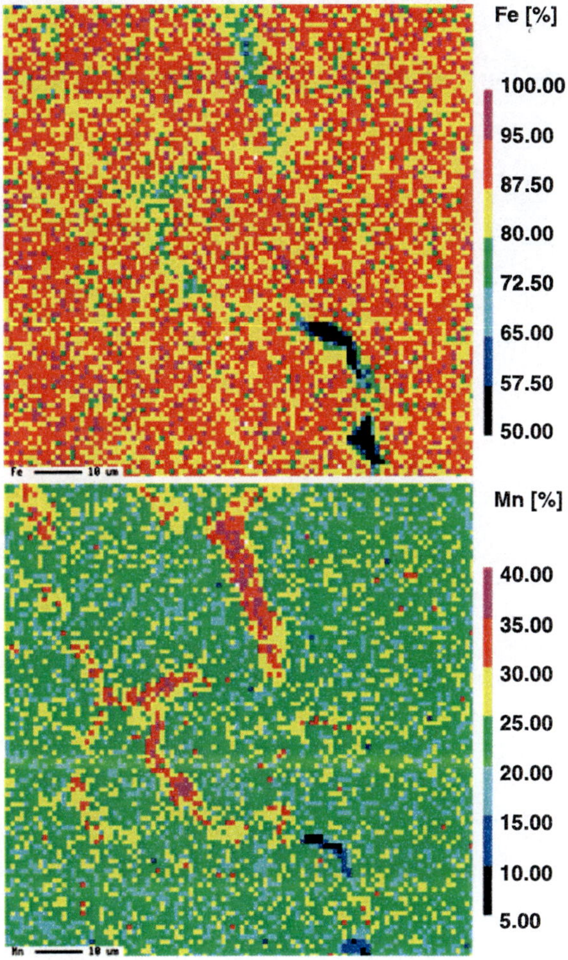

Fig. 8 EPMA analysis (Fe, Mn) of the area around the first solidification cracks in FeMn-4/DZ01 (see also Fig. 8)

With electron probe microanalysis (EPMA) measurements of a sample welded with 3.0 kJ/cm and a critical distance to the edge of 8 mm, high concentrations of manganese could be detected near the crack tip (Fig. 8). The increase of the manganese content goes hand in hand with a decrease of the iron content. Hot cracking promoting elements such as phosphorous or sulphur could not be detected.

The comparative welds with the G 18 8 Mn filler metal showed the same macroscopic behaviour as the welds with the FeMn filler metal. The cracks occurred longitudinally in the welding direction on the underside of the lower sheet. But the analysis of the cross sections revealed a different behaviour. In contrast to the welds with the Fe-Mn filler metal the cracks were clearly filled with weld metal (Figs. 7 and 9). In large cracks filled with weld metal the formation of different

(a)

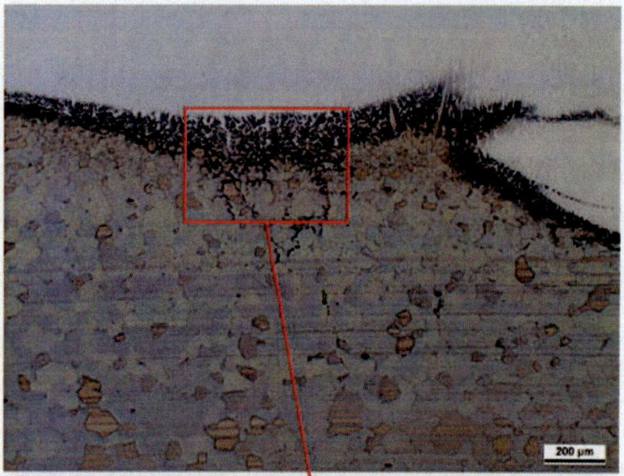

(b)

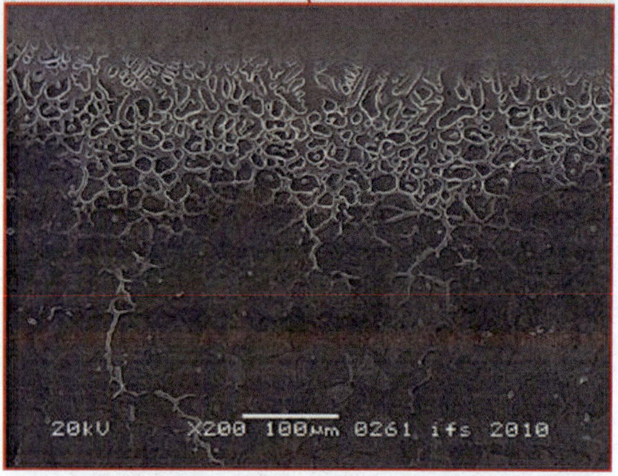

Fig. 9 (**a**) Optical micrograph of the HAZ of FeMn-4/G 18 8 Mn with filled cracks (**b**) SEM micrograph of the area around the fusion line. Note the difference in magnification

phases was metallographic detectable. By means of energy-dispersive X-ray analysis (EDX) very high concentrations of manganese (35%) and phosphorous (7%) could be measured in the centre of the filled crack (Fig. 10, Table 7 Spot B). This result led to the suspicion that a low-melting eutectic composition type Mn-Mn_4P with a solidus temperature of 960°C [14] had formed. Further measurements on specimens welded with a lower heat input and less hot cracking in the HAZ could not confirm the formation of the eutectic composition. Only in large cracks filled with weld metal the eutectic composition was detectable. Since the base metal shows the HAZ cracking regardless of the used filler metal, it appears that the manganese/phosphorous eutectic is in this case not the cause for the hot cracking. It seems more plausible that the formation of the eutectic is a result of the cracking and in particular due to the filled cracks with the specific filler metal. Nevertheless such increasing phosphorous contents should be considered to be critical. Note that the P-content in the base metal is 0,036 wt-%.

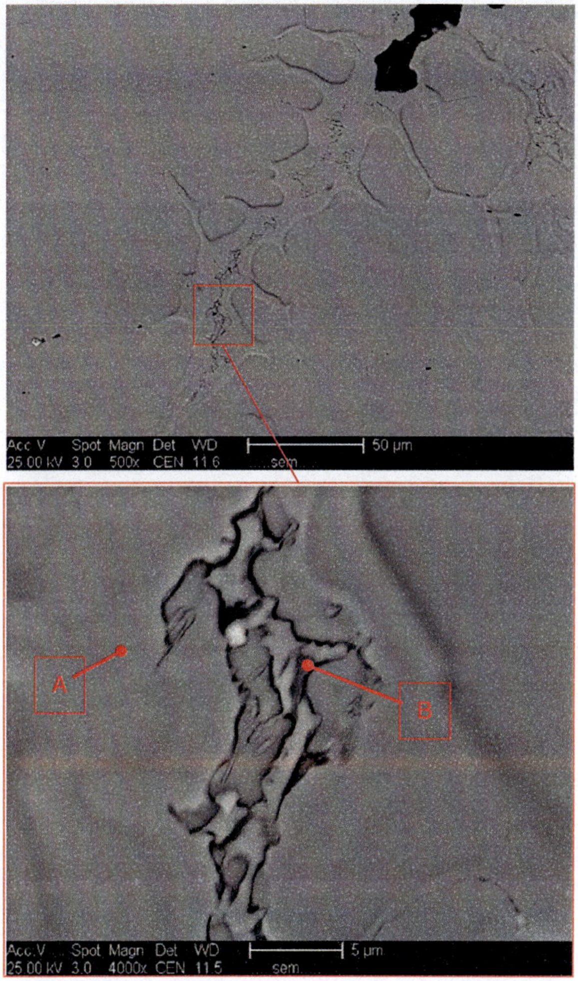

Fig. 10 Back-scattered electrons (BSE) micrographs of the filled liquation cracks; A and B mark the spots where the EDX-analysis was performed (Table 7)

Table 7 Results of the EDX analysis of the crack-filling weld metal (Fig. 10)

EDX-analysis	Fe		Mn		Cr		Ni		P	
	wt-%	at-%	wt-%	at-%	wt-%	at-%	wt-%	at-%	wt-%	at-%
Spot A	70.29	68.46	24.10	23.93	2.87	3.01	0.96	0.89	–	–
Spot B	45.55	42.43	35.09	33.23	10.41	10.42	–	–	7.26	12.19

The series of tests conducted with the self-restrained specimens confirmed the trend exhibited by the materials in the PVR-test in that the Fe-Mn-steels are characterised by a considerably higher hot cracking susceptibility than the Cr-Ni-steels. The occurrence of the first hot crack as liquation cracks in the HAZ of the FeMn-4 confirms the results of the flat tensile test.

Conclusions

The PVR-tests discussed show that the novel high-manganese TWIP-steels exhibit a considerably greater hot cracking susceptibility than the Cr-Ni-steels tested. Their hot cracking susceptibility is comparable to that of fully-austenitic Ni-based alloys or special stainless steels. Among the Fe-Mn-steels, the X70MnAlSi15-3-3 alloy was the most susceptible to hot cracking.

The self-restrained hot crack test for GMA-welding confirmed the results obtained from the PVR-tests. Irrespective of the filler material used, the hot cracks occurred mainly in the coarse-grained high temperature region of the HAZ of the Fe-Mn-steels along the weld seam. The combination of a large solidification temperature range, poor heat conduction, high thermal expansion coefficients and extreme grain growth in the HAZ of the Fe-Mn-steels played a significant role, particularly for the self-restrained specimens for arc welding processes tailored to real weldment geometries.

The high hot cracking susceptibility exhibited by the novel high manganese content steels requires, as with various Al-alloys, the employment of suitable technological and structural measures in component manufacturing.

Acknowledgements These tests were funded by the Federal Ministry of Economics and Technology (BMWi) assisted by the Arbeitsgemeinschaft industrieller Forschungsvereinigungen (AiF - industrial research consortium) "Otto von Guericke" e.V. (AiF-No. 15.201 B/DVS-Nr. 1.058) and supported by the DVS Forschungsvereinigung Schweißen und verwandte Verfahren e. V. (DVS institute for welding and related processes). The authors gratefully acknowledge the support.

References

1. Schröder T (2004) Ausgekochter Stahl für das Auto von morgen. MaxPlanckForschung. 3: 36–41.
2. Kim S, Cho J, Kwak W, Kim G, Kwon O (2007) Development of TWIP Steel for Automotive Application. InSteelCon, New Developments in Metallurgical Process Technology 3: 690–697.
3. Salzgitter AG (2009) Corus und Salzgitter kooperieren bei der Entwicklung eines neuen Stahls mit herausragenden Materialeigenschaften. press release 18.02.2009.
4. Gräßel O (2000) Entwicklung und Charakterisierung neuer TRIP/TWIP Leichtbaustähle auf der Basis Fe-Mn-Al-Si. Technische Universität Clausthal-Zellerfeld, Dissertation.
5. Bleck W (2000) Weiterentwickelte Stahlwerkstoffe eröffnen neue Anwendungsfelder. Materialwiss Werkst. 31:162–168.
6. Folkhard E (1984) Metallurgie der Schweißung nichtrostender Stähle, Springer, Wien, New York.

7. Masumoto I, Tamaki K, Kutsuma M (1972) Hot Cracking of Austenitic Steel Weld Metal. J. Jpn Weld. Soc. 41(11):1306–1314.
8. Thier H (1976) Deltaferrit und Heißrisse beim Schweißen chemisch beständiger austenitischer Stähle. DVS-Report 41:100–104.
9. Kirchheiner R, Köhler M, Heubner U (1992) Alloy 59, A New Highly Corrosion Resistant Material for the Chemical Process Industry, Environmental Pollution Control and Related Applications. Mater Corros 43:388–395.
10. ISO/TR 17641-3:2004: Destructive tests on welds in metallic materials – Hot cracking tests for weldments – Arc welding processes – Part 3: Externally loaded test.
11. Herold H, Streitenberger M, Pchennikov A (2001) Hot Cracking Theory by Prokhorov and Modelling of the PVR-test. Welding in the World 45:3/4:pp. 17–22.
12. Ploshikhin V, Prikhodovsky A, Makhutin M, Llin A, Zoch H (2005) Integrated mechanical-metallurgical approach to modeling of solidification cracking in welds. In: Böllinghaus T. Herold H. (eds), Hot Cracking Phenomena in Welds, Springer, Berlin Heidelberg, pp. 223–244.
13. Fink C, Zinke M, Keil D, Pries H (2010) Untersuchungen an neuartigen Fe-Mn-Stählen mit TWIP-Effekt mittels thermo-mechanischer Zyklen. ISBN: 978-3-87155-260-1, DVS-report 258:187–192.
14. Wiechmann E (1937) Beiträge zur systematischen Verwandtschaftslehre. 77. Über die niederen Phosphide des Mangans Thermische Analyse des Systems Mangan/Manganmonophosphid. Z Anorg Allg Chem. doi:10.1002/zaac.19372340204.

Solidification Cracking Susceptibility in C-Mn Steel CO_2 Laser Welds

M.F. Gittos, S.M.I. Birch, and R.J. Pargeter

Introduction

High power CO_2 laser welding of C-Mn steel has been a competitive process for many applications, including shipbuilding and structural steelwork, for many years [1]. There are many advantages, mostly related to precision, quality, ease of automation and cost. One problem which remains unpredictable, and therefore difficult to control, however, is solidification cracking. It is evident that some steels are more susceptible to solidification cracking in autogenous laser welds than others, and this is most likely to be due to differences in chemical composition, although other differences, such as strength, may play a part. It is also evident that process parameters affect the risk of cracking; high travel speeds increase the risk, as demonstrated in earlier work [2, 3], and variations in other parameters, particularly focus position which affect weld shape [4], are also relevant. However, welding power been found to be a variable influence on crack sensitivity [3, 5].

The effects of chemical composition have been explored, using a 10 kW laser in previous work at TWI [2]. In this, it was found that cracking increased as the carbon level fell from about 0.15 to 0.09%, the latter corresponding to the maximum freezing temperature range. It was felt that it would be of interest to explore the effects of carbon content below 0.09%. However, since the time of ref. 2, the 10 kW laser had been de-commissioned and therefore the opportunity was taken in the present work not only to explore the effect of carbon content below 0.09% but also the robustness of equations generated using welds made with the 10 kW laser, on welds made with a different laser (25 kW) over a wider range of compositions. The opportunity has also been taken to investigate the effect of strength on cracking, independent of composition, and also to vary focus position.

M.F. Gittos (✉)
TWI Ltd, Cambridge, UK
e-mail: mike.gittos@twi.co.uk

J. Lippold et al. (eds.), *Hot Cracking Phenomena in Welds III*,
DOI 10.1007/978-3-642-16864-2_13, © Springer-Verlag Berlin Heidelberg 2011

Background

Solidification cracking occurs when solidifying weld metal cannot sustain the thermal strain during the final stages of weld cooling. Weld metal solidifies over a range of temperatures, and on solidifying, contracts considerably. It has been suggested that cracking occurs when the supply of liquid metal has been cut off, or the strain is too high, or because the volume of the region containing the liquid film is larger than the volume of liquid available to fill the void created [6]. Indeed, cracking was prevented in a crack-sensitive high sulphur (0.2%) free-cutting steel by using induction heating to reportedly reduce the weld strain during solidification [7].

Matsuda and Ueyama [8] suggested a process by which cracking occurs in laser welds. This indicated that cracks formed in the brittle temperature range during solidification, and that due to the shape of the molten weld pool, different regions of the weld cracked at different times. Since the "well" region is the first to solidify, cracks occur here before the nail head solidifies. The schematic from this paper illustrates this theory (Fig. 1) and the shape of the weld pool is illustrated by the longitudinal section taken from a laser weld made in earlier work at TWI (Fig. 2).

It has been postulated that the occurrence and extent of solidification cracking is controlled by three principal factors, namely composition, weld solidification pattern and strain on the solidifying weld metal. For submerged arc welds, it is generally accepted that elements which increase the freezing range of the steel (such as C, S and P) increase the susceptibility to solidification cracking. Previous work in this programme indicated that C was the dominant element with S acting to a lesser extent [2, 9] and a composition based cracking index was derived from the results of experimental trials:

$$L = -1123C + 3096C2 + 816S + 90 \text{ (for 0.09–0.22\%C)} \quad (1)$$

Where L = summed length of cracks in the Russell test specimen.

The studies also suggested cracking susceptibility decreased with increasing carbon content between 0.09 and 0.18%, before increasing again above this level, with the most crack resistant steels having carbon contents in the range 0.15–0.18%. Nevertheless, in the generality of fusion welding, there is reason to believe that other elements can affect the solidification cracking susceptibility, such as P and Nb (detrimental) and Si, Mn and Mo (beneficial).

The occurrence of solidification cracking in steels is somewhat complicated by the peritectic reaction from delta ferrite to austenite. Certain elements (in particular sulphur) have different solubilities in the two phases – S is less soluble in austenite than in ferrite. Therefore elements that promote formation of ferrite reduce the risk of solidification cracking by reducing the amount of S available to form low melting point phases.

However, it has been speculated by Ohshita et al. [10] that a very low carbon steel weld metal (0.04–0.09%C) that solidifies completely to ferrite will experience extreme contraction strains on transformation to austenite (due to its associated volume contraction) so that the risk of solidification cracking is actually increased.

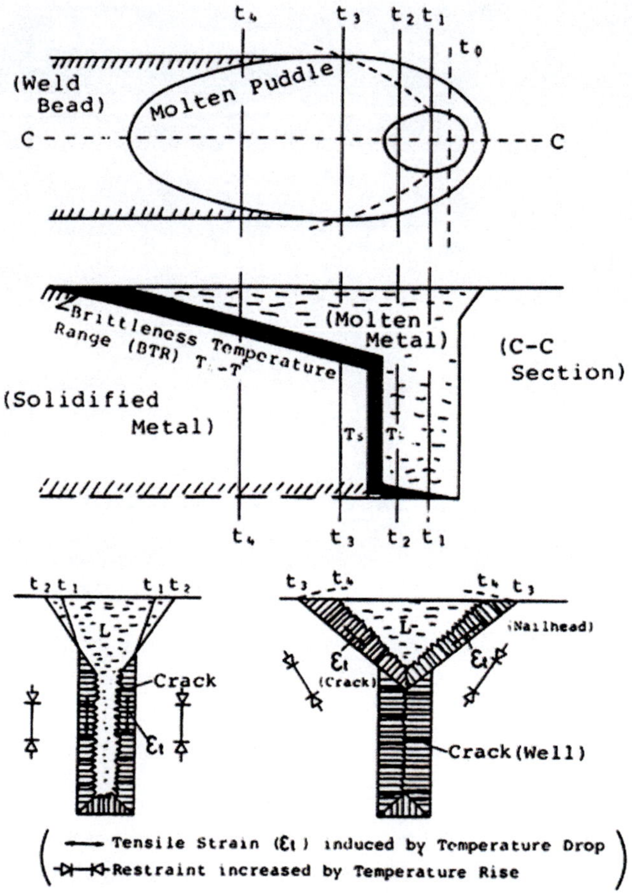

Fig. 1 Relationship between solidification process and crack generation. From Matsuda and Ueyama [8]. Reproduced by permission Taylor & Francis group

The solidification structure of the weld metal is particularly influential on the occurrence of solidification cracking, and is largely controlled by the weld bead shape. In turn, this is influenced by factors such as welding speed, joint preparation and fit-up. It is generally accepted that reductions in welding speed reduce the solidification cracking susceptibility. High speed welding typically gives a solidification structure of columnar grains perpendicular to the welding direction which meet at a single centre line boundary. This boundary generally has a limited ability to accommodate liquid films and strain [1].

Reduction of the welding speed results in the columnar grains being less perpendicular to the welding direction, and the formation of a complex central region consisting of finer columnar grains parallel to the welding direction, and equiaxed grains, which is more beneficial to the prevention of cracking. The previous report in this programme [2] indicated that this was indeed the case. It was discovered in

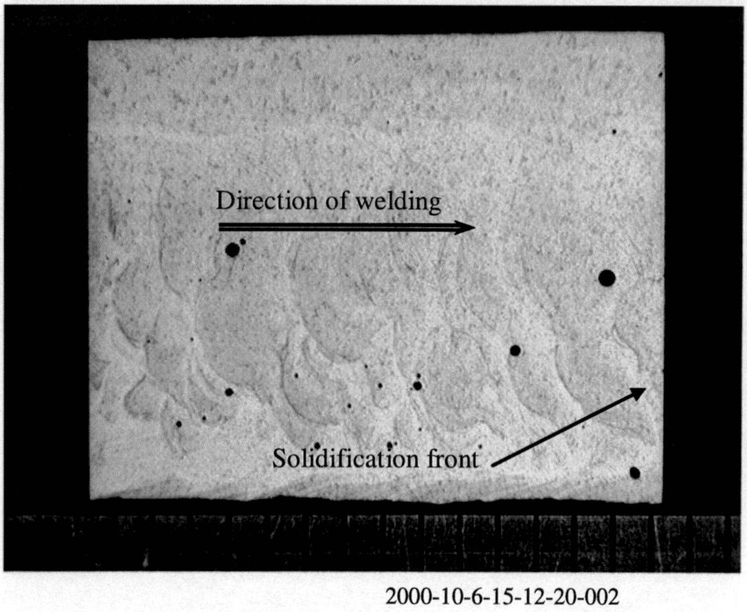

Fig. 2 Longitudinal section through a laser melt run clearly showing the solidification fronts, similar to that shown in Fig. 1. A millimetre scale is shown

this work that a reduction in welding speed from 1 to 0.8 m/min, when performing melt runs in 11.6 mm thick plate with a 10 kW laser at full power, significantly reduced the occurrence of solidification cracking in all steels tested. Although no examples of the solidification structure were given in the report, it has subsequently been observed that the amount of equiaxed grains at the centre line increased with reducing welding speed (Fig. 3).

An additional consideration in the prevention of solidification cracking is the weld profile, and hence the depth to width ratio. A low depth to width ratio reduces the risk of cracking. This again was demonstrated by the weld microstructures from the previous work (Fig. 4). Also straight sided welds are considered to be more desirable than welds which bulge in through thickness direction. It is suggested by Kristensen and Krarup [11] that under some circumstances, the molten pool at the upper surface of the work-piece cannot feed sufficient molten metal into the cavity below, as there is no continuous molten channel available.

Fraser and Metzbower [12] performed a detailed study on laser weld solidification structure and their effects on fatigue crack propagation. The observations on solidification structure are considered to be pertinent in the current study. Quoting the observation of these investigators "The direction of maximum grain growth is normal to the fusion boundary and parallel to the direction of heat flow. As the contour of the fusion boundary changes so does the direction of grain growth. In the

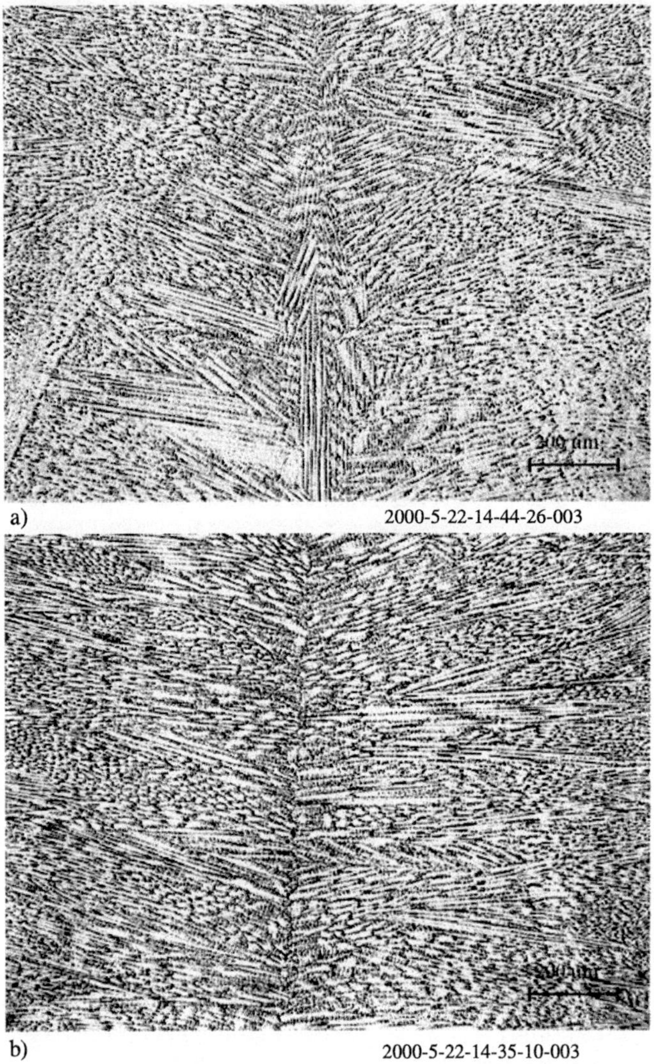

Fig. 3 Photomicrographs showing the plan view of the centre line of laser melt runs in steel. (**a**) Welding speed = 0.8 m/min. The microstructure at the centre line consists of a region of equiaxed grains and columnar grains parallel to the welding direction. (**b**) Welding speed = 1 m/min. The microstructure at the centre line consists of columnar grains meeting head on, perpendicular to the welding direction

broad shallow pool at the top of the fusion zone, the columnar grains extend upward as they grow from the base plate toward the weld centre line. In the remainder of the fusion zone, the fusion boundary is relatively straight and, for the most part, the columnar grains grow parallel to each other as the solidification front advances toward the weld centre line". These observations would seem to be borne out by

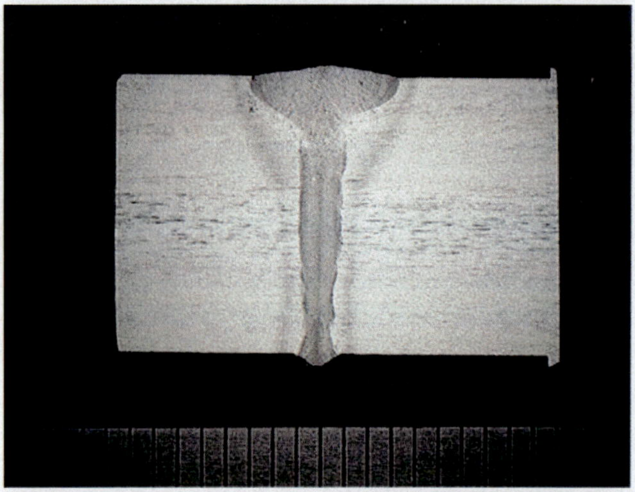

a) 2000-5-19-15-5-57-002

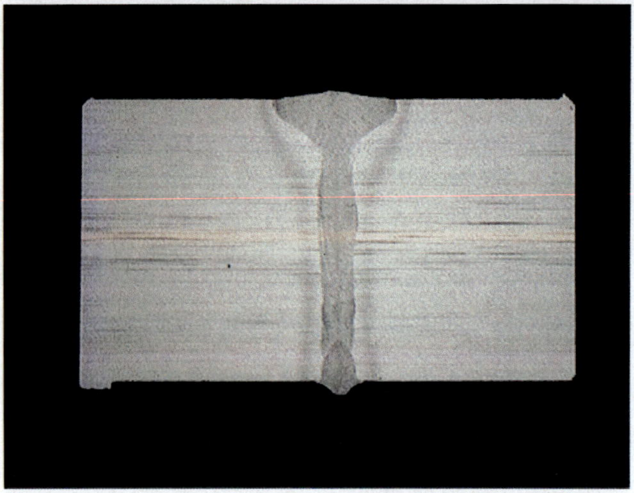

b) 2000-5-22-11-23-32-002

Fig. 4 Photomicrographs showing the transverse view of the laser melt runs in steel. (**a**) Welding speed = 0.8 m/min. (**b**) Welding speed = 1 m/min

those made from the previous work. In work performed by David and Liu [13] on the laser welding of thorium-doped iridium alloys, it was observed that, in the two dimensions at least, the weld pool forms one of two distinct shapes depending on speed, with a certain degree of transition between the two (Fig. 5).

Slower speeds result in an elliptical weld pool which exhibits a continual variation in the direction of maximum thermal gradient from the fusion boundary to the weld centreline. Therefore no grains experience favoured growth direction for an extended period, and hence a structure forms in which the grains trend to curve

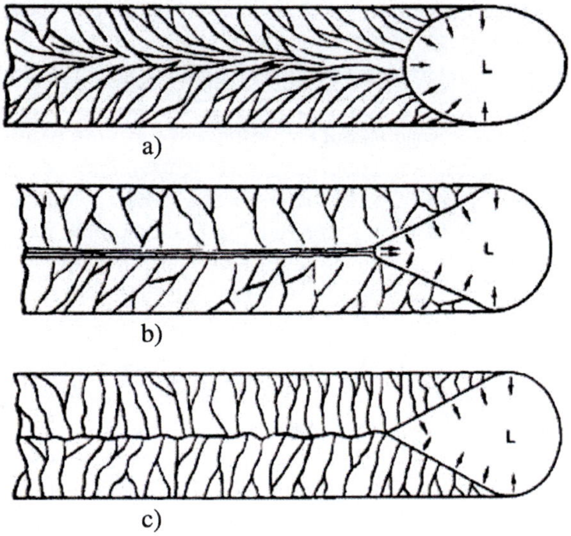

Fig. 5 Figure from ref. 13 showing fusion zone grain structure developed in: (**a**) elliptical weld puddle; (**b**) a transitional weld puddle; (**c**) teardrop weld puddle. Reproduced by permission AWS

towards the direction of welding. Conversely, at faster speeds the teardrop pool that forms provides almost an invariant direction of maximum thermal gradient at all points on the pool edge from the fusion boundary to the weld centreline. Thus there is favourable growth of grains in certain orientations at the fusion boundary at the expense of other grains, leading to a coarse columnar fusion zone structure with grains orientated towards the centre of the fusion zone.

These observations go some way to explaining the reason why reduced centreline solidification cracking was experienced in the welding trials from the previous work which used the slightly reduced welding speed. Where the grains grow towards the weld centreline, perpendicular to the fusion boundary (as in Fig. 5c) the potential for a continuous film of low melting point liquid to form at the weld centreline is higher than it would be where the grain solidification front, and hence the growth direction, is inclined to the fusion line, or even where a structure of fine equiaxed grains forms at the weld centreline, as observed in Fig. 3a. It is suggested that the fusion zone is fed by three of the five feeding mechanisms described by Campbell [14] for the feeding of castings. The five mechanisms identified are liquid feeding, mass feeding, interdendritic feeding, burst feeding and solid feeding. The last two of these mechanisms rely heavily on significant pressures present in casting that are absent from welding, developed from the weight of the reservoir of liquid metal and the restraint/clamping of the casting mould. Therefore, it is believed that the first three mechanisms, which rely less heavily on pressure, may influence the feeding of welds.

Liquid feeding generally precedes the other forms of feeding, and involves low viscosity liquid filling a relatively wide feed path. Mass feeding refers to the movement of a slurry of solidified metal and residual liquid, although in casting and therefore probably in welding, this mechanism is considered to be only of minor importance, but may cause difficulties by blocking the area requiring feeding, with the solid phase. Interdentric feeding describes the flow of residual liquid through a zone of interlaced dendrites. This mechanism does require a certain amount of pressure to be successful. By this time, it is likely that the low melting point phases will have segregated out and films will form in the interdendric regions. One could speculate that the solidification structure of the weld could either help or hinder the feeding process. In the situations where columnar grains meet head on, liquid feeding would be possible right up to the point at which the grains meet at the centreline. However, this presents the opportunity for low melting point phases to form a continuous film along the weld centreline. Where no favourable growth direction is observed, the amount of liquid feeding may be reduced, due to the formation of an interlaced dendritic structure. However, the shape of the liquid weld pool would provide an appropriate supply of liquid metal to aid both mass feeding and interdentric feeding. It is also possible that the orientation of the grains (end on to the weld pool) would make for more efficient interdendritic feeding. This may result in the persistence of liquid films along the columnar grain structure, leading to high susceptibility to flare type solidification cracking. It would, therefore, seem that the reduction of low melting point liquid phases by the control of the composition, would be the key to reducing solidification cracking.

The focus position of the laser with respect to the work piece can affect the weld bead profile. With the laser focused on the surface of the work piece the laser beam is at its narrowest where it impinges on the work piece. Therefore, the energy input per unit area is maximised, penetration potential is at its greatest and the weld is at its narrowest. The narrow waist of a focused laser beam can extend over several mm, so the focus position may have to be moved significantly for welding to be done with a defocused laser. Such welding would have the effect of reducing the heat input per unit area. Thus, penetration of the weld would also be reduced, and the width of the weld bead increased. Jones [15] observed that moving the focus position away from the material surface tended to produce welds with a narrower top bead but a wider weld central region, and, in general, reduced the amount of cracking in the samples tested.

Experimental Procedure

Test Method and Equipment

The externally restrained Russell test equipment (Fig. 6 and Ref. 16 and 17) was used to apply a load to the test specimen, sufficient to cause cracking in the most resistant steels initially tested. The strain gauges on top of the rig in Fig. 6 were calibrated against the load developed across the loading pins on the right, as a result of tightening the bolt on the left. The test weld was made along the central portion of the test specimen, stressed by the loading pins, from left to right, into a decreasing stress field. The primary assessment was by radiography. The laser used in the trials

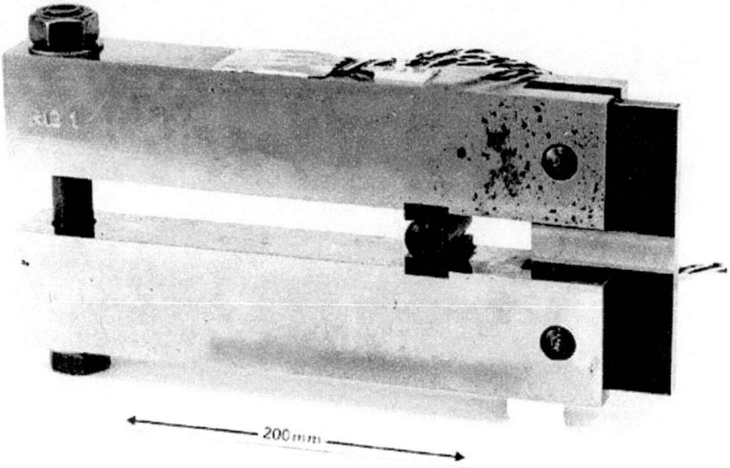

64780

Fig. 6 Photograph of the Russell test apparatus

was a UTIL 25 kW CO_2 laser. The maximum power output was actually estimated as 18 kW, and the power at the workpiece was approximately 66% of the output power, largely due to the energy absorption by the mirrors which reflected the beam towards its target. At full power, the laser emitted a beam which had a power of approximately 12 kW at the workpiece.

Materials

The compositions of the steels tested in this programme are given in Table 1. Material was sourced which covered the carbon range 0.04–0.25%. Most of the steels were C-Mn steels, with or without microalloying, covering a strength range of ~300–500 MPa yield. Two were weathering steels, with relatively high Ni, Cr, V and Cu levels, and there was one armour plate steel, with a yield strength of ~940 MPa, and relatively high Ni, Cr and Mo levels (Table 1).

The specimens tested were 150 mm long and 80 mm wide. The plates from which the samples were removed varied in thickness from 12 to 30 mm. Where necessary, material was removed equally from both sides of the plate so that the specimens were a maximum of 15 mm thick, and would fit in the testing rigs. A central strip 20 mm wide was machined across the width of the specimens, removing material equally from both sides of the plate, to make the area for the melt run 11.6 mm thick for all specimens. This also had the benefit of removing all scale from the surface of the specimens.

Table 1 Chemical compositions of the steels tested

Steel	Used in phases	Element, wt% C	Mn	Si	S	P	Ni	Cr	Mo	V	Nb	Cu	Al	O	N	Crack type [a]
1B4017	III	0.18	*0.63*	0.21	*0.024*	*0.029*	*0.005*	*0.008*	<0.003	<0.001	<0.002	0.011	0.005	*0.0055*	0.0085	A/B
1B4050	III, IV	0.20	*0.48*	0.21	*0.049*	*0.035*	0.180	*0.100*	*0.019*	0.002	<0.002	0.200	0.029			B (L)
1B4051	III	0.20	*0.47*	0.20	*0.050*	*0.029*	0.170	0.051	*0.016*	0.002	<0.002	0.250	0.021			A/B (L)
1B4052	III	0.20	0.93	0.24	*0.021*	0.021	*0.004*	*0.005*	<0.003	0.001	<0.002	0.007	0.028			B
1B4053	III	0.17	0.86	0.26	*0.020*	0.021	*0.004*	*0.006*	<0.003	0.001	<0.002	0.007	0.026			A
1B4054	III	0.14	0.94	0.22	*0.018*	*0.030*	*0.004*	*0.007*	<0.003	0.002	<0.002	0.008	0.020			A/B
1B3200	I, II, III, IV, V	0.22	0.84	0.32	0.012	0.015	0.020	0.040	<0.005	<0.002	<0.002	0.030	0.007	0.0021	0.0032	(B)
1B689	III	0.11	*1.35*	0.34	*<0.002*	0.010	0.300	0.070	*0.020*	<0.002	*0.016*	0.230	0.023	0.0010	0.0070	A
1B469	III	*0.08*	*1.80*	0.41	0.002	0.015	0.020	0.020	0.010	<0.002	*0.039*	0.020	0.029	0.0004	0.0023	A
1A287	III	0.19	0.89	0.20	*0.045*	0.009	0.030	*0.010*	0.005	<0.002	<0.002	0.030	0.005	*0.0131*	0.0036	A/B
1B833	III	0.16	*1.36*	0.32	0.011	0.015	0.020	0.020	<0.005	<0.002	*0.023*	0.020	0.028	0.0010	0.0033	A/(B)
1B834	III	0.15	1.04	0.34	0.009	0.009	0.230	*0.680*	<0.005	*0.070*	<0.002	*0.340*	0.031	0.0011	0.0051	A
1B317	III	*0.25*	0.49	0.34	0.020	0.027	*0.640*	*1.360*	*0.320*	*0.010*	*0.004*	0.150	0.014	*0.0053*	0.0086	A/B (L)
1B819	III, V	0.16	1.11	0.23	0.006	0.014	0.220	*0.540*	<0.005	*0.050*	<0.002	0.290	0.038	0.0010	0.0030	A/B
1B681	III	*0.04*	*1.45*	0.24	0.002	0.007	*0.510*	*0.010*	0.005	<0.002	*0.034*	*0.530*	0.036	*0.0030*	0.0050	A
1B877	III, V	*0.05*	1.26	0.25	0.005	0.011	*0.510*	*0.005*	<0.005	<0.002	*0.019*	<0.005	0.034	*0.0040*	0.0046	A
1B707	III	*0.06*	*1.45*	0.30	0.000	0.010	0.230	0.020	0.010	*0.050*	*0.044*	0.010	0.043			A
1B702	III	*0.07*	*1.47*	0.29	*<0.002*	0.017	0.030	0.040	<0.005	*0.030*	*0.028*	0.010	0.034	0.0020	0.0050	A
2A221	III	*0.08*	*1.40*	0.24	0.002	0.007	0.020	*0.010*	0.005	0.002	0.002	0.005	0.024			A
L4294	IV	0.10	1.03	0.01	0.007	0.010	0.021	0.015	<0.003	<0.001	<0.002	0.001	0.026			A
Range used to develop equation in Ref. 2		0.09 0.22	0.74 0.131	0.01 0.47	0.002 0.016	0.006 0.028	0.02 0.35	0.02 0.09	<0.002 0.010	<0.002 0.005	<0.002 0.002	0.005 0.33	0.004 0.047	0.0008 0.0025	0.0022 0.0107	

TWI analysis reference numbers: S/00/33, S/98/116, S/99/43, S/92/296, S/92/287, S/85/117, S/90/12, S/89/40, S/89/96, S/89/307

Note: 1B834 and 1B819 are weathering steels; 1B317 is armour plate (yield strength 943 MPa)

Numbers in italics lie outside the range of compositions used to develop Birch's equation (Ref. 2). See last two lines of table.

[a] See Fig. 9. L indicates presence of liquation cracks in the HAZ.

Cracking Tests

The work investigated the following:

1. The effects of steel composition on centreline solidification cracking susceptibility in C-Mn steels with C contents in the range 0.04–0.25% welded with a 25 kW laser.
2. The effects of steel heat treatment condition, and hence inherent material strength, on the solidification cracking susceptibility of C-Mn steels.
3. The effects of laser focus position, relative to the workpiece surface on solidification cracking susceptibility of C-Mn steels.

Compositional Effects on Cracking Susceptibility

Melt runs had not previously been made with the 25 kW laser in Russell Test Specimens. Therefore, it was necessary to identify the welding parameters for the test welds. Initially, melt runs were made in steel 1B4050 with increasing welding speed (from 0.6 to 2.2 m/min) using the maximum laser power output, to identify a welding speed which gave an acceptable weld bead profile. Once this had been established, the restraint condition required to give cracking in selected steels (1B4017, 1B4051 and 1B4054), chosen to give a range of cracking susceptibilities on the basis of previous work [2], was determined by applying different loads (between 6 and 14 kN) and making melt runs across the width of the specimen (W4-x series of welds).

Having identified the welding parameters and restraint conditions, triplicate full penetration melt runs were made in 19 different steels with the laser at full power (approximately 12 kW at the work piece), a welding speed of 1.2 m/min and an applied load of 12 kN (W5-x and W7-x series of welds).

The majority of the test specimens (W4-x and W5-x) were welded in the same direction as used originally by Russell [17] for exploring solidification cracking in electron beam welds, namely from the most highly stressed end into a decreasing stress field, as described above. The W7-x series, however, were welded in the opposite direction.

Effects of Parent Plate Strength

Two samples from each of three selected steels (L4294, 1B4050 and 1B3200) were heat treated differently. One of each was heated to 890°C and then quenched in water, whereas the other was annealed by heating to 950°C followed by slow cooling. Samples of these materials were subjected to a through thickness Vickers hardness survey using an indenting load of 10 kg in order to estimate any differences in strength. The heat treated specimens were then tested by applying 12 kN load and making a full penetration melt run in each using the laser at full power travelling at 1.2 m/min. All these test welds (W6-x and the comparator W5-x) were welded into a decreasing stress field.

Effects of Laser Focus Position

Three steels were selected from previous trials which gave different amounts of cracking (1B819, 1B877 and 1B4051). These were welded as previous samples had been (triplicate full penetration melt runs, 12 kW power, 12 kN load and 1.2 m/min welding speed), but with varying laser focus position with respect to the specimen surface (5 mm below the surface, 5 mm above the surface and 8 mm above the surface). All this series of welds (W8-x) were welded into an increasing stress field, by contrast with the similar sets of welds made at 0 mm focus position (W5-x series).

Evaluation

Test specimens, were examined by X-radiography so that the extent of centreline cracking could be determined by totalling the length of all cracks observed in each test piece. This allowed materials tested using the same condition to be ranked according to the relative total crack lengths.

A number of welded specimens were selected for sectioning so that comparisons could be made between different steels and different welding conditions. Sections were taken in both the transverse and longitudinal directions, prepared using standard metallographic techniques, etched in 2% nital to reveal the transformed microstructure, or a saturated aqueous solution of picric acid with a wetting agent (SASPA NANSA) to reveal the solidification structure, and examined using light microscopy. Photomicrographs were taken to record the observations of the light microscope examinations.

Results

Compositional Effects on Cracking Susceptibility

Figure 7 shows the transverse metallographic sections from the melt runs to identify the suitable welding speed. Those pictured were performed at 1.0 m/min, 1.2 m/min and 1.4 m/min. It can be seen that the slowest of the three had a large uneven underbead, whereas the fastest exhibited undercut in both the top and underbead. The weld made at 1.2 m/min was considered to be most representative of a weld shape acceptable to industry.

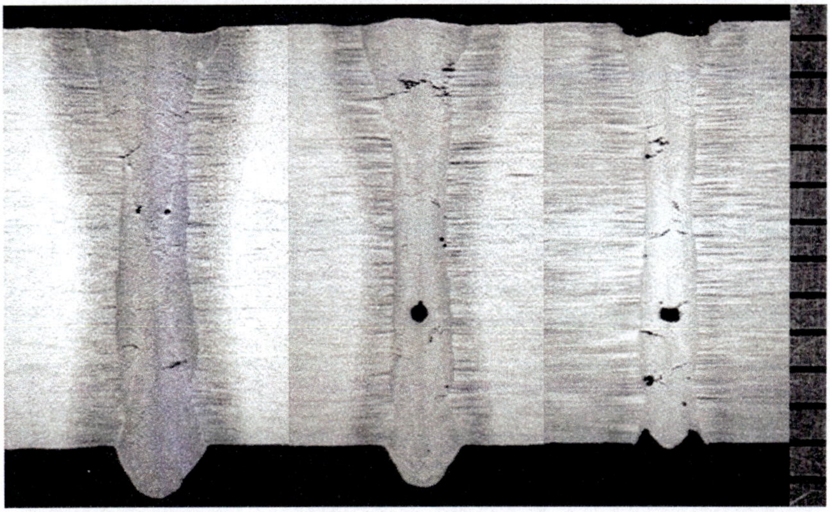

2000-7-21-15-27-56-002 a) W3.3 – 1.0m/min
2000-7-21-14-51-55-002 b) W3.4 – 1.2m/min
2000-7-21-15-10-59-002 c) W3.5 – 1.4m/min

Fig. 7 Photomicrographs of full penetrating melt runs made at varying welding speeds, as indicated. Scale in mm

Table 2 shows the results from the cracking tests. Welds W4-1 to W4-15 are those which determined the load to be applied for the test welds. The results are also shown in Fig. 8. A load of 12 kN was chosen for remaining tests as it gave cracking in 2 of the selected steels, and was also the load used for the majority of the welding trials in the previous work.

Welds W5-1 to W5-42 and W7-1 to W7-15 are the compositional based cracking tests using a laser power of 12 kW at the work piece, a load of 12 kN and a welding speed of 1.2 m/min. It can be seen that cracking was commonly intermittent along the centreline of the meltrun, that individual crack lengths ranged between 1.0 mm (the lower limit of detection) and 27.5 mm, and that total crack lengths ranged between 0 and 29.0 mm. In mean percentage terms, values ranged from 0 to 36.25%.

At least one sample from each steel tested was sectioned, transverse to the welding direction, through a crack (where present). This revealed two types of cracking, as indicated schematically and illustrated in Fig. 9. Type A cracking is purely centreline cracking, lying within the central portion of the weld. In severe cases this has propagated into the top portion of the weld. Type B cracking is "flare cracking", typically, but not exclusively lying within the top portion of the weld. In severe cases, this has also developed into a centreline crack in the top portion. The types of cracking observed were consistent for a given steel, and have been included in Table 1. Both types of cracking were interdendritic but Type B cracking (in a flare position or centreline) was generally intercolumnar, and Type A cracking was between the

Table 2 Results of the cracking tests

Weld no	Steel		Individual crack lengths, mm				Total crack length, mm	Cracking weld length %	Mean crack length, mm	Mean weld length %	Start or finish at hole end	Width ratio	
												Type A	Type B
Effect of load (6, 8, 10, 12, 14 kN)													
		Load, kN											
W4.1	1B4017	6	0.0				0.0	0.00	–	–	S		
W4.2	1B4017	8	0.0				0.0	0.00	–	–	S		
W4.3	1B4017	10	0.0				0.0	0.00	–	–	S		
W4.4	1B4017	12	0.0				0.0	0.00	–	–	S		
W4.5	1B4017	14	2.0				2.0	2.50	–	–	S		
W4.6	1B4051	6	16.0				16.0	20.00	–	–	S		
W4.7	1B4051	8	14.5	6.5			21.0	26.25	–	–	S		
W4.8	1B4051	10	20.0	2.0	1.0		23.0	28.75	–	–	S		
W4.9	1B4051	12	38.0				38.0	47.50	–	–	S		
W4.10	1B4051	14	25.0	5.5	3.0		33.5	41.88	–	–	S		
W4.11	1B4054	6	0.0				0.0	0.00	–	–	S		
W4.12	1B4054	8	1.5				1.5	1.88	–	–	S		
W4.13	1B4054	10	0.0				0.0	0.00	–	–	S		
W4.14	1B4054	12	1.5				1.5	1.88	–	–	S		
W4.15	1B4054	14	0.0				0.0	0.00	–	–	S		
Effect of composition at 12 kN load													
W5.1	1B4050		15.0	4.0	1.5	2.0	22.5	28.13	22.5	28.1	S		
W5.2	1B4050		18.0	3.0	4.0		25.0	31.25			S		1.2
W5.3	1B4050		14.0	4.0	2.0		20.0	25.00			S		
W5.4	1B4051		20.5	3.0			23.5	29.38	25.3	31.7	S		
W5.5	1B4051		27.5	1.5			29.0	36.25			S		1.0
W5.6	1B4051		15.0	4.0	2.5	2.0	23.5	29.38			S		
W5.7	1B4052		0.0				0.0	0.00	0.0	0.0	S		1.2

Table 2 (continued)

Weld no	Steel	Individual crack lengths, mm								Total crack length, mm	Cracking weld length %	Mean crack length, mm	Mean weld length %	Start or finish at hole end	Width ratio	
															Type A	Type B
W5.8	1B4052	0.0								0.0	0.00			S		
W5.9	1B4052	0.0								0.0	0.00			S		
W5.10	1B4053	1.5								1.5	1.88	0.5	0.6	S	1.4	
W5.11	1B4053	0.0								0.0	0.00			S		
W5.12	1B4053	0.0								0.0	0.00			S		
W5.13	1B4054	3.0	3.0	2.0	1.5					9.5	11.88	6.7	8.3	S	1.2	
W5.14	1B4054	4.5	1.0							5.5	6.88			S		
W5.15	1B4054	2.5	1.5	1.0						5.0	6.25			S		
W5.16	1B4017	5.0								5.0	6.25	4.0	5.0	S	1.3	
W5.17	1B4017	3.0								3.0	3.75			S		
W5.18	1B4017	3.0	1.0							4.0	5.00			S		
W5.19	1B3200	0.0								0.0	0.00	0.0	0.0	S		
W5.20	1B3200	0.0								0.0	0.00			S		1.2
W5.21	1B3200	0.0								0.0	0.00			S		
W5.22	1B689	5.0	1.0	1.0						7.0	8.75	5.7	7.1	S	1.3	
W5.23	1B689	3.5								3.5	4.38			S		
W5.24	1B689	4.5	2.0							6.5	8.13			S		
W5.25	1B469	3.0	6.5	2.5	1.5					13.5	16.88	13.7	17.1	S		
W5.26	1B469	6.5	2.0	1.0	1.5	1.0	1.0	2.0	2.0	17.0	21.25			S	1.2	
W5.27	1B469	4.5	1.0	1.5	3.5					10.5	13.13			S		
W5.28	1A287	4.5								4.5	5.63	1.5	1.9	S	1.3	1.3
W5.29	1A287	0.0								0.0	0.00			S		
W5.30	1A287	0.0								0.0	0.00			S		
W5.31	1B833	0.0								0.0	0.00	0.8	1.0	S		
W5.32	1B833	2.5								2.5	3.13			S	1.2	
W5.33	1B833	0.0								0.0	0.00			S		

Table 2 (continued)

Weld no	Steel	Individual crack lengths, mm						Total crack length, mm	Cracking weld length %	Mean crack length, mm	Mean weld length %	Start or finish at hole end	Width ratio	
													Type A	Type B
W5.34	1B834	0.0						0.0	0.00	2.0	2.5	S		
W5.35	1B834	0.0						0.0	0.00			S		
W5.36	1B834	1.0	1.0	1.5	1.5	1.0		6.0	7.50			S	1.4	
W5.37	1B819	2.5	3.5	1.0	5.0			12.0	15.00	12.5	15.6	S		
W5.38	1B819	4.5	5.5	2.0	3.0	2.5	2.0	19.5	24.38			S	1.9	
W5.39	1B819	3.5	1.5	1.0				6.0	7.50			S		
W5.40	1B317	6.5						6.5	8.13	7.2	9.0	S		
W5.41	1B317	7.0						7.0	8.75			S		
W5.42	1B317	8.0						8.0	10.00			S	1.3	1.3
W7.1	1B-681	2.0	1.0	2.0				5.0	6.25	3.3	4.2	F	1.1	
W7.2	1B-681	4.0	1.0					5.0	6.25			F		
W7.3	1B-681	0.0						0.0	0.00			F		
W7.4	1B–877	15.0	3.0					18.0	22.50	10.2	12.7	F	1.1	
W7.5	1B–877	9.0						9.0	11.25			F		
W7.6	1B–877	3.5						3.5	4.38			F		
W7.7	1B-707	3.0						3.0	3.75	3.7	4.6	F		
W7.8	1B-707	4.0						4.0	5.00			F		
W7.9	1B-707	4.1						4.1	5.13			F	1.3	
W7.10	1B-702	5.0						5.0	6.25	4.3	5.4	F		
W7.11	1B-702	3.5						3.5	4.38			F		
W7.12	1B-702	1.5	3.0					4.5	5.63			F	0.9	
W7.13	2A-221	3.5	2.0					5.5	6.88	5.2	6.5	F		
W7.14	2A-221	2.0	5.0					7.0	8.75			F	1.2	
W7.15	2A-221	3.0						3.0	3.75			F		

Table 2 (continued)

Weld no	Steel		Individual crack lengths, mm				Total crack length, mm	Cracking weld length %	Mean crack length, mm	Mean weld length %	Start or finish at hole end	Width ratio	
												Type A	Type B
Effect of strength (Quenched, as received and Annealed)													
W6.1	L4294	Annealed	1.0	1.0	1.5	2.0	5.5	6.88	–	–	S	1.5	
W6.2	L4294	Quenched	0.0				0.0	0.00	–	–	S		
W6.3	1B4050	Annealed	10.5	2.5	4.0		17.0	21.25	–	–	S		1.0
W5.1-3	1B4050	As Recd.							22.5	28.1	S		
W6.4	1B4050	Quenched	16.0	2.0			18.0	22.50	–	–	S		1.0
W6.5	1B3200	Annealed	0.0				0.0	0.00	–	–	S		
W5.19-21	1B3200	As Recd.							0.0	0.0	S		
W6.6	1B3200	Quenched	0.0				0.0	0.00	–	–	S		
Effect of focus position, (–5 mm, 0 mm, +5 mm, +8 mm)													
		Focus, mm											
W8.1	1B-4051	–5	11.0				11.0	13.75	8.2	10.2	F	1.1	1.1
W8.2	1B-4051	–5	7.0				7.0	8.75			F		
W8.3	1B-4051	–5	6.5				6.5	8.13			F		
W8.4	1B-819	–5	6.0				6.0	7.50	6.0	7.5	F	1.1	
W8.5	1B-819	–5	2.0				2.0	2.50			F		
W8.6	1B-819	–5	4.0	2.0	4.0		10.0	12.50			F		
W8.7	1B-877	–5	5.0	8.0			13.0	16.25	9.3	11.7	F	1.2	
W8.8	1B-877	–5	7.0				7.0	8.75			F		
W8.9	1B-877	–5	2.0	6.0			8.0	10.00			F		
		Focus, mm											
W5.4-6	1B-4051	0							25.3	31.7	S		
W5.37-39	1B-819	0							12.5	15.6	S		
W7.4-6	1B-877	0							10.2	12.7	F		
W8.10	1B-4051	5	20.0				20.0	25.00	30.3	37.9	F		
W8.11	1B-4051	5	28.0	3.0	23.0		54.0	67.50			F		1.3

Table 2 (continued)

Weld no	Steel		Individual crack lengths, mm			Total crack length, mm	Cracking weld length %	Mean crack length, mm	Mean weld length %	Start or finish at hole end	Width ratio	
											Type A	Type B
W8.12	1B-4051	5	17.0			17.0	21.25			F		
W8.13	1B-819	5	2.0			2.0	2.50	3.0	3.8	F		
W8.14	1B-819	5	4.0			4.0	5.00			F	1.1	
W8.15	1B-819	5	3.0			3.0	3.75			F		
W8.16	1B-877	5	12.0			12.0	15.00	10.7	13.3	F	1.2	
W8.17	1B-877	5	3.0	8.0		11.0	13.75			F		
W8.18	1B-877	5	5.0	3.0	1.0	9.0	11.25			F		
W8.19	1B-4051	8	15.0			15.0	18.75	15.3	19.2	F		
W8.20	1B-4051	8	7.0	4.0		11.0	13.75			F		
W8.21	1B-4051	8	20.0			20.0	25.00			F		1.0
W8.22	1B-819	8	3.5			3.5	4.38	4.8	6.0	F		
W8.23	1B-819	8	3.0	5.0		8.0	10.00			F	1.2	
W8.24	1B-819	8	1.0	2.0		3.0	3.75			F		
W8.25	1B-877	8	4.0			4.0	5.00	9.8	12.3	F		
W8.26	1B-877	8	11.0			11.0	13.75			F		
W8.27	1B-877	8	8.0	2.0	4.5	14.5	18.13			F	1.1	

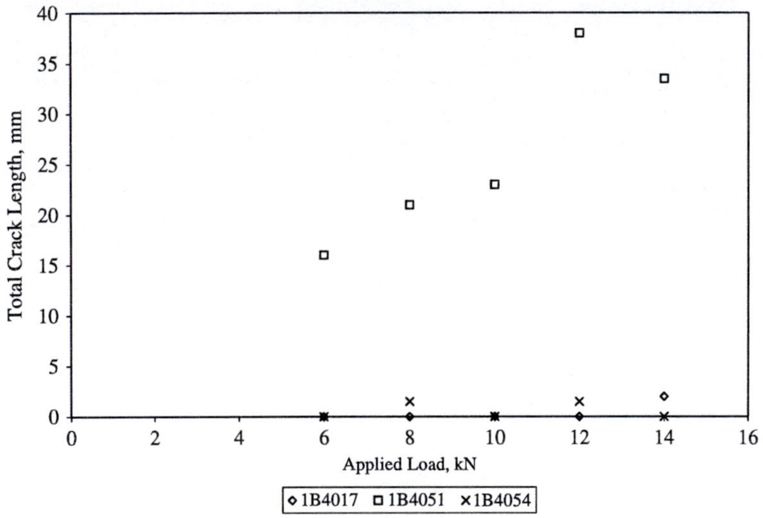

Fig. 8 Comparison of total crack length as a function of applied load

ends of dendrites, approaching from opposite sides of the crack. Examples are show in Fig. 10. It was not unusual for "flare cracks" to be aligned with pearlite bands in the parent steel, and in some cases, small liquation cracks were found adjacent to the ends of the solidification cracks (Fig. 11). Such liquation cracks were only found in steels with compositions known to be susceptible to liquation cracking in arc welding [18], as shown in Fig. 12.

Figures 13, 14 and 15 show comparisons in both plan and transverse section between steels exhibiting different amounts of cracking. Figure 13 shows both transverse and plan sections through weld W5-5, the most severely cracked sample. The plan section was taken at approximately 8 mm from the weld cap, and the figure showing this section is made up of three adjacent photographs. It can be seen that the weld exhibits both centreline and flare solidification cracking. The centreline crack appears, from the transverse section, to be concentrated in the upper third of the weld where the fusion zone is at its widest, and all the cracking is Type B, although the plan section reveals the occurrence of some centreline cracking lower down in the weld bead, which may be Type A.

Figure 14 shows both transverse and plan sections through weld W5-24. The transverse section has failed to intersect any of the cracking reported from radiographic inspection, but careful inspection of the plan section revealed the presence of fine intermittent weld centreline cracks. No flare cracking was present, and the cracking in this sample was all Type A.

Figure 15 shows the transverse and plan sections of weld W5-20 in which no cracking was found from radiographic inspection, and indeed the absence of centreline cracking was confirmed by examination of the metallographic sections. Small flare cracks (Type B) were present on the transverse section, however.

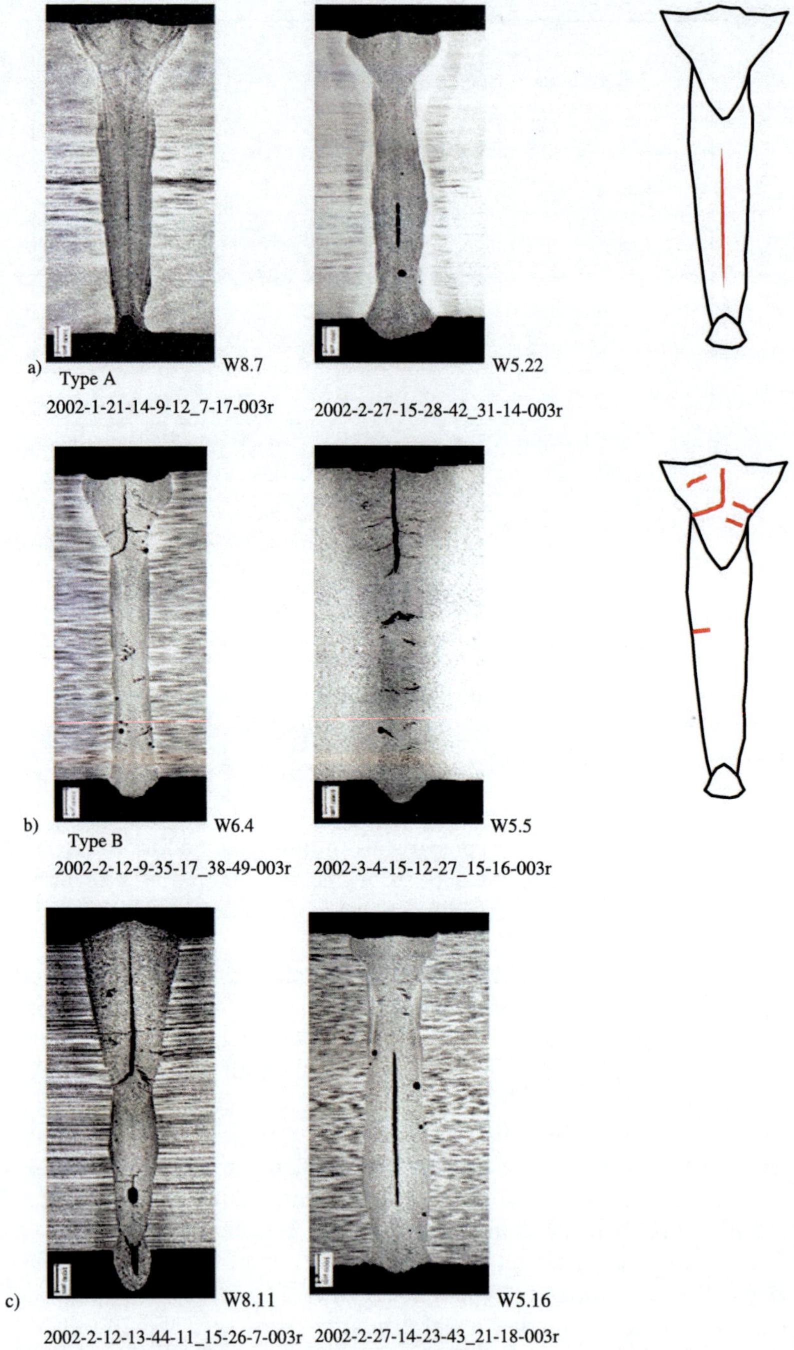

Fig. 9 Different types of cracking all etched in SASPA-NANSA. (**a**) Type A; (**b**) Type B. Magnifications given by micron marks: (**c**) Mixed types A & B

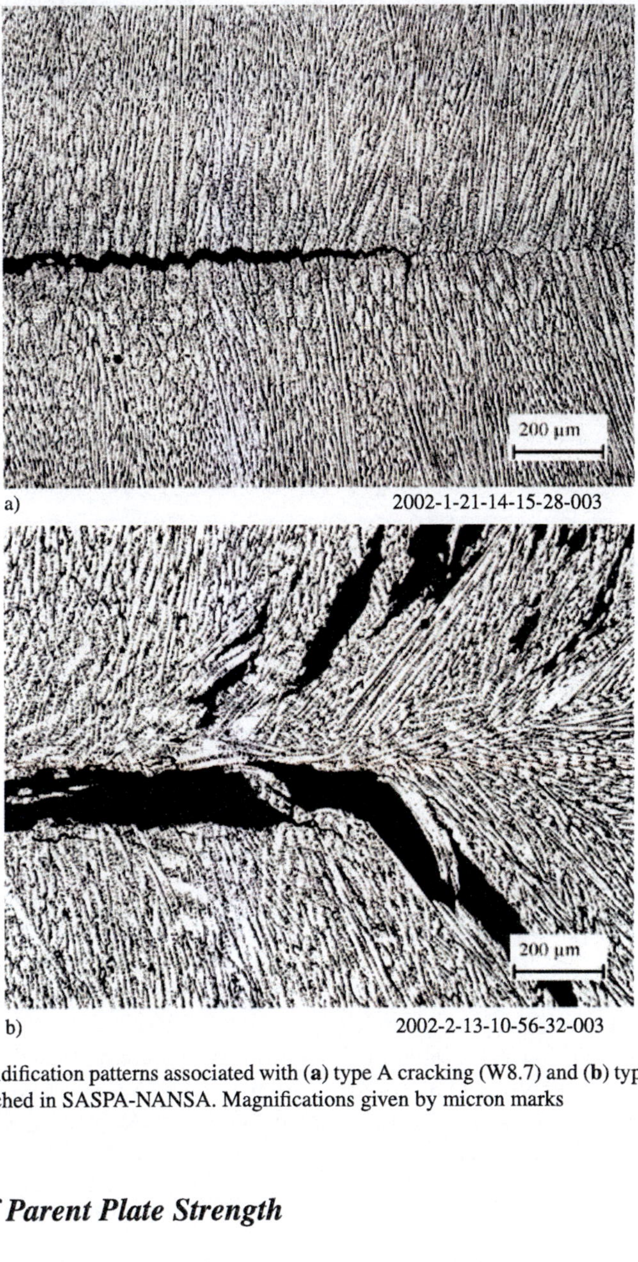

Fig. 10 Solidification patterns associated with (**a**) type A cracking (W8.7) and (**b**) type B cracking (W8.11). Etched in SASPA-NANSA. Magnifications given by micron marks

Effect of Parent Plate Strength

Table 3 shows the results of the hardness surveys performed on the heat treated samples, along with the assumed yield strength, calculated using the correlation reported by Hart [19]:

$$\sigma y = 3.25\ \mathrm{HV} - 221 \tag{2}$$

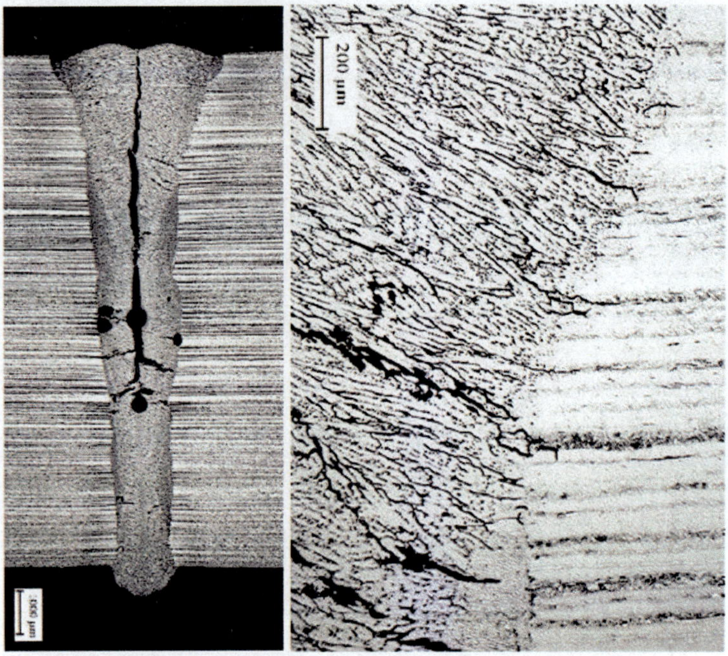

2002-2-12-8-47-47_51-2-003r 2002-2-13-11-33-18-003r

Fig. 11 Example of association of Type B cracking with HAZ liquation cracks. W8.1. Etched in SASPA NANSA, magnifications given by micron marks

Where

σy = yield strength (MPa)

HV = Vickers Hardness Number.

It can be seen that the heat treatment has resulted in significant differences in the hardness results, and hence in the assumed yield strength of the steels (up to about 300 MPa difference for 1B3200). Steel 1B4050 exhibited little or no difference in the amount of cracking as a result of this strength change, the total range of crack length being 21–28% (Table 3). Photomacrographs are presented in Fig. 16. No cracking was produced in any samples from 1B3200. Steel L4294 did exhibit some difference, with the annealed sample containing cracks totalling 6.9% of its length and the quenched sample free from cracking altogether. This is opposed to the suggestion that the lower yield point steels would experience some relief in the strain and reduce the cracking in the sample. Overall, no clear picture of the effect of strength on cracking has emerged.

Effects of Laser Focus Position

The results of tests made at different focus position (W8-x) are included in Table 2. Data from welds made in the same steels at 0 mm focus may be added to this set,

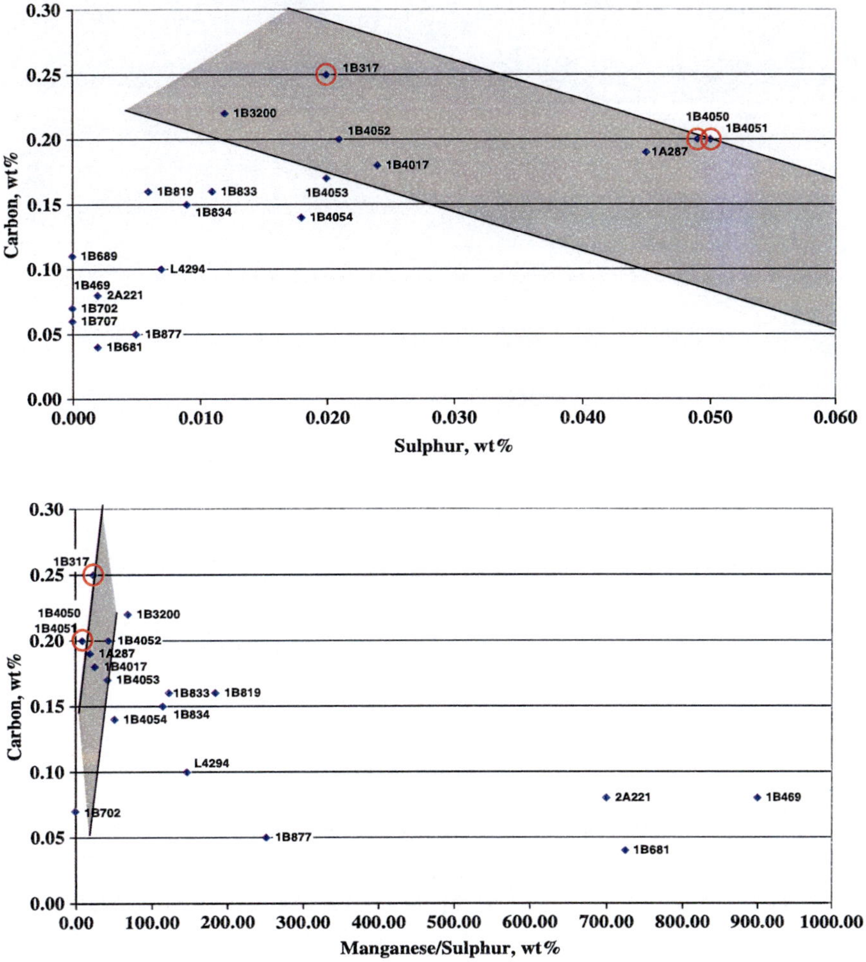

Fig. 12 Test steel compositions plotted on diagrams indicating the susceptibility to liquation cracking in arc welding (from Ref. 18), with steels which demonstrated liquation cracks in the Russell tests *circled*

but care must be exercised with regard to steels 1B4051 and 1B819, as the 0 mm welds were made in the opposite direction (into a decreasing stress field) to the W8-x welds.

Crack length has been plotted as a function of focus positions in Fig. 17. It can be seen that for the two steels which exhibit Type A cracking, there is no measurable effect of changing focus position on crack length, whereas for 1B4051, which exhibited Type B cracking, the –5 mm focus position does appear to be marginally beneficial.

Photomacrographs of transverse sections through welds made with different focus positions are presented in Figs. 18, 19 and 20. For the material in which there was an apparent effect of focus position on measured crack length (1B4051,

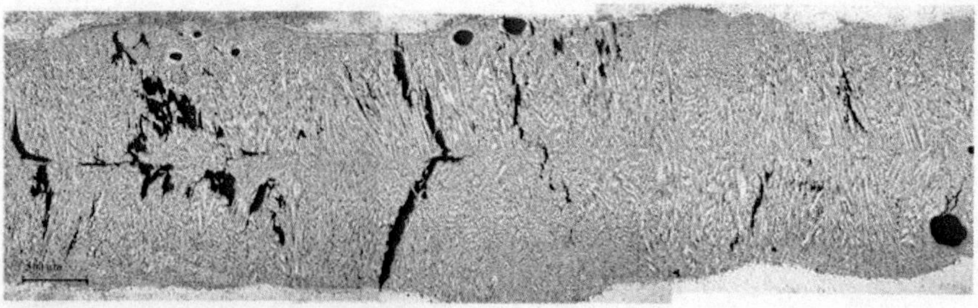

2000-7-14-14-15-5-003, 2000-7-14-14-17-45-003 and 2000-7-14-14-19-17-003

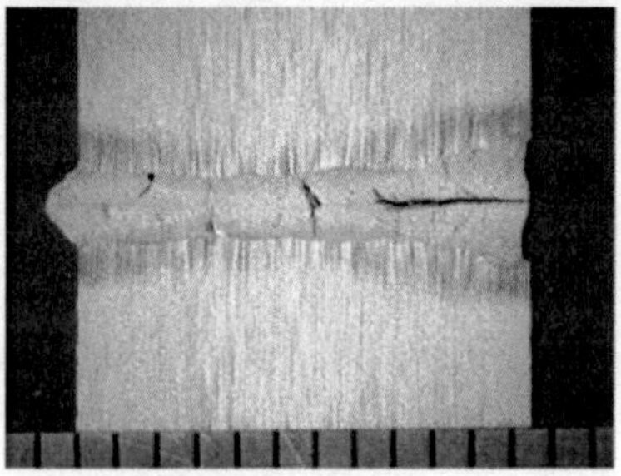

2000-7-7-13-36-1-002

Fig. 13 Plan and transverse sections of weld W5-5, etched in 2% nital. Magnifications given by micron mark and a mm scale

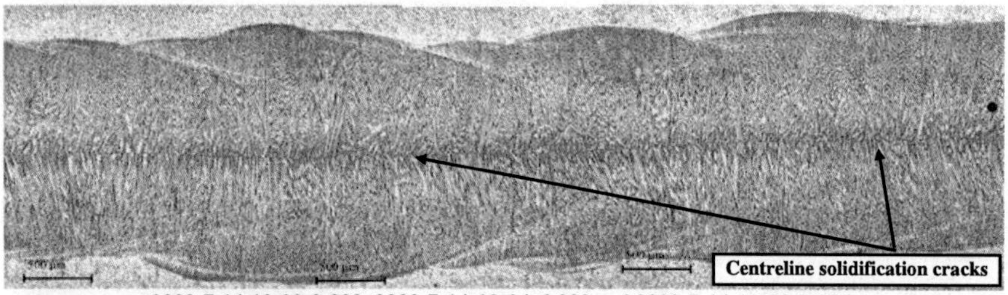

2000-7-14-13-22-0-003, 2000-7-14-13-24-6-003 and 2000-7-14-13-28-3-003

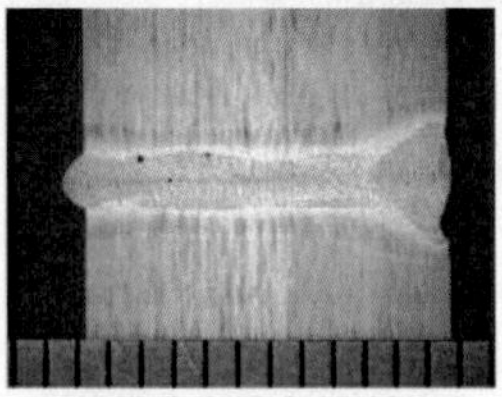

2000-7-7-12-8-12-002

Fig. 14 Plan and transverse sections of weld W5-24, etched in 2% nital. Magnifications given by micron mark and a mm scale. The plan section shows evidence of centreline solidification cracking, as marked

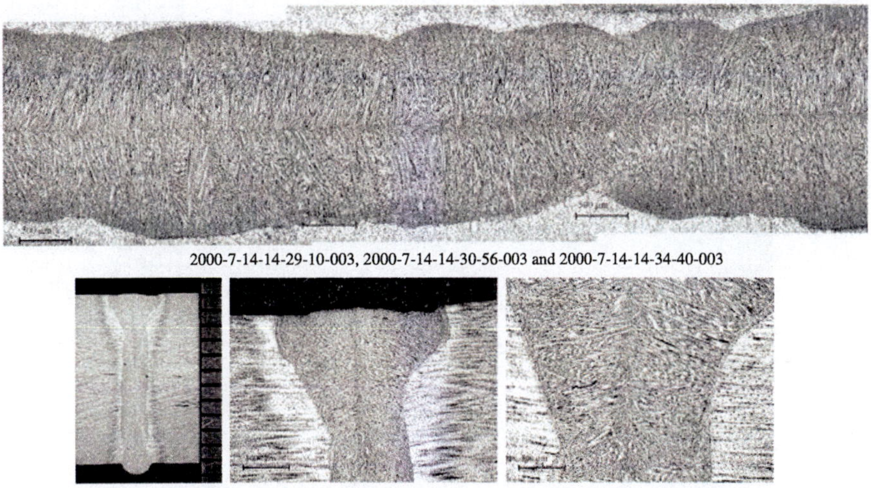

Fig. 15 Plan and transverse sections of weld W5-20, etched in 2% nital. Magnifications given by micron marks and a mm scale. No centreline cracking is evident in either plan or transverse sections. Some flare cracking is present on the transverse section

Fig. 18), focus position does not appear to have had a marked effect on either weld shape, or the through thickness extent of cracking. (This is based on only one section per weld, however). By contrast, for the two materials in which there was no measurable effect of focus position on length of cracking (1B819 and 1B877, Figs. 19 and 20), the weld width was more waisted at 0 mm and +5 mm, and had a smoother taper at –5 mm and +8 mm, with the best profile at +8 mm. The through thickness extent of cracking was also greatest at +8 mm in both these materials, however.

Table 3 Results of hardness and cracking tests performed in the heat treated steels

Steel ID	Average hardness	Assumed yield strength (MPa)[a]	Measured cracking (%)
L4294 (annealed)	110	140	6.9
L4294 (quenched)	171	340	0
1B4050 (annealed)	132	212	21
1B4050 (as received)	138	232	28
1B4050 (quenched)	176	356	23
1B3200 (annealed)	144	251	0
1B3200 (as received)	143	248	0
1B3200 (quenched)	234	547	0

[a]Calculated from Hart (Ref. 19).

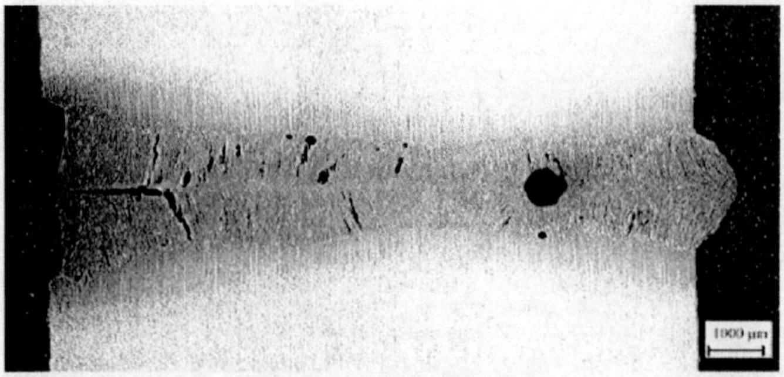

a) 2002-1-8-14-50-9_55-36-003

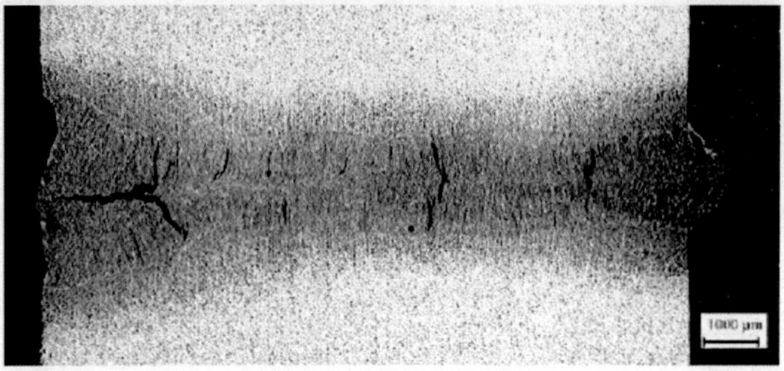

b) 2002-1-8-14-16-1_20-30-003

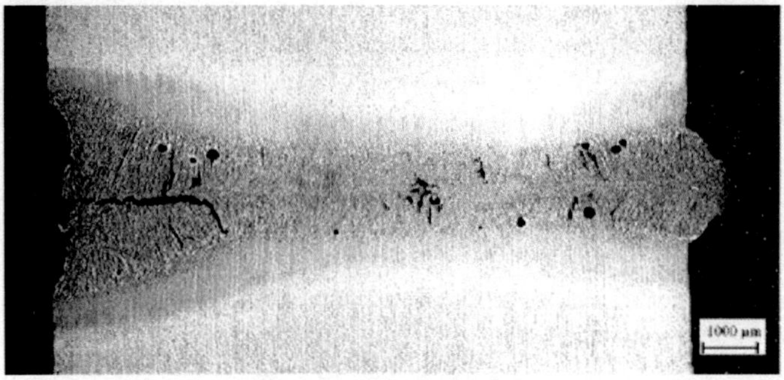

c) 2002-2-12-9-35-17_38-49-003

Fig. 16 Effect of steel strength on welds in steel 1B4050. Magnifications given by micron marks. (**a**) W6-3 annealed, average parent steel hardness = 132HV. (Etched in 2% nital.). (**b**) W5-2 as received, average parent steel hardness = 137HV. (Etched in 2% nital.). (**c**) W6-4 austenitised and quenched, average parent steel hardness = 176HV. (Etched in SASPA-NANSA.)

Discussion

The overall aim of the work was to enable combinations of steels and laser welding conditions to be selected which are resistant to solidification cracking. Following on from experience with arc and electron beam welding, it was hoped that a compositional parameter could be derived from a quantitative weldability test, and that this

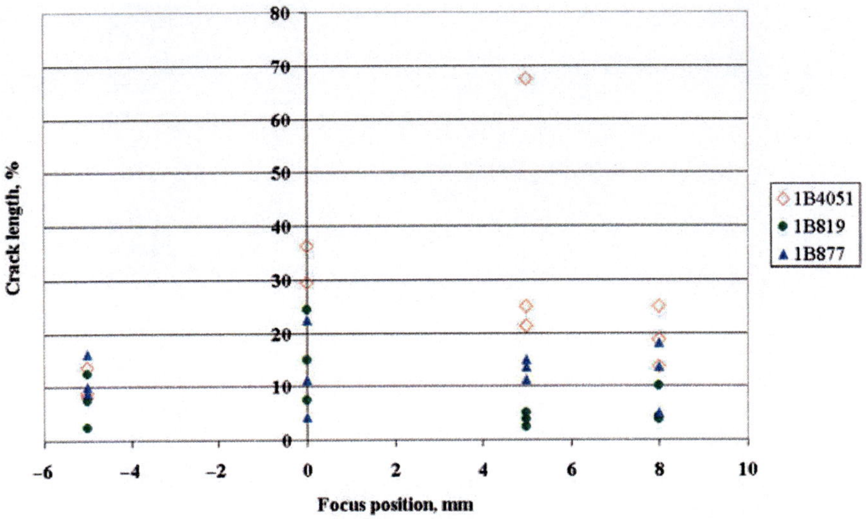

Fig. 17 Effect of focus position on cracking. 1B4051, Type B cracking (with a small amount of type A). 1B819, Type A cracking (with a small amount of type B). 1B877, all type A cracking. (Note, 0 mm focus, 1B4051 and 1B819 welded in the opposite direction to all other welds in this series)

could be incorporated in an overall scheme for the determination of safe welding conditions.

Rees and Birch have generated regression equations which have provided some measure of the effect of composition on solidification cracking in similar (lower power) laser welds, but with quite large uncertainty. The test which has been used in all this work is the Russell test, which was developed to explore solidification cracking in electron beam welds. Measurement of cracking in the test is by radiography, and thus only the total length of favourably orientated (centreline) cracking is recorded. However, in the recent work reported in this document, metallographic sectioning has revealed that there are two types of solidification cracking occurring, one of which does not always result in cracks which would be favourably orientated for radiographic inspection. It is possible that the controlling parameters for the two types of cracking are different, and thus one compositional parameter may be inadequate to predict the overall risk of solidification cracking. Since, in many steels, both types of cracking have occurred, derivation of two separate equations is complicated.

In an attempt to understand the effects of composition on the two types of cracking, the crack length results have been plotted against predicted crack length according to equations by Russell [17] (for electron beam welds) and Birch [2] (for laser welds), and also against Solidification Flaw Index (SFI) for laser welds [20], and the Unit of Crack Susceptibility (UCS) formula for submerged arc welding [21]. Each datum point has been identified as Type A or B, or A&B.

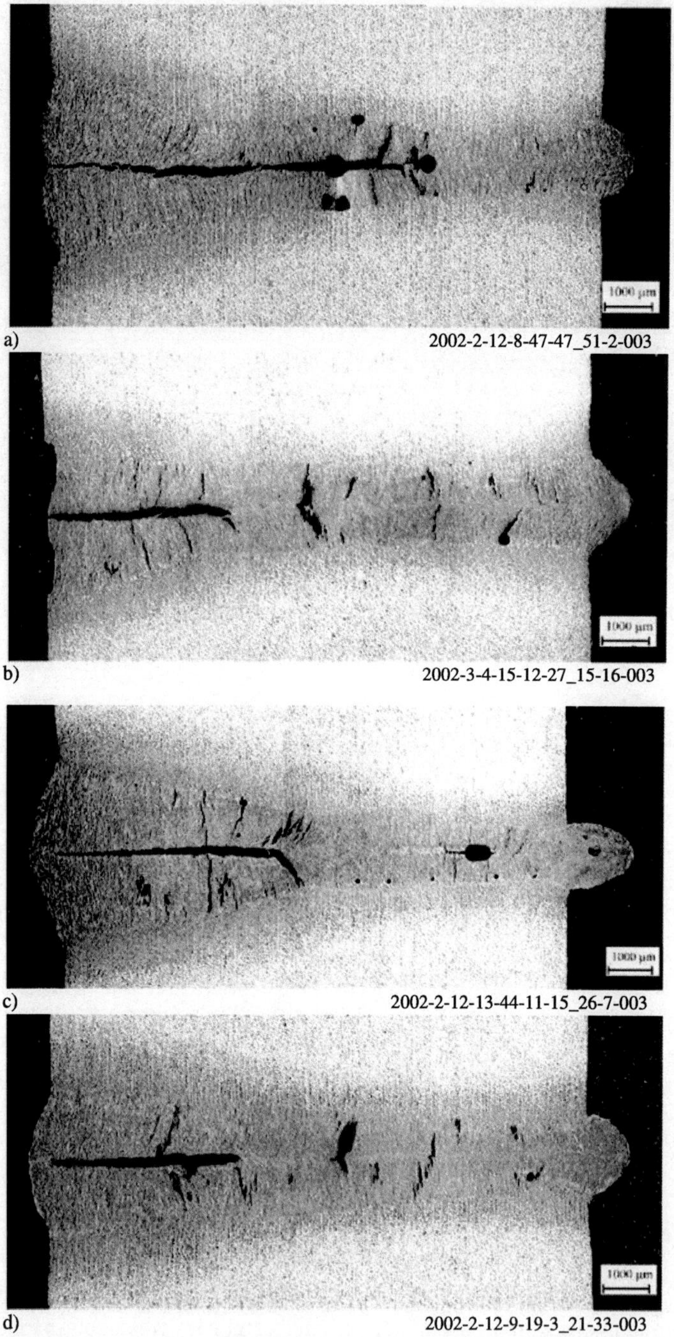

Fig. 18 Effect of focus position on welds in steel 1B4051. All etched in SASPA-NANSA. Magnifications given by micron marks. **a** W8-1, −5 mm. **b** W5-5, 0 mm. **c** W8-11, +5 mm. **d** W8-21, +8 mm

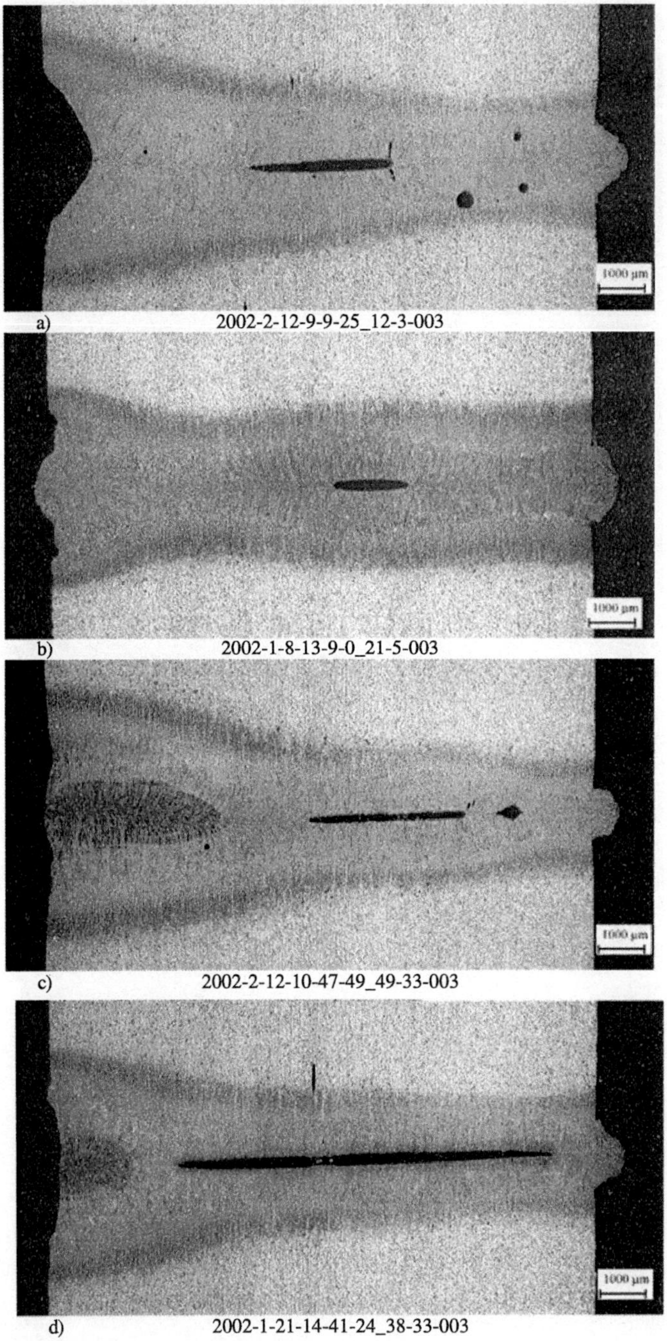

Fig. 19 Effect of focus position on welds in steel 1B819. (**a**) (**c**) & (**d**) etched in SASPA-NANSA. **b** etched in 2% nital. Magnifications given by micron marks
a) W8-4; b) W5-38, 0 mm c) W8-14, +5 mm d) W8-23, +8 mm

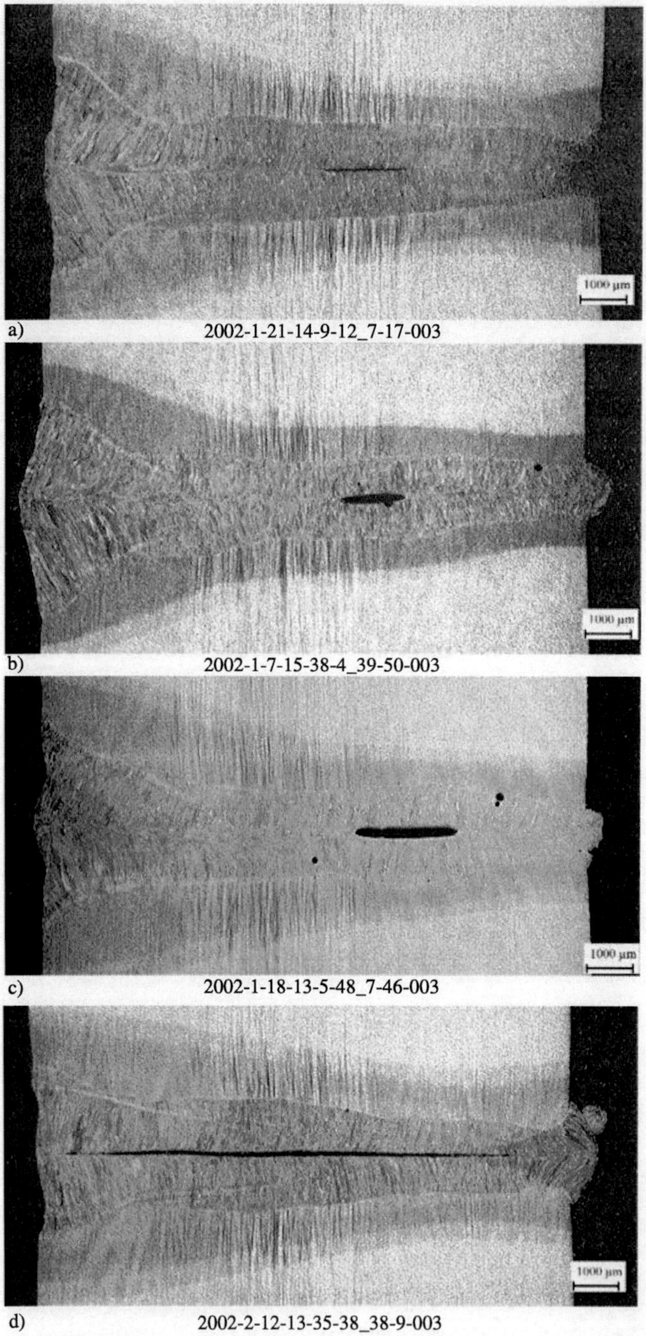

Fig. 20 Effect of focus position on welds in steel 1B877. (**a**) etched in SASPA-NANSA (**b**) and (**c**) etched in 2% nital. Magnifications given by micron marks
a) W8-7, −5 mm; b) W7-4, 0 mm; c) W8-16, +5 mm; d) W8-27, +8 mm

$$\mathrm{SFI} = \mathrm{C}\,(10\mathrm{Mn} - 200\mathrm{S} - 400\mathrm{P}) + 0.3 \tag{3}$$

$$\mathrm{UCS} = 230\mathrm{C}^{*} + 190\mathrm{S} + 75\mathrm{P} + 45\mathrm{Nb} - 12.3\mathrm{Si} - 5.4\mathrm{Mn} - 1 \tag{4}$$

where C^* = carbon content or 0.08, whichever is higher.

A further complication in assessing the present results, however, is a difference in welding direction. Weld series W7-x and W8-x were inadvertently welded into an increasing stress field (from right to left in Fig. 6), and this would possibly have had some effect on measured crack length. In order to explore this, data generated using just the W5-x series of weld were regressed, and predicted crack length was plotted as a function of measured crack length for both these welds and the W7-x series of welds. The equation was:

$$\mathrm{L} = 12.59 - 131\mathrm{C} + 331\mathrm{S} - 378\mathrm{P} + 21.8\mathrm{Ni} \tag{5}$$

The results are plotted in Fig. 21, and the crack sizes for the W7-x series of welds extended outside the scatterband for the W5-x series, for which the regression was produced. The fit would be improved if the W7-x crack sizes were larger, and this suggests that welding into an increasing stress field has resulted in shorter cracks. The correlation coefficient for the regression, which only used 14 data points, was

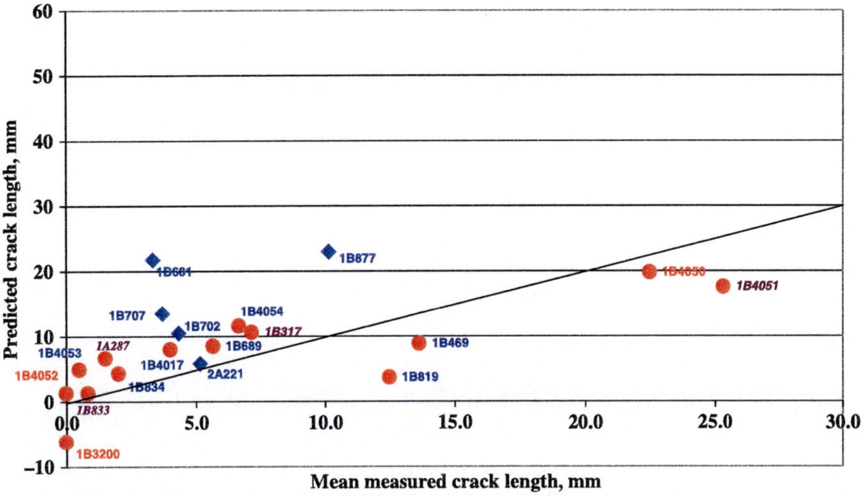

Fig. 21 Comparison of measured crack length vs. crack length predicted using a regression against chemical composition for W5-x welds (*red circles*), welded into a decreasing stress field. (L = 12.59 – 131C + 331S – 378P + 21.8Ni). All ≤0.08%C. *Blue* diamonds are data points for W8-x welds, welded into an increasing stress field. *Line* shows 1:1 predicted: measured relationship; *blue* material designations, type A cracking; *violet* italic material designations, type A & B cracking; *red* material designations, type B cracking

0.80 and it is not considered to be sufficiently reliable to use for predictive purposes, but does serve to indicate the possible effect of welding direction.

The steels used for the W7-x series, however, were all low carbon (0.04–0.08%), and another possibility is that the lower crack sizes of these welds were due to the low carbon content. The range of carbon contents was insufficient to be confident of trends within the group, however, and it is not possible to be certain whether either welding direction or carbon content below 0.09% have had a significant effect.

Figure 22 shows the present data plotted against predicted crack lengths using the equation developed by Russell [17] for electron beam welds, and that generated by Birch [2] for lower power laser welds. At first sight, neither equation provides a good fit to the current data. Looking at the test welds made into an increasing stress field, these are all unconservatively predicted by the Russell equation, and all conservatively predicted by the Birch equation. There is no clear distinction between Types A&B cracking. However, if the results from two weathering steels and the armour plate (1B834, 1B317 and 1B819) which fall well outside the compositional range used to develop the Birch equation, are omitted, a reasonable prediction is given by Fig. 22b, with one outlier. This is for steel 1A287, which has an unusually high oxygen content (0.0131%). This is over five times higher than the maximum used for the Birch equation, and the effect on fluidity could account for the apparently greater crack resistance of this steel. Overall, for steels within or close to the range used to develop the Birch equation, it gives a good prediction of cracking risk for the 25 kW laser welds (Fig. 22c).

Figure 23 shows the present data plotted against other compositional parameters for susceptibility to solidification cracking. No limiting value of Solidification Flaw Index (SFI) is recommended by Ref. 20, but it is stated that as the SFI value increases, the steel susceptibility to solidification flaws increases. No such relationship is apparent for welds containing purely Type A cracking, or Type A with also Type B. There is an apparent trend, however, for welds containing Type B cracking, with or without Type A cracking.

No relationship with Unit of Crack Susceptibility (UCS) is apparent, but Type B cracking (which is more similar to solidification cracking observed in arc welds) is restricted to compositions giving UCS values of 28 or over.

→

Fig. 22 Present data plotted as a function of crack lengths predicted by (**a**) Russell's equation for electron beam welds: $L = 52.6C + 1972S + P\,(4268C - 285) - Mn\,(1135S\text{-}9) - 21.2$ and (**b**) Birch's equation for lower power laser welds: $L = -1123C + 3096C^2 + 816S + 90$. **c** Present data, omitting W8-x welds, welded into an increasing stress field, and welds in weathering steels and armour plate, plotted as a function of crack lengths predicted by Birch's equation for lower power laser welds. Line shows 1:1 predicted: measured relationship. W5-x welds (*red circles*), welded into a decreasing stress field. W8-x welds, (*blue diamonds*), welded into an increasing stress field. *Blue* material designations, type A cracking; *violet* italic material designations, type A & B cracking; *red* material designations, type B cracking

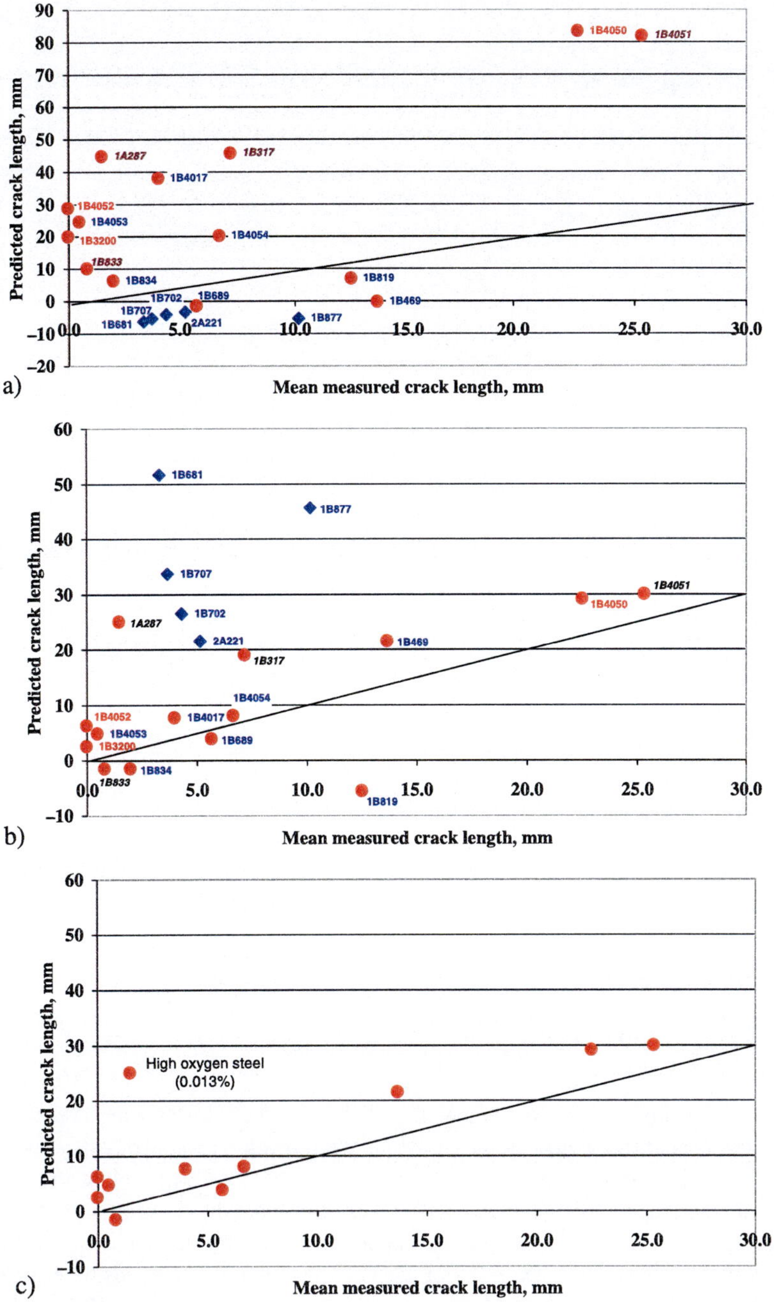

Fig. 22 (continued)

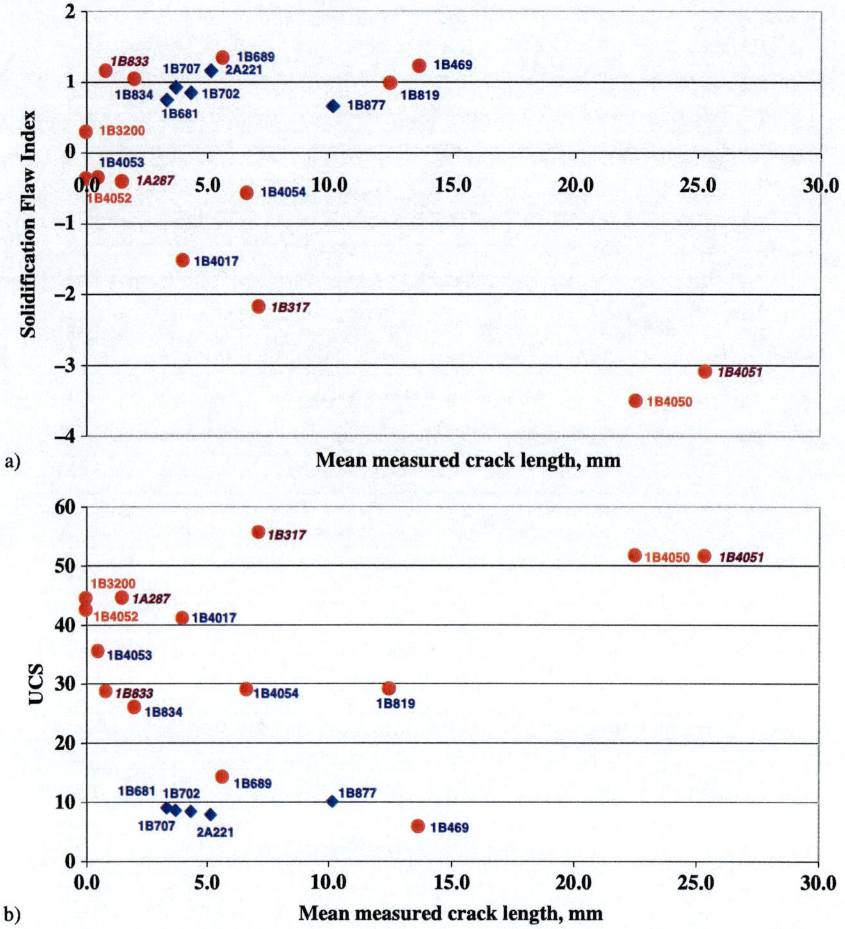

Fig. 23 Present data plotted against compositional factors: (**a**) Solidification flaw index for laser welds. (**b**) Unit of crack susceptibility for arc welds

W5-x welds (*red circles*), welded into a decreasing stress field W8-x welds, (*blue diamonds*), welded into an increasing stress field. *Blue* material designations, type A cracking; *violet* italic material designations, type A & B cracking; *red* material designations, type B cracking

Type A cracking has been connected with welding conditions which results in a narrow neck near the weld cap, and a wider weld bead near mid thicknesses [20]. Since uncracked welds have not been sectioned, it is not possible to relate width ratio to risk of cracking rigorously. However, Fig. 24 shows the distribution of width ratio (defined as the maximum width below the neck, divided by the neck width) for all cracked welds which have been sectioned. This consisted of 31 welds, of which 20 had almost entirely Type A cracking, 8 had no Type A cracking, and 3 had mixed Type A and Type B cracking. As expected, the distribution for welds exhibiting Type A cracking is skewed with respect to the overall distribution towards higher width ratios, and Type B cracking shows a very flat distribution.

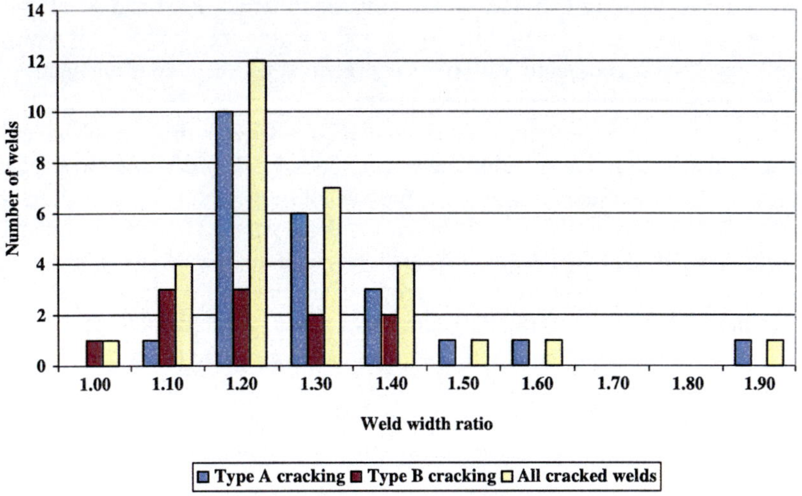

Fig. 24 Distribution of width ratio for cracked welds which have been sectioned (including welds made with different focus positions)

Looking at the transverse sections, it was evident that there was some variation in width ratio within weld series made under constant welding conditions, and possibly therefore, due to an effect of parent material composition. It is possible that chemical composition is having an indirect effect on risk of Type A cracking through an effect on width ratio. Although only one section has been taken through selected welds, an attempt was made to explore this for the W5-x, W6-x and W7-x series of welds, all of which were made using the same laser conditions. Multiple regression analysis

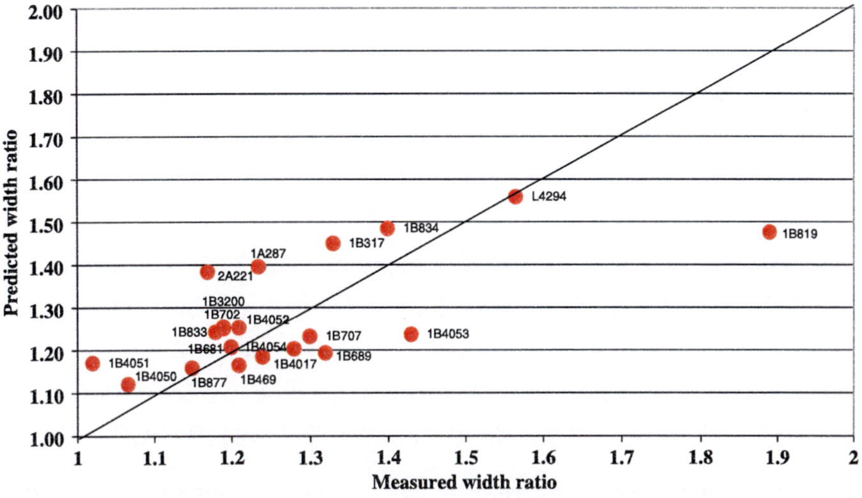

Fig. 25 Results of regression of width ratio as a function of chemical composition

did not indicate a strong correlation, but suggested that Cr, P, Si and Ni might be significant elements (C, S, P, Si, Mn, Ni, Cr and Al were included to start with). A multiple correlation coefficient of 0.67 was produced for these four elements. The regression equation has been simplified to give:

Width Ratio = 1.66 – 9.6P – 0.86Si – 0.36Ni + 0.42Cr [6] and measured vs predicted values using this equation have been plotted in Fig. 25. The physical significance of these four elements to width ratio is not immediately apparent, and it is not suggested that this particular correlation should be pursued further. However, the approach of exploring the effect of chemical composition on weld shape, using multiple sections through repeat welds to improve the database, is worth taking forward. It would also be worth exploring longitudinal variations in weld width. If some credence is given to the above equation, then it would be apparent that phosphorus and Ni are both promoting a low width ratio, which would reduce the risk of cracking. This is contrary to the metallurgical effects of these elements, and thus the overall effect on solidification cracking in laser welds may be complex.

Practical Implications

With regard to the selection of materials suitable for laser welding, the equation developed in previous work, using a 10 kW CO_2 laser, has been found to still give a reasonable indication of risk of cracking in the present work, using a higher power, although selection of a limiting value of L for any given set of welding conditions, is not straightforward. It is evident that use of the equation must be limited to steels of composition and strength within or close to the range for which the equation was developed, specifically; yield strength up to ~500 MPa, and C-Mn composition, as detailed in Table 1.

It is also apparent that compositions which give high values of UCS should be avoided, with cracking being experienced in the present work with UCS >28.

Previous work has demonstrated the importance of travel speed, with only a slight decrease in speed proving capable of eliminating cracking in tests. In the present work, the generally accepted benefits of adjusting focus position to improve weld shape have been found to have a less predictable or beneficial effect on risk of cracking.

Conclusions

A programme of weldability testing has been carried out on CO_2 laser welds in C-Mn steels using the Russell test. The following conclusions have been drawn:

1. Solidification cracking in laser welds may occur in two forms, namely Type A, restricted to the weld centreline, within the centre of plate thickness, and Type B "flare cracking", also associated with some centreline cracking in severe cases, typically, but not exclusively, lying within the top part of the weld.

2. The compositional index developed in previous work [2] has been found to be successful at predicting the susceptibility of a slightly wider compositional range of steels (than used for derivation of the index) to solidification cracking during laser welding.
3. It is not possible to identify the effect of carbon content below 0.09% with confidence from the present work.
4. There is no apparent effect of strength per se over a range of up to ~300 MPa yield on the risk of solidification cracking in C-Mn steel laser welds.
5. It is likely that there is an effect of steel composition on weld bead shape, which may influence risk of Type A cracking.
6. Type B cracking was only observed in steels with a UCS value of 28 or above.
7. Compositions susceptible to liquation cracking in arc welding are also likely to suffer liquation cracking, from which Type B cracking may grow, in laser welds.
8. Focus position only apparently had an effect on cracking in materials exhibiting Type B cracking, when the optimum position was –5 mm.
9. Welding direction in the Russell test has a small effect on the total crack length, with slightly shorter crack lengths being produced by welding into an increasing stress field.

Acknowledgements This work was funded by Industrial Members of TWI as part of the Core Research Programme. The authors acknowledge valuable input from colleagues, particularly Peter Hart on metallurgical issues and the late Derek Russell on process issues.

References

1. Russell JD (1996) High Power CO_2 Laser welding for high quality fabrication of steel structures. Proceedings of the 37th Laser Material Processing Conference, Nagoya, October 1996, pp. 19–34.
2. Birch SMI (1999) Effects of composition and welding speed on solidification cracking in C-Mn steel laser welds. TWI Members' Report 681/1999, July 1999.
3. Sugihashi A, Kido M, Kawi Y and Yano Y (2002) Study of CO_2 laser welding to suppress defects in thick steel plates welding bead. ICALEO 2002 Proceedings, Scotsdale, AZ, October 2002.
4. Engstrom H, Nilsson K, Flinkfeldt J, Skirfors A and Miller M (2001) Laser welding of thick structural steels. 8th Nordic Conference, Laser Materials Processing, Copenhagen Denmark, August 2001, pp. 93–104.
5. Harrison PL (2005, April) High power laser welding of Nb microalloyed steel plates. Science and Technology of Welding and Joining 10(2):211–219.
6. Bailey N, Jones SB (1977) Solidification cracking of ferritic steels during submerged-arc welding. The Welding Institute.
7. Gobel G and Brenner B (2006) Avoiding hot by induction based change of thermal strains during laser welding. ICALEO 2006 Proceedings, Scotsdale, AZ, October 2006.
8. Matsuda F and Ueyama T (1993) Solidification crack susceptibility of laser weld metal in 0.2C-Ni-Cr-Mo steels: Effects of bead configuration and S and P contents. Welding International 7(9):686–692. http://www.informaworld.com
9. Rees GI (1997) Compositional factors controlling solidification cracking in C-Mn steel laser welds. TWI Members' Report 623/1997, September 1997.

10. Ohshita O, Yurioka N, Mori N and Kimura T (1983, May) Prevention of solidification cracking in very low carbon steel welds. Welding Journal 62:129s–136s.
11. Kristensen JK and Krarup P (1996) Testing of laser welds in shipyards and structural steelwork. Select International Conference: Exploitation of Laser Processing in Shipyards and Structural Steelwork, Glasgow, UK, pp. 30–31, May 1996.
12. Fraser FW and Metzbower EA (1983) Solidification structure and fatigue crack propagation in lb welds. Lasers in Materials Processing – Conference Preceedings, Los Angeles, 24–26 January 1983, Paper 8301–019, pp. 196–207.
13. David SA and Liu CT (1982, May) High-power laser and arc welding of thorium-doped iridium alloys. Welding Journal 61(5):157s–163s.
14. Campbell J (1993) Castings. Publ Butterworth-Heinemann Ltd, ISBN 07506 1696 2, Ch5.
15. Jones IA (1999) Procedures for reducing solidification cracking in CO_2 laser welds in structural steels. TWI Members' Report 674/1999, March 1999.
16. Rees GI (1996) The development of a solidification cracking test for carbon manganese steel laser welds. TWI Members' Report 573/1996, October 1996.
17. Russell JD The development of an electron beam weld solidification cracking test. IIW document IV-298-81.
18. Baker RG, Dolby RE and Watkinson F (1970) The assessment of cracking problems. Conference on "Weldability of structural and pressure vessel steels". The Welding Institute.
19. Hart PHM (1975, June) Yield strength from hardness data. The Welding Institute Research Bulletin 16(6).
20. Lloyds Register of Shipping (2001) LR Draft Guidelines for Approval of CO_2 Laser Welding, Rev1, May 2001.
21. Bailey N and Jones SB (1977) Solidification cracking of ferritic steels during submerged-arc welding. The Welding Institute.

Part III
Nickel-Base Alloys

Application of Solidification Models for Controlling the Microstructure and Hot Cracking Response of Engineering Alloys

John N. DuPont

Introduction

The solidification cracking susceptibility of engineering alloys is controlled, in large part, by the solidification temperature range and amount of solute-rich liquid that exists at the end of solidification. The distribution of solute-rich liquid within the low temperature region of the solid + liquid mushy zone is particularly important for controlling cracking susceptibility. These features are, in turn, controlled by the solidification conditions and alloy composition. Solidification models can provide useful insight into the cracking behavior of engineering alloys by revealing how alloy composition affects the final amount of solute-rich liquid, the solidification temperature range, and the distribution of liquid at the edge of the mushy zone where cracking is known to occur. The application of solidification models for understanding the cracking behavior of various nickel base alloy systems is described in this article. Examples are shown for a range of model complexities, starting with a simple binary approach and working towards ternary and multicomponent solidification models.

Use of Binary Models – Application to Ni-Cr-Mo-Gd Alloys

Research is currently underway to develop thermal neutron absorbing structural materials for spent nuclear fuel applications. These materials will be used for the internal baskets that separate spent fuel assemblies and are required for structural support, spent nuclear fuel geometry control, and nuclear criticality safety. Recent research [1–3] has focused on development of Gadolinium (Gd) enriched Ni base alloys for this application. Gd can serve as an effective alloy addition for this application because it has a very high neutron absorption cross section [1, 4], thus providing the capability of handling the highly enriched fuel that cannot currently

J.N. DuPont (✉)
Department of Materials Science and Engineering, Lehigh University, Bethlehem, PA, USA
e-mail: jnd1@lehigh.edu

J. Lippold et al. (eds.), *Hot Cracking Phenomena in Welds III*,
DOI 10.1007/978-3-642-16864-2_14,

be handled with boron containing stainless steel alloys. The current plan for fusion welding of this alloy involves the use of a Gd-free commercial filler metal such as alloy UNS N06455 or UNS N06059. With this approach, the Gd content in the fusion zone will vary with weld metal dilution, and the dilution level is strongly affected by the welding parameters [5, 6]. Considering the case in which a base metal containing 2 wt% Gd is welded with a Gd–free filler metal, the concentration of Gd in the fusion zone will vary linearly with dilution and can be close to 0 wt% Gd (at low dilution values) to 2 wt% Gd (for an autogenous weld in which the dilution is 100%). The welding parameters are typically selected based on the required weld size and joint design requirements, and a wide range of welding parameters can be expected in practice. This, in turn, can produce a wide range of fusion zone Gd concentration values. The Gd concentration can also vary throughout the fusion zone of a given weld in multiple-pass welding applications.

Recent research has shown that Gd controls the solidification behavior of this alloy. In particular, the solidification temperature range and amount of terminal eutectic type constituents that form at the end of solidification are essentially dominated by the Gd concentration [3]. In addition, it has also been well established that the solidification cracking susceptibility is also strongly affected by the solidification temperature range and amount of terminal eutectic type constituent that form during solidification of the fusion zone [7, 8]. However, the influence of Gd concentration on fusion zone solidification cracking susceptibility has not yet been investigated.

Six alloys were investigated in this research. Their compositions are summarized in Table 1. The first alloy (Alloy 0) is a Gd-free UNS N06455 composition used as a basis for comparison. The next five alloys (Alloys 1–5) have increasing additions of Gd from ~0.5 to ~2.5 wt% Gd in ~0.5 wt% increments. The P and S contents of all the alloys were less than 0.001 wt%. The alloys were produced by vacuum induction melting of elemental components and cast into graphite molds approximately 10 × 10 × 30 cm. The ingots were then held at 16 h at 1,200°C and subsequently hot rolled at the same temperature to a thickness of ~12 mm. They were then air-cooled, annealed at 1,200°C for 4 h, and water quenched. Differential thermal analysis (DTA) was conducted on a Netzsch STA 409 unit using samples weighing 250–500 mg. The samples were heated and cooled through the solidification temperature range at a rate of 10°C/min, and reaction temperatures were

Table 1 Compositions of Ni-Gd-Mo-Cr alloys used for varestraint testing, differential thermal analysis, and microstructural characterization. All values shown in weight percent

Alloy	Ni	Mo	Cr	Fe	Gd	Si	C
0	Bal.	15.9	16.1	0.088	0.00	0.088	0.0065
1	Bal.	15.67	16.00	0.016	0.46	0.070	0.0052
2	Bal.	15.53	16.08	0.017	1.01	0.081	<0.005
3	Bal.	15.56	15.91	0.015	1.49	0.070	0.0064
4	Bal.	15.16	15.95	0.016	1.90	0.070	0.0065
5	Bal.	15.12	15.56	0.016	2.45	0.071	0.0066

taken as deviations from the local baseline. Samples for Varestraint testing were machined to dimensions of 165 × 25.4 × 6.4 mm. Varestraint testing was conducted with GTAW autogenous welds made at 150 A, 25 V, and a travel speed of 2 mm/s. An augmented strain of 5% was used during the tests, and two samples were tested of each alloy. The maximum crack length was used as the indicator of cracking susceptibility and was measured with a stereomicroscope. Samples from the DTA and Varestraint specimens were prepared for microstructural characterization using standard metallographic procedures. Microstructural imaging was conducted by light optical microscopy (LOM) and scanning electron microscopy (SEM), while qualitative chemical analysis was conducted by energy dispersive spectrometry.

Typical DTA curves are shown in Fig. 1. The figures show heating and cooling curves of Alloy 4 with 1.90 wt% Gd. This alloy is typical of all the alloys examined and exhibits one major peak and one minor peak during both heating and cooling. On heating, the minor peak at 1,289°C is associated with liquation of the γ/$GdNi_5$ eutectic type constituent [3]. The major peak is associated with melting of the primary γ-austenite, and the liquidus is reached at 1,397°C. During cooling, the primary γ phase exhibits significant undercooling before it begins to solidify at ~1,300°C, and solidification terminates with formation of the γ/$GdNi_5$ eutectic type constituent at ~1,250°C.

Table 2 summarizes the DTA heating and cooling data. The average values are reported from two or three tests conducted on each alloy. (The wide variation and large standard deviation values associated with the liquidus temperatures measured on cooling are attributed to undercooling.). The liquidus temperature generally decreases with increasing Gd concentration while the variation in the eutectic temperature is not as significant. Figure 2 shows an LOM photomicrograph of the DTA sample for Alloy 4 with 1.90 wt% Gd that was typical for all the alloys. As expected [3], these alloys exhibit γ dendrites and an interdendritic γ/Ni_5Gd eutectic type constituent.

Figure 3 shows the influence of Gd on the solidification cracking susceptibility. The cracking susceptibility is low for the Gd-free base UNS N06455 alloy. The cracking susceptibility then increases with increasing Gd concentration up to ~1 wt% Gd. The cracking susceptibility then decreases to a level that is similar to the base UNS N06455 alloy at Gd concentrations of 1.5–2.5 wt%. Typical microstructures of the Varestraint samples are shown in Figs. 4 through 7. Figure 4 shows that no significant crack healing due to back-filling of solute rich liquid occurred in the alloy with 1.01 wt% Gd. Similar results were obtained with the alloy that had 0 and 0.46 wt% Gd. However, a significant amount of crack healing was observed in the alloys that had additions of Gd at the 1.49 wt% level and above. Examples of this are shown in the LOM photomicrographs of Figs. 5 and 6. A typical example of the relatively large amount of γ/Ni_5Gd eutectic type constituent observed in the back-filled region is shown in the SEM photomicrograph in Fig. 7b.

The Varestraint weldability results show that the cracking susceptibility reaches a maximum at ~1 wt% Gd, and decreases with both higher and lower Gd additions. Previous research [7, 8] has shown that the solidification cracking susceptibility of engineering alloys is strongly affected by the solidification temperature range and

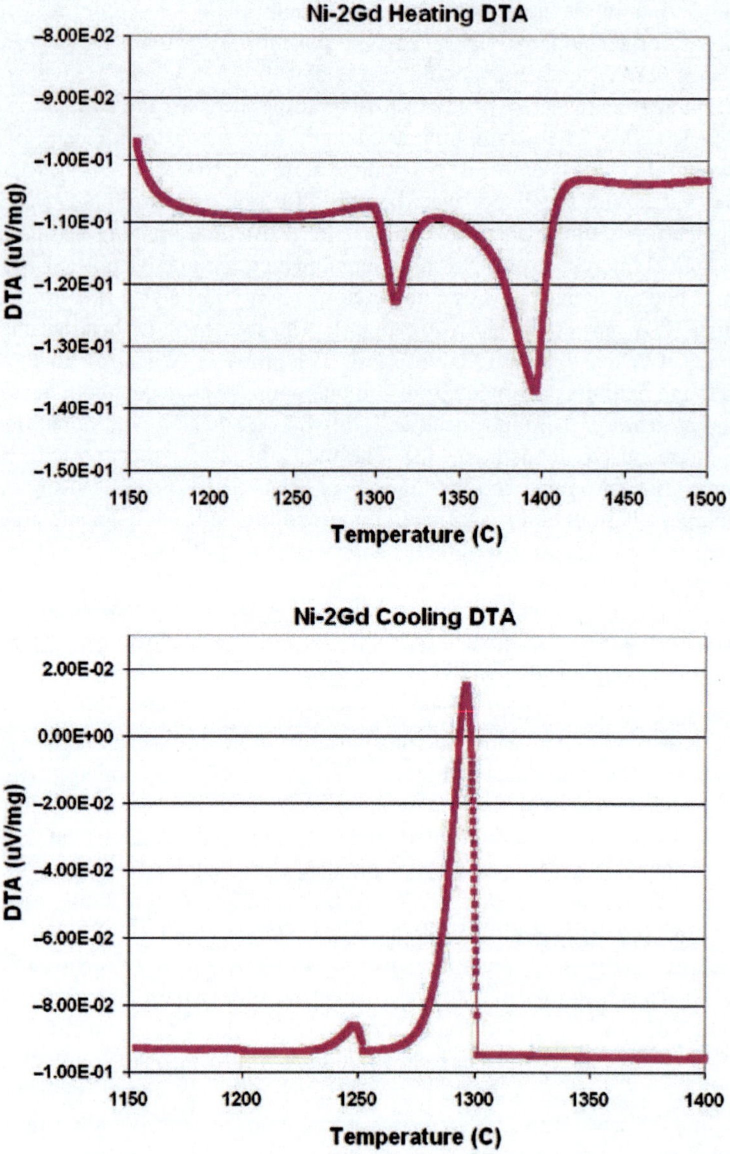

Fig. 1 Typical DTA results. Heating curve (*top*) and cooling curve (*bottom*) for alloy 4 with 1.90 wt% Gd

Table 2 Summary of liquidus and eutectic temperatures measured using differential thermal analysis. All values reported in °C

		Heating		Cooling	
Alloy	Gd Conc. (wt%)	1st Reaction	Liquidus	Liquidus	1st Reaction
1	0.46	1,298 ± 2	1,413 ± 4	1,365 ± 14	1,265 ± 4
2	1.01	1,300 ± 1	1,410 ± 3	1,313 ± 4	1,260 ± 2
3	1.49	1,293 ± 1	1,397 ± 3	1,331 ± 20	1,258 ± 4
4	1.90	1,289 ± 1	1,397 ± 2	1,345 ± 28	1,251 ± 5
5	2.45	1,291 ± 2	1,392 ± 1	1,342 ± 16	1,256 ± 6

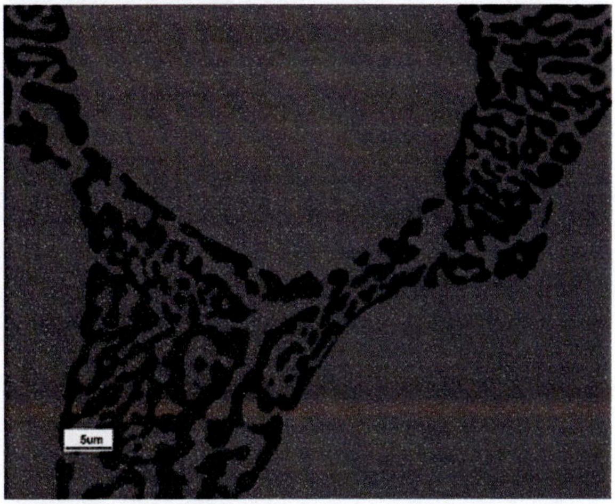

Fig. 2 LOM photomicrograph of DTA sample of alloy 4 with 1.90 wt% Gd

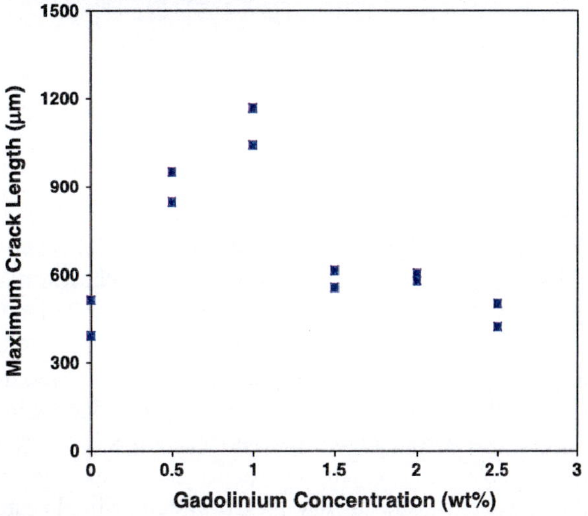

Fig. 3 Varestraint results showing maximum crack length as a function of Gd concentration

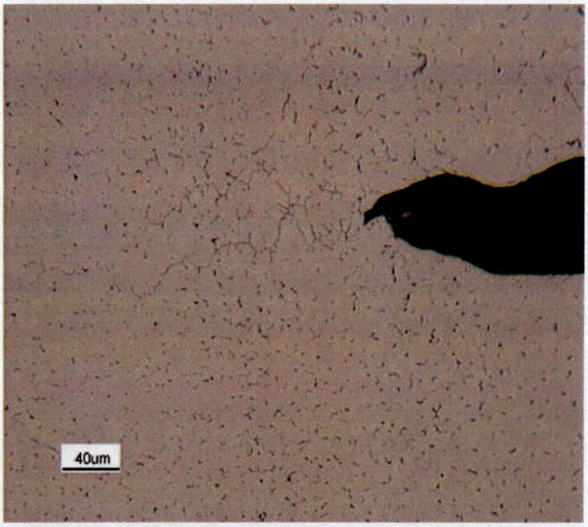

Fig. 4 LOM Photomicrograph of solidification crack induced in Varestraint sample of alloy 2–1.01 wt% Gd

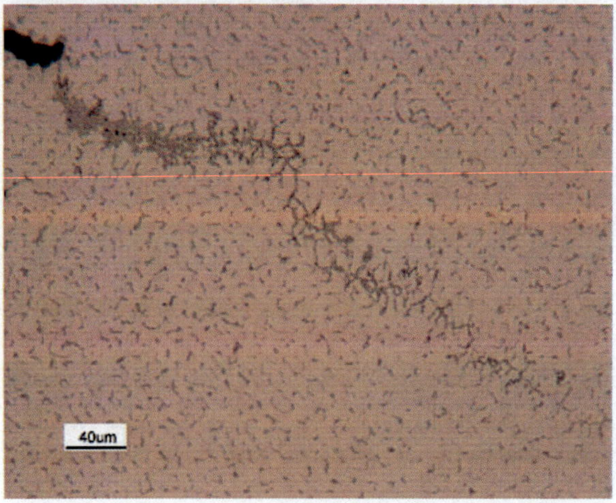

Fig. 5 LOM Photomicrograph of solidification crack induced in Varestraint samples of alloy 3 with 1.49 wt% Gd

amount of terminal eutectic type constituent that forms at the end of solidification. In light of this, it is worth summarizing the influence of Gd on the solidification behavior of these alloys. The solidification behavior has recently been described in detail [3] and will only be summarized here in relation to the weldability results.

Solidification of these alloys initiates at the liquidus temperature by the formation of primary γ-austenite. Essentially no Gd is dissolved in the austenite matrix. Thus, as solidification proceeds, the liquid becomes increasingly enriched in Gd until the Liquid $\rightarrow \gamma + Ni_5Gd$ eutectic type reaction is reached, at which point solidification is terminated. This reaction sequence and temperature range is generally similar to that expected in the binary Ni-Gd system. Simple binary Ni-Gd alloys with less than about 13 wt% Gd exhibit a similar two step solidification sequence consisting of primary austenite formation followed by a terminal eutectic reaction involving

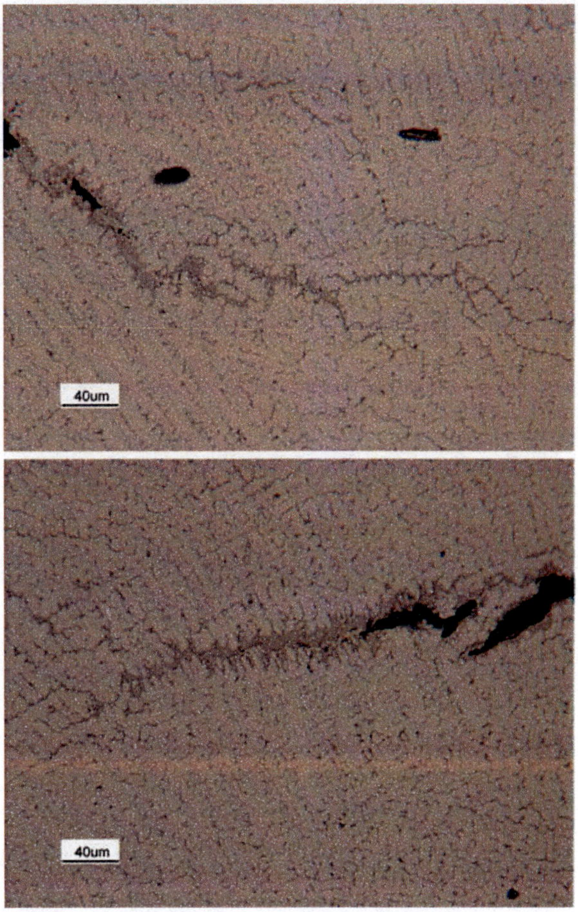

Fig. 6 LOM Photomicrographs of solidification cracks induced in Varestraint samples (*top*) alloy 4–1.90 wt% Gd, (*bottom*) alloy 5–2.45 wt% Gd

the $Ni_{17}Gd_2$ intermetallic at 1,275°C [9]. By comparison, the multi-component Ni-Cr-Mo-Gd alloys examined here complete solidification at ~1,258°C by a terminal eutectic type reaction involving the Ni_5Gd intermetallic. Thus, although the secondary phase within the terminal eutectic constituent is different in each case, the terminal reaction temperatures are very similar. In fact, as shown in Fig. 8, a pseudo binary solidification diagram has recently been developed [3] for this alloy that is similar to the phase diagram of a binary eutectic alloy. In this case, the "solvent" is represented by the Ni-Cr-Mo solid solution γ-austenite phase and Gd is treated as the solute element. The similarity of this γ-Gd binary system to a binary eutectic system is readily evident in several ways, including: the as-solidified microstructure consists of primary γ dendrites surrounded by an interdendritic eutectic type constituent in which the secondary phase in the eutectic is solute rich; the amount of eutectic type constituent increases with increasing solute content; and the proportional amount of each phase within the eutectic constituent is relatively insensitive to nominal solute content [3]. Also note that the eutectic temperature is not strongly dependent on the nominal Gd concentration. Key points of the

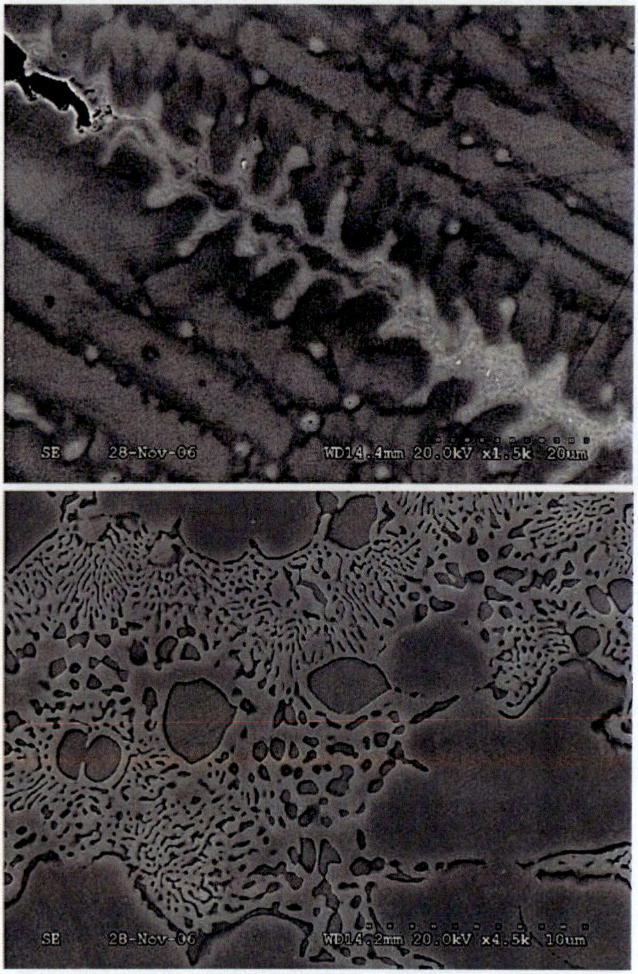

Fig. 7 *Top*: SEM photomicrograph near crack tip region of alloy 0 with no Gd. *Bottom*: alloy 3 with 1.49 wt% Gd

diagram shown in Fig. 8 were determined with a combination of thermal analysis and quantitative microstructural characterization techniques.

The pseudo binary diagram developed for these alloys is useful for interpreting the weldability results because it permits direct determination of the solidification temperature range and amount of terminal γ/Ni_5Gd eutectic constituent as a function of Gd concentration. The fraction of terminal γ/Ni_5Gd eutectic constituent (f_e) that forms during solidification can be calculated with the Scheil equation [10] via:

$$f_e = \left[\frac{C_e^{Gd}}{C_o^{Gd}}\right]^{\frac{1}{k-1}} \qquad (1)$$

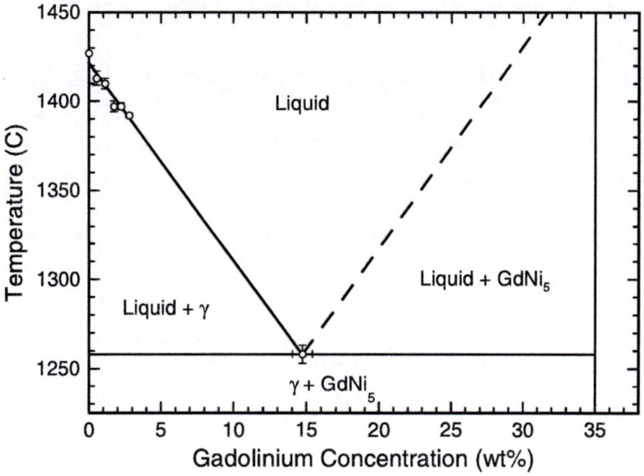

Fig. 8 Pseudo binary solidification diagram for (Ni-Cr-Mo)-Gd alloys

Where C_e^{Gd} is the concentration of Gd in the liquid at the eutectic reaction (14.7 wt% Gd, Fig. 8), C_e^{Gd} is the nominal Gd concentration, and **k** is the distribution coefficient for Gd. Equation (1) is valid for conditions in which the diffusivity of solute (Gd) in the solvent (γ) is insignificant. Previous work [2] has shown that essentially no Gd is dissolved in the γ matrix, which results in the equivalent condition of negligible solute diffusivity in the solvent. The lack of Gd solubility in austenite also indicates that **k** for Gd is 0, and equation (3) reduces simply to:

$$f_e = \frac{C_o^{Gd}}{C_e^{Gd}} \tag{2}$$

The solidification temperature range of fusion welds is best represented by the separation between the on-heating liquidus temperature (T_L) and on-cooling eutectic temperature (T_e) because solidification initiates epitaxially at the fusion line without the need for undercooling [11, 12]. As mentioned previously and discussed in more detail elsewhere [3], the eutectic temperature (T_e) does not vary significantly with Gd concentration and can be represented by an average value of $T_e = 1{,}258°C$. The liquidus line in Fig. 8 can be expressed by a simple linear equation of the form

$$T_L = T_o + m_L C_o^{Gd} \tag{3}$$

Where T_o is the melting point of the Ni-Cr-Mo "solvent" and m_L is the liquidus slope. Linear regression analysis of the phase diagram information provided in Fig. 8 leads to values of $T_o = 1{,}422°C$ and $m_L = -11.2°C/wt\%$ Gd. Thus, the solidification temperature range (ΔT) can be directly determined as a function of Gd concentration via

$$\Delta T = T_o + m_l C_o^{Gd} - T_e \tag{4}$$

The solidification temperature range is important from a weldability perspective because it controls the size of the solid + liquid "mushy" zone that trails the fully molten weld pool. Assuming a constant temperature gradient in the solid + liquid region (i.e., fixed welding parameters and sample size/geometry), the size of the mushy region is given simply by the ratio of the solidification temperature range to temperature gradient. Thus, alloys with narrow solidification temperature ranges are generally crack resistant because the mushy zone is relatively small. The influence of the amount of terminal eutectic liquid is a bit more complicated. At very low amounts of terminal liquid, the cracking resistance is generally not adversely affected because there is not enough liquid to cover the solidification grain boundaries and interdendritic regions. Thus, solid-solid boundaries can be easily established without interference from the liquid, and the cracking resistance is generally good. (The presence of surface active elements that modify the solid/liquid surface energy and promote extensive wetting of the boundaries by the liquid at even small liquid fractions can alter this general trend.) As the fraction of terminal eutectic liquid increases, the liquid more actively wets the boundaries, thus interfering with the formation of solid-solid boundaries and causing an increase in the cracking susceptibility. At considerably higher fractions of liquid, back-filling of cracks can occur by the liquid and heal the cracks as they form. This phenomenon is similar to liquid feeding by risers used to control solidification shrinkage defects in castings.

With this background in mind, the influence of Gd concentration on the cracking susceptibility, amount of terminal eutectic constituent, and solidification temperature range are summarized in Fig. 9. The lines in the bottom two figures represent the fraction eutectic and solidification temperature range, respectively, calculated with the equations above, and there is good agreement between the calculated and measured values. This supports the use of a pseudo binary analog for modeling the solidification behavior of these alloys. More importantly, the results provide a basis for developing a detailed understanding of the weldability results. At low Gd concentrations, the solidification temperature range is high, but the fraction of terminal eutectic liquid is very low. Thus, the cracking susceptibility is low. As the Gd concentration increases, there is only a slight decrease in the solidification temperature range, but a rather significant increase in the fraction of terminal eutectic liquid. This leads to the increase in cracking susceptibility shown in Fig. 9 that reaches a maximum at ~1 wt% Gd. With still increasing amounts of Gd, the amount of terminal liquid increases to the point where cracking susceptibility decreases due to back-filling of solidification cracks. The reduction in the solidification temperature range with increasing Gd concentration also assists in decreasing the cracking susceptibility. The amount of terminal liquid required to promote back-filling has been suggested to be in the range of 7–10 volume percent [13]. The results in Fig. 9 support this, where the cracking susceptibility is observed to decrease when the fraction of terminal eutectic reaches this range. This occurs when the Gd concentration reaches ~1.5 wt%. This phenomenon is also supported by direct observation of back-filling that became appreciable when the Gd reached 1.5 wt%, as shown in Figs. 4 through 7.

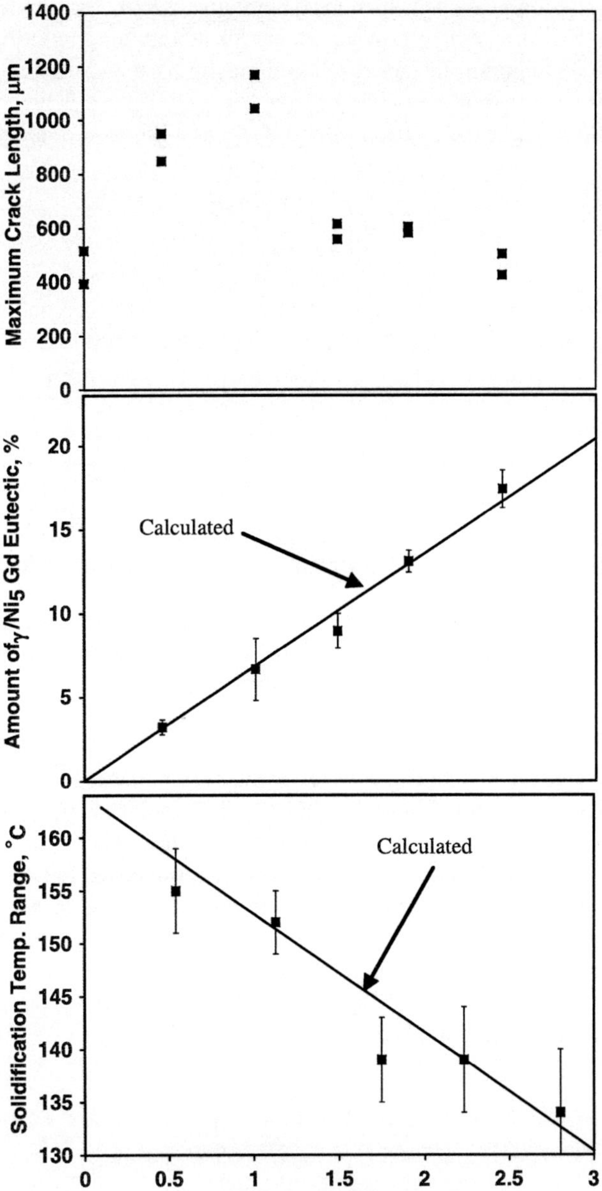

Fig. 9 Influence of Gd concentration on (*top*) maximum crack length, (*middle*) amount of γ/Ni_5Gd eutectic, and (*bottom*) solidification temperatures range

These effects can be evaluated in more quantitative detail by combining solidification theory with simple heat flow equations for determining both the size of the crack susceptible region of the solid + liquid mushy zone and variation in fraction liquid with distance within the mushy region. The well known Scheil equation can also be used to calculate the fraction of liquid (f_L) as a function of temperature via

$$f_L = \left[\frac{T_o - T}{T_o - T_L}\right]^{\frac{1}{k-1}} \tag{5}$$

Where, as before, T_o is the melting point of the Ni-Cr-Mo "solvent" and T_L is the liquidus temperature of the alloy (as affected by Gd concentration). T is the actual temperature. Noting that **k** for Gd is 0, equation (6) reduces simply to

$$f_L = \frac{T_o - T_L}{T_o - T} \quad \text{for k} = 0 \tag{6}$$

The variation in temperature within the solid + liquid region can be estimated using the Rosenthal heat flow solution [14] which, for three dimensional heat flow, is given by

$$T = T_p + \left(\frac{\eta P}{2\pi hr}\right) \exp\left[\frac{-S(r - x)}{2\alpha}\right] \tag{7}$$

Where T is the actual temperature, T_p is the preheat temperature, η is the heat source transfer efficiency, P is the arc power, **h** is the thermal conductivity of the base metal, α is the thermal diffusivity of the base metal, S is the heat source travel speed, **r** is the radial distance from the heat source, and **x** is the distance behind the heat source. The weld centerline is of most interest here because this is the location where the temperature gradient is the lowest and, as a result, where the crack susceptible solid + liquid region is the largest. This accounts for the experimental observation of the maximum crack length occurring in the weld centerline region. At the weld centerline, r = x and equation (7) reduces to

$$T = T_p + \left(\frac{\eta P}{2\pi hx}\right) \tag{8}$$

Equation (8) permits an estimate of the temperature variation (T) with position (x) behind the heat source. Thus, these equations can be combined to obtain an expression between the fraction liquid (f_L) and distance along the centerline of the weld (x)

$$f_L = \frac{-m_l C_o^{Gd}}{T_o - T_p + \frac{\eta P}{2\pi hx}} \tag{9}$$

This expression is useful because it permits direct estimation of both the variation in fraction liquid with location and the size of the crack susceptible region of the mushy zone as a function of alloy parameters (T_o, m_L, h, k, C_e^{Gd}) and welding parameters (T_o, P). The solid + liquid region will exist where the temperature is between T_L and T_e, and the fraction liquid will vary from 1 at $T = T_L$ to f_e at $T = T_e$.

Figure 10 shows $f_L - x$ curves for the Gd containing alloys evaluated here. The curves have all been shifted so that the reference point is taken as distance from the solid/liquid interface where $f_L = 1$ (instead of distance from the heat source). This permits more direct comparison between curves of different nominal Gd concentration. The terminal fraction eutectic values (f_e) calculated to form at the edge of the mushy zone for each alloy are also noted in the figure. The heat source transfer efficiency was taken from the data for GTA welding under similar currents [15], and the value for thermal conductivity was taken as an average value near the high temperature regime for an alloy with a similar Ni content [Inconel X, Ref. 16]. The power (P) used during the Varestraint test was 3,750 W, and no preheat was used so that $T_p = 25°C$. Other values were previously defined. It should be noted that the variation in temperature with position calculated through equation (8) is not expected to be highly accurate due to the assumptions invoked to arrive at the Rosenthal solution, most notably that the thermal properties are assumed constant with temperature

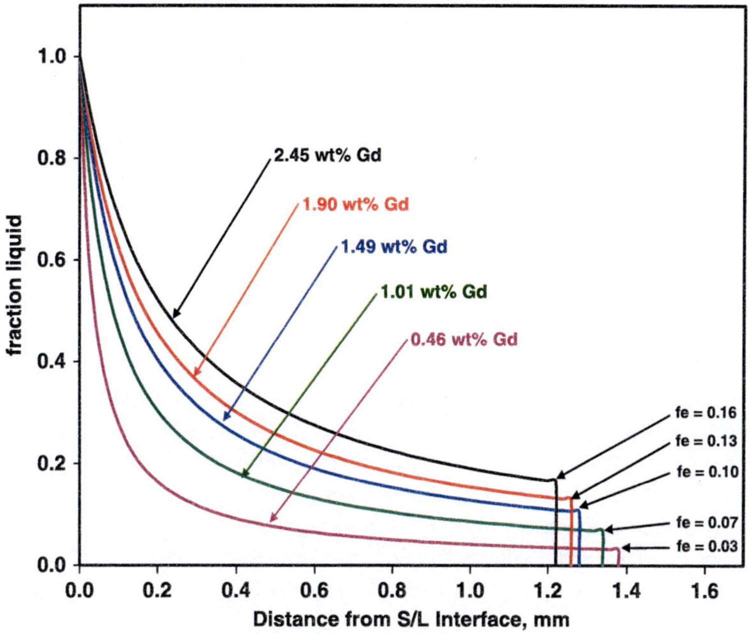

Fig. 10 Variation in fraction liquid as a function of distance from the solid/liquid interface for the B-free alloys with various concentrations of Gd

and the heat flow via convection in the melt pool is ignored. However, the objective here is to evaluate the influence of nominal composition on the solid + liquid zone characteristics and resultant weldability *under identical heat flow conditions.* In view of this, it is only required to have an estimate of the functional form of the temperature variation with the mushy zone. As demonstrated by the good comparison between experimental and calculated values of ΔT and f_e shown in Fig. 9b, c, the influence of nominal composition on mushy zone characteristics can be accurately captured with these equations. This is also confirmed by noting that the experimentally determined maximum crack lengths shown in Fig. 9a and the calculated mushy zone sizes shown in Fig. 10 are of similar size. Solidification cracks are not expected to propagate through the high temperature region of the mushy zone where the liquid fraction is high because solid-solid boundaries have not yet begun to form in these regions, but rather are constrained to a region within the mushy zone. Thus, the maximum crack length values should be less then the size of the mushy region, and this trend is certainly observed by the maximum crack lengths shown in Fig. 9a and the mushy zone sizes shown in Fig. 10.

More importantly, the calculated $f_L - x$ curves, combined with the experimental weldability results, aid in the development of a more detailed understanding on the influence of Gd on the mushy zone characteristics and resultant cracking susceptibility. Qualitatively, at the lowest Gd concentration (0.46 wt%), the mushy zone is the largest because the solidification temperature range is the widest. In addition, much of the mushy zone is occupied by relatively low f_L values. For example, the fraction liquid only varies from 0.03 at the edge of the mushy zone (at $x \approx 1.4$ mm) to 0.07 at a distance of 0.5 mm behind the solid/liquid interface, (i.e., ~65% of the mushy zone is occupied by $0.03 < f_L < 0.07$). This liquid fraction range is high enough to cause moderate cracking, but too low for appreciable back-filling and healing of the cracks. The mushy zone size for the 1.01 wt% Gd alloy is slightly reduced due to the smaller solidification temperature range. However, the higher amount of liquid present in the trailing edge of the mushy zone aggravates cracking by preventing the formation of solid/solid boundaries, and the terminal fraction liquid value of 0.07 is too low to permit healing of cracks by back-filling. This suggests that, at a terminal value of $f_e \approx 0.07$ and below, the liquid from the molten pool is cut-off from cracks that form in the mushy zone because the solid dendritic morphology is well developed.

At Gd concentrations of 1.49 wt% and higher, the f_L values never drop below 0.10 in the mushy zone. Also note that the f_L values do not decrease as substantially with increasing distance for these higher Gd alloys. Each of these factors would help promote a more continuous liquid phase between the fully molten weld pool and the solid + liquid mushy zone, thus permitting the back-filling that was observed experimentally. The decrease in solidification temperature range that occurs due to a reduction in the liquidus temperature also contributes to the improved cracking susceptibility of the alloys with Gd concentration higher than 1.49 wt%.

Use of Ternary Models – Application to IN 718

The solidification paths (i.e., variation in liquid composition during solidification) in binary alloys are relatively easy to understand because the phase diagram is planar and the liquid composition is therefore restricted to follow the liquidus line. Thus, the solidification path is easily interpreted by direct inspection of the binary diagram. However, with ternary alloys, the addition of a third element changes the planar phase diagram into a volume and the liquidus line transforms into a surface. Here, the solidification path and resultant solidification microstructure cannot be determined by direct inspection of the liquidus surface because the path will depend on the relative solute redistribution behavior of each solute in the system [6, 7].

Relatively simple solidification path equations can be derived for limiting cases of solute redistribution in ternary alloys. These equations, which are analogous to those already known and utilized for binary alloys, form a set of limiting conditions for making quantitative estimates of microstructural development in ternary alloys. For ternary solidification, three limiting cases can be identified based on the diffusivity of solute in the solid phases: 1) negligible diffusion of each solute in the solid phases, referred to as non-equilibrium solidification, 2) negligible diffusion of one solute in the solid phases and infinitely fast diffusion of the other solute in the solid phases, referred to here as intermediate equilibrium, and 3) infinite diffusion of each solute in the solid phases (equilibrium). These models are briefly reviewed below, followed by application to a commonly used Ni base superalloy, IN 718.

Solute Redistribution Equations

Two sets of solute redistribution equations are needed to fully describe the solidification paths of ternary alloys. The first set describes the variation in liquid composition and fraction liquid during the primary stage of solidification (e.g., $L \rightarrow \alpha$). These expressions can be used to quickly identify if the liquid composition is enriched to a mono-variant eutectic-type reaction and, if so, what type of reaction will occur and the fraction of total eutectic constituent that will form in the microstructure. The second set of expressions describes the variation in liquid composition and fraction liquid during the mono-variant eutectic reaction. These expressions can be used to determine if the liquid composition is enriched to the ternary eutectic reaction (e.g., $L \rightarrow \alpha + \beta + \gamma$) and the fractions of both the mono-variant (i.e., α/β or α/γ) and ternary eutectic ($\alpha/\beta/\gamma$) constituents. Only the primary solidification path expressions are discussed here. More detailed information on the full ternary model can be found elsewhere [17].

The assumptions for equilibrium condition are identical to those implemented for equilibrium binary solidification: infinite diffusion of each solute in the liquid and solid phases; equilibrium is maintained at the solid/liquid interface, and there is no undercooling during growth. Two additional assumptions are invoked here, namely that the equilibrium distribution coefficient for each element in each phase is constant and that solute redistribution of each element occurs independently. The relation between fraction liquid and liquid composition for each solute (A and B) for this condition is

$$f_L = \frac{C_{oA} - k_{\alpha A} C_{LA}}{(1 - k_{\alpha A}) C_{LA}} \tag{10a}$$

$$f_L = \frac{C_{oB} - k_{\alpha B} C_{LB}}{(1 - k_{\alpha B}) C_{LB}} \tag{10b}$$

Where f_L is the fraction liquid, C_{oj} is the nominal concentration of element **j**, C_{Lj} is the concentration of element **j** in the liquid, and k_{ij} is the equilibrium distribution coefficient for element **j** in phase **i**. An expression for the primary solidification path (i.e., relation between C_{LA} and C_{LB}) can be obtained by setting expressions (10a) and (10b) equal and solving for C_{LA}

$$C_{LA} = \frac{C_{oA}}{\left(\frac{1 - k_{\alpha A}}{1 - k_{\alpha B}}\right)\left(\frac{C_{oB} - k_{\alpha B} C_{LB}}{C_{LB}}\right) + k_{\alpha A}} \tag{11}$$

Equation (11) is the solidification path for primary solidification for equilibrium conditions and can be plotted directly on the liquidus surface. The plot is initiated at $C_{LA} = C_{oA}$ and $C_{LB} = C_{oB}$. The corresponding value of f_L at any liquid composition can be determined from equation (10a) or (10b).

When the liquid composition given by equation (11) intersects a mono-variant eutectic line, the remaining fraction liquid transforms to a binary type eutectic and possibly the ternary eutectic. Thus, the value of f_L at the intersection point of the primary solidification path and mono-variant eutectic line defines the total amount of eutectic, f_e^t, in the microstructure. Details on the calculation procedure for determining this value are provided elsewhere [17].

The assumptions invoked for intermediate equilibrium are similar to equilibrium except that one solute now displays negligible diffusion in the solid phases. The solidification path for this condition is given by

$$C_{LA} = C_{OA}\left(\frac{C_{OB} - k_{\alpha B} C_{LB}}{(1 - k_{\alpha B}) C_{LB}}\right)^{k_{\alpha A} - 1} \tag{12}$$

This expression is analogous to equation (11) and is used in a similar manner. The assumptions for non-equilibrium are similar to the conditions previously considered except that both solutes now display negligible diffusion in the solid phases, and the corresponding solidification path is given by

$$C_{LA} = C_{oA}\left(\frac{C_{LB}}{C_{oB}}\right)^{\frac{k_{\alpha A} - 1}{k_{\alpha B} - 1}} \tag{13}$$

The difference among the three cases considered here is governed by the diffusivity of solutes in the solid phases. Thus, the condition which most closely describes the solidification behavior for an actual application can be assessed by considering factors which affect the ability of the solute to diffuse in the solid. This includes the inherent diffusivity of the solute, the solidification time (as controlled by the cooling rate and solidification temperature range), and the length of the concentration gradient (~half the dendrite spacing during dendritic growth). Brody and Flemings [18] considered the influence of back diffusion in the solid during solidification and showed that diffusion in the solid phase is insignificant when the dimensionless parameter $\alpha = (D_s t_f/L^2) << 1$, where D_s is the diffusivity of solute in the solid, t_f is the local solidification time, and L is half the dendrite spacing. Clyne and Kurz [19] have shown that the Brody-Flemings model is only valid under limited solute diffusion when $\alpha < \sim 0.1$. The model breaks down at α values above this value since, in the presence of a fast diffusing element, solute is no longer conserved within the volume element considered. Clyne and Kurz proposed an alternate form of the α parameter, although they recognized this alternate form is not based on a physical model. More recently, Kobayashi [20] and Ohnaka [21] have provided a more rigorous analysis of back diffusion during solidification. These models can be used to assess the significance of solid state diffusion during solidification and to select a condition which most closely represents the application of interest.

Recent results indicate that some general comments can be made in this regard for moderate cooling rate conditions typical of casting and arc welding (these general trends may not hold under higher cooling rate conditions typical of, for example, high energy density welding and splat quenching). It is well known that diffusion rates for interstitial elements in cubic crystal structures such as BCC and FCC are orders of magnitude higher than substitutional diffusion rates, and that substitutional diffusion rates in BCC are significantly higher than those in FCC. For example, it has been shown that solid state diffusion can be assumed to occur infinitely fast during solidification for interstitial diffusion of carbon in BCC iron based alloys [19] and FCC nickel based alloys [22]. Significant substitutional solid state diffusion can occur during primary solidification of BCC phases in ferritic Fe-Cr-Al alloys [23] and stainless steels [24]. In contrast, substitutional diffusion has generally been shown to be negligible during solidification of FCC iron based [23] and nickel based alloys [22].

Application to IN 718

IN 718 is a Nb-bearing superalloy that solidifies in a manner similar to the ternary Ni-Nb-C system. This alloy forms two binary-type eutectics at the terminal stages of solidification – γ/NbC and γ/Laves. A liquidus surface that represents this system that is very similar to the Ni-Nb-C ternary system is shown schematically in Fig. 11 and displays three primary phase fields of interest – γ, NbC, and Laves. IN 718 has a nominal Nb and C content that is located within the primary γ phase field. During solidification, Nb and C are rejected to the liquid, which results in increasing Nb and C concentrations in the liquid as solidification proceeds. This occurs until the primary solidification path intersects the γ/NbC phase boundary, at which

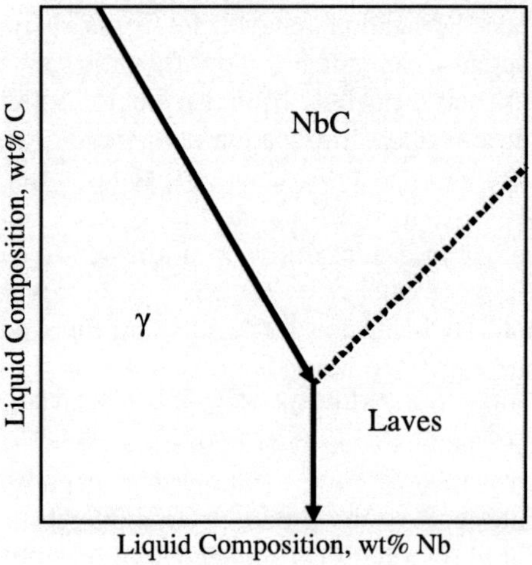

Fig. 11 Schematic illustration a liquidus projection for IN 718

point γ and NbC form by a eutectic reaction. The formation of the high C NbC phase causes C depletion in the liquid, and solidification terminates by formation of Laves phase through the L → γ + Laves reaction. Previous work [8] has demonstrated that relatively small variations in Nb and C to these types of alloys can have a significant effect on the solidification behavior and resultant weldability. In particular, C additions have been shown to increase the L → γ + NbC start temperature. This has the effect of decreasing the solidification temperature range and improving resistance to solidification cracking. These effects can be assessed with the ternary model described above.

Table 3 Summary of alloy compositions and measured and calculated L → γ + NbC) reaction start temperatures. Compositions in weight percent, temperatures in °C

Element	Heat 1	Heat 2	Heat 3	Heat 4	Heat 5
Ni	Bal.	Bal.	Bal.	Bal.	Bal.
Al	0.46	0.41	0.28	0.46	0.42
Cr	17.65	17.15	17.68	17.32	17.19
Fe	19.36	20.56	19.47	19.49	19.19
Mo	2.90	2.92	2.87	2.88	2.86
Nb	5.17	5.02	2.97	6.38	5.07
Ti	0.90	0.87	0.84	0.88	0.90
C	0.04	0.02	0.05	0.06	0.09
Measured L → γ + NbC Temp. (°C)	1,260 +/– 12	Not Detected	1,290 +/– 9	1,283 +/– 9	1,296 +/– 9
Calculated L → γ + NbC Temp. (°C)	1,260	1,237	1,297	1,264	1,294

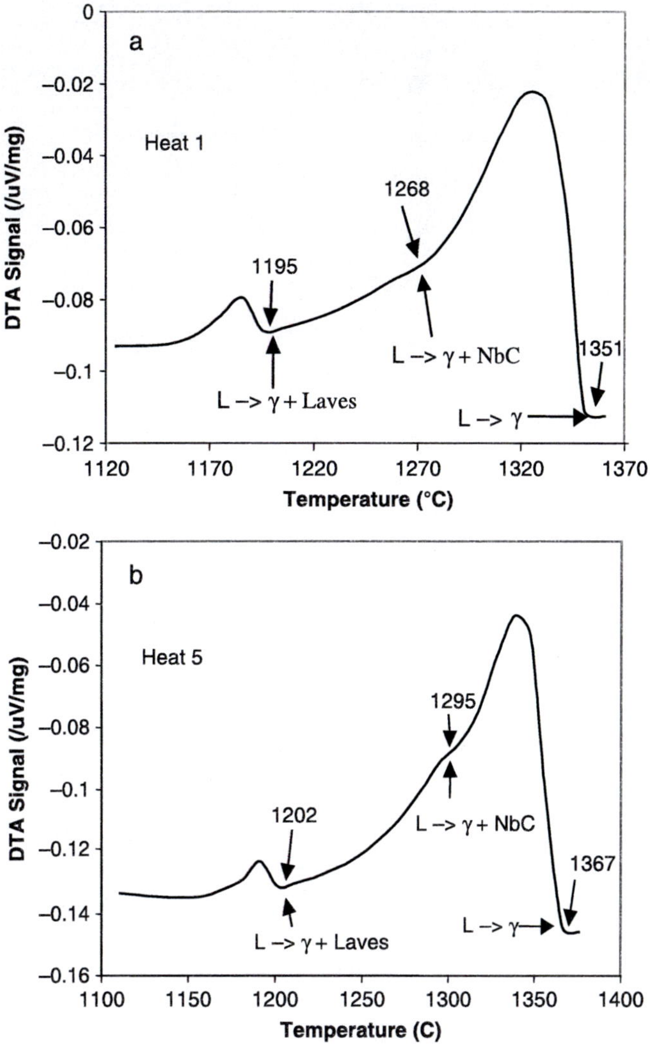

Fig. 12 DTA curves for alloys 1 (*top*) and 5 (*bottom*)

Table 3 lists the chemical composition of five heats of IN 718 that contain variations in the Nb and C concentrations. Figure 12 shows typical DTA cooling curves for Heats 1 and 5, where the three step L $\rightarrow$ γ, L $\rightarrow$ γ + NbC, L $\rightarrow$ γ + Laves reaction sequence is displayed for each alloy. However, note that a seemingly small change in C content from 0.04 to 0.09 wt% increases the L $\rightarrow$ γ + NbC start temperature by an average of 36°C. These experimentally determined reaction start temperatures are summarized in Table 3. In agreement with previous results [8], the L $\rightarrow$ γ + Laves reaction occurs near 1,200°C over a relatively small temperature

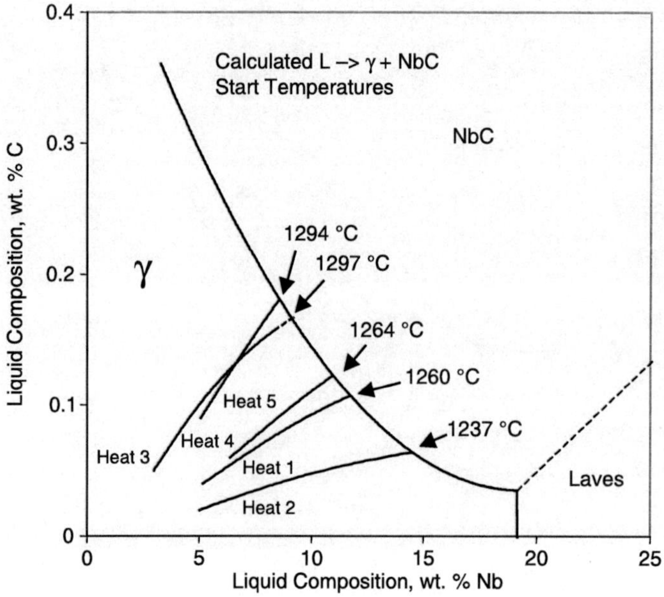

Fig. 13 Calculated primary solidification paths of five experimental alloys superimposed on the calculated liquidus projection

interval. Figure 13 shows a calculated ternary-like liquidus projection for IN 718. This diagram was computed with Thermocalc by determining the position of the lines of two-fold saturation which separate the γ, NbC, and Laves primary phase fields. Although the diagram is displayed in ternary-like fashion, it accounts for the presence of eight elements (Ni-Fe-Cr-Mo-Al-Ti-Nb-C) by approximating "γ" as an elemental constituent. The liquid composition at which the $L \rightarrow \gamma + NbC$ reaction is replaced by $L \rightarrow \gamma + Laves$ is calculated at 19.1 wt% Nb and 0.03 wt% C. These values are in excellent agreement with those previously reported as 19.1 wt% Nb and 0.04 wt% C [8].

Superimposed on the diagram are the primary solidification paths as calculated with equation (12) for intermediate equilibrium conditions in which C diffuses infinitely fast and Nb diffusion is negligible, which has been shown to be valid for this system [22]. The intersection of the primary solidification path with the line of two fold saturation separating the γ and NbC phase fields provides a predicted value for the start temperature of the $L \rightarrow \gamma + NbC$ reaction. These predicted values are summarized in Table 3, and good agreement is observed between the calculated and measured reaction temperatures. These results also explain, in quantitative fashion, the observed effect of C increasing the $L \rightarrow \gamma + NbC$ reaction temperature. With increasing nominal C concentration, the primary solidification path is directed relatively high into the C-rich side of the diagram where the $L \rightarrow \gamma + NbC$ reaction temperature is relatively high.

Multicomponent Modeling

The retaively simple models described above are useful for assessing the solidification behavior of alloys that behave in a manner analogous to binary or ternary alloys. Multicomponent thermodynamic and kinetic software are now available that can also be used for such purposes, and a simple example is provided here.

Mo-enriched super austenitic stainless steels are often used in a wide variety of environments where good corrosion resistance and toughness is required. Welds in these alloys typically exhibit inferior corrosion resistance relative to the base metal due to microsegregation of Mo, and this problem is often mitigated with the use of Ni base filler metals that are high in Mo. (e.g., IN622 and IN686). With this approach, the nominal Mo concentration of the weld is increased to help compensate for microsegregation. The nominal Mo concentration of the weld can vary considerably depending on the dilution level, which is, in turn, controlled by the welding variables. Welds in these alloys are known [25] to form sigma phase at the end of solidification, and the toughness and solidification cracking resistance have been shown to be largely controlled by the sigma phase content [25, 26]. Considering that the sigma phase is Mo rich, changes in weld metal composition that occur with changes in welding parameters (dilution) are expected to potentially have a significant effect on the sigma phase content and resultant toughness and cracking susceptibility. These potential effects are explored below with the use of multicomponent solidification modeling tools.

Welds were prepared on Alloy CN3MN (cast version of Alloy AL6XN) over a wide range of dilution levels using IN686 filler metal, and the sigma phase content was measured for each weld metal dilution level using image analysis techniques. The composition of the base metal and filler metal are provided in Table 4. Figure 14 shows a light optical photomicrograph of a weld prepared at the 21% dilution level where the interdendritic sigma phase is readily observed. (Backscattered diffraction confirmed the presence of this phase.)

Table 4 Compositions (in weight percent) of alloy CN3MN and filler metal IN686 used to make dissimilar welds

Element	CN3MN	IN686
C	0.02	0.00
Mn	0.40	0.23
Fe	48.03	0.06
S	0.00	0.00
P	0.03	0.00
Si	0.40	0.01
Cu	0.10	0.01
Ni	24.00	58.99
Al	0.00	0.27
Ti	0.00	0.05
Cr	20.50	20.38
Mo	6.30	16.13
W	0.00	3.87
N	0.22	0.00

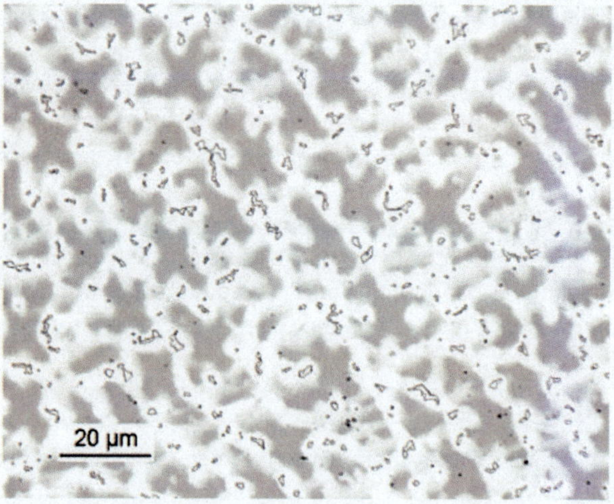

Fig. 14 LOM photomicrograph of a weld prepared at the 21% dilution level where the interdendritic sigma phase is readily observed

The solidification behavior of these welds can be qualitatively understood by reference to the Ni-Cr-Mo ternary liquidus projection shown in Fig. 15 [27]. The nominal composition of the fusion zones results in primary solidification to γ-austenite. Past research has demonstrated that the partition coefficient of Mo is less than unity in welds of very similar composition [25]. This indicates that the liquid composition will be increasingly enriched in Mo as primary solidification proceeds. The partition coefficient for Cr has been shown to be only slightly less than unity for similar weld compositions, indicating that Cr will only slightly segregate to the liquid. Thus, primary $L \rightarrow \gamma$ solidifcation will occur and result in Mo enrichment in the liquid. This stage of solidifcation continues until the the primary solidification path intersects a monovariant eutectic line. Reference to Fig. 15 indicates that NiMo, σ, P, and/or α can potentially form at the final stages of solidification. The diffraction results indciated that only σ phase formed in these welds, so the primary solidification path intersected the σ/γ monovariant eutectic line, and solidification terminated along this line.

Figure 16 shows the variation in sigma phase content as a function of dilution level. The increase in weld metal Mo concentration that occurs with decreasing dilution would typically be expected to produce a concomitant increase in the amount of sigma phase in the weld (since sigma is a Mo-rich phase). However, the amount of sigma phase does not show a significant correlation to the dilution and corresponding Mo concentration of the weld. This immediately suggests that other factors besides the nominal concentration of Mo are also affecting the sigma phase content.

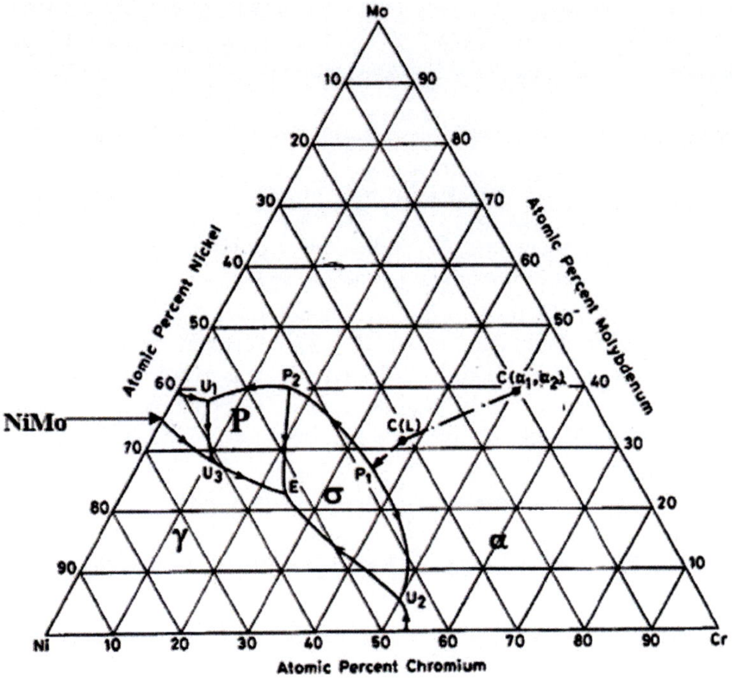

Fig. 15 Ni-Cr-Mi liquidus projection

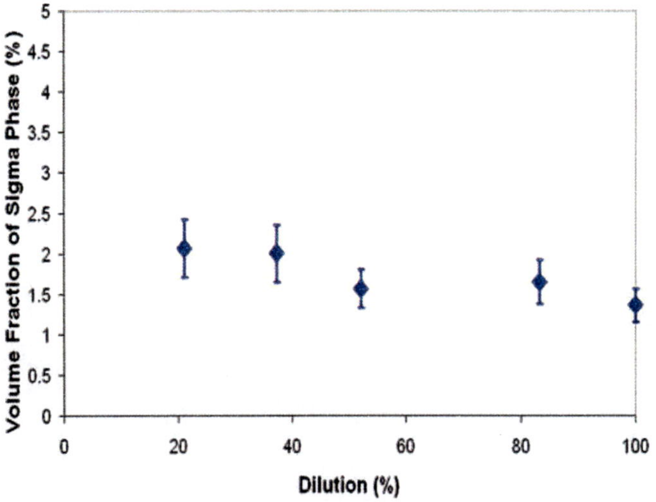

Fig. 16 Sigma phase content as a function of dilution for dissimilar welds between alloy CN3MN and IN 686 filler metal

The sigma phase content is controlled by the amount of eutectic constituent that forms during the end of solidification. The fraction of eutectic that forms during solidification is given through the Scheil equation by

$$f_e = \left(\frac{C_{ej}}{C_{oj}}\right)^{\frac{1}{k_j - 1}} \qquad [14]$$

where f_e is the fraction eutectic, C_{ej} is the concentration of element j in the liquid at the intersection point of the primary solidification path and monovariant eutectic line (i.e., the eutectic composition), C_{oj} is the nominal concentration of element j, and k_j is the partition coefficient for element j. Thus, the amount of sigma phase that forms will depend on three factors: k, C_o, and C_e. It is also important to note that the Ni and Fe contents in the weld are also changing appreciably with changes in dilution, and these changes to Ni and Fe may be affecting the values of k and Ce. This can be assessed with multicomponent models.

As an example, Fig. 17 shows the variation in liquid composition during solidification of a 100% dilution level weld as simulated using Thermocalc. Solidification starts at the nominal composition (C_o), which is controlled by the dilution level. The eutectic composition (C_e) is given by the inflection point where the eutectic reaction $L \rightarrow \gamma + \sigma$ begins. It is important to note that changes in the nominal weld composition can induce changes to all three factors that affect the sigma phase content (C_o, C_e, and k). Figure 18 shows similar calculated curves for all the dilution levels of interest. It is apparent that the eutectic composition decreases appreciably with increasing dilution. Figure 19 shows the variation in the Mo partition coefficient

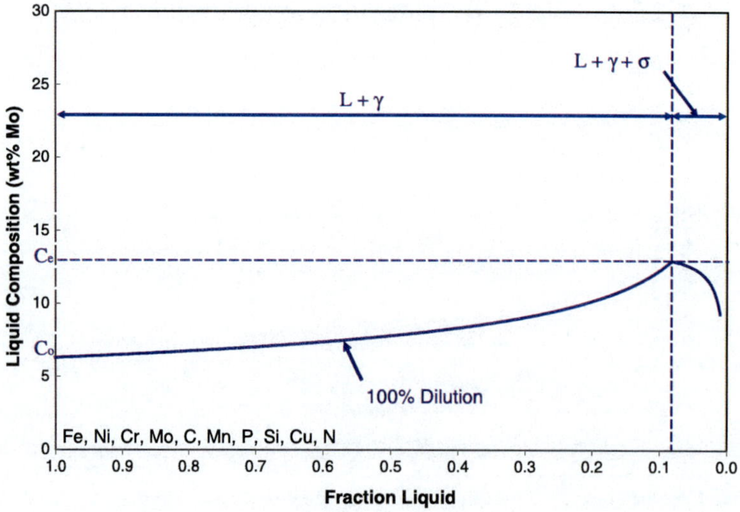

Fig. 17 Variation in liquid composition during solidification of a 100% dilution level weld as simulated using Thermocalc

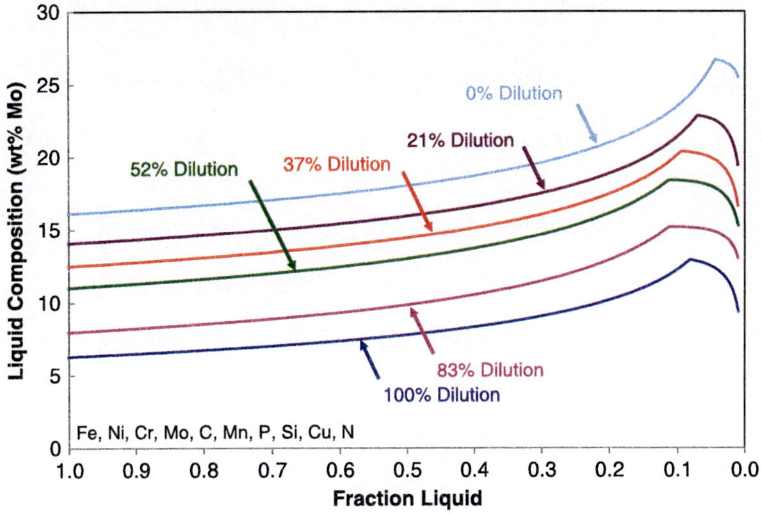

Fig. 18 Calculated solidification curves for all the dilution levels

for the same dilution levels of interest. The partition coefficient for a given dilution does not vary significantly during solidification, and it increases slightly (from ~0.67 to 0.84) with decreasing dilution. Figure 20 summarizes the variations in C_o, C_e, and k as a function of dilution. According to equation [14], an increase in C_o and a decrease in C_e and k will increase the fraction eutectic and resultant amount of secondary phase. The data in Fig. 20 demonstrate there are offsetting effects in

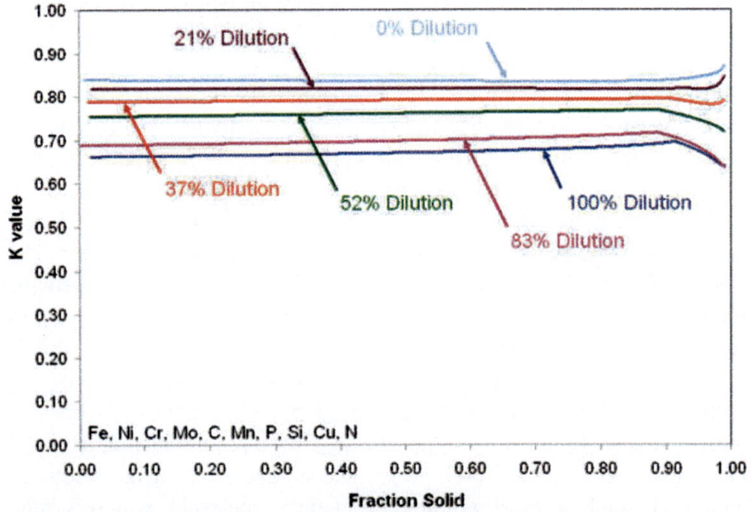

Fig. 19 Variation in the Mo partition coefficient for all dilution levels

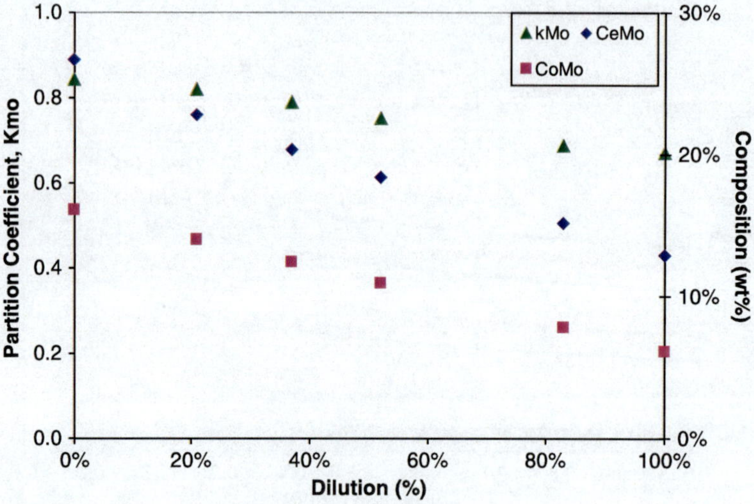

Fig. 20 Variation in C_o, C_e, and k as a function of dilution

these parameters that keep the sigma phase content relatively independent of the weld metal composition. In other words, C_e and k each decrease with increasing dilution, which would, in itself, lead to an increase in fraction eutectic and amount of sigma phase. However, these changes are offset by the decrease in nominal Mo concentration that occurs with increasing dilution. This accounts for the relatively constant sigma phase content observed in these welds.

It is recognized that the final sigma phase content will also depend on the maximum solid solubility of Mo in the austenite and sigma phases. All of these factors can be accounted for with a Scheil simulation. The Scheil simulation is justified here because all the substitutional elements of interest are known to exhibit insignificant diffusion rates during solidification in austenite [28]. As an example, Fig. 21 illustrates the variation in austenite and sigma phase contents during solidification for a weld at 100% dilution. In this case, primary $L \rightarrow \gamma$ solidification begins at 1,385°C, the $L \rightarrow \gamma + \sigma$ reaction begins at 1,305°C, and solidification terminates at 1,232°C with the formation of 1.9 weight percent sigma. This solidification sequence matches the experimental results. Table 5 shows the amount of sigma phase calculated for all the dilution levels of interest. (It is recognized that the sigma phase is measured on a *volume* percent basis, while the Scheil simulation calculates the sigma phase content on a *weight* percent basis. However, the density of these phases is not significantly different, so a direct comparison can be made between the units of measure.) There is reasonable agreement between the measured and calculated amounts of sigma phase. These results also carry important practical implications, since IN686 filler metal can be used at various dilution levels with minimal changes to the sigma phase content in the weld, which is beneficial from a solidification cracking and toughness standpoint.

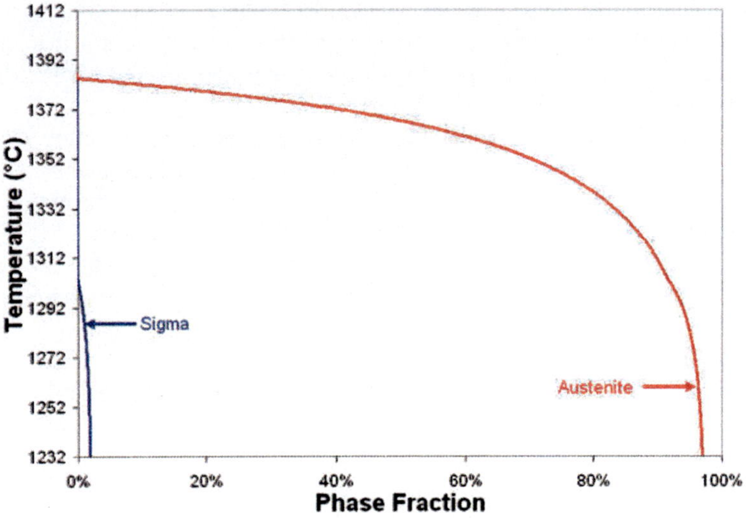

Fig. 21 Variation in austenite and sigma phase contents during solidification for a weld at 100% dilution

Table 5 Comparison of measured and calculated sigma phase contents

Dilution	Calculated percent sigma Wt%	Measured sigma content, Vol %
100	1.9	1.4 ± 0.2
83	2.4	1.7 ± 0.3
52	2.3	1.6 ± 0.2
37	1.9	2.0 ± 0.4
21	1.4	2.1 ± 0.4

Summary

Binary, ternary, and multicomponent solidification models are useful tools for understanding and controlling the solidification behavior and resultant weldability of engineering alloys. These models have been applied to several alloys of practical interest. For the Ni-Cr-Mo-Gd alloys examined, it was demonstrated that these alloys solidify in a manner similar to a binary eutectic system. Solidification initiates with a primary $L \rightarrow \gamma + \rightarrow \gamma$ reaction and terminates at ~1,258°C with a eutectic type $L + \gamma + Ni_5Gd$ reaction. The solidification cracking susceptibility is a maximum at ~1 wt% Gd, and decreases with both higher and lower Gd additions. Low cracking susceptibility at Gd concentrations below ~1 wt% is attributed to a relatively small amount of terminal liquid that exists over much of the mushy zone. Low cracking susceptibility at Gd concentrations above ~1 wt% is qualitatively attributed to a reduced solidification temperature range and back-filling of solidification cracks. Simple heat flow equations were combined with solidification theory to develop a relation between the fraction liquid (f_L) and distance

(x) within the mushy zone. The calculated $f_L - x$ curves can be combined with details on the phenomenology of crack formation to provide useful insight into composition-solidification-weldability relations of the Gd containing alloys. Models for predicting the primary solidification path in ternary alloys were described and applied to understanding the influence of Nb and C additions on alloy IN 718. Results of that work demonstrated that C additions raise the start temperature of the $L \rightarrow \gamma + NbC$ reaction. This occurs because the primary solidification path is directed relatively high into the C-rich side of the diagram where the $L \rightarrow \gamma + NbC$ reaction temperature is relatively high. Finally, multicomponent solidification models were applied to understand the influence of weld metal composition on phase formation in dissimilar welds between alloy CN3MN and filler metal IN686. Results from these calculations were useful to reveal why the sigma phase fraction did not vary significantly with weld metal composition, which was attributed to offsetting changes in equilibrium distribution coefficient, nominal composition, and eutectic composition that occurred with variations in dilution.

Acknowledgements Part of this work was supported by the United States Department of Energy, Assistant Secretary for Environmental Management, under DOE Idaho Operations Office Contract No. DE-AC07-99ID13727. This work was performed at Lehigh University through support from the National Spent Nuclear Fuel Program. Sandia is a multiprogram laboratory operated by Sandia Corporation, a Lockheed Martin Company, for the United States Department of Energy's National Nuclear Security Administration under contract DE-AC04-94AL85000.

References

1. Hull L, Pace M, Lessing, P, Rogers R, Mizia R, Propp A, Shaber E and Taylor L (2000) Advanced neutron absorbers for doe snf standardized canisters – feasibility study. Idaho National Engineering and Environmental Laboratory Report DOE/SNF/REP-057 Rev. 0.
2. DuPont JN, Robino CV, Michael JR, Mizia RE and Williams DB (2004, November) Physical and welding metallurgy of Gd-Enriched austenitic alloys for spent nuclear fuel applications – Part II: Nickel Base Alloys. Welding Journal 83:289s–300s.
3. Susan DF, Robino CV, Minicozzi MJ and DuPont JN (2006) A solidification diagram for Ni-Cr-Mo-Gd alloys estimated by quantitative microstructural characterization and thermal analysis. Metallurgical and Materials Transactions 37A:2817–2825.
4. Van Konynenburg RA, Curtis PG and Summers TSE (1998) Scoping corrosion tests on candidate waste package basket materials for the yucca mountain project, Report UCRL-ID-130386, Lawrence Livermore National Laboratory.
5. Mizia RE, Robino CV and DuPont JN (2006, August) Development and testing of an advanced neutron absorbing gadolinium alloy for spent fuel storage. Nuclear Technology 155.
6. DuPont JN and Marder AR (1996) Dilution in single pass arc welds. Metallurgical and Material Transactions B 27B:481–489.
7. Cieslak MJ (1991) The welding and solidification metallurgy of Alloy 625. Welding Journal 70:49s–56s.
8. DuPont JN, Robino CV and Marder AR (1998) Solidification and weldability of Nb-Bearing superalloys. Welding Journal 77:417s–431s
9. ASM International (1992) Binary Alloy Phase Diagrams, vol. 3.
10. Scheil E (1942) Z. Metallk 34:70.
11. David SA and Vitek JM (1989) International Materials Review 34:213–245.
12. Kou S (2003) Welding Metallurgy, 2nd edition. Wiley, Hoboken, NJ, pp. 170–172.

13. Clyne TW and Kurz W (1982) The effect of melt composition on solidification cracking of steel, with particular reference to continuous casting. Metallurgical and Material Transactions B 13A:259–266.
14. Rosenthal D (1941) Welding Journal 20:220s.
15. Fuerschbach PW and Knorovsky GA (1991) Welding Journal 70:287s–297s.
16. Lucks CF and Deem HW (1958) Thermal Properties of Thirteen Metals, STP No. 227. ASTM, Philadelphia, PA.
17. DuPont JN (2006, June) Mathematical modeling of solidification paths in ternary alloys: Limiting cases of solute redistribution. Metallurgical and Materials Transactions A 37A:1937–1947.
18. Brody HD and Flemings MC (1966) Transactions AIME 236:615–623.
19. Clyne TW and Kurz W (1981) Metallurgical and Material Transactions A 12A:965–971.
20. Kobayashi S (1988) Journal of Crystal Growth 88:87–94.
21. Ohnaka I (1986) Transactions of the Iron and Steel Institute 26:1045–1051.
22. DuPont JN, Robino CV and Marder AR (1998) Acta Materialia 46:4781–4790.
23. Banovic SW, DuPont JN and Marder AR (1999) Welding Journal 78:23s–30s.
24. Brooks JA and Thompson AW (1991) International Materials Review 36:16–43.
25. Banovic SW, DuPont JN and Marder AR (2003) Dilution and microsegregation in dissimilar metal welds between super austenitic stainless steels and ni base alloys. Science & Technology of Welding and Joining 6(6):374–383.
26. Talbot AM and Furman DE Transactions of the American Society for Metals 45:429–440.
27. Gupta KP (1990) Phase Diagrams of Ternary Nickel Alloys: Part I. Indian Institute of Metals, Calcutta, India.
28. DuPont JN, Lippold JC and Kiser SD (2009) Welding Metallurgy and Weldability of Nickel Base Alloys. Wiley, Hoboken, NJ, pp. 59–70.

In Situ Observations of Ductility-Dip Cracking Mechanism in Ni-Cr-Fe Alloys

J.S. Unfried, E.A. Torres, and A.J. Ramirez

Introduction

Ductility-dip cracking (DDC) is an intergranular solid state cracking phenomenon characterized by an important reduction of ductility in a temperature range between 0.5 and 0.8 homologous temperature of the material [1–3]. Several face centered cubic (FCC) pure metals and alloys undergo DDC [4–6]. Among the most important materials suffering DDC are Ni-based alloys, Cu-based alloys, some Ti-based alloys, and Stainless steels. Even though DDC has been the focus of countless research programs for several decades and several technological measures have been proposed and used to reduce its occurrence, a comprehensive understanding of the fundamental metallurgical mechanism behind this phenomenon has not been fully explained. Due to the recent interest regarding DDC of solid solution strengthened Ni-base alloys within the nuclear power generation industry, several possible explanations to DDC mechanism have been proposed, which may divided in three main groups, depending on the dominating mechanism: (i) Grain boundary sliding (GBS) [2–3, 22, 31, 34–35], (ii) Grain boundary embrittlement due to impurity (P, S, and H) segregation[25–27], and (iii) Intergranular fracture enhanced by intermediate temperature intergranular precipitation of Cr-rich carbides [30].

The first report on DDC in the open literature at the beginning of the past century involving Cu and its alloys concluded that second phase precipitation did not play an important role on DDC at intermediate temperatures [4]. Soon after Ni-based alloys came into service in 1931 [7], two important theories about the role of GBS in high temperature intergranular cracking were published by *Rachinger* [8] and *Lifshitz* [9]. These studies clarified when the high temperature intergranular cracking is or is not dominated by creep conditions. Figure 1 shows a non-linear timeline emphasizing important advances on DDC research and solid solution strengthen Ni-base alloys.

Initially, the DDC mechanism was explained as a diffusion creep phenomenon involving low strain rates at intermediate temperatures [5]. Later the relationship

J.S. Unfried (✉)
Brazilian Synchrotron Light Laboratory (LNLS), Campinas, Brazil; School of Mechanical Engineering, State University of Campinas, Campinas, Brazil
e-mail: junfried@lnls.br

J. Lippold et al. (eds.), *Hot Cracking Phenomena in Welds III*,
DOI 10.1007/978-3-642-16864-2_15,

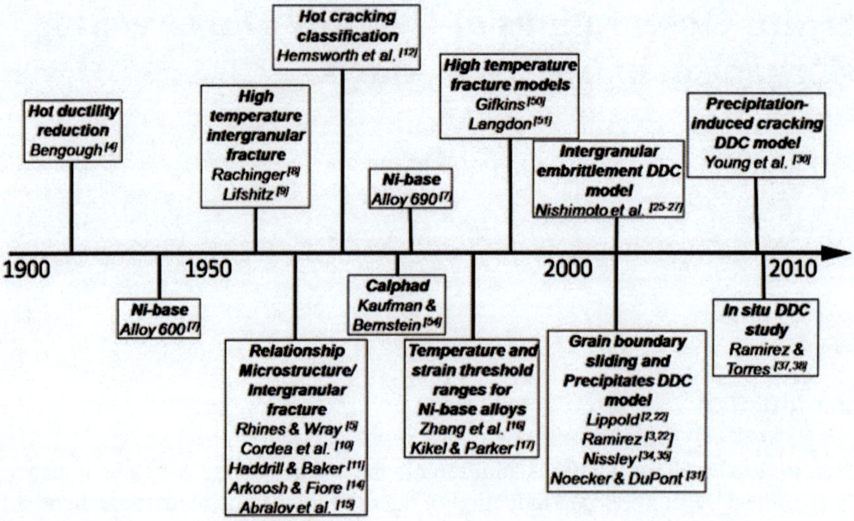

Fig. 1 Non linear timeline of intergranular hot cracking and DDC research. It was also included information about solid solution strengthened Ni-base alloys and the Calphad method

among DDC, stacking fault energy, microsegregation, and microstructure (grain size, morphology, and second phase distribution) were outlined [10, 11]. Afterward, the classification and clear differentiation of high temperature cracking associated with welding in different alloys systems was established. This classification differentiated solidification and liquation cracking phenomena from those occurring in the solid state during high temperature forming and welding [12]. Most recent DDC research effort has concentrated on as-welded Ni-based alloys, especially ERNiCrFe-7 (FM-52), ERNiCr-3 (FM-82), and Alloys 600 and 690 [13–15]. On the other hand, relevant information linking thermo-mechanical history, high temperature formability, and microstructure has been neglected.

Several DDC aspects involving as-welded Ni-based alloys, which are well-known today, were established after Alloy 690 and its filler metal (ERNiCrFe-7) were introduced. Chromium depletion in the lattice of material, massive carbide precipitation, shear stress along the grain boundaries (GB) [14], and cracking along grain boundaries normal to the externally applied stress [15] were related to DDC in these types of alloys. Likewise, there were documented the ductility-dip temperature range (DTR<100°C), minimum threshold strain (E_{min}<15%) [16]; and DDC temperature range for 600 and 690 alloys (800–1,300°C) [17]. Initially, experimental efforts to reproduce and study DDC in as-welded Ni-based alloys were based on Varestraint-like tests [18, 19]. Later, the strain-to-fracture (STF) test [20], which is carried out in a Gleeble® thermo-mechanical simulator, made possible important advance on the unveiling of the DDC phenomena fundamentals.

Most recent works on the understanding of the DDC mechanism highlights GBS as the controlling factor. Work based on STF measurements studied the nucleation and growth of cavities at GB triple points and along GBs. Grain size, intergranular second phases and GB morphology, and impurity segregation are important factors controlling the material high temperature loss of ductility. The DDC temperature range is limited at lower temperatures by the reduced GBS and at higher temperatures by the onset of recrystallization [2–3, 5, 21–22].

Other research based on Varestraint testing explains the material loss of ductility, leading to DDC, solely to impurity segregation to GBs [23–27]. However, DDC has been observed in Ni-based alloys with very low levels of impurities. Nevertheless, this may be an important mechanism controlling DDC when as-welded microstructures are repeatedly reheated as a result of multipass welding. Finally, some other work using hot-ductility testing explains DDC as a precipitation-induced cracking, which favors nucleation and propagation of cavities that are conditioned by grain boundary sliding and GB morphology [28–31].

Based on the previously described results and discussions, some efforts have been made to increase the DDC resistance of some Ni-base alloys, especially alloy ERNiCrFe-7, which is very important for the nuclear power generation and petrochemical industries. The addition of primary carbide formers (Nb, Zr, Ta, and Ti) and solid solution hardening (Mo) elements to as-welded Ni-base alloy ERNiCrFe-7 have promoted outstanding increases in DDC resistance [32–36]. It has been verified that profuse primary carbides precipitation precludes GB migration, producing tortuous GBs, which reduces GBS, strain concentration at triple points, and void formation, consequently increasing DDC resistance [21]. Despite the good correlation between the experimental observations and mechanistic descriptions for DDC presented above, it is yet unclear the relationship among GBS, alloy modification, second phase precipitation, and the material DDC susceptibility. Moreover, the proposed mechanisms do not explain DDC in pure metals, ultra-clean (very low impurities) and precipitate free alloys. Therefore, more comprehensive and controlled studies are necessary to understand the fundamentals behind this phenomenon.

Recently, fundamental DDC research based on SEM *in situ* thermo-mechanical testing at temperatures up to 1,050°C and forces up to 4,000 N has been conducted [37]. Figure 2 shows the *in situ* experimental setup, which uses a high temperature tensile mini-stage with heating and deformation control systems coupled to a scanning electron microscope (SEM). This experimental setup has enabled the real time observation of DDC and the associated GBS in Ni-base alloys ERNiCrFe-7 and ERNiCr-3, providing quantitative measurements for micro-threshold strain, as shown in Fig. 3. Different from the STF test, for a given temperature, the micro-threshold strain can be determined just from one test. Such results are qualitatively comparable with previously reported STF data. In addition, quantitative GBS information retrieved from the *in situ* tests have been directly linked with GB morphology, GB orientation to the external applied force, and the DDC resistance [38]. The great advantage of such *in situ* SEM approach is the ability to study the microstructural behavior, and crack nucleation and growth in real time, providing a direct link between several metallurgical aspects of the DDC phenomenon that have not been fully considered in previous research.

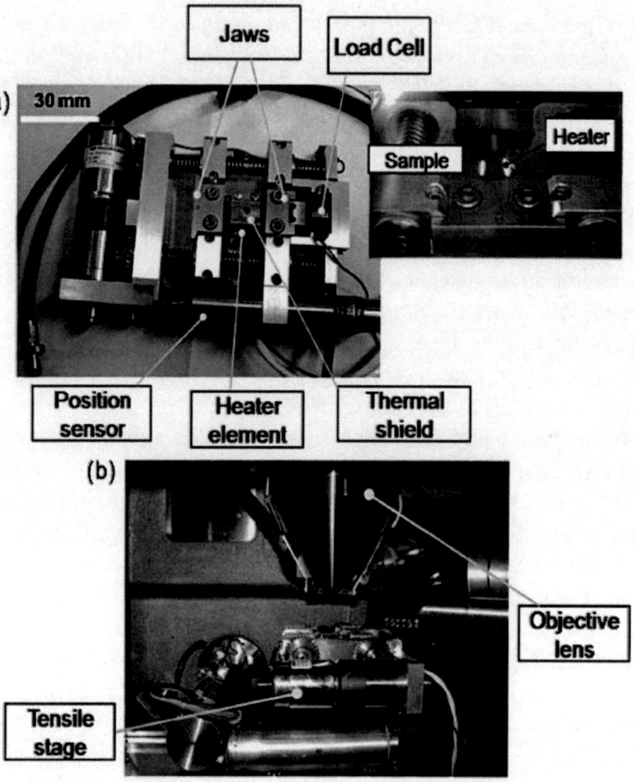

Fig. 2 *In situ* experimental setup. (**a**) Tensile mini-stage, including the furnace and the cooling system. (**b**) Device mounted inside the SEM vacuum chamber

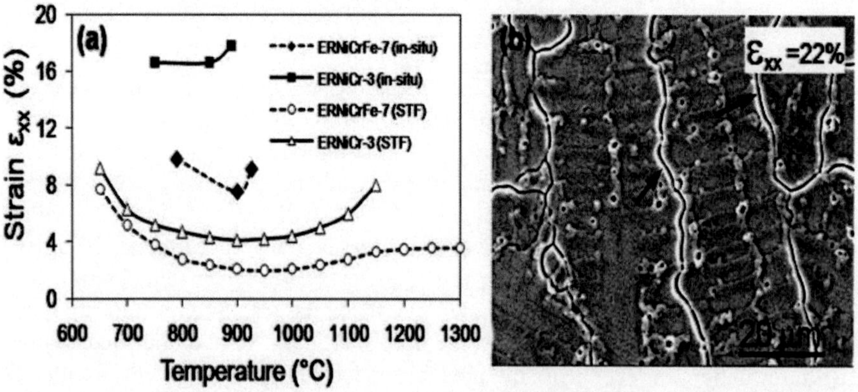

Fig. 3 DDC susceptibility study using high temperature *in situ* SEM and strain-to-fracture (STF) measurements for alloys ERNiCr-3 and ERNiCrFe-7 [21]. (**a**) Comparison of threshold-strain for DDC versus temperature measured using *in situ* SEM and STF tests. (**b**) Intergranular DDC opening in alloy ERNiCr-3 during in situ SEM test performed at 927°C [38]

Therefore, here are addressed the relationships among several microstructural features (GB morphology, chemical distribution and partitioning, and primary and second phase precipitation) and their effects on the high temperature ductility and DDC behavior in a set of experimental alloys based on ERNiCrFe-7 alloy with Nb, Mo, and Hf additions. The high temperature ductility and DDC have been evaluated using the thermo-mechanical *in situ* SEM test, with special attention to the microstructural behavior during high temperature plastic deformation.

Experimental Procedures

Materials

Buttons (20 g) of Ni-base alloys based on ERNiCrFe-7 (master alloy) with additions of high purity Nb, Mo, and Hf pellets were arc melted within an Argon atmosphere. Optimized additions ranges to the master alloy were obtained using Calphad-based numerical modeling. The thermodynamic and kinetic calculations were performed using Thermo-Calc® and Dictra® software along with Ni-data, Mob-2, and Mob-Ni databases. A welding cooling rate of 100 $K.s^{-1}$ was used for the modeling in accordance with a methodology described elsewhere [32]. In addition, a high purity and interstitial elements-free alloy with the same Ni, Cr, and Fe content of ERNiCrFe-7 alloy was melted using high purity pellets.

The experimental alloys ingots were homogenized and recurrently cold rolled and annealed to obtain ~1 mm thick sheet. The homogenization and annealing heat treatments were carried at 1,200°C during 12 h using an electric furnace with inert gas atmosphere. Specimens for the *in situ* SEM experiments, with geometry and dimensions shown in Fig. 4, were electric discharge machined from the cold rolled sheets. Using a similar procedure as the one proposed for the STF test, as-welded microstructure with radial GB distribution was obtained by autogenously spot welding on the center of the *in situ* SEM specimens using manual GTAW process (Argon shielding, ranging between 15 and 30 A during 3 and 5 s).

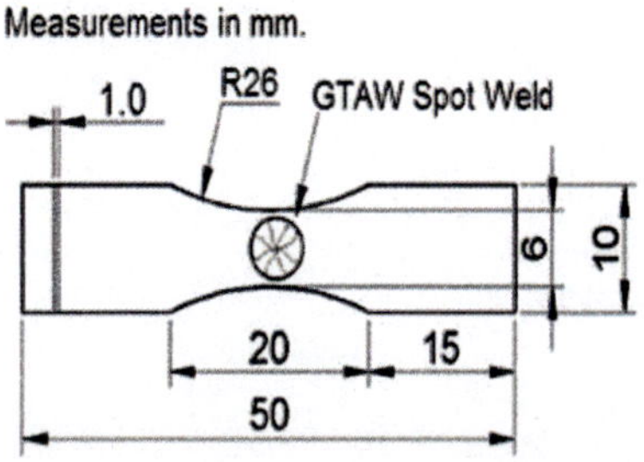

Fig. 4 *In situ* SEM dog-bone specimen with as-welded radial GB distribution [37]

Chemical and Microstructural Characterization

The experimental alloys chemical composition was measured using x-ray fluorescence and combustion/fusion mass spectrometry techniques. As-welded microstructure was characterized using an optical microscope (OM) Olympus® BX51M coupled to PaX-It image analysis software, scanning electron microscope (SEM) JEOL® JSM-5900 LV operating between 10 and 25 kV, and transmission electron microscope (TEM) JEOL® JEM-2100 ARP operating at 200 kV, both coupled with Noran® Energy-Dispersive X-ray (XEDS) spectrometers. Crystallographic and grain boundary analyses were performed on an SEM using an HKL Channel 5® backscattered electron diffraction (EBSD) system.

In situ SEM specimens were prepared using conventional grinding and polishing procedures. Final polishing was performed using 1 μm diamond paste and 0.05 μm colloidal silica on a vibratory polishing system. Electrolytic etching was performed using 10%-vol. aqueous chromic acid solution (H_2CrO_3) at room temperature and applying DC voltage between 1.5 and 2.5 V during 60–150 s. TEM foils from as-welded microstructure were prepared by grinding down to 100 μm, followed by Argon ion-milling.

High Temperature Deformation *in situ* SEM Experiments

Polished and etched dog-bone-shaped specimens in the as-welded condition, as presented in Fig. 4, were subjected to *in situ* SEM high temperature deformation. The parameters used for these experiments are summarized in Table 1. Two sets of experiments were performed for each experimental alloy to provide at least 5 different measurements within the interest temperature range (500–1,000°C).

Digital video and high quality still images from the high temperature deformation process were recorded at selected regions of the as-welded microstructure. Afterwards, the videos and still images were digitally processed and analyzed to study microstructural changes, crack nucleation and growth, and strain quantification. Finally, GBS was quantified at specific regions using a methodology discussed elsewhere [38].

Table 1 Parameters used for the high temperature deformation *in situ* SEM experiments

Variable	Values
Test temperature range	500–1,000°C
Limit force	3.8 kN
Stroke rate	0.05 mm s^{-1}
Sample thickness range	0.6–1.0 mm
Heating time range	2–5 min

Results

Microstructural Characterization and Numerical Modeling

Table 2 shows the experimental alloys chemical composition. Alloy-1 is a solid solution strengthened Ni-base alloy with similar Ni, Cr and Fe contents of ERNiCrFe-7. Alloy-2 is the master ERNiCrFe-7 alloy. Nb, Mo, and Hf additions were such that Alloy-3 (with Nb and Mo additions) had similar composition to modified ERNiCrFe-7 DDC resistant alloy, which has been previously studied [34–36]. Alloy-4 composition range ($2.0 < Nb < 3.0 - 3.0 < Mo < 3.5 - 0.2 < Hf < 0.4$ %wt) has been optimized using Calphad-based modeling to maximize primary intergranular and intragranular carbonitride precipitation, preserving other fundamental metallurgical characteristics [39, 40]. Previous work has shown that small Hf additions to wrought Ni-base alloys increases high temperature ductility, promotes Hf-rich round-shape primary carbonitrides precipitation, reduces second phase precipitation, increases waviness of GBs, and reduces grain size [41–44].

Figure 5 shows the most relevant thermodynamic and kinetic calculations results for the studied Ni-base alloys. The simulation reveals that Nb, Mo, and Hf additions strongly promote primary carbonitrides precipitation (MX), reduce the precipitation temperature and mass fraction of $M_{23}C_6$ secondary carbides, and also expands the alloy solidification temperature range. Experimental MX mass fraction data has been included in Fig. 5, which was calculated from measured volume fraction using optical microscopy techniques and the theoretical density of carbonitrides. Table 3 summarizes the main features of primary and secondary carbides precipitation expected with Calphad calculations and kinetic conditions.

Figure 6, shows SEM micrographs of the experimental alloys in the as-welded condition and Table 4 summarizes the main microstructural characteristics observed using OM and SEM. Nb, Mo, and Hf additions have promoted a change in the solidification mode from cellular to columnar dendritic. Likewise, a significant increase

Table 2 Measured chemical composition (wt%) of Ni-base experimental alloys

Element	Alloy-1 Ni-Cr-Fe	Alloy-2 ERNiCrFe-7	Alloy-3 ERNiCrFe-7 + Nb + Mo	Alloy-4 ERNiCrFe-7 + Nb + Mo + Hf
Ni	61.5	59.2	54.9	55.8
Cr	28.5	29.1	27.4	27.4
Fe	10	10.3	9.6	9.4
Mo	–	–	4.4	3.2
Nb	–	–	2.3	2.4
Hf	–	–	–	0.36
Mn	–	0.28	0.30	0.26
Ti	–	0.48	0.45	0.50
Al	–	0.45	0.67	0.69
C	–	0.027	0.022	0.025

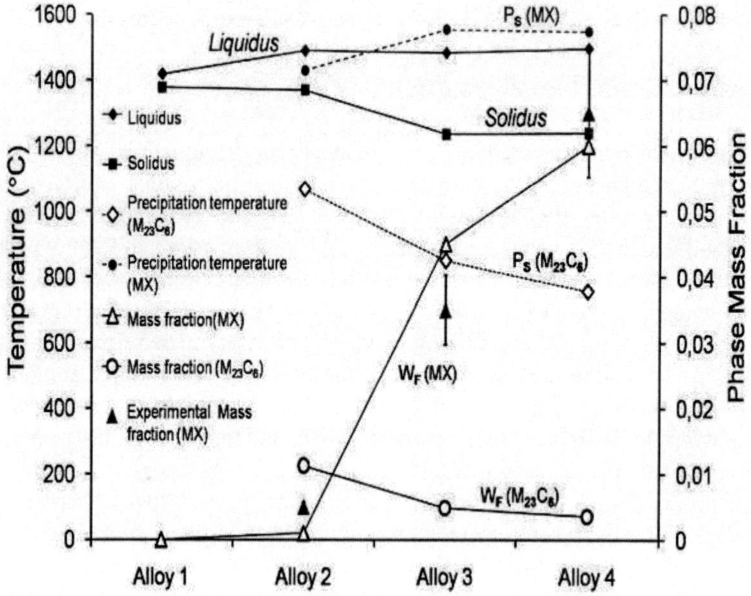

Fig. 5 Thermodynamically and kinetically calculated phase transformation temperatures and phase mass fraction for the experimental alloys. Experimental MX mass fraction MX-W_F, which was calculated from volumetric fraction using theoretical density. Phase precipitation start temperature (P_S), phase mass fraction (W_F). MX-W_F and MX-P_S were calculated using kinetic modeling. $M_{23}C_6$-W_F, $M_{23}C_6$-PS, *liquidus*, and *solidus* temperatures were calculated using thermodynamic modeling

in the interdendritic segregation and second phase precipitation (intergranular and intragranular) was observed.

The final grain boundary morphology has been conditioned by type, frequency, and distribution of primary carbonitrides. Thus, the predominately flat migrated

Table 3 Calculated carbides ($M_{23}C_6$) and carbonitrides (MX) precipitation. Kinetic Calphad-based calculations

	MX – primary carbides			$M_{23}C_6$ – carbides		
					Mass fraction ($\times 10^{-3}$)	
Alloy	Type	Average size (nm)	Mass fraction ($\times 10^{-3}$)	Type	Interdendritic	Intradendritic
1	No precipitates			No precipitates		
2	(Ti,Cr,Fe)(C,N)	50	1.0	$(Cr,Fe)_{23}C_6$	7.8	3.8
3	(Nb,Ti)(C,N)	250	45	$(Cr,Mo,Nb,Fe)_{23}C_6$	3.1	1.4
4	(Hf,Ti)(C,N) (Nb,Hf,Ti)(C,N)	200	60	$(Cr,Mo,Nb,Fe)_{23}C_6$	1.5	2.1

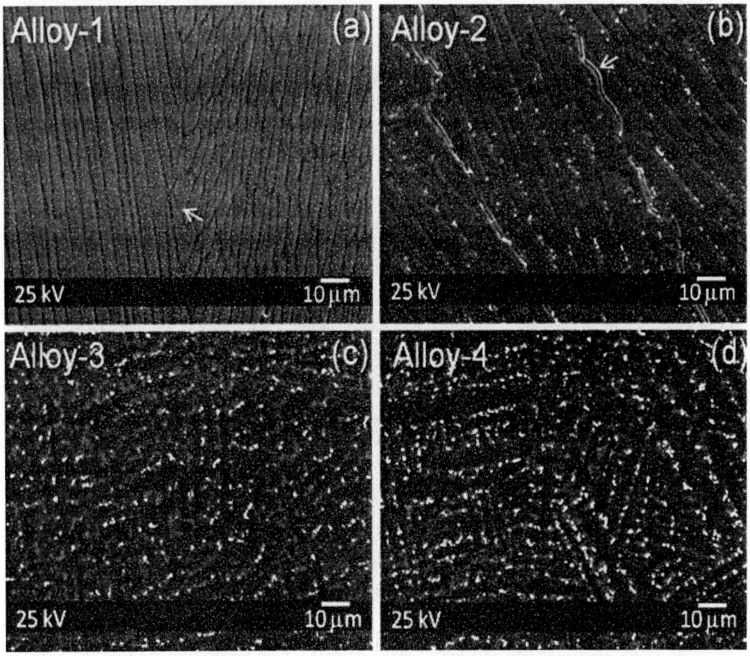

Fig 6 Experimental alloys microstructure in the as-welded condition. Migrated grain boundaries are shown for (**a**) alloy 1 and (**b**) alloy 2 (*white arrows*). The extensive intragranular precipitation and important segregation are observed on (**c**) alloy 3 and (**d**) alloy 4. Secondary electron SEM images

GBs observed in alloys 1 and 2, were the result of the absence of precipitates in alloy-1 and the sporadic Ti-rich carbonitrides observed in alloy-2 (master alloy). On the other hand, the GB pinning caused by the extensive precipitation of Nb-, Ti-, and Hf-rich primary carbides observed in alloys 3 and 4 have promoted the

Table 4 Microstructural characteristics of the experimental alloys in the as-welded condition. Analyses by optical and scanning electron microscopy

		Predominant elements			
Alloy	Solidification structure	Matrix (solid solution)	Segregation	Precipitates	Grain boundary morphology
1	Cellular	Ni-Cr-Fe	Cr (slight)	No precipitates	Flat
2	Cellular	Ni-Cr-Fe	Cr, Ti (strong)	Ti-rich primary carbides	Flat
3	Columnar dendritic	Ni-Cr-Fe-Mo	Mo, Nb (strong)	Nb- and Ti-rich primary carbides, Laves	Undulated
4	Columnar dendritic	Ni-Cr-Fe-Mo	Mo, Nb (strong)	Nb- and Hf-rich primary carbides	Undulated

formation of undulated migrated GBs [3, 21–22, 32, 35]. Electron backscattered diffraction (EBSD) analyses shown in Fig. 8 confirmed the GBs character in the as-welded experimental alloys. These EBSD analyses were especially important for alloys 3 and 4 where the GBs were difficult to identify, as evident in Fig. 6.

The secondary electron images and XEDS maps in Figs. 7 and 8 show in detail the microstructural characteristics and chemical distribution of the as-welded

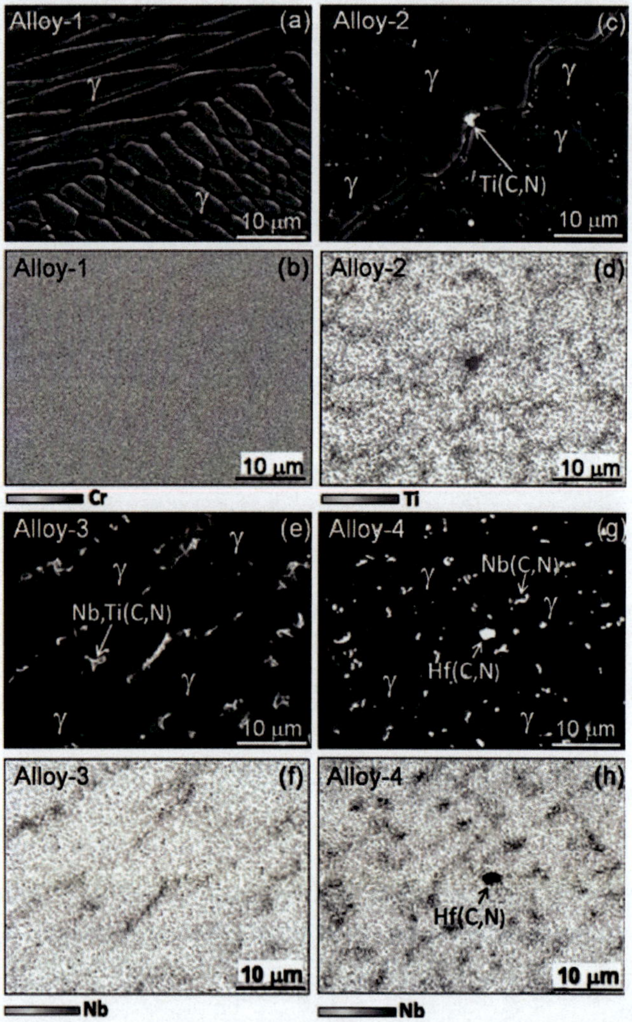

Fig. 7 Characterization of the as-welded experimental alloys. (**a**) and (**b**) Microstructure and Cr-distribution of Alloy-1; (**c**) and (**d**) Microstructure and Ti-distribution of Alloy-2; (**e**) and (**f**) Microstructure and Nb-distribution of Alloy-3 (**g**) and (**h**) Microstructure and Nb-distribution of Alloy-4. The different frequency and type of precipitates are shown in the secondary electron SEM images. The chemical segregation is shown in the correspondent XEDS maps from the same regions shown in the images above

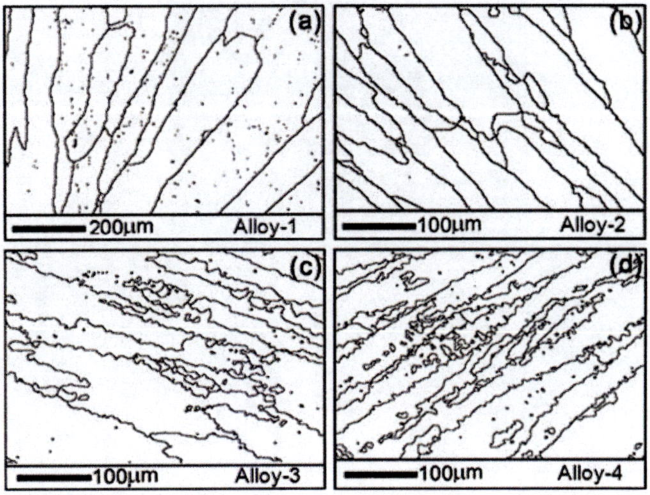

Fig. 8 Images show EBSD analysis of grain boundary character. (**a**) Alloy-1 and (**b**) alloy-2 exhibit large grain size and straight boundaries. Alloy-1 presents the largest grain size. (**c**) Alloy-3 and (**d**) alloy-4 exhibit wavy grain boundaries

experimental alloys. Evidence of interdendritic segregation and second phase precipitation are highlighted in these images. In accordance with the kinetic carbonitrides precipitation calculations presented in Table 3, the images in Fig. 7 reveal the precipitation increase with the addition of carbonitride forming elements. The strong Nb and Mo segregation observed on the modified alloys 3 and 4 also agrees with results of kinetic calculations (results not shown here).

Figure 9 presents representative results of the transmission electron microscopy (TEM) analyses performed in alloys 2, 3 and 4. Primary carbides and intermetallic phases have been identified by XEDS microanalysis and selected area electron diffraction (SAD). Table 5 summarizes the macroscopic (OM) and microscopic (SEM and TEM) relevant characteristics of the second phases observed in the experimental alloys. The Ti, Nb, and Hf additions to the ERNiCrFe-7 master alloy are expected to increase the primary precipitation with specific spatial distribution and particle morphology [32]. The addition of Ti produced carbonitrides with faceted morphology. Nb additions produced carbonitrides with elongated and complex morphologies (Chinese-script like and branched) [32–33, 35]. On the other hand, simultaneous Nb and Hf additions produced more rounded fine carbonitrides. Detailed characterization of these precipitates is presented elsewhere [45].

High Temperature in situ Experiments

The high temperature *in situ* SEM experiments allowed the controlled simulation of DDC conditions and the direct observation of intergranular cracking phenomenon. Figure 10 shows the DDC susceptibility measurements for experimental alloys in

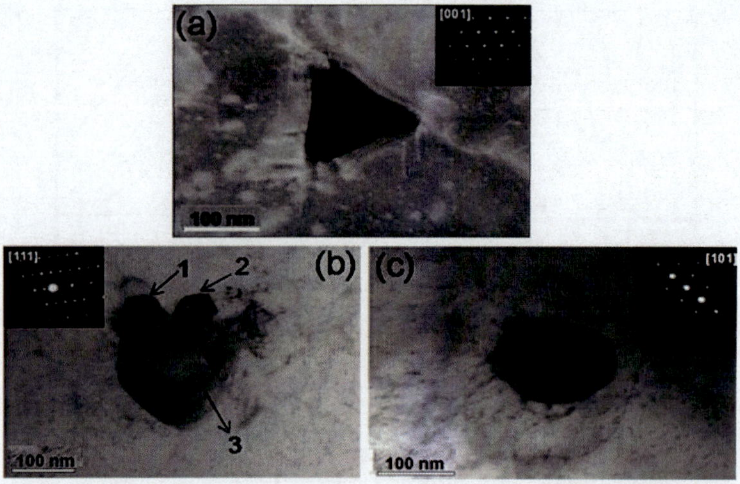

Fig. 9 TEM Bright field images and selected area electron diffraction patterns of MX-primary carbonitrides and intermetallic phases observed in the experimental alloys. (**a**) Typical polygonal primary Ti(C,N) on Alloy-2. (**b**) Laves phase (3) and Nb(C,N) precipitates (1 and 2) observed in Alloy-3. (**c**) Round Hf(C,N) primary carbonitride observed in Alloy-4. Diffraction patterns were performed on MX-primary carbonitrides. 200 keV

Table 5 Microstructural characterization of precipitates observed in the experimental alloys

Alloy	Type and frequency	Localization	Average size range (μm)	Average distance (μm)[a]	Morphology
1	No precipitates		No precipitates		
2	MX – Sporadic	Intergranular, intragranular	0.01 – 2.0	> 50	Polygonal and faceted
	$M_{23}C_6$ – Sporadic	Intergranular	0.05 – 0.1[b]	Numerous[b]	Round and square[b]
3	MX – Profuse	Intergranular, intragranular	0.02 – 10	5–50	Elongated
	$M_{23}C_6$	Not observed	Not observed	Not observed	Not observed
4	MX – Profuse	Intergranular, intragranular	0.01 – 5.0	5–10	Round
	$M_{23}C_6$	Not observed	Not observed	Not observed	Not observed

[a]Observed distance between intergranular precipitates.
[b]Extracted from [3].

the temperature range between 500 and 1,000°C. The cracking susceptibility has been measured using the threshold-strain for DDC versus test temperature [38]. The results in Fig. 10 show that Alloys 1 and 2 are more susceptible to DDC than Alloys 3 and 4. In accordance to *Zhang's* criteria [16], the threshold-strain for cracking for alloys 1 and 2, in the temperature range between 650 and 1,000°C, exhibit DDC. On the other hand, alloys 3 and 4 exhibited good ductility during the tests,

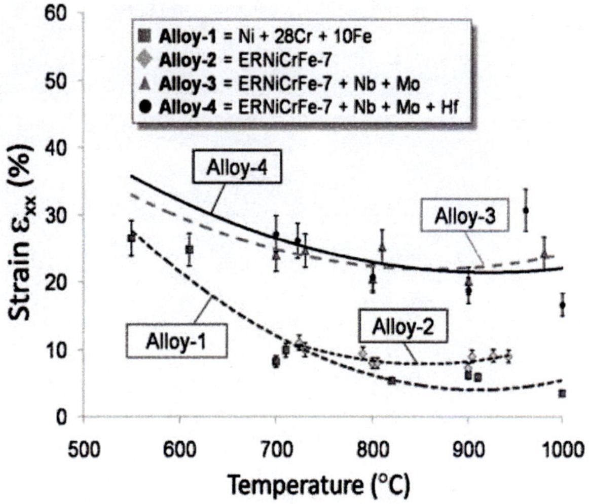

Fig. 10 DDC susceptibility quantification using the high temperature deformation *in situ* SEM test. Experiments within a temperature range between 500 and 1,000°C. ε_{xx} is the strain along the direction of external load application

showing a slight reduction within the tested temperature range. The minimum strain for cracking (ε_{min}) shown by alloys 1 and 2 was 3.5% at 980°C and 7.3% at 910°C, respectively. The ε_{min} measured for alloy 2 using *in situ* SEM is higher than one measured using the Gleeble®-based strain-to-fracture test (~2.0% at 900°C), which has been previously discussed [37].

Sequences of images recorded during high temperature deformation *in situ* tests are shown in Fig. 11. Deformation sequence for alloy-2 (master alloy) has been presented elsewhere [37, 38]. Three different stages of the high temperature deformation process are presented: (1) test beginning, with $\varepsilon_{xx} = 0$, showing the selected critical region presenting GBs normal to the external load; (2) intermediate stage of test, showing the emergence of deformation bands which have been activated by thermo-mechanical conditions; and (3) at the moment before crack opening.

The *in situ* tests show the buildup of deformation bands inside grains in the selected critical region and the beginning of an accommodation process of the of polycrystalline aggregate, which involves deformation, translation and rotation of grains. The behavior shown in Fig. 11 has been observed at different temperatures within the evaluated temperature range.

The average imposed horizontal (x direction) strain is presented for each stage image in Fig. 11 reveals the large differences in the materials ductility just before cracking (third stage). Alloys 1 and 2 withstand low strains before reaching the cracking stage. On the other hand, alloys 3 and 4 do not show evidence of deformation bands buildup or grain accommodation when similar levels of strain, as the ones necessary to onset the intergranular cracking on alloys 1 and 2, are applied. These observations are very important for the understanding of the relationships between the microstructural features (including crystallographic orientation, GBs, precipitates, etc) and plastic deformation.

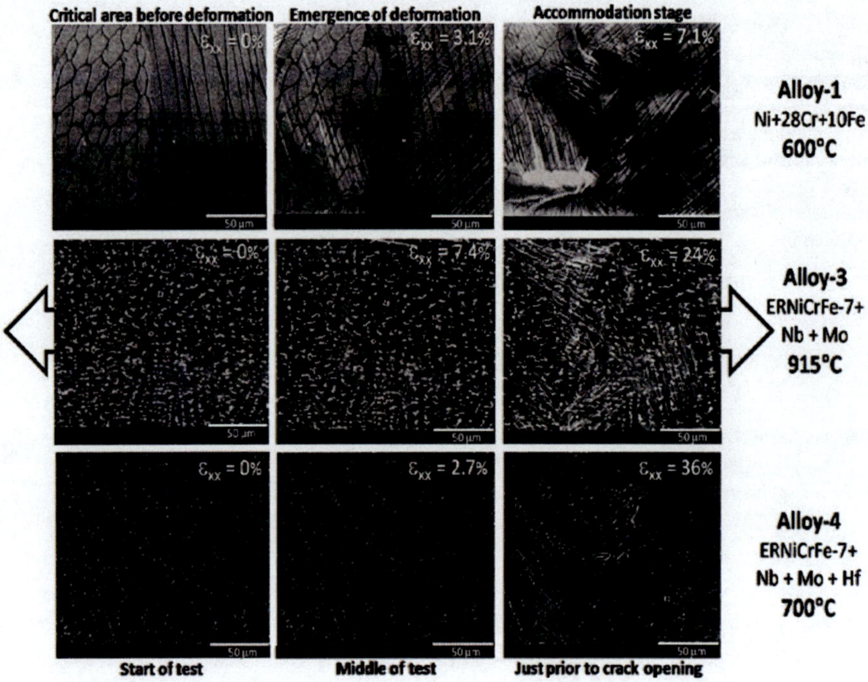

Fig. 11 Image sequences from the high temperature deformation *in situ* tests for alloys 1, 3, and 4

The *in situ* induced ductility-dip cracks morphology for the experimental alloys is shown in Fig. 12. As can be seen, all the cracks are localized along of the migrated grain boundaries. A higher incidence of cracking at triple points was not observed. While alloys 1 and 2 show straight cracks, alloys 3 and 4 show wavy cracks. However, despite the above mentioned morphological differences, all cracks exhibit a wedge or W-type appearance [11, 46].

Figure 13 shows the methodology used for the grain boundary sliding (GBS) measurement, which has been presented in detail elsewhere [38]. Table 6 shows the results of GBS measurements, which have been linked with the imposed strain, temperature, GB morphology, and GB orientation to the externally applied load. There are reported two types of GBS, pure sliding (S_P), which is related to the relative movement of grains along the boundary, and deformation sliding (S_d), which is related to internal deformation of each grain [38]. Table 6 also reports the angle formed between the analyzed grain boundary and the tensile axis.

GBS measurements suggest a decrease of pure sliding (S_P) when carbide promoting (Nb and Hf) and solid solution strengthening (Mo) elements are added. Such additions also increased the allowed strain (ε_{xx}) in the alloys before the cracking onset within the entire test temperature range. Deformation sliding (S_D) values indicates that alloy 4 withstands the highest intragranular deformation. However,

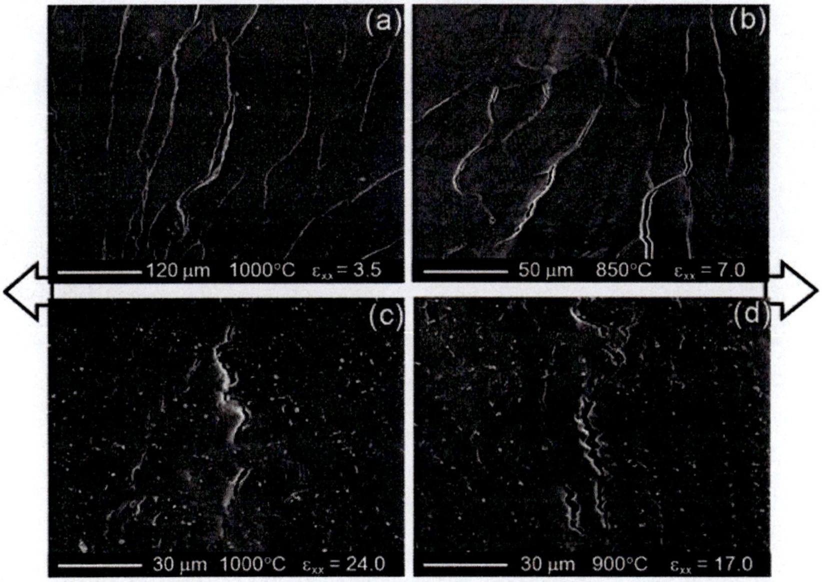

Fig. 12 Images show the DDC cracks morphology in as-welded microstructure at final of *in situ* high temperature deformation tests. There are observed deformation bands and the crack path along grain boundaries. All cracks are wedge type

this last result must be carefully analyzed because the measurements have been performed in a selected region, which involved a reduced number of grains. Finally, the microstructure that withstands the highest average deformation before cracking is the one containing undulated GBs, which agrees with previous experimental observations and discussions [1–2, 22, 32–33, 35].

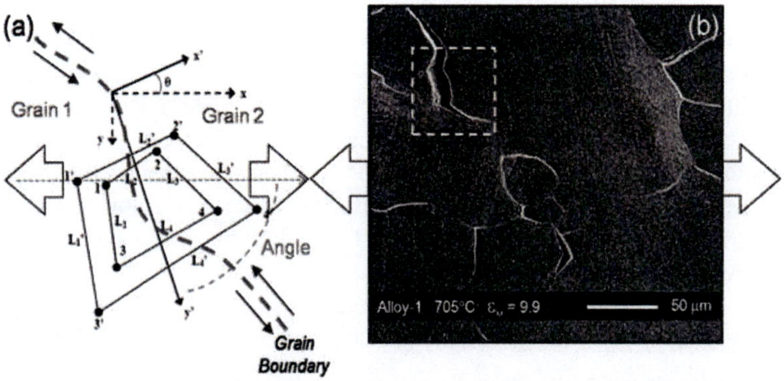

Fig. 13 Quadrilateral method for grain boundary sliding quantification. **a** Schematic representation of quadrilaterals before (1,2,3,4 positions) and after (1',2',3',4' positions) deformation. **b** Example of selected area for GBS measurement

Table 6 Grain boundary sliding calculation results using quadrilateral method [38]

Alloy	Pure sliding (S_P)	Deformation sliding (S_D)	Strain (ε_{xx})	Angle (°)	Temperature (°C)	GB morphology
1	0,98	1,66	9,1	67	705	Flat
	0,31	0,16	3,5	73	1,000	
2	–	–	–	–	–	Flat
	0,85	0,88	14	79	912	
3	1,03	1,34	24,3	57	730	Undulated
	0,55	0,05	26,5	62	1,000	
4	0,11	1,98	26,1	62	702	Undulated
	0,25	3,32	18,4	64	995	

Discussion

Role of Element Additions

The role of Nb and Mo additions on the alloy ERNiCrFe-7 microstructure and DDC resistance has been widely discussed in the open literature [32–33, 35–36]. Nb-addition to alloy ERNiCrFe-7 promotes Nb-rich primary carbides precipitation, which pins GBs favoring the development of their undulated morphology. Recent work has shown the strong Mo segregation toward the interdendritic regions and has suggested its possible influence on the Chinese script morphology of the Nb-rich precipitates [33]. The Mo in solid solution within the FCC matrix plays an important role on the grains internal deformation due to promoting low stacking fault energy, pinning of dislocations by jogs, and plastic deformation increasing at high temperatures [52].

Calphad modeling and microstructural characterization have shown that 2.0–3.0%-wt Nb addition to ERNiCrFe-7 promotes strong Nb (up to 25%-wt) segregation toward interdendritic regions, favoring Laves phase, Nb-rich, and Nb+Ti-rich primary carbide precipitation. Such Nb-rich precipitates and laves phase exhibit elongated and Chinese-script-like morphologies (Figs. 6 and 7), making it very difficult to differentiate such phases by OM or SEM. Only TEM analyses using SAD and XEDS have made possible the unequivocal identification of such phases.

Mo additions to alloys 3 and 4 have promoted strong interdendritic segregation of this element and homogeneous distribution within the matrix. However, Mo-rich precipitates have not been observed. Calphad modeling has indicated severe interdendritic Mo segregation, achieving 10–30%-wt Mo content in such region when 3.0–4.0%-wt Mo additions have been made to ERNiCrFe-7 [48]. Such interdendritic Mo enrichment is influencing the primary carbides precipitation kinetics and morphology [33].

The Hf addition to alloy 4 has influenced the morphology and chemistry of the Nb- and Nb+Ti-rich MX (M stand for metal and X for C and/or N) primary precipitates (Fig. 7). Such precipitates have incorporated some of this Hf addition and became predominately round-like. Nb and Hf additions have reduced or suppressed $M_{23}C_6$ formation in the as-welded microstructure of alloys 3 and 4, as previously reported [32, 35]. Thermodynamic and kinetic calculations have shown the $M_{23}C_6$ solvus temperature and final mass fraction reduction in as-solidified condition, as shown in Fig. 5 and Table 3.

MX primary precipitates nucleation and growth are key factors to understand GB pinning during the as-solidified microstructure cooling. Calphad calculations show that in alloy 2 the MX precipitates nucleate as TiN within the liquid metal. While the metal solidification proceeds, C atoms become part of the structure, forming Ti(C,N) carbonitride, which growths cooperatively with the γ-matrix as part of an eutectic reaction. A similar process occurs with Nb and Hf, however in such case these two elements are also incorporated into the MX structure at specific temperatures [32, 48]. The MX formation before solidification is complete explains such precipitation strong effect on the GBS pinning. Therefore, changes in the MX precipitation start temperature, fraction at the end of solidification, morphology, and distribution will have an important effect on the GBs morphology, as shown by the EBSD measurements shown in Fig. 8. Exhaustive precipitate characterization and their effect on GB morphology in the Ni-Cr-Fe modified alloys is presented elsewhere [45].

Nb, Mo, and Hf additions have expanded the solidification range of the experimental alloys (Fig. 5), which has been shown to increase the solidification cracking susceptibility [49].

DDC Mechanism

DDC has been recognized as a high temperature GBS phenomenon [22, 35]. However, it is well-known that several high temperature intergranular fracture phenomena in polycrystalline materials are dominated by GBS [51]. Figure 14 shows a generalized classification of GBS in polycrystalline materials as: (a) Rachinger

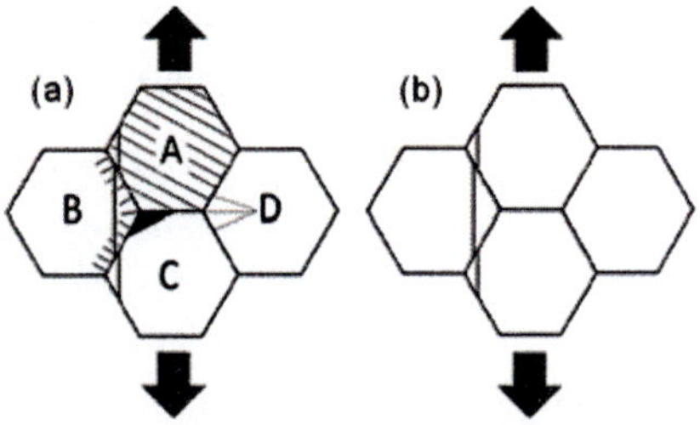

Fig. 14 Schematic representation of high temperature GBS flow mechanisms: (**a**) Rachinger mechanism, showing deformation bands (grain A), strain buildup along the GB (grain B), W-type crack opening at triple point (between grains A, B, and C), and triple point fold (grain-D). (**b**) Lifshitz mechanism related to high temperature diffusion creep. *Offset lines* indicate GBS and *arrows* show external tension axis [52]

mechanism, which is independent of diffusion creep; and (b) Lifshitz mechanism, which is dominated by diffusion creep [50, 51].

Several important characteristics of Rachinger GBS have been observed on different stages the ductility-dip cracks produced using the high temperature deformation *in situ* SEM approach. Deformation bands buildup, strain accumulation at GBs, and W-type cracks along GBs and at triple points suggest that DDC is a high temperature Rachinger type GBS phenomenon. In addition, the grains accommodation to compensate the polycrystalline aggregate deformation, which increases the number of grains along the tensile axis, while maintaining the average grains shape, especially those oriented normal to the tensile axis, confirms that DDC is a GBS phenomena, which is different from diffusion creep.

From Rachinger mechanism viewpoint, microstructural aspects as solid solution hardening, intragranular precipitation, and GB waviness are very important because of their effect on the grains internal deformation process, delaying strain accumulation along the GBs. The DDC susceptibility measurements shown in Fig. 10 suggest that the above mentioned factors play an important role delaying strain accumulation at initial stages of the modified alloys (alloys 3 and 4) deformation. The GB morphology acts blocking pure GBS at the last stages of the deformation. The massive MX precipitation may be interfering with the dislocation movement within the grains. Mo-addition promotes the increase of plastic deformation at high temperature by interaction with dislocations movement [52]. Nevertheless, ongoing research on the elemental effects on the stacking fault energy [53] in combination with high temperature deformation *in situ* studies coupled with EBSD [54] are expected to provide a more complete and fundamental understanding of the DDC mechanism.

Conclusions

A study DDC performed on four experimental alloys based on ERNiCrFe-7 alloy using thermodynamic and kinetic modeling, high temperature deformation *in situ* SEM experiments, and detailed materials characterization was developed. This has made possible the reproduction under controlled conditions of the solid state intergranular failure mechanism, called DDC and study the failure mechanism fundaments, allowing reaching the following conclusions:

1. The Nb and Hf additions promoted extensive MX primary precipitation. Nb-rich MX precipitates exhibited an elongated or Chinese-script-like morphology, while Hf-rich primary MX precipitates were predominately rounded.
2. The Nb- and Hf-rich primary MX precipitates promoted GB pinning, promoting tortuous GBs and a slight grain size reduction.
3. Mo additions in Ni-base alloys have promoted matrix hardening by solid solution and increased the plastic deformation at high temperature by interaction with dislocation movement.
4. The Nb, Hf, and Mo reduced the $M_{23}C_6$ precipitation, but exacerbated interdendritic segregation and expanded the alloy solidification range.

5. The alloys 1 and 2, without significant MX precipitation and flat GBs were more susceptible to DDC cracking than alloys 3 and 4, with Nb, Mo and Hf additions, which contained extensive MX precipitation and wavy GBs.
6. The observed characteristics of the high temperature deformation process suggest that DDC is a high temperature intergranular fracture type which is governed by Rachinger grain boundary sliding.

Acknowledgements The authors would like to thank: the Brazilian Synchrotron Light Laboratory (LNLS), FAPESP, and CNPq by founding this research; Prof. Rubens Caram Jr. from UNICAMP for allowing the use of the arc melting capabilities; Dr. Adelino Coelho from UNICAMP for allowing the use of the heat treatment furnaces; and Technological Institute of Sao Paulo – IPT staff for performing the cold rolling operations.

References

1. Noecker FF II and DuPont J.N. (2009) Metallurgical investigation into ductility dip cracking in Ni-based alloys: Part I. Welding Journal 88:7s–20s.
2. Lippold JC and Nissley NE (2008) Ductility-dip cracking in high chromium Ni-base filler metals. Hot Cracking Phenomena in welds II, Part V: Ductility-dip cracking, pp. 409–426.
3. Ramirez AJ and Lippold JC (2005) New insight into the mechanism of ductility-dip cracking in Ni-base weld metals. Hot Cracking Phenomena in welds, Part I: Phenomena and mechanism, pp. 19–41.
4. Bengough GD (1912) A study of the properties of alloys at high temperatures. Proceedings of Annual General Meeting of the Institution of Mechanical Engineers, pp. 123–174.
5. Rhines FN and Wray PJ (1961) Investigation of the intermediate temperature ductility minimum in metals. Transactions of the ASM 54:117–128.
6. Nippes EF et al. (1955) An investigation of the hot ductility of high temperature alloys. Welding Journal 23:183s–196s.
7. Hodge FG (2006, September) The history of solid-solution-strenghtened Ni alloys for aqueous corrosion service. Journal of Materials, Metallurgical and Mineralogical 58:28–31.
8. Rachinger J (1952–1953) Relative grain translations in the plastic flow of aluminum. Journal of the Institute of Metals 81:33–41.
9. Lifshitz IM (1963) On the theory of diffusion-viscous flow of polycrystalline bodies. Soviet Physics JETP 17:909–920.
10. Cordea JN et al. (1962) Investigation to determinate causes of fissuring in stainless steel and nickel-based alloy weld metals. Technical documentary report No. ASD-TDR-62-317, pp. 1–41.
11. Haddrill DM and Baker RG (1965) Microcracking in austenitic weld metal. British Welding Journal 12:411–419.
12. Hemsworth B et al. (1969, February) Classification and definition of high temperature welding cracks in alloys. Metal Construction and British Welding Journal 1:5–16.
13. Yeniscavich W (1966) A correlation of Ni-Cr-Fe alloy weld metal fissuring with hot ductility behavior. Welding Journal 45:344s–356s.
14. Arkoosh MA and Fiore NF (1972) Elevated temperature ductility minimum in Hastelloy alloy-X. Metallurgical Transactions 3:2235–2240.
15. Abralov MA and Abdurakhmanov RU (1974) Research into the especial features of the structure of Nickel with different purities, and its susceptibility to ductility-dip cracking when welded. Automation Welding 27(10):7–10.
16. Zhang YC et al. (1985) Weldability of Fe-36%Ni Alloy (report III). Transactions of JWRI 14(1):107–114.

17. Kikel MJ and Parker DM (1998) Ductility dip cracking susceptibility of Inconel filler metal 52 and Inconel alloy 690. MAO-T-98-0233. Proceedings of the 5th International Conference of Trends in Welding Research. BWX Technologies Inc., Pine Mountain, GA, pp. 1–19.
18. Arata Y et al. (1977) Solidification crack susceptibility in weld metals of fully austenitic stainless steels (Report II): Effect of ferrite, P, S, C, Si and Mn on ductility properties of solidification brittleness. Transactions of JWRI 6(1):105–116.
19. Arata Y et al. (1977) Solidification crack susceptibility in weld metals of fully austenitic stainless steels (Report III): Effect of Strain rate on cracking threshold in weld metal during solidification. Transactions of JWRI 6(2):197–206.
20. Nissley NE (2002) Development of the strain-to-fracture test to study ductility dip cracking in austenitic alloys. Master thesis, Ohio State University, Columbus, OH, p. 240.
21. Collins MG, Ramirez AJ, Lippold JC (2004) An investigation of ductility dip cracking in nickel-based weld metals- Part III. Welding Journal 83:39s–49s.
22. Ramirez AJ, Lippold JC (2004). High temperature behavior of Ni-based weld metal part II: Insight into the mechanism for ductility dip cracking. Materials Science and Engineering A 380:259–271.
23. Yamaguchi S et al. (1980) The effect of minor elements on the hot-workability of nickel-based superalloys. Philosophical Transactions of the Royal Society of London. Series A, Mathematical and Physical Sciences 295(1413):122.
24. Matsuda F (1984) Weldability of Fe-36%Ni Alloy II: Effect of chemical composition on reheat hot cracking in weld metal. Transactions of JWRI 13(2):241–247.
25. Nishimoto K, Saida K, Okauchi H (2006) Microcracking in multipass weld metal of alloy 690 part 1: Microcracking susceptibility in reheat weld metal. Science and Technology of Welding and Joining 11(4):455–461.
26. Nishimoto K, Saida K, Okauchi H (2006) Microcracking in multipass weld metal of alloy 690 part 2: Microcracking mechanism in reheat weld metal. Science and Technology of Welding and Joining 11(4):462–479.
27. Nishimoto K, Saida K, Okauchi H (2006) Microcracking in multipass weld metal of alloy 690 part 3: Prevention of microcracking in reheat weld metal by addition of La to filler metal. Science and Technology of Welding and Joining 11(4):471–479.
28. Zhang YC (1985) Weldability of Fe-36%Ni Alloy (report V). Transactions of JWRI 14(2):119–124.
29. Zhang YC (1985) Weldability of Fe-36%Ni Alloy (report VI). Transactions of JWRI 14(5):125–134.
30. Young GA et al. (2008) The mechanism of ductility dip cracking in Nickel-Chromium alloys. Welding Journal 87:31s–43s.
31. Noecker II FF and DuPont JN (2009) Metallurgical investigation into ductility dip cracking in ni-based alloys: Part II. Welding Journal 88(3):62s–77s.
32. Ramirez AJ and Garzón CM (2008) Thermodynamic and kinetic approach to ductility dip cracking resistance improvement of Ni-base alloy ERNiCrFe-7: effect of Ti and Nb additions. Hot Cracking Phenomena in welds II, Part V: Ductility-dip cracking, pp. 427–453.
33. Afonso CRM, Lippold JC and Ramirez AJ (2010) Influence of Mo and Nb additions on the ductility dip cracking susceptibility of high chromium Nickel base weld metal. Submitted: Metallurgical transactions A.
34. Nissley NE and Lippold JC (2008) Ductility-Dip cracking susceptibility of nickel-based weld metals Part I: Strain-to-Fracture testing. Welding Journal 87:257s–264s.
35. Nissley NE and Lippold JC (2009) Ductility-Dip cracking susceptibility of nickel-based weld metals Part II: Microstructural characterization. Welding Journal 88:131s–140s.
36. Kiser SD, Zhang R and Baker BA (2009) A new welding material for improved resistance to ductility dip cracking. Proceedings of the 8th International Conference on Trends in Welding Research. Pine-Mountain, GA, pp. 639–644.
37. Torres EA, Paternella FG, Caram R and Ramirez AJ (2009) In situ high temperature straining tests to study welding solid state cracking phenomena. Proceedings of the 8th International Conference on Trends in Welding Research, Pine-Mountain, GA, pp. 354–357.

38. Torres EA, Caram R and Ramirez AJ (2010) Grain boundary sliding phenomenon and its effects on high temperature ductility of Ni-base alloys. Materials Science Forum 638–642:2858–2863.
39. J Unfried S et al. (2009) Numerical modeling and experimental analysis during solidification of arc weld in Ni-Cr-Fe alloys with Hf-additions. In: Proceedings of the 9th International Seminar on Numerical Analysis of Welding. Leibnitz, Austria.
40. J Unfried S et al. (2009) Modeling of carbides precipitation during weld solidification of solid solution strengthened Ni-Cr-Fe alloys with Nb, Mo and Hf additions. In: Proceedings of the 11th International Conference on Advanced Materials. Rio de Janeiro, Brazil.
41. Zimina LN, Burova NN and Makushok OV (1986) Effect of hafnium on the structure and properties of wrought nickel-base alloys. Metal Science and Heat Treatment 28(2):130–135.
42. Kotval PS, Venables JD and Calder RW (1972) The role of Hafnium in modifying the microstructure of cast nickel-base superalloys. Metallurgical and Materials Transactions 3:453–458.
43. Dahl JM, Danesi WF and Dunn RG (1973) The partitioning of refractory metal elements in hafnium-modified cast nickel-base superalloys. Metallurgical and Materials Transactions 4:1087–1096.
44. Duhl JM, Sullivan CP (1971) Some effects of hafnium additions on the mechanical properties of a columnar-grained nickel based superalloy. Journal of Metals 23:38–40.
45. J Unfried S and Ramirez AJ (2010) Grain boundary morphology and precipitates characterization of Ni-Cr-Fe alloys modified with Nb, Mo and Hf additions. Unpublished research.
46. Fields RJ and Ashby MF (1980) Observation of wedge cavities in the SEM. Scripta Metallurgica 14:791–796.
47. Kraft T and Exner HE (1998) Restricted equilibrium phase diagram of Nickel rich corner of quaternary system Ni-Cr-Fe-Mo. Materials Science and Technology 14(5):377–381.
48. J Unfried S et al. (2010) Numerical Modeling and experimental analysis of carbides precipitation during Arc-Welding Of Ni-Cr-Fe alloys modified with Nb, Mo And Hf additions. 65° Congresso anual da Associação Brasileira de Metalurgia, Materiais e Mineração – ABM.
49. Ramirez AJ, Sowards JW and Lippold JC (2006) Improving the ductility-dip cracking resistance of Ni-base alloys. Journal of Materials Processing Technology 179(1–3):212–218.
50. Gifkins RC (1994) Grain boundary participation in high-temperature deformation: An historical review. Materials Characterization 32:59–77.
51. Langdon TG (2006) Grain boundary sliding revisited: Developments in sliding over four decades. Materials Science 41:597–609.
52. Tiearney TC and Grant NJ (1982) Measurement of structural parameters important in creep of Ni-Mo and Ni-W solid solutions. Metallurgical and Material Transactions A 13:1827–1836.
53. J Unfried S et al. (2010) Stacking fault energy measurements in Ni-Cr-Fe solid solution strengthened superalloys using synchrotron radiation. Submitted to: Materials Science and Engineering.
54. J Unfried S and Ramirez AJ (2010) A new approach to fundamental mechanism of Ductility-dip cracking in Ni-Cr-Fe alloys using in situ experiments. Unpublished research LNLS.

Further Development of the Cast Pin Tear Test for Evaluating Solidification Cracking in Ni-Base Alloys

B.T. Alexandrov and John Lippold

Introduction

The original cast pin tear test is attributed to Frederick C. Hull [1] in 1960s as a method for ranking the susceptibility of stainless steels to hot cracking during welding. This test was based on levitation melting and casting of small charges of the tested alloy in copper molds to produce cast pins with varying geometry and a constant mass (19 g). The level of restraint applied to the cast pin during solidification was controlled by the pin and mold geometry. Plots of the total length of circumferential cracks versus the mold number were used to generate arbitrary rankings of the susceptibility to hot cracking in tested alloys. The main advantages of the CPTT over other hot cracking tests are the small amount of material needed for testing (less than 500 g) and the applicability of CPTT for alloy development.

A second generation CPTT was developed in the Welding and Joining Metallurgy Laboratory at The Ohio State University [2–4]. The complex levitation melting technique of the original CPTT was replaced by gas tungsten arc (GTA) melting of the tested charge on a water-cooled open copper hearth. The varying mold and pin geometries in Hull's design were replaced by a cylindrical mold and pin with constant diameter and varying length. This allowed replacing the arbitrary mold numbering in the Hull's solidification susceptibility ranking with the actual mold/pin length. An optimized testing procedure was developed with the second-generation CPTT that provided better test reproducibility and improved test sensitivity in alloys highly susceptible to solidification cracking.

This paper presents the application of the CPTT in the development of a chromium-free welding consumable for stainless steels, and specifically to evaluate the effect of base metal dilution on the susceptibility of this consumable to solidification cracking. Development of CPTT rankings for a number of high-chromium Ni-base filler metals that are currently in use or under consideration by the nuclear industry is also described. The sensitivity to solidification cracking in these alloys is

B.T. Alexandrov (✉)
Welding Engineering Program, The Ohio State University, Columbus, OH, USA
e-mail: alexandrov.1@osu.edu

J. Lippold et al. (eds.), *Hot Cracking Phenomena in Welds III*,
DOI 10.1007/978-3-642-16864-2_16, © Springer-Verlag Berlin Heidelberg 2011

discussed in relation to their composition and solidification behavior. The latter was studied utilizing a button melting technique with simultaneous single sensor differential thermal analysis [5, 6]. The solidification cracking susceptibility rankings generated by CPTT were validated by the transverse Varestraint test.

Cast Pin Testing Procedure

The second-generation CPTT apparatus is shown in Fig. 1. A charge of the tested alloy is melted in a gas-shielded chamber over a water-cooled copper hearth by a GTA torch. The hearth (Fig. 2) has a central opening that connects the gas-shielded chamber to a copper-beryllium mold (Fig. 3). The molten charge is transferred through the hearth opening into the mold where it solidifies as a cast pin, Fig. 4. The mold cavity is designed to produce a cylindrical cast pin with restraining head and foot. The parameters of the CPTT procedure are listed in Table 1. The charge mass and corresponding cast pin lengths are shown in Table 2.

The cast pins are examined for cracks using a binocular microscope at magnifications from 10× to 70×. The projected crack length in a plane that is perpendicular to the pin axis is measured in degrees and is converted into percentage of circumferential cracking. Based on the CPTT results of particular alloy, a response curve of Maximum Circumferential Cracking (MCC) is plotted as a function of pin length, Fig. 5. The maximum pin length of no cracking and the minimum pin length of 100% circumferential cracking are used as quantitative criteria for ranking the susceptibility to solidification cracking.

Fig. 1 Cast pin tear test apparatus

Fig. 2 Open water cooled copper hearth

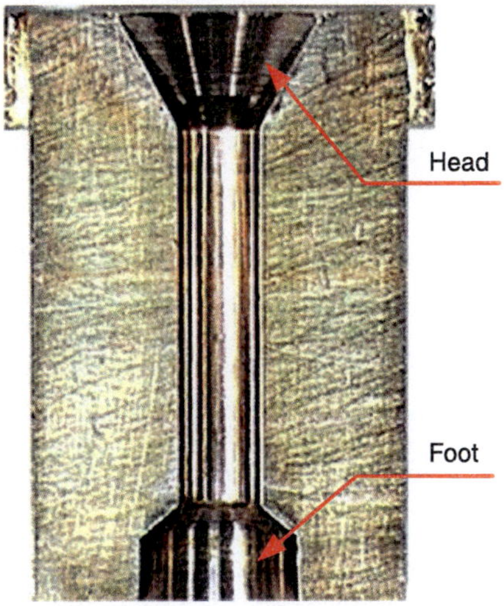

Fig. 3 Cast pin mold made of Cu-Be-Co alloy C17000

Application of the Cast Pin Tear Test

Development of a Chromium-Free Welding Consumable

The CPTT was applied in the development process of a chromium-free consumable for welding stainless steels. This consumable has been developed as a replacement of the conventional Type 308, 309, and 316 consumables in order to eliminate the hexavalent chromium (Cr(VI)) that is generated in the fume during welding of stainless steels [7–9]. The motivation for development of this consumable was the recent

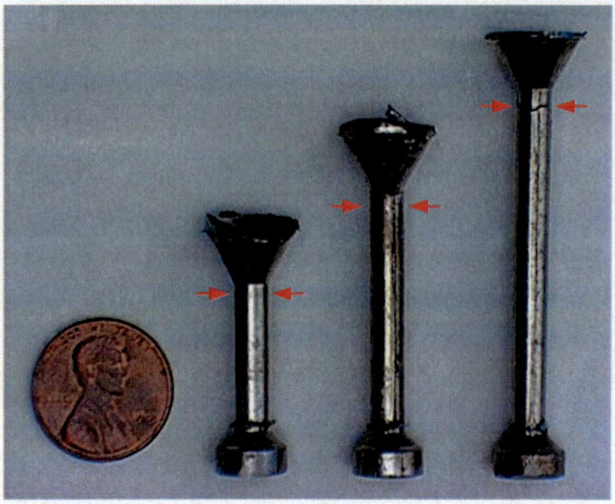

Fig. 4 Typical cast pin samples. *Arrows* show the typical location of solidification cracks

Table 1 Cast pin tear test procedure

Parameter	Value	
Mold material	Cu; Cu-Be	
Shielding gas	Ar	
Gas flow rate, ml/s	70–80	(9–10 ft^3/ h)
Arc current, A	250	
Arc length, mm	15	(0.6 in.)
Arc time, s	5–7	
Over-pressure, MPa	0.007–0.014	(1–2 psi)
Cast pin diameter, mm	9.525	(0.375 in.)
Cast pin length, mm	12.7–50.8	(0.5–2 in.)
Length increments, mm	3.175	(0.125 in.)

Table 2 Range of mold (cast pin) lengths and corresponding pin/charge volume and mass (for pure Ni samples)

Mold length, mm (in.)		Volume, mm^3	Mass, g
12.700	(0.500)	1,122.6	10.0
15.875	(0.625)	1,178.7	10.5
19.050	(0.750)	1,234.8	11.0
22.225	(0.875)	1,291.0	11.5
25.400	(1.000)	1,347.1	12.0
28.575	(1.125)	1,403.2	12.5
31.750	(1.250)	1,459.4	13.0
34.925	(1.375)	1,515.5	13.5
38.100	(1.500)	1,571.6	14.0
44.450	(1.750)	1,683.9	15.0
50.800	(2.000)	2,806.5	16.0

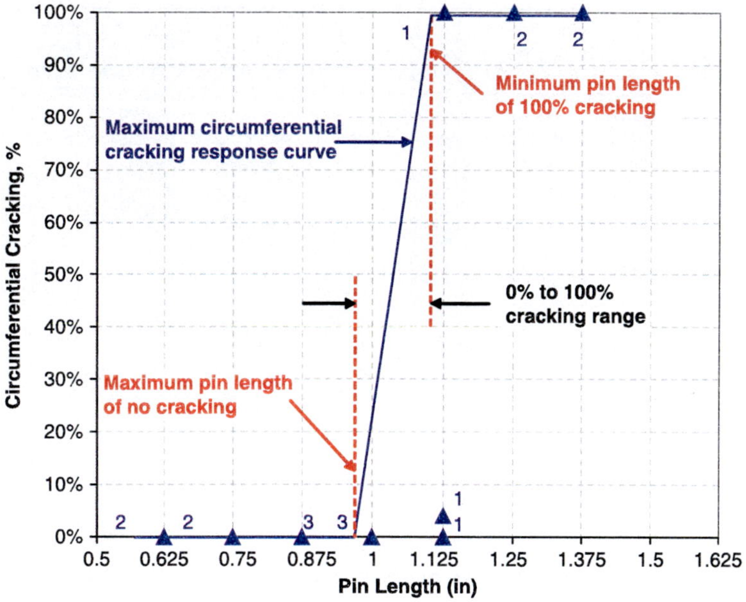

Fig. 5 Example of quantitative criteria for evaluation of susceptibility to solidification cracking by CPTT. The numbers on the plot (1, 2, or 3) indicate the number of samples tested for each mold length

reduction of the permissible exposure level of Cr(VI) from 52 to 5 micrograms/m^3 by the Occupational Safety and Health Administration (OSHA) in the United States.

Nominal compositions of Ni-7.5%Cu-1%Pd and Ni-7.5%Cu-1%Ru were evaluated during the development of this consumable. These compositions were selected to meet the corrosion resistance and mechanical properties of Type 304 steel base metal and the conventional stainless steel welding consumables used to join Type 304. As a part of the consumable development process, buttons, filler wires, and shielded metal arc electrodes with compositions close to the nominal compositions listed above were subjected to weldability testing. Since solidification cracking is a typical weldability problem in most Ni-base alloys, the effect of dilution with the 304L base metal on the solidification behavior and susceptibility to solidification cracking was studied.

In the initial stage of consumable development, the solidification ranges in alloys with compositions Ni-7.5%Cu-1%Pd and Ni-10%Cu-1%Ru were determined by the SS DTA technique using a button melting procedure [6]. The measured solidification ranges are shown in Fig. 6. The undiluted Ni-Cu-Pd alloy had significantly wider solidification range than the Ni-Cu-Ru alloy (114°C vs. 46°C). Dilution of the Ni-Cu-Pd alloy with 15% 304L base metal did not have a significant effect, but at 30% dilution the solidification range was increased to 146°C with formation of a eutectic phase. By increasing the dilution to 30%, the solidification range in the Ni-Cu-Ru alloy progressively increased to 117°C. In contrast, the solidification

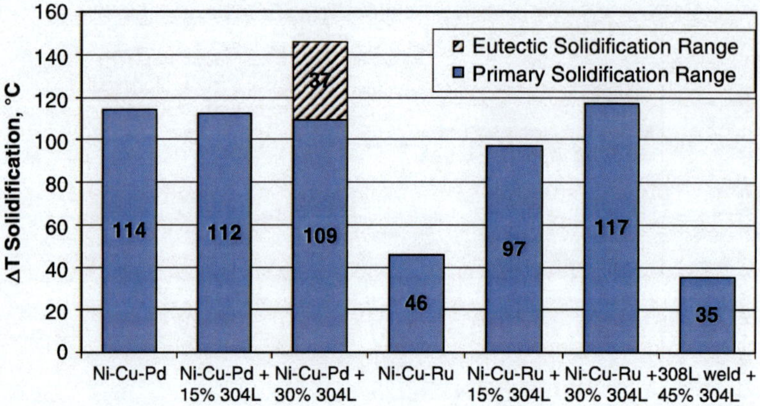

Fig. 6 Solidification ranges in Ni-7.5%Cu-1%Ru and Ni-10%Cu-1%Ru alloys at different dilutions with Type 304L base metal compared to diluted 308L weld metal. The solidification ranges are determined using the SS DTA button melting technique [3]

range of a conventional 308L consumable at 45% dilution with 304L base metal was only 35°C.

This data suggests an increase in solidification cracking susceptibility of the Cr-free consumables relative to Type 308L. The solidification range data also suggests an increase in cracking susceptibility of these consumables with increased dilution by Type 304L. While no solidification cracking was observed during sample fabrication during the development program, solidification ranges from 100 to 150°C generally indicate a "moderate" susceptibility to solidification cracking.

At a later stage of consumable development, the CPTT was utilized to test and rank the susceptibility to solidification cracking of a Ni-7.5Cu-1Ru-0.5Al-0.5Ti filler wire [7, 8]. The composition of this electrode is shown in Table 3. The maximum cracking response curves generated by CPTT for non-diluted Ni-Cu-Ru wire and for 25% and 50% dilution with steel 304L base metal are compared to three Ni-base alloys in Fig. 7.

Table 3 Chemical composition (wt%) of Ni-Cu-Ru filler wire

Element	Content
Ni	89.30
Cu	8.20
Ru	1.36
Al	0.56
Ti	0.53
C	0.014
N	<0.001
O	0.0031
S	<0.001
P	<0.005

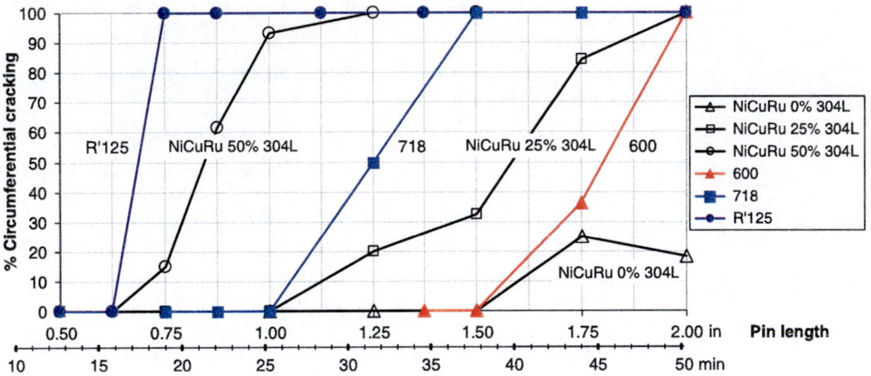

Fig. 7 Effect of dilution with steel 304L base metal on the solidification cracking susceptibility in a Ni-Cu-Ru filler wire; CPTT ranking [7, 8]

The undiluted Ni-Cu-Ru weld metal had better resistance to solidification cracking than Inconel 600. The minimum pin length for crack initiation was 1.5 in. and 100% cracking was not achieved within the available range of pin lengths up to 2 in. The maximum circumferential cracking response curves at 25% and 50% dilution were located, respectively, between the cracking response curves of Alloys 600 and 718 and between these of Alloy 718 and René 125.

The fracture surface in a completely separated pin (100% circumferential cracking) at 25% dilution is shown in Fig. 8. The apparent interdendritic cracking is consistent with a typical solidification cracking morphology. The primary dendrite

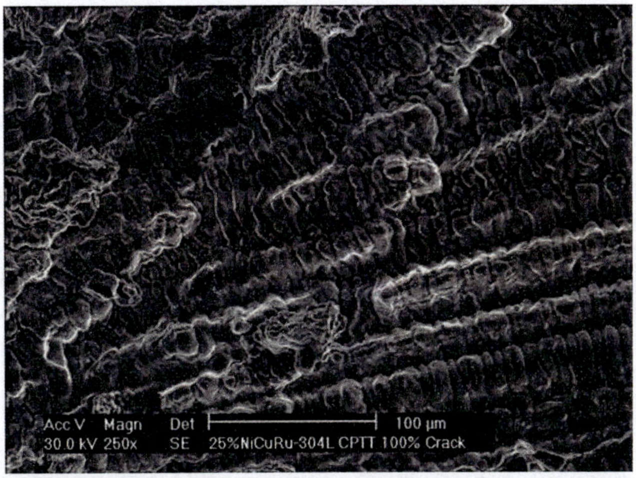

Fig. 8 Fracture surface in fully separated 2 in. long cast pin of Ni-Cu-Ru filler wire diluted 25% by Type 304L [7, 8]

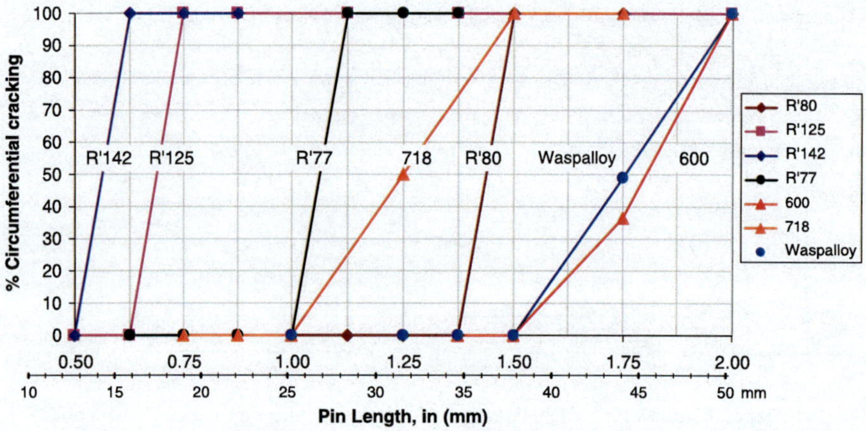

Fig. 9 CPTT ranking of solidification cracking susceptibility in several Ni-base superalloys and Alloy 600 [4]

arm spacing was determined to be approximately 25 μm, which was similar to the solidification substructures observed in weld deposits of this consumable [7, 8].

A previously published CPTT ranking for Ni-base alloys (Fig. 9, [4]) ranked Alloy 600 as fairly resistant, Alloy 718 as moderately susceptible, and René 125 as highly susceptible to solidification cracking. Compared to this ranking, the weld metal of Ni-Cu-Ru filler wire can be considered as: (1) fairly resistant to solidification cracking when undiluted with Type 304 base metal, (2) moderately susceptible at low dilution levels, and (3) susceptible to solidification cracking at high dilution (50%). This behavior is in good agreement with the trend of the solidification range data shown in Fig. 6. This increase in cracking susceptibility with dilution by Type 304 is attributed, in part, to the increase in Fe content of the weld deposit.

Use of the CPTT to Evaluate High-Cr, Ni-Base Filler Metals

High-chromium, Ni-base filler metals 82 (ERNiCr-3), 52 (ERNiCrFe-7), and 52M (ERNiCrFe-7A) are used for cladding and dissimilar metal welding applications in commercial nuclear power plants. Alloys 52 and 52M have superior resistance to stress corrosion cracking relative to Alloy 82, but are susceptible to ductility-dip cracking (DDC) during welding under high restraint conditions. In order to improve the resistance to DDC, a new high chromium nickel-base filler metal 52MSS (ERNiCrFe-13) with higher Nb and Mo content was recently developed [6]. Due to its higher Nb content, alloy 52MSS was expected to be more susceptible to solidification cracking than Alloys 52 and 52M which it is designed to replace.

The solidification behavior in two heats of Alloy 52MSS (52MSS-A and 52MSS-B), a single heat of Alloy 52M, and in a modified heat of Alloy 82 (MLTS-2) was studied using SS DTA. The CPTT was utilized to rank the solidification

Table 4 Chemical composition of the high-chromium Ni-base filler metals

Element, wt%	52M	52MSS-A	52MSS-B	MLTS-2
Al	0.09	0.07	0.24	0.06
B	0.0004	–	–	0.0002
C	0.02	0.03	0.018	0.031
Co	0.011	<0.001	0.003	–
Cr	30.11	29.92	29.20	26.88
Cu	0.03	0.06	0.055	–
Fe	8.87	8.31	8.63	3.00
Mn	0.72	0.19	0.70	3.16
Mo	0.05	3.83	3.68	–
Nb	0.87	2.57	2.4	2.65
Ni	59.21	54.67	54.67	63.84
P	0.002	<0.001	0.016	0.003
S	0.0005	0.001	0.0006	0.0006
Si	0.11	0.12	0.15	0.15
Ta	<0.01	0.017	0.013	–
Ti	0.16	0.19	0.21	0.19
Zr	0.001	–	–	0.0002

cracking susceptibility in these filler metals and to compare cracking behavior to other Ni-base alloys and filler metals. The CPTT ranking was validated using the Transvarestraint test. The chemical compositions of the filler metals tested are presented in Table 4.

The solidification behavior of these filler metals was conducted utilizing a new SS DTA button-melting procedure that accurately replicates the solidification process in GTA welds of Ni-base alloys [6]. This procedure includes melting by GTA pulse of a small (1 g) button of the alloy over a 20 g base button of the same composition as shown in Fig. 10. The cooling thermal history in the test button is measured by a Type C (W/W-Re) thermocouple and recorded by a fast sampling rate data acquisition system. The solidification range in the tested filler metal is determined by processing the acquired thermal data using the SS DTA software.

Typical solidification curves in the tested alloys determined by SS DTA are presented in Fig. 11. These curves show a varying evolution rate of latent heat of fusion, which changes the slope on the SS DTA curve (indicated as “kinetics change” in Fig. 11). The rate of evolution of latent heat of fusion in alloys MLTS-2 and 52MSS significantly decreases between the points designated in Fig. 11 as “kinetics change” and “eutectic start”. The solidification process finishes with a large thermal effect associated with formation of eutectic phase, Fig. 11a–c. There is only a slight change in the solidification kinetics of alloy 52M without a separate thermal effect of eutectic solidification, Fig. 11d.

The solidification parameters in the filler metals determined by SS DTA are presented in Table 5. Alloy MLTS-2 has the widest solidification range and eutectic reaction temperature range, which can be attributed to its potential to form a gamma-NbC eutectic (highest carbon and niobium content). Alloys 52MSS-A and

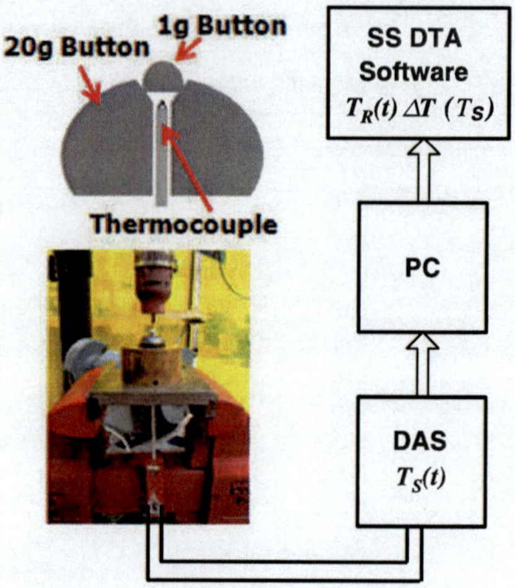

Fig. 10 SS DTA device for phase transformation analysis in GTA welds

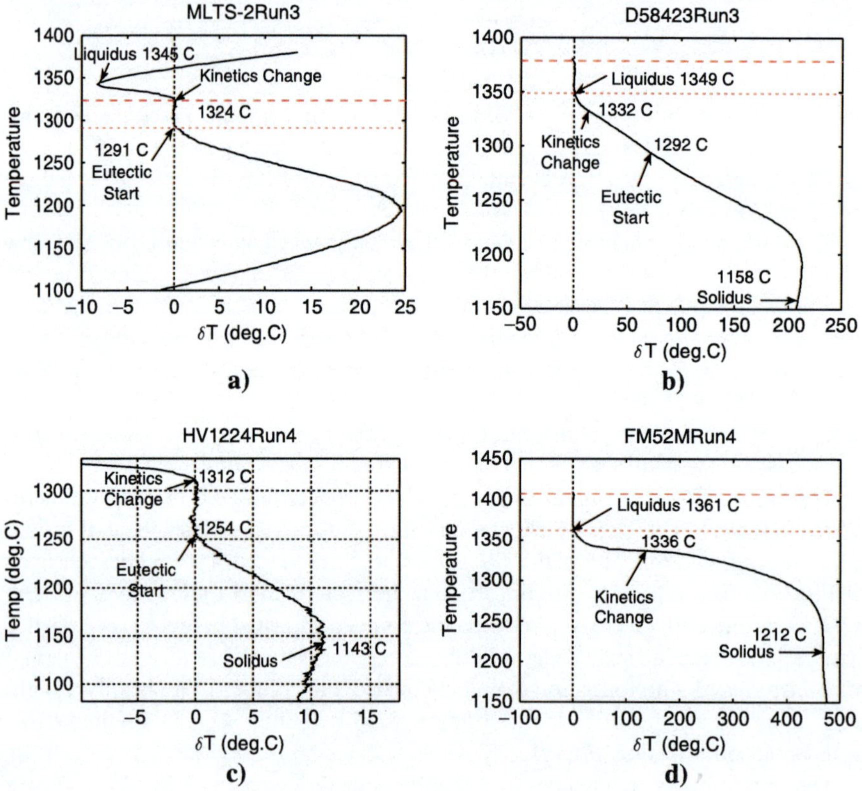

Fig. 11 Solidification behavior in alloys MLTS-2, 52MSS-A, 52MSS-B, and 52M determined by SS DTA

Table 5 Solidification parameters in the tested alloys determined by SS DTA (average of five measurements)

Solidification parameters, °C	52MSS-A	52MSS-B	MLTS-2	FM52-M
Liquidus	1351	1340	1343	1358
Kinetics change	1325	1319	1317	1340
Eutectic start	1290	1261	1286	–
Solidus	1158	1148	1112	1214
Eutectic range	132	113	174	–
Solidification range	192	192	232	144

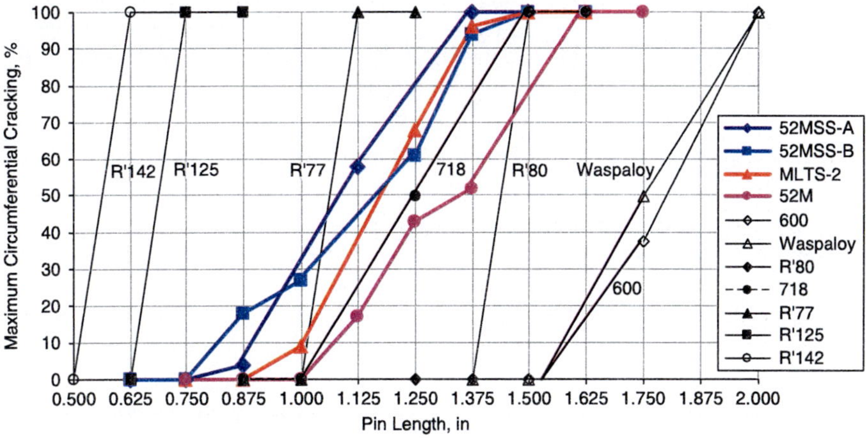

Fig. 12 Comparative ranking of filler metals 52M, MLTS-2, 52MSS-A, and 52MSS-B to some Ni-base superalloys and Alloy 600

52MSS-B have equal solidification ranges, but the former has a wider eutectic temperature range that can be explained by its higher carbon content and slightly higher niobium content. Alloy 52M has the smallest solidification range with no definitive thermal effect of eutectic reaction on its SS DTA curve (Fig. 12d). Due to its low carbon and niobium content, this alloy is not expected to form any significant fraction of eutectic phase.

The results of cast pin testing in the four high-chromium, Ni-base filler metals are presented in Table 6 and Fig. 12. Table 6 summarizes the maximum pin length with no cracking, the minimum pin length with 100% cracking, and the zero to 100% cracking in each tested alloy. In Fig. 12, the susceptibility to solidification cracking in these alloys is compared to some Ni-base superalloys and filler metals previously tested by CPTT [4].

Based on the CPTT results, 52M is the least susceptible to solidification cracking, followed by MTLS-2, 52MSS-B, and 52MSS-A. The CPTT ranking of alloys 52M, 52MSS-B, and 52MSS-A correlates well with the solidification behavior reported in Table 5. Due to their wider solidification ranges and formation of eutectic phases, alloys 52MSS-A and 52MSS-B are more susceptible than alloy 52M. The CPTT

Table 6 CPTT solidification cracking parameters in the tested alloys

Alloy	Max. pin length with no cracking, in.	Min. pin length with 100% cracking, in.	Zero to 100% pin length range, in.
52MSS-A	0.750	1.375	0.625
52MSS-B	0.750	1.500	0.750
MLTS-2	0.875	1.500	0.625
FM52-M	1.000	1.625	0.625

resolved the effect of small composition variations in the tested heats of alloy 52MSS. The higher carbon and niobium content that extended the eutectic range in alloy 52MSS-A resulted in a higher susceptibility to solidification cracking (shorter pin length for 100% cracking, Table 6 and Fig. 12) as compared to alloy 52MSS-B. The lower susceptibility to solidification cracking in alloy MLTS-2 than in 52MSS-A and 52MMS-B is believed to be the result of crack healing by eutectic liquid. Since MLTS-2 would be expected to form the highest fraction of eutectic phase at the end of solidification, crack healing is facilitated.

Figure 12 also shows that Alloy 52M has better resistance to solidification cracking than the Ni-base superalloy 718 and is more susceptible than Waspaloy and Alloy 600. The susceptibility to solidification cracking in alloys 52MSS-A, 52MSS-B, and MLTS-2 is in the range of some moderately susceptible Ni-base superalloys as René 77 and René 80. These results also confirm that the sensitivity of CPTT is sufficient to differentiate the resistance to solidification cracking in various high-chromium Ni-base filler wires and in different heats of the same consumable (52MSS-A and 52MSS-B).

The fracture surface in the cast pins of all the filler metals exhibited the typical solidification cracking "egg-crate" morphology, as shown in Fig. 13. This is additional proof that the CPTT accurately replicates the solidification cracking process in GTA welds of these filler metals. The precipitates found on the fracture surface of MLTS-2 (Fig. 14b) were identified as Nb-rich, suggesting the presence of NbC from the eutectic reaction at the end of solidification. It should also be noted that the MLTS-2 fracture morphology is different than that of the other alloys and perhaps suggests the presence of more liquid along the fracture path at failure. This supports the theory of eutectic healing which may have improved the solidification cracking resistance of this filler metal relative to 52MSS (Fig. 12).

For comparison and validation of the CPTT results, Transvarestraint testing of the four high-chromium Ni-base filler metals was also conducted and is summarized Fig. 14 as a plot of the Maximum Crack Distance (MCD) versus the applied strain. The Transvarestraint test provided the same solidification cracking ranking as the CPTT: alloy 52M is the most resistant to solidification cracking followed by MLTS-2, 52MSS-B, and 52MSS-A.

Note that the MCD of alloys 52MSS-A and 52MSS-B is larger than alloy MLTS-2 at lower strain levels (below 1% augmented strain). This behavior replicates the larger maximum circumferential cracking (MCC) in alloys 52MSS-A and

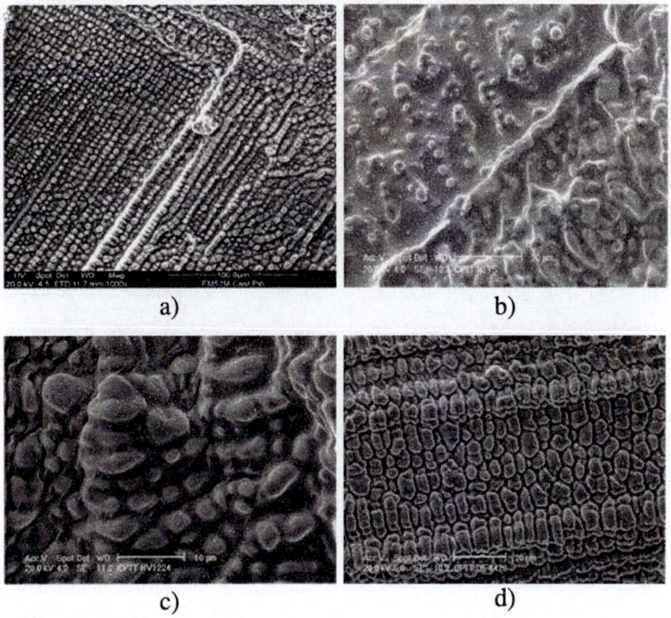

Fig. 13 Fracture surfaces in fully cracked pins of high-chromium Ni-base filler metals tested by CPTT: (**a**) 52M, (**b**) MLTS-2, (**c**) 52MSS-A, (**d**) 52MSS-B. Note different magnifications

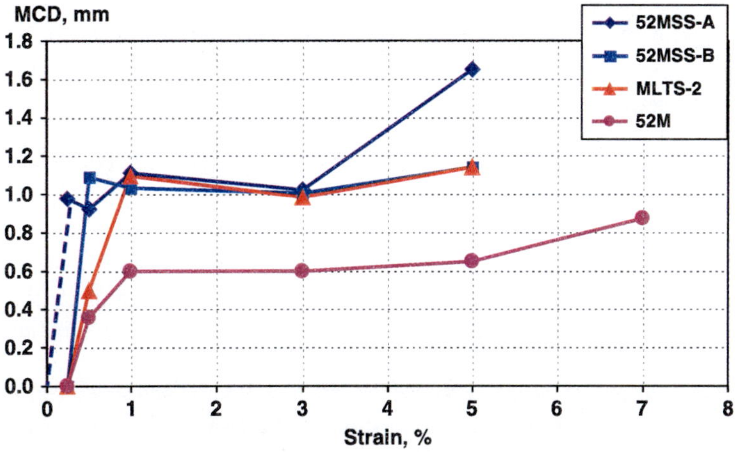

Fig. 14 Transvarestraint test results from high chromium Ni-base filler metals: Maximum crack distance (MCD) as function of the applied strain

52MSS-B relative to MLTS-2 at lower CPTT restraint levels (pin lengths between 19.05 and 28.575 mm), as shown in Fig. 13.

At 5% augmented strain, the MCD of alloy 52MSS-A is larger than in alloys 52MSS-B and MLTS-2, which corresponds to the larger MCC in alloy 52MSS-A

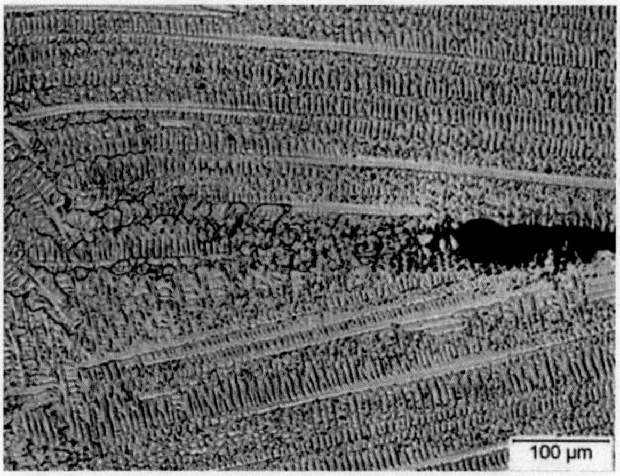

Fig. 15 Crack backfilling in Transvarestraint test sample of alloy MLTS-2

as compared to alloys 52MSS-B and MLTS-2 at 1.375 in. pin length. The slightly better resistance to solidification cracking at lower strain levels in alloy MLTS-2 as compared to alloys 52MSS-A and 52MSS-B could again be explained by more effective crack healing resulting from the larger amount of eutectic phase formed in this alloy. A crack backfilling effect in Transvarestraint test sample of alloy MLTS-2 is shown in Fig. 15.

In summary, the Transvarestraint test results for the four high-chromium Ni-base filler metals validates the solidification cracking susceptibility rankings in these alloys generated by the CPTT and the sensitivity of the CPTT to resolve the effect of relatively small heat-to-heat composition variations.

Conclusions

1. The modified cast pin tear test provides a reliable and efficient tool for evaluation and ranking the susceptibility to solidification cracking in precipitation strengthened and solid solution strengthened Ni-base alloys and welding filler metals.
2. The CPTT successfully differentiated the effect of dilution and small heat-to-heat composition variations on the solidification cracking susceptibility of several Ni-base filler metals.
3. Simultaneous application of the CPTT and the technique for single sensor differential thermal analysis allows the correlation of solidification temperature range and solidification cracking susceptibility.
4. The effect of filler metal dilution by the base metal was also demonstrated by the combination of SS DTA and CPTT of Ni-Cu consumables diluted with Type 304 stainless steel. The solidification ranges of the undiluted Ni-7.5%Cu-1%Pd and Ni-10%Cu-1%Ru filler wires are 114°C and 46°C respectively, but increase

to 146°C and 117°C at 30% dilution. Using CPTT, the Ni-7.5Cu-1Ru-0.5Al-0.5Ti filler wire was ranked as fairly resistant to solidification cracking at 0% dilution, as moderately susceptible to craking at 25% dilution, and as susceptible to cracking at 50% dilution.

5. The CPTT and the Transvarestraint test provided equivalent ranking of the solidification cracking susceptibility in the high-chromium Ni-base filler metals. Filler metal 52M was the most resistant to cracking followed by the modified filler metal 82 (MLTS-2), 52MSS-B and 52MSS-A.
6. The solidification cracking susceptibility in the high-chromium Ni-base filler metals directly corresponds to their solidification behavior. Alloy 52M has the smallest solidification range with no evidence of a eutectic reaction. Alloy MLTS-2 has the widest solidification range and the largest thermal effect of a eutectic solidification. Its better resistance to cracking than alloy 52MSS is related to crack healing by the eutectic phase healing. Both heats of alloy 52MSS had the same solidification range, but the more susceptible one (52MSS-A) exhibited a wider eutectic temperature range.

Acknowledgements The research presented in this paper has been supported by the Strategic Environmental Research and Development Program (SERDP) and the Electric Power Research Institute (EPRI). We would also like to thank Dr. Jeffrey Sowards, Mr. Adam Hope, and Mr. Ben Sutton who conducted many of the tests that were reported here.

References

1. Hull F (1959) Cast-Pin Tear Test for Susceptibility to Hot Cracking. Welding Journal 38: 176–181.
2. Ryan D (2003) Development of Modified Cast-Pin Tear Test to Evaluate the Solidification Cracking Susceptibility of Fully Austenitic Materials. The Ohio State University: MS Thesis.
3. Alexandrov B Nissley N Norton S Lippold J (2006) Development of a Weldability Test for High Performance Base and Filler Materials, Edison Welding Institute: Report No. MR0606.
4. Alexandrov B Lippold J Nissley E (2008) Evaluation of Weld Solidification Cracking in Ni-Base Superalloys Using the Cast Pin Tear Test. Springer-Verlag Hot Cracking Phenomena in Welds II:193–214.
5. Alexandrov B Lippold J (2009) Method and Device for Investigation of Phase Transformations in Metals and Alloys. United States Patent No. 7,473,028.
6. Alexandrov B Hope A Sowards J Lippold J McCracken S (2009) Susceptibility to Solidification Cracking in High Chromium Nickel-Base Weld Filler Metals. EPRI First International Conference Welding and Fabrication Technology for New Power Plants.
7. Development of Chromium-Free Welding Consumables for Stainless Steels (2009) The Ohio State University: SERDP Project PP-1415 Final Report..
8. Sowards J (2009) Development of a Chromium-free Consumable for Joining Austenitic Stainless Steels, The Ohio State University: PhD Dissertation.
9. Sowards J Liang D Alexandrov B Frankel G Lippold J (2010) Solidification Behavior and Weldability of Dissimilar Welds between a Cr-free, Ni-Cu Welding Consumable and Type 304L Austenitic Stainless Steel. Metallurgical and Materials Transactions: submitted for publication.

Behavior and Hot Cracking Susceptibility of Filler Metal 52M (ERNiCrFe-7A) Overlays on Cast Austenitic Stainless Steel Base Materials

Steven L. McCracken and Richard E. Smith

Background

The application of weld overlays (WOLs) to mitigate SCC in both pressurized water reactors (PWRs) and boiling water reactors (BWRs) has been effectively used for nearly three decades. Initially WOLs were designed using stainless steel filler metals to mitigate IGSCC in stainless steel to stainless steel pipe weldments. Subsequently, WOLs were applied to dissimilar metal weld (DMW) joints where stainless steel piping joined low alloy steel (LAS) components such as vessel nozzles and valve bodies. Nickel-base weld filler metals are typically used to join these dissimilar metal components. Because stainless steel filler metals are not suitable for welding over nickel-base metal due to weld solidification cracking issues, a 30% chromium nickel-base filler metal similar in composition to Inconel 690 is used. This high chromium composition is resistant to stress corrosion cracking and is the filler metal of choice for DMW overlays.

Previous 52M Hot Cracking Experience on Wrought Piping

In the fall of 2006 an attempt was made to install a 52M WOL on the pressurizer surge nozzle at the San Onofre Unit 3 Nuclear Generating Station [1]. Welding was done with the machine gas tungsten arc (GTA) welding process with water in the surge piping. Welding was in the horizontal (2G) position. The WOL on the surge nozzle was not successful due to 52M hot cracking. Weld overlays on the spray nozzle, relief nozzle, and three safety nozzles on top of the pressurizer were successfully welded with no hot cracking problems.

The surge nozzle geometry comprised a short section of SA376 Type 316 (0.017% S, 0.25% P, and 0.40% Si) stainless steel surge line pipe, a short reducing safe end of SA351 CF8M cast stainless steel, and a short portion of the SA508 Class 2 surge nozzle. The stainless pipe to cast safe end was welded with 308L filler

S.L. McCracken (✉)
Electric Power Research Institute, Charlotte, NC, USA
e-mail: smccracken@epri.com

J. Lippold et al. (eds.), *Hot Cracking Phenomena in Welds III*,
DOI 10.1007/978-3-642-16864-2_17, © Springer-Verlag Berlin Heidelberg 2011

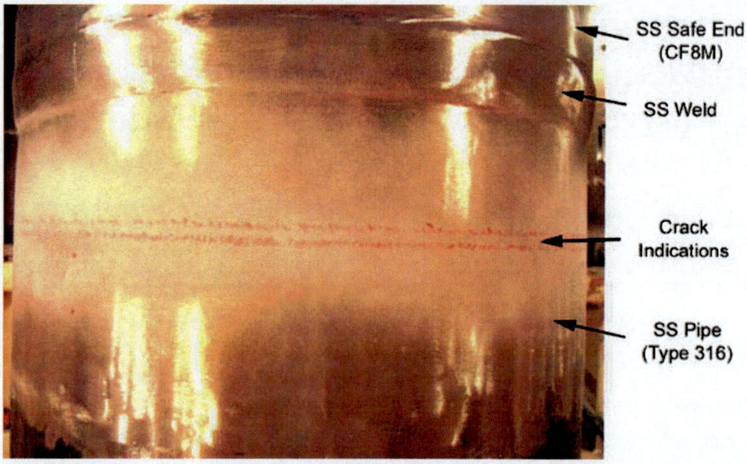

Fig. 1 Liquid penetrant indications observed after initial three beads of 52M deposit on Type 316 stainless steel surge pipe [1]

metal and the safe end to nozzle was welded with 82/182 filler metal. The WOL design covered all or a portion of each of these materials.

Hot cracking occurred (Fig. 1) with the initial 52M weld beads on the SA376 Type 316 wrought pipe. A second attempt just above the initial location also resulted in cracking. Due to schedule constraints the WOL work on the surge nozzle was halted and all 52M weld metal was subsequently removed pending root cause investigation. The unit was returned to service after the five nozzles on top of the pressurizer were successfully overlaid with 52M weld metal.

Root cause investigations and weld demonstrations determined that the high levels of S and P (0.017% and 0.25% respectively) in the SA376 Type 316 stainless steel pipe were responsible for the 52M hot cracking. It was determined that a single ER308L stainless steel buffer layer could be deposited on the surface of the stainless steel pipe to minimize dilution of impurity elements into the 52M deposit. Moreover, the ER308L weld metal is resistant to hot cracking since primary solidification is delta ferrite. This buffer layer technique (Fig. 2) was demonstrated on a

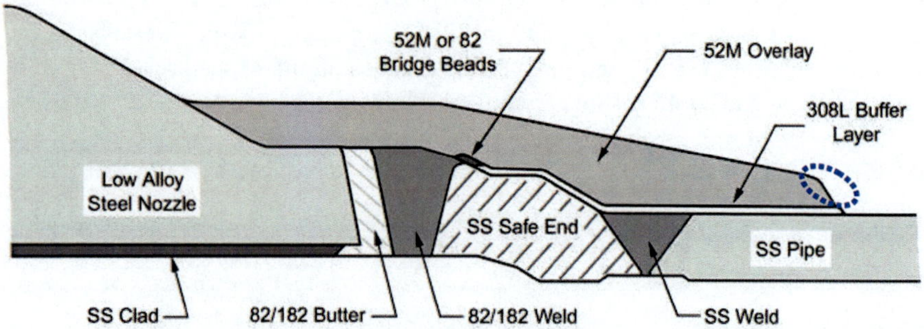

Fig. 2 Schematic of typical Weld Overlay (WOL) with the area of 52M hot cracking over austenitic stainless steel piping noted by the dashed ellipse

full scale mockup and subsequently used to install weld overlays on the pressurizer surge nozzles at San Onofre Unit 2 and Unit 3.

To date more than 150 weld overlays using 52M have been successfully installed using the stainless steel (ER308L or ER309L) buffer layer approach to control and mitigate the potential for 52M hot cracking due to high S and P levels in the stainless steel pipe or elbow.

52M Hot Cracking on Cast CF8A Piping

During the spring of 2007 weld overlays with 52M were attempted on the large diameter reactor vessel hot leg nozzles in a PWR nuclear power plant (Fig. 3). The weld overlays were not successful due to 52M hot cracking. The hot leg nozzle materials were similar to the San Onofre surge nozzle except the pipe was centrifugally cast SA351 CF8A stainless steel pipe and the safe end was wrought SA376 Type 304 stainless steel. The nozzles in both cases were forged SA508 Class 2 low alloy steel. Stainless steel E308L welds joined the pipe to safe end and 82/182 filler metal was used for the dissimilar metal weld (DMW). The CF8A cast pipe and weld filler metals in vicinity of the most severe hot cracking are listed in Table 1. Each of the listed materials is in compliance with the applicable specification composition ranges and limits.

The unexpected hot cracking was located in the area of the 1st and 2nd 52M weld beads deposited on the ER308L buffer layer (Fig. 4). The ER308L stainless

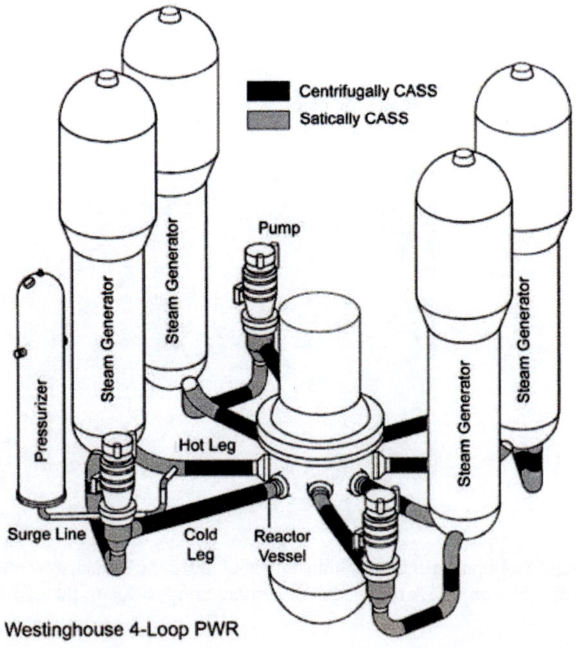

Fig. 3 Schematic of typical PWR reactor coolant system layout

Table 1 CF8A pipe, ER308L buffer layer, and 52M weld metal compositions

	Cr	Ni	Fe	Mn	Si	Nb	Al	Ti	Mo	C	S	P
CF8A	20.5	8.4	69.3	0.65	1.03	0.01	0.006	0.001	0.02	0.040	0.015	0.022
ER308L	20.1	10.5	66.9	1.90	0.52	–	–	–	0.08	0.018	0.001	0.014
52M	29.0	60.8	8.2	0.68	0.09	0.8	0.130	0.200	0.01	0.020	0.005	0.005

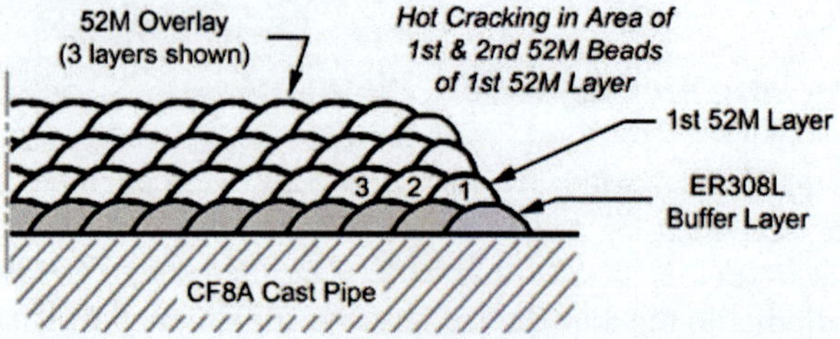

Fig. 4 Schematic of ER308L buffer layer with three layers of 52M weld metal (detail of dashed area in Fig. 1)

steel buffer layer was examined by liquid penetrant prior to installing the 52M weld metal and no cracking was observed. Moreover, the welding procedure used for the 52M deposit was similar to procedures that had been used successfully with the San Onofre surge nozzle overlays and for many other overlays applied using the stainless steel buffer layer technique. Similar hot cracking was replicated on a mockup that simulated the welding parameters, configuration, and pipe material in the plant (Fig. 5).

Unfortunately, due to schedule constraints and the inability to resolve the 52M hot cracking, the weld overlay work was halted. Four small boat samples of the 52M

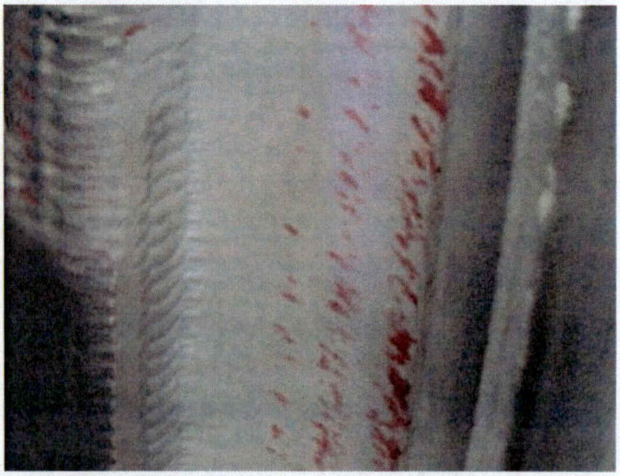

Fig. 5 Liquid penetrant indications of 52M hot cracks welded on an ER308L buffer layer deposited on CF8A base metal of similar composition to the cast pipe in the plant

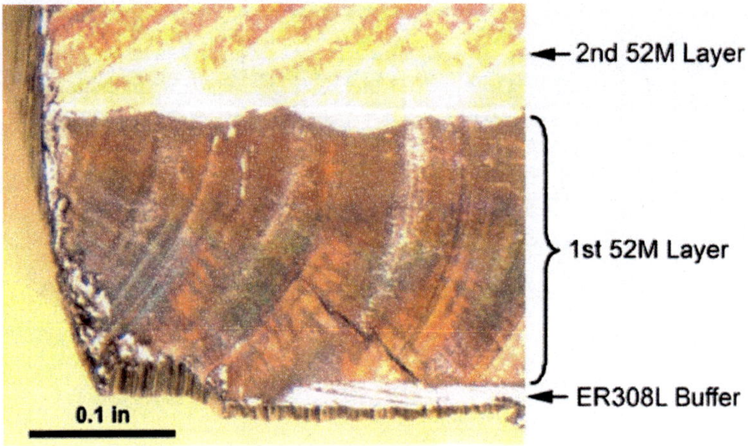

Fig. 6 Boat sample with hot cracks in the *1st* 52M layer

overlay deposits were taken in areas where hot cracking occurred (Fig. 4). The 52M and ER308L WOL weld metal were removed and the affected CF8A base material was examined to ensure the surface was free of rejectable indications.

Visual examination of the boat samples showed that the cracking was contained in the 1st layer of 52M (Fig. 6) and was generally in the 1st and 2nd beads of the 1st 52M layer (Fig. 4). There was no evidence of cracking in the ER308L buffer layer or 52M beads beyond the 1st layer.

52M Hot Cracking Investigation and Root Cause Analyses

A multidisciplinary team was assembled to collect pertinent data and investigate the 52M hot cracking. The following sections detail this investigation and present data and information obtained.

ER308L and 52M Welding Parameters

Weld overlays on the hot leg nozzles were done with the machine GTA welding process with water in the piping. Weld parameters used for the ER308L buffer and 1st 52M layer were in the 70 kW/in^2 power ratio range based on previous experience and mockup testing. The power ratio is a relationship developed at EPRI to control dilution in machine GTA welds where the wire feed is independent of the welding energy [2]. The power ratio equation is given by:

$$\mathit{Power\ ratio} = \frac{\mathit{Power}}{\begin{matrix}\mathit{Cross \cdot sectional\ area}\\ \mathit{of\ deposited\ metal}\end{matrix}} = \frac{\mathit{Amperage} \times \mathit{Voltage}}{\left[\frac{\mathit{Wire\text{-}feed\ speed}}{\mathit{Travel\ speed}} \times \begin{matrix}\mathit{Cross\text{-}sectional}\\ \mathit{area\ of\ filler}\end{matrix}\right]}$$

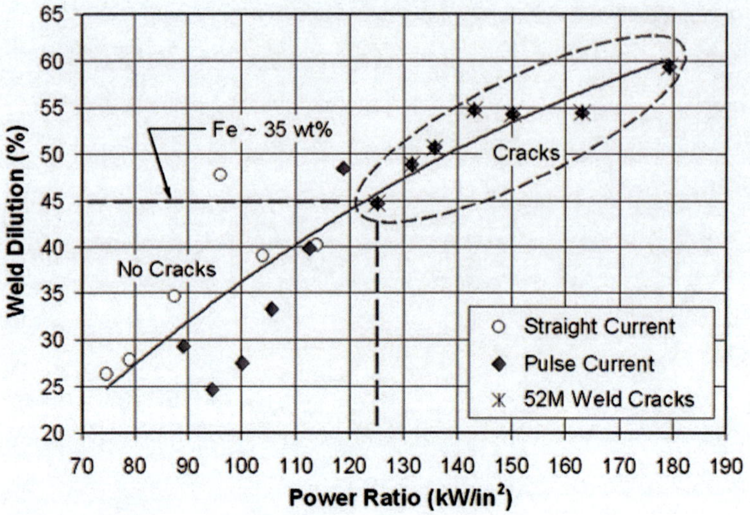

Fig. 7 Relationship of dilution versus power ratio for 52M deposits on Type 304L austenitic stainless steel base metal welded in the horizontal 2G position (dilution is ~10% higher for welding in the vertical up 5G position)

It can be seen that for a given heat input, an increase in filler wire volume produces a lower power ratio with a corresponding decrease in dilution. The parameter is useful for controlling the proportion of the weld arc energy used to melt the substrate versus the portion of arc energy used to melt the weld wire.

Figure 7 shows the relationship of power ratio versus dilution for 52M weld metal deposited on Type 304L stainless steel in the horizontal (2G) position. These data were developed by EPRI as part of the root cause evaluation and GTA welding process development following the 52M hot cracking experience on the Type 316 wrought surge pipe at San Onofre. These data show no cracks below 125 kW/in^2 power ratio; where dilution is generally less than 45% and the calculated Fe is below 35 wt% (the significance of Fe will be discussed later). It is important to note, that dilution is also a function of weld position. Experimental testing shows that dilution in the pipe vertical up (5G) position is approximately 10% higher than dilution in the horizontal (2G) position at the same power ratio. Even with an increase of 10% dilution for welding in the 5G position, a power ratio of 70 kW/in^2 is well below the threshold where hot cracking would be expected.

Dilution of ER308L and 52M by CF8A Cast Pipe

Testing on Type 304L base metal (Fig. 7) indicates that 52M welded in the vertical up (5G) position with the prescribed 70 kW/in^2 power ratio would result in single bead dilution in the range of 35%. Dilution for a layer with multiple beads with approximately 50% overlap would be about 20% due to adjacent bead dilution effects. In the field the observed ER308L dilution on the CF8A pipe and the 1st 52M layer dilution on the ER308L buffer layer were much higher.

The measured dilution of ER308L test beads on a CF8A cast pipe section with 1.29% Si, 0.019% S and 0.032% P using the same welding parameters as prescribed in the field was as high as 70% (Fig. 8). Comparison of ER308L single beads deposited on two 304 stainless steel plates and on the CF8A pipe showed that dilution was significantly higher on the CF8A cast pipe (Fig. 9). The data also suggests that stainless steel base metals that contain high Si and S promote an increase in dilution.

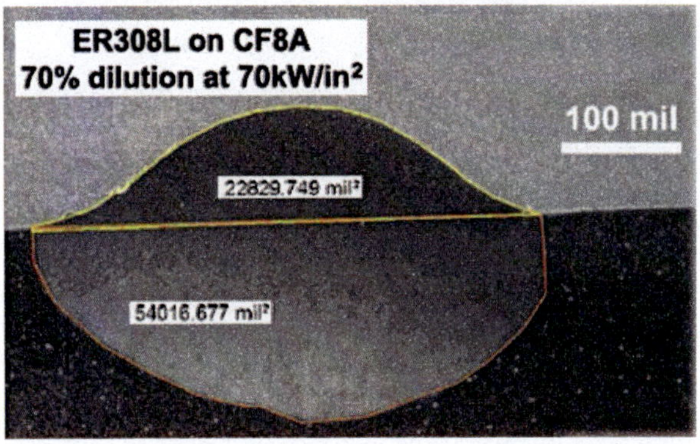

Fig. 8 Cross-section of single ER308L bead deposited on CF8A cast pipe with 70% dilution

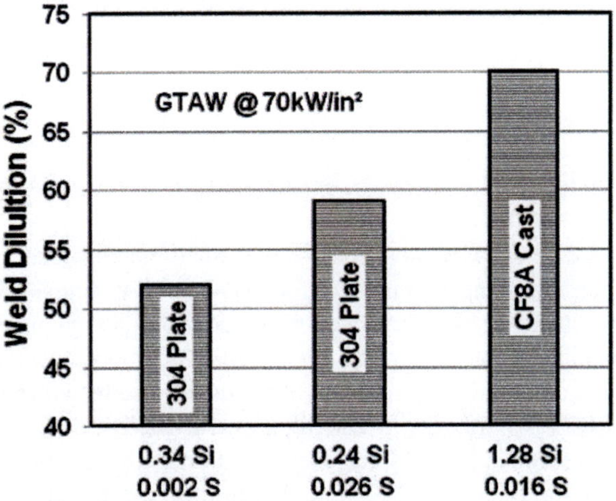

Fig. 9 ER308L single bead dilution on 304 rolled plate and CF8A cast pipe (welding position was 5G vertical up)

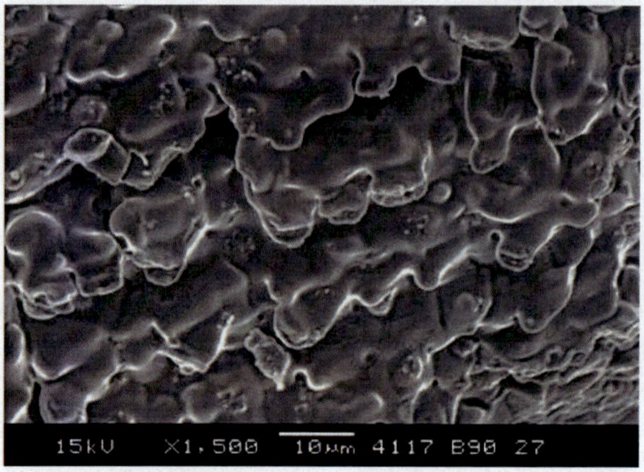

Fig. 10 SEM fracture surface of 52M weld metal crack from boat sample

Fracture Surface Analyses of 52M Hot Cracks

A crack in one of the boat samples was opened and the exposed surface examined by SEM to identify the mode of cracking in the 52M weld metal. Figure 10 shows the fracture surface of the crack. The dendritic egg crate type fracture morphology with evidence of liquid film on the surface is indicative of weld solidification cracking, which is a form of hot cracking. Composition analyses in the SEM of the fracture surface were also performed to investigate the presence of elements that are known to form low melting constituents. These analyses indicated elevated levels of Nb and Ti. No elevated level of Si was detected on this sample. Note that elevated levels of S, P, or B, if present, would likely not be detected by SEM.

Compositional analyses were performed on the surface of several other cracks opened for examination. These analyses consistently identified the presence of elevated Nb and in a few cases elevated Si. A MnS particle was identified on one crack surface.

EDS Compositional Analyses of Boat Sample Cross-Sections

Figures 11, 12, and 13 show photographs and polished metallurgical cross sections of three of the four boat samples removed from weld overlay regions where 52M hot cracking occurred (see Fig. 4). The location of each EDS sample area is noted on the figures by number. Tables 2, 3, and 4 list the EDS composition analyses of the sample areas for each of the boat samples. These tables also list an estimate of 52M dilution by the ER308L buffer layer. The dilution estimate is calculated from the Fe and Ni content determined by EDS analyses. Dilution of the ER308L buffer by the CF8A pipe was assumed to be 50%. Note that in this case a variation in ER308L buffer layer dilution has only a minor effect on the diluted 52M composition due to the similar composition of the ER308L and CF8A. Table 1 lists the CF8A (B hot

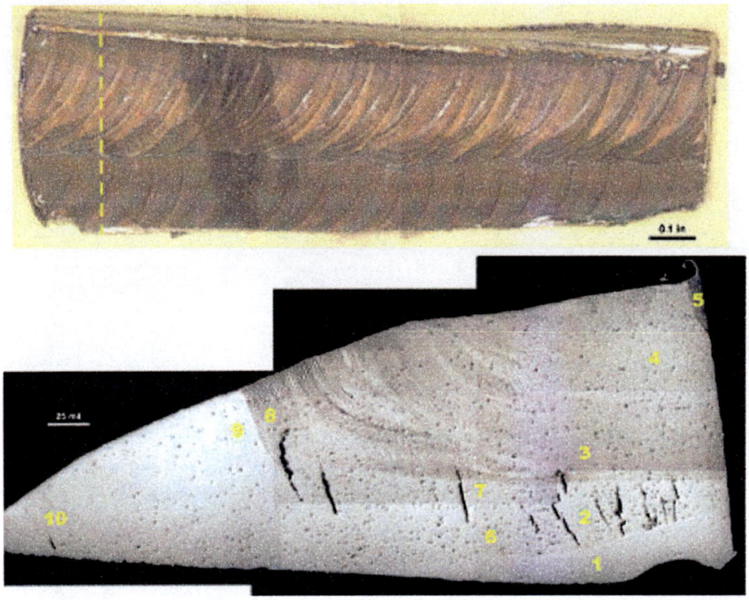

Fig. 11 Boat sample B-180-2 removed from B hot leg CF8A pipe at 180° position. *Dashed line* (*top left*) shows location of cross-section (*bottom*). Composition at each numbered sample area is listed in Table 2 below

Fig. 12 Boat sample D-180 removed from D hot leg CF8A pipe at 180° position. *Dashed line* (*top right*) shows location of cross-section (*bottom*). Composition at each numbered sample area is listed in Table 3 below

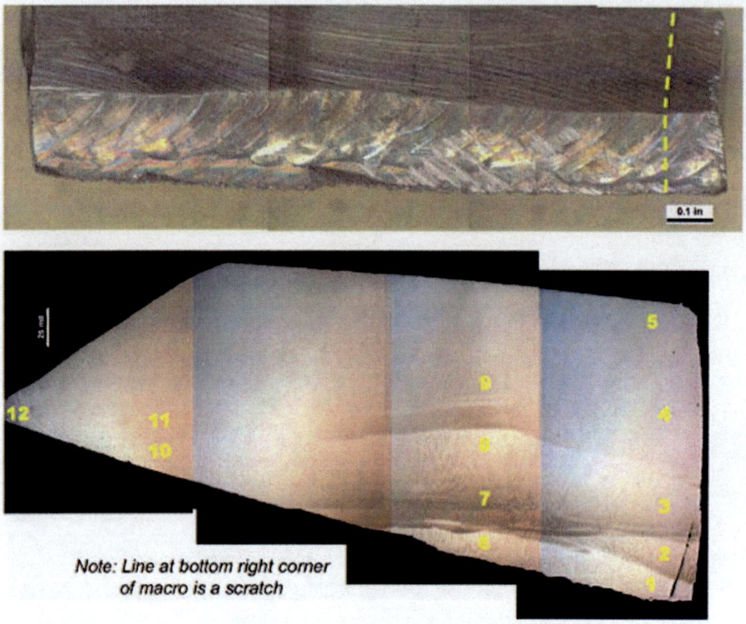

Fig. 13 Boat sample D-90 removed from D hot leg CF8A pipe at 90° location. *Dashed line* (*top right*) shows location of cross-section (*bottom*). Composition at each numbered sample area is listed in Table 4 below

leg pipe), the ER308L, and the 52M compositions used for the dilution calculations. Below are general observations for each of the three boat samples.

- *Boat sample B-180-2 (Fig. 11 and Table 2)*: Sample B-180-2 was removed from the 180° position on the B hot leg CF8A pipe. Sample area 1 is in the ER308L buffer layer which exhibits no cracking. Sample areas 2, 6, and 7 are in regions

Table 2 EDS composition analyses of boat sample B-108-2. Arrows indicate sample areas with cracking

	Si	Cr	Mn	Fe	Ni	Nb	52M[a] Dilution	
1[b]	0.96	21.42	1.39	67.36	8.78	0.09	–	
2	0.73	25.13	1.28	44.81	27.67	0.38	63	←
3	0.58	26.52	0.91	36.92	34.66	0.42	49	
4	0.62	26.32	1.21	36.23	35.07	0.54	48	
5[c]	0.68	26.91	1.38	31.43	38.79	0.82	41	
6	0.72	24.60	1.47	46.09	26.73	0.39	65	←
7	0.98	25.15	1.21	44.98	27.33	0.35	63	←
8	0.61	26.70	1.16	36.15	35.15	0.24	48	
9	0.71	24.64	1.39	45.44	27.43	0.39	64	
10	0.65	25.06	1.03	45.53	27.36	0.37	64	

[a] *Estimate of 52M dilution by ER308L buffer based on Fe and Ni content. Dilution of ER308L by CF8A case assumed to be 50% for calculations.*
[b] *Sample area 1 is in the ER308L buffer layer.*
[c] *Smaller EDS sample area in order to fit in area of interest.*

Table 3 EDS composition analyses of boat sample D-180. *Arrows* indicate sample areas with cracking

	Si	Cr	Mn	Fe	Ni	Nb	52M[a] Dilution	
1	0.43	27.03	1.09	34.34	36.67	0.45	45	←
2	0.40	27.70	1.46	30.98	39.04	0.42	40	←
3	0.18	29.56	1.14	18.08	50.38	0.67	18	
4	0.20	30.47	1.02	13.79	53.81	0.70	11	
5	0.50	27.59	1.11	30.88	39.68	0.22	40	←
6	0.18	29.69	0.86	18.22	50.42	0.62	18	
7	0.42	27.76	0.71	32.63	37.85	0.64	43	←
8	0.47	28.03	1.18	27.36	42.10	0.87	34	
9	0.21	29.23	1.25	18.82	49.68	0.81	20	
10	0.10	30.78	1.18	12.81	54.44	0.70	10	

[a]*Estimate of 52M dilution of ER308L buffer based on Fe and Ni content. Dilution of ER308L by CF8A cast assumed to be 50% for calculations.*

with extensive cracking and are located in the 1st 52M layer since they are adjacent to the ER308L buffer layer and because the Fe is in the 45–46 wt% range. The dilution of 52M weld metal by the ER308L buffer layer in these sample areas with cracking is estimated to be between 63 and 65%.

- *Boat sample D-180 (Fig. 12 and Table 3)*: Sample D-180 was taken from the 180° position on the D hot leg CF8A pipe. This boat sample contains all 52M weld metal with no ER308L buffer layer. The pattern of results is similar to the D-180-2 sample except the dilution levels are slightly lower and the extent of cracking is less. Cracks in sample areas 1, 2, 5 and 7 appear to be in the 1st 52M layer where Fe is in the 31–34 wt% range. The dilution of 52M weld metal by the ER308L buffer layer in these crack areas is estimated to be between 40 and 45%.
- *Boat sample D-90 (Fig. 13 and Table 4)*: Sample D-90 was taken from the 90° position on the D hot leg CF8A pipe. This boat sample contains all 52M weld

Table 4 EDS composition analyses of boat sample D-90. No evidence of cracking in this sample

	Si	Cr	Mn	Fe	Ni	Nb	52M[a] Dilution
1	0.35	28.17	1.16	24.96	44.51	0.85	30
2	0.26	29.24	0.98	20.12	48.33	1.06	22
3	0.21	29.61	1.24	17.20	50.99	0.76	17
4	0.21	30.56	0.87	12.75	54.70	0.92	10
5	0.22	31.15	0.68	10.74	56.19	1.02	7
6	0.45	28.46	1.05	25.22	44.29	0.54	30
7	0.31	29.82	0.89	17.30	50.79	0.89	17
8	0.23	29.63	0.98	16.61	51.60	0.95	16
9	0.25	30.61	1.15	12.90	54.19	0.90	10
10	0.29	30.46	0.93	16.99	50.87	0.46	17
11	0.18	30.58	0.71	12.49	55.14	0.90	9
12	0.39	29.79	0.99	17.33	50.46	1.01	18

[a] *Estimate of 52M dilution by ER308L buffer based on Fe and Ni content. Dilution of ER308L by CF8A cast assumed to be 50% for calculations.*

metal with no ER308L buffer layer. There is no evidence of cracking in this sample. The maximum estimated dilution of 52M by the ER308L buffer layer in this sample is 30% where Fe is 25 wt%.

Review of composition analyses of all the sample areas with cracking showed Ni contents between 27 and 40 wt% and Fe between 31 and 46 wt%. Based on these levels of Ni and Fe, the dilution of 52M weld metal by the ER308L buffer layer in areas of cracking is estimated to be between 40 and 65%. These results are in general agreement with other experimental testing by EPRI (Fig. 7) which showed increased potential for 52M hot cracking when dilution by Type 304L stainless steel is above 45%.

Discussion

Analyses of the boat samples removed from the B and D hot leg overlay welds and subsequent mockup testing clearly identified the mode of cracking as weld metal solidification cracking. Solidification cracking occurs when liquid films are present along the weld metal grain boundaries at the terminal stage of solidification. Liquid films are typically associated with low melting point constituents or eutectics such as FeS, (Nb,Ti)C or Laves which lower the solidus temperature and expand the solidification temperature range. The following sections discuss a likely mechanism for the 52M solidification cracking observed in the 1st 52M layer and present a plausible cause for the unexpectedly high dilution observed in the ER308L buffer and 1st 52M layers.

Computational Modeling of Solidification Behavior

Computational modeling using Dictra^{TM} was performed at The Ohio State University to investigate the solidification behavior of the 1st 52M layer [3]. Dilution of the ER308L buffer layer by the CF8A pipe was assumed to be 30% and dilution of the 52M by the diluted ER308L buffer layer was modeled at 20%, 30%, and 40%. In addition, the Si content was artificially increased to 1.0 wt% at the 40% dilution level to investigate the influence of Si on the solidification behavior. Table 5 lists the compositions used in the model.

Dictra^{TM} computational modeling predicts an increase in solidification temperature range from 204 to 298°C as dilution increases from zero to 40% and predicts formation of the (Nb,Ti)C low melting carbide eutectic at the terminal stage of solidification (Fig. 14 and Table 6). The modeling confirms that the elevated Nb detected by EDS on the boat sample fracture surface is likely the low melting (Nb,Ti)C or NbC eutectic that wets the weld metal grain boundaries at the terminal stage of solidification. Increasing the Si content up to 1 wt% with 40% dilution reduced the solidification temperature range. This prediction substantiates the fact that elevated

Table 5 Dilution compositions used for Dictra™ computational modeling

	CF8A	308L	52M	ER308L 30%[a]	Diluted 52M 20%	30%	40%	40%[b]	
Cr	20.5	20.1	29.0	20.2	27.3	26.4	25.5	25.5	Cr
Ni	8.4	10.5	60.8	9.9	50.6	45.5	40.4	39.7	Ni
Fe	69.3	66.9	8.2	67.6	20.1	26.0	32.0	32.0	Fe
Mn	0.65	1.90	0.68	1.53	0.85	0.93	1.02	1.02	Mn
Si	1.03	0.52	0.09	0.67	0.21	0.26	0.32	1.00	Si
Nb	0.010		0.800	0.003	0.641	0.561	0.481	0.481	Nb
Al	0.006		0.130	0.002	0.104	0.092	0.079	0.079	Al
Ti	0.001		0.200		0.160	0.140	0.120	0.120	Ti
Mo	0.020	0.080	0.010	0.062	0.020	0.026	0.031	0.031	Mo
C	0.040	0.018	0.020	0.025	0.021	0.021	0.022	0.022	C
S	0.015	0.001	0.005	0.005	0.005	0.005	0.005	0.005	S
P	0.022	0.014	0.005	0.016	0.007	0.008	0.010	0.010	P

[a] *Assumed 30% dilution for Dictra™ calculations.*
[b] *Si increased to 1% to investigate influence on solidification.*

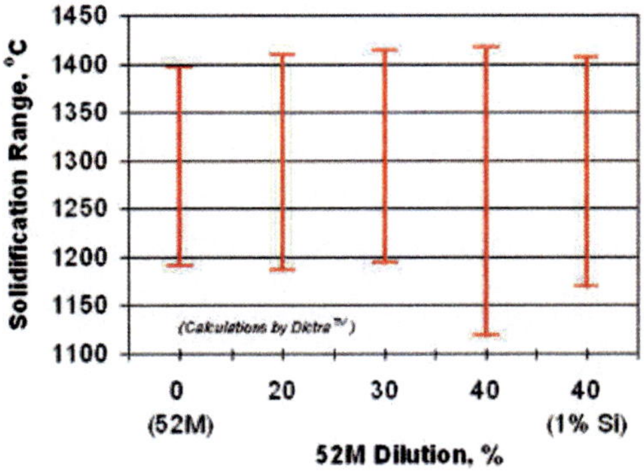

Fig. 14 Solidification temperature ranges predicted by Dictra™

Table 6 Dictra™ prediction of 52M solidification behavior when diluted by the ER308L buffer layer [3]

Solidification parameter	52M	20%	30%	40%	40% (1 wt% Si)
Liquidus, °C	1,396	1,409	1,414	1,416	1,406
Solidus, °C	1,192	1,187	1,194	1,118	1,169
Eutectic temp, °C	1,264 (Nb, Ti)C	1,275 (Nb, Ti)C	1,279 (Nb, Ti)C	None	None
Solidification range, °C	204	222	220	298	237
Eutectic, Vol%	5.8	6.2	6.1	None	None

Si was not detected on the crack surface by EDS and also implies Si up to 1.0 wt% does not directly promote solidification cracking in the diluted 52M weld metal. Finally, it is worth noting that S content is constant and P increases only slightly as dilution increases (Table 5). This implies that the cracking at higher dilution is not due to FeS, NiS, Fe_3P or Ni_3P low melting phases since S and P levels are essentially constant with increasing dilution.

It should be noted that the predictions by Dictra™ are based on the Ni alloy thermodynamic database where Ni is the major alloying element. Inspection of Table 5 shows that the Ni drops below 50 wt% with just 30% dilution which indicates the Dictra™ predictions may be outside the range of the Ni alloy database. The predicted trends are likely correct but the absolute values may not be accurate. In fact, single sensor differential thermal analysis (SS DTA™) testing indicates that the solidification temperature range for 52M is closer to 145°C [4] compared to the 204°C predicted by Dictra™ in this study.

Influence of Fe Content on Propensity for Hot Cracking

Dupont et. al. studied the solidification behavior of Nb-bearing nickel-based super alloys [5] with nominal composition of 68Ni-19Cr-10Fe (close to Inconel 82 and 600 compositions) and 45Fe-30Ni-19Cr (close to Incoloy composition). In order to understand the effect of Nb on solidification behavior the Nb content was varied from 1.82 to 4.92 wt% in the 68Ni-19Cr-10Fe alloy group and 1.66–4.88 wt% in the 45Fe-30Ni-19Cr group. Table 7 lists the partition coefficients measured for Fe, Ni, Cr, Si, and Nb for these two groups of alloys with the given Nb composition ranges. A partition coefficient of unity indicates the element does not partition to the liquid during solidification. In contrast, an element with a partition coefficient that is less than unity will segregate to the liquid.

Figure 15 shows the relationship of the Nb partition coefficient (k_{Nb}) and nominal Fe content for Ni-19Cr alloys. This relationship indicates that Nb segregation to the liquid will increase as the Fe content increases above 30 wt%. This can be understood by considering the maximum solubility of Nb in Ni and Fe. The maximum solubility of Nb in γ-Ni is 18.2 wt% (at 1286°C). In contrast, the maximum solubility of Nb in γ-Fe is 1.5 wt% (at 1210°C) which explains the increasing rejection of Nb to the liquid with higher Fe and lower Ni content.

Table 7 Partition coefficient for elements in Nb bearing nickel-base alloys [5]

10Fe 68Ni 19Cr 2-5 Nb (nominal)					45Fe 30Ni 19Cr 2-5Nb (nominal)				
Fe	Ni	Cr	Si	Nb	Fe	Ni	Cr	Si	Nb
1.00	1.02	1.06	0.71	0.45	1.06	1.00	1.02	0.58	0.25
11 experimental heats					9 experimental heats				

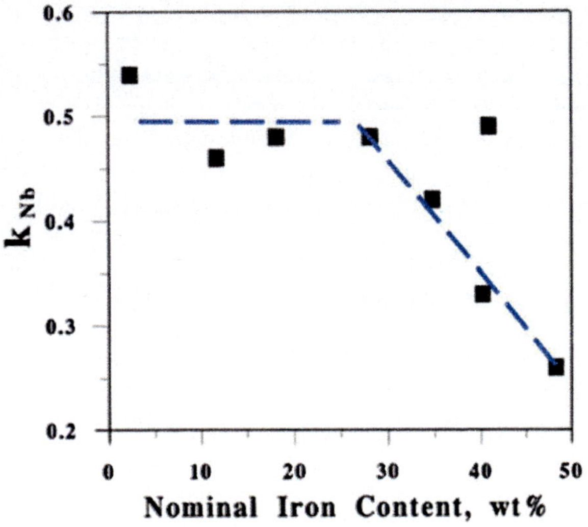

Fig. 15 Relationship for the general effect of Fe content on the Nb partition coefficient (k_{Nb}) in nickel-base alloys [5]

The implication of this research is that dilution of 52M by the ER308L buffer layer increases the Fe content in the weld puddle. Increasing Fe above 30 wt% in turn leads to a more pronounced segregation of Nb to the liquid during solidification. The higher segregation of Nb to the liquid expands the solidification temperature range by forming the low melting NbC eutectic during the terminal stage of solidification. Both the wider solidification temperature range and the higher fraction of NbC eutectic increase the propensity for solidification cracking.

Influence of Cast CF8A Composition on Dilution

The power ratio approach to control dilution and 52M hot cracking for installation of weld overlays has been used commercially for a number of years with success. Unfortunately this approach was ineffective when welding over the CF8A cast stainless steel piping. This was due to the unexpected and significant difference in melting and solidification behavior between 304L austenitic stainless steels used for weld process development and the vintage CF8A cast stainless piping encountered in the field. This difference was observed by welders that reported the weld puddle on the CF8A cast material was excessively fluid and "hot" compared to welds on 304L wrought austenitic stainless steel with identical weld parameters. The important questions to be answered are:

(a) "Why the difference in the weld puddle and unexpectedly high dilution?" and subsequently,
(b) "How can the excessive dilution be mitigated?"

It was noted that although castings had been successfully overlaid in the field, most of the weld overlay application experience had been on wrought product forms (i.e. pipe or forgings). Comparisons of wrought to cast product forms of austenitic stainless steels show that the most significant difference is Si levels. Many cast specifications for stainless steels permit up to 2.0% Si, or several times that permitted for similar type wrought products. Silicon is an element known to increase the fluidity of the molten puddle and is added to improve fluidity of the molten metal during casting.

Sulfur is known to have a strong influence on the weld bead shape by affecting the surface tension forces on the weld pool [6, 7]. Low S < 0.008% promotes a negative surface tension to temperature gradient (dγ/dT) and high S > 0.015% promotes a positive dγ/dT gradient. A negative surface tension gradient creates an outward fluid flow with a wide shallow weld shape and a positive gradient creates an inward fluid flow with narrow deep weld shape (Fig. 16). This surface tension driven fluid flow is well understood and is known as the Marangoni effect. Other forces such as electromagnetic (Lorentz), buoyancy, and aerodynamic drag, can also affect the weld pool shape and flow. Recent modeling and research, however, indicate that the Marangoni surface tension forces dominate the other forces under normal GTA welding conditions [8].

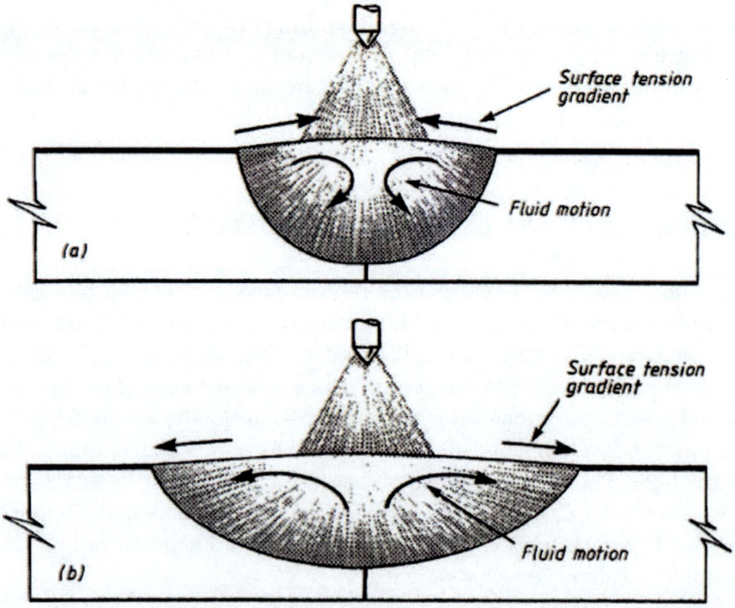

Fig. 16 Schematic of (**a**) fluid motion and bead shape for > 0.015% S substrates, and (**b**) fluid motion and bead shape for <0.008% S substrate [9]

Cross-Sectional Area

Treatment combination	Effect	Best value
Si	31.3	17.3
Ti	18.1	37.9
S-Si	17.7	35.0
Ti-Al	−16.2	6.4
Al	−15.4	4.4
P	14.1	17.7
Mean	–	27

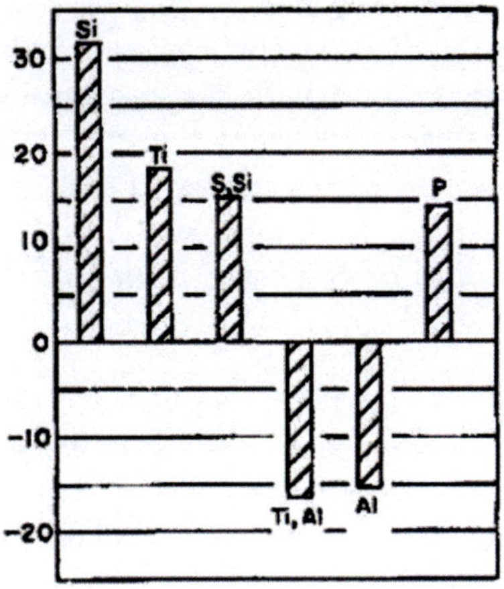

Fig. 17 Influence of minor element additions on cross-sectional area of autogenous Inconel 600 welds (bar graph plots the 'Effect' column values) [11]

Sulfur may also have a strong influence on the dilution of nickel-base alloys. Walsh et. al. investigated the effect of subtle changes of S, N_2 and O_2 on the autogenous GTA weld pool of Inconel 718 [10]. An increase of S from 0.0032 to 0.016% resulted in a 46% increase in depth-to-width ratio, 18% increase in volume, and 32% increase in depth.

Savage et. al. studied the effects of minor additions of Al, Mn, P, S, Si, and Ti on the autogenous weld pool shape and cross-sectional area of Inconel 600 [11]. This statistical analysis, which included 64 samples, indicated that the average cross-section area of pure Inconel 600 increased by 27% with minor additions of the given elements (i.e., the mean increase was 27%). Figure 17 shows the cross-sectional area results of the study. The 'Effect' column indicates an increase (or decrease) in cross-sectional area when only trace amounts of the select element or element pair is intentionally added. The 'Best Value' column indicates the influence on cross-sectional area as a percentage of the 27% mean value. Minor additions of Ti and S-Si are the only treatments that increase the cross-sectional area above the mean of 27%. The influence of Ti is offset by the Ti-Al combination which acts to decrease the cross-sectional area. This implies that the S-Si combination has the overall greatest influence on increasing cross-sectional area. Note that the 'Best Value' for Si is 17.3%, which is less that the 27% mean, whereas S-Si is 35%. This indicates that Si acting independently does not increase the cross sectional area but does increase the cross sectional area when combined with S. An explanation for this combined S-Si effect is that Si reduces the weld puddle viscosity which enhances the narrow downward Marangoni flow forces that occur with high S content.

The research by Walsh and Savage indicate that S and S-Si can have a significant influence on dilution of autogenous nickel-base welds. Though not conclusive, this information points to an S-Si synergy capable of producing an unexpectedly high dilution in 52M weld metal when welding on stainless steel metals with high S-Si levels. This synergistic effect of high base metal S-Si levels on the dilution of austenitic weld metals was observed with a limited number of ER308L single bead experiments on stainless steel base materials of various S-Si levels (Figs. 8 and 9).

Mitigation of High Dilution Influence of Cast CF8A Material

The discussion above leads to the conclusion that a fundamental key to minimize susceptibility to 52M hot cracking when welding on cast austenitic base materials is dilution control. In this case study the specific welding parameters with defined power ratio were successfully developed and demonstrated on Type 304L plate material. Unfortunately the cast CF8A pipe composition altered the weld puddle fluidity and flow characteristics which ultimately led to high dilution of 52M by the CF8A diluted ER308L buffer layer. As discussed above, this high dilution ultimately resulted in solidification cracking of the 52M weld deposit. It is therefore critical that the relationship between the welding techniques, parameters, and power ratio be tested, proven and well understood for the specific base metals and weld metal combinations to ensure proper dilution control and adequate margin against solidification cracking.

Development of specific welding parameters and techniques are outside the scope of this paper, however, there are a number of basic techniques and methods that can, if properly implemented, control dilution and minimize susceptibility to solidification cracking in 52M overlay welds. Following are three such methods:

(a) Development of low power ratio parameters on representative base metal which control dilution within an acceptable range. Definitive experiments and high fidelity mockups are necessary to properly identify the optimum welding parameters and techniques that minimize dilution for solidification crack resistance, but also maintain adequate margin against other weld defects such as lack of fusion.
(b) Use of one or more buffer layers to reduce the deleterious elements that are diluted into the 52M weld deposit (Fig. 18). One or more ER308L, ER309L, or

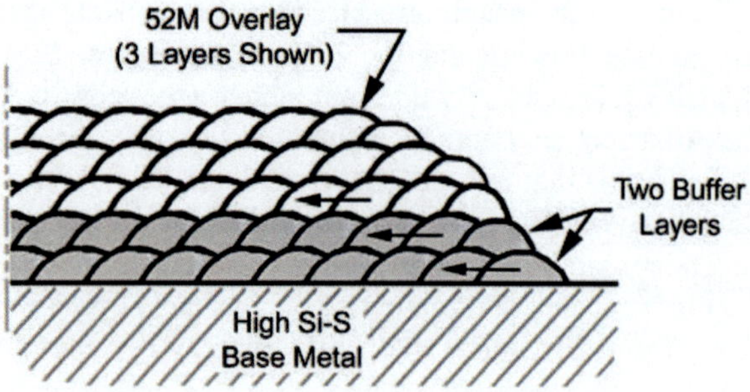

Fig. 18 Example of weld overlay with two buffer layers to minimize 52M dilution with base metal that increases potential for solidification cracking

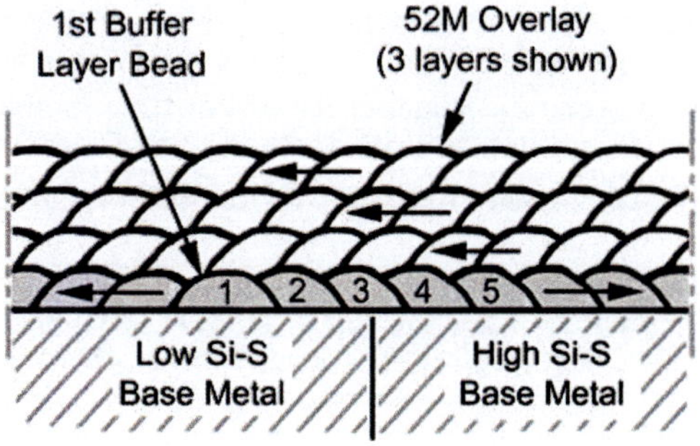

Fig. 19 Example of buffer layer bead progression to minimize 52M dilution with base metal that increases potential for solidification cracking

possibly a crack resistant Inconel 82 layer(s) may be considered. Note that the use of buffer layer(s) still requires an understanding of the base metal influence on the weld puddle flow characteristics and dilution.

(c) Special weld bead placement, sequencing, and progression can be used to minimize localized high dilution areas such as occurs with single beads without overlap. For example, placement of weld beads on low Si-S stainless steel base material (e.g., SS safe end or SS weld in Fig. 2) with subsequent bead placement stepping towards the high S-Si base material (e.g., CF8A pipe) takes advantage of bead overlap effects to reduce dilution with the deleterious base metal (Fig. 19).

Conclusions

1. Metallurgical analyses of cracks in the boat samples with 52M weld deposits diluted by the ER308L buffer layer and CF8A base metal confirm that the cracking mechanism was solidification cracking.
2. Close examination of metallurgical cross sections of the boat samples confirmed that the solidification cracking was isolated to the 1st 52M layer where the weld deposits were diluted with between 31 and 46 wt% Fe. These results are in close agreement with previous dilution studies by EPRI that showed a similar solidification cracking threshold at approximately 35 wt% Fe.
3. The documented research on Nb-bearing Ni-based alloys combined with the Dictra^{TM} computational modeling and the metallurgical analyses of the diluted 52M weld metal provide strong evidence that the Ni-NbC eutectic reaction is promoting solidification cracking in the 52M weld deposit.
4. The synergistic influence of high Si and S in the CF8A cast pipe is the likely cause of the unexpectedly high dilution in the ER308L buffer and 1st 52M layers. Additional research and work in this area is necessary to better understand and quantify the threshold and influence of Si and S on the weld pool properties and dilution behavior of 52M weld metal.

5. The key to avoiding 52M solidification cracking on cast austenitic stainless steels with high Si and S, or other deleterious elements, is to control dilution. The power ratio equation is an effective method to control dilution. However, the relationship between the power ratio and dilution must be tested and proven for specific base metal and weld metal combinations.

Acknowledgements The authors wish to recognize Dr. John Lippold, OSU Welding Engineering Department, for his insight and expert guidance during the root cause investigation. Sue Anderson, Duke Energy Manager – Metallurgy & Welding Services, performed and supervised the metallurgical analyses of the boat samples and also provided support for experimental dilution investigations. Greg Frederick, EPRI Program Manager – Welding & Repair Technology Center, provided expert support for much of the mockup and experimental dilution investigations. James Puzan, Manager of Projects – Aquilex-WSI Nuclear Services, provided key support and assistance with weld process experiments supporting the root cause investigation.

References

1. N.G. Cofie, R.E. Smith, R.L. Bax, C.S. Lohse, B. Hermanns, J. Valsvig, L.D. Yepez, and D. Parker, (2008) Effectiveness of Stainless Steel Buffer Layer to Address Hot Cracking During Weld Overlay Repair of Dissimilar Metal Alloy 82/182 Welds with Stainless Steel Piping. *Proceedings ASME PVP Conf.*, July 27–31, Chicago, IL, Paper No. PVP2008-61411.
2. D.W. Gandy, R.E. Smith, S.J. Findlan, and W.J. Childs, (August 1992), A Better Way to Control GTA Weld Dilution, *Welding Design & Fabrication*, pp. 41–43.
3. J.C. Lippold, (2009) Private Correspondence, Catawba Root Cause Analysis, June 8, 2009.
4. S.L. McCracken, B.T. Alexandrov, J.C. Lippold, J.W. Sowards, and A.T. Hope, (2010) Hot Cracking Study of High Chromium Nickel-Base Weld Filler Metal 52MSS (ERNiCrFe-13) for Nuclear Applications, *Proceedings ASME PVP Conf.*, July 18–22, 2010, Bellevue, WA, Paper No. PVP2010-25787.
5. J.N. DuPont, C.V. Robino, J.R. Michael, M.R. Notis, and A.R. Marder, (1998) Solidification of Nb-Bearing Superalloys: Part I. Reaction Sequences. *Metall. Mater. Trans. A,* 29A: 2797–2806.
6. C.R. Heiple and J.R. Roper, (April 1982) Mechanism for Minor Element Effect on GTA Fusion Zone Geometry. *Welding Journal.* , pp. 97s–102s.
7. C.R. Heiple and J.R. Roper, (1981) Effects of Minor Elements on GTAW Fusion Zone Shape. *Trends in Welding Research Conf. Proceedings*, November 16–18, New Orleans, LA, pp. 489–520.
8. K.C. Mills, B.J. Keene, R.F. Brooks, and A. Shirali, (1998) Marangoni Effects in Welding, *Phil. Trans. R. Soc. Lond.* A 356: 911–925.
9. M.J. Tinkler, I. Grant, G. Mizuno, and C. Gluck, (1983) Welding 304L Stainless Steel Tubing Having Variable Penetration Characteristics," *International Conf. London England*, November 15–17, Paper No. 29.
10. D.W. Walsh, A. Demmons, and D. Gibbs, (1998) The Effects of Minor Elements on Bead Shapes in Inconel 718, *5th International Trends in Welding Research Conference Proceedings*, June 1–5, Pine Mountain, GA, pp. 352–356.
11. W.F. Savage, E.F. Nippes, and G.M. Goodwin, (April 1977) Effect of Minor Elements on Fusion Zone Dimensions of Inconel 600. *Welding Journal* pp. 126s–132s.

Investigation of Weld Crack Mitigation Techniques with Advanced Numerical Modeling and Experiment – Summary

Y.P. Yang, S.S. Babu, J.M. Kikel, and F.W. Brust

Introduction

Hot cracking has been a subject of intensive studies over the last few decades. In general, two basic approaches are usually taken to control hot cracking: (1) improving the material ductility of weld and heat-affected zone and (2) improving the thermomechanical conditions during welding. This paper summarizes published methodologies to design these conditions using computational models. Studies that consider alterations in the welding-induced transient stress and strain through (a) trailing cooling, (b) side rolling, (c) controlling heat input, and (d) weld sequencing have been considered for aluminum, nickel and iron-alloys.

Trailing cooling [1] and side rolling techniques [2] were investigated to control solidification and liquation cracking with thermal-elastic-plastic modeling and experimental trials. Numerical analysis helped understand the mechanism and found the best process parameters of trailing cooling and side rolling. Cooling devices and mechanical rolling equipment were designed for experimental trials. Experimental results show that both trailing cooling and side rolling can effectively prevent solidification and liquation cracking during welding of 2024 high strength aluminum alloy.

Controlling heat input [3] was used to mitigate ductility-dip cracking and solidification cracking during cladding and buttering a low alloy steel pipe with Ni-Cr filler wire. Experimental mockup and thermo-mechanical modeling were performed to understand the mechanical reasons for the cracks. Experimental and modeling results suggest that reducing the heat input in regions susceptible to cracking could be a viable mitigation technique which has been proven during experiment.

Weld sequence [4] was studied with numerical modeling to understand the cracking reasons in bimetallic welds that join the hot leg to the reactor pressure vessel nozzle in a US Nuclear Power Plant. Numerical results showed that weld sequence

Y.P. Yang (✉)
Edison Welding Institute, Columbus, OH, USA
e-mail: yyang@ewi.org

J. Lippold et al. (eds.), *Hot Cracking Phenomena in Welds III*,
DOI 10.1007/978-3-642-16864-2_18, © Springer-Verlag Berlin Heidelberg 2011

affected the stress distributions on the inner surface of the hot leg. Tensile weld residual stresses, in addition to service loads, contribute to the crack and its growth rates.

Thermomechanical Conditions for Hot Cracking

The necessary conditions for hot cracking are the presence of tensile strains in the region behind the weld pool as shown in Fig. 1. Figure 1 illustrates the stress and strain distribution near a weld region. Before the weld pool, a plastic compressive zone is formed during thermal expansion and the constraint of surrounding low-temperature area. During melting, stress and strain in the weld pool is zero. The region behind the weld pool is subjected to tensile stress and strain which could induce hot cracking. Further behind the weld pool, plastic tensile stress and strain exists, but no solidification cracking occurs since material has recovered the full ductility after solidification. Some materials could have ductility dip cracking (DDC). Ductility dip cracking can also occur between weld pass due to reheating and re-melting.

In this study, mechanical strains were used as a measure of the driving force for hot cracking instead of stresses, since in the region behind the weld pool, the transient stress level is usually low due to significantly reduced material yield strength at high temperature. As depicted in Fig. 2, if the tensile strain rate (with respect to temperature) exerting on the region behind the weld pool becomes smaller than "the critical strain rate for temperature drop (CST)" as depicted by the tangent line (Curve B) to the ductility curve, or

$$\frac{d\varepsilon}{dt} < CST \tag{1}$$

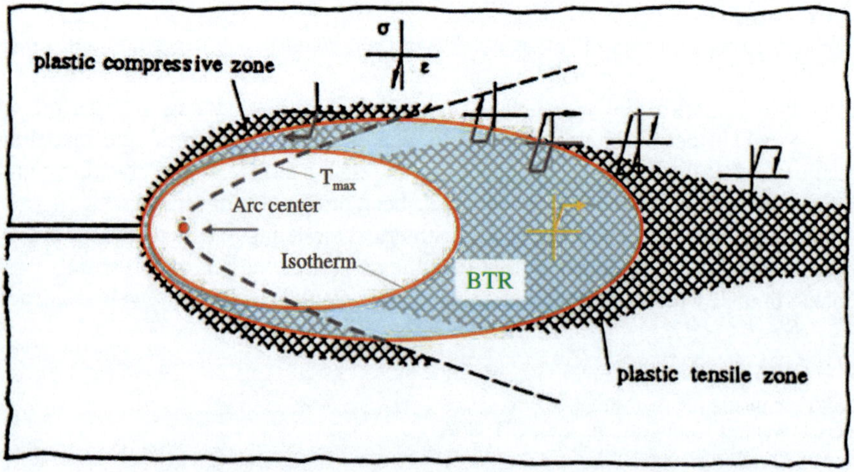

Fig. 1 Stress and strain distributions during welding [5]

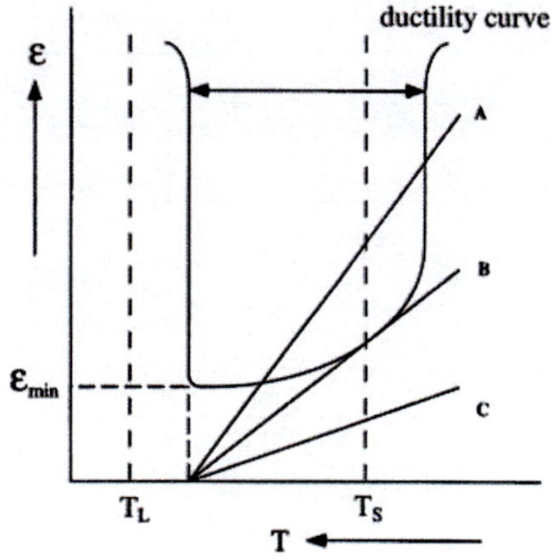

Fig. 2 Thermomechanical conditions associated with hot cracking

as depicted by the tangent line (Curve B) to the ductility curve, or in theory, hot cracking can be avoided, as depicted by line C. Decomposing $d\varepsilon/dT$, one obtains the following:

$$\frac{d\varepsilon}{dT} = \frac{\partial\varepsilon/\partial t}{\partial T/\partial t} \tag{2}$$

The first term $\partial\varepsilon/\partial t$ in Equation (2) becomes the standard expression of strain rate determined by the thermomechanical response during welding. The second term $\partial T/\partial t$ represents cooling rate typically controlled by the heat flow characteristics of the work piece under consideration. For a given material, both the strain rate and the cooling rate can be altered by either modifying welding parameters or introducing local heating/cooling mechanisms, or applying mechanical means.

Trailing Heat Sink

The trailing cooling technique [1] was developed on aluminum alloy 2024. As shown in Fig. 3, a heat sink was introduced behind the GTA welding torch. The heat sink could dramatically change the temperature distributions as illustrated in Fig. 3b. Compared to the temperature distributions of conventional welding (Fig. 3a), a low temperature zone is formed in the location of heat sink. This low temperature zone will contract as illustrated with arrows in Fig. 3b. The contraction will apply an auxiliary compressive pressure to the area in the BTR and behind the BTR.

Assuming a simple additive relationship between welding-induced strain (ε_W) and the strain (ε_C) generated by the heat sink, the total mechanical strain at any moment in time becomes,

$$\varepsilon = \varepsilon_W + \varepsilon_C \tag{3}$$

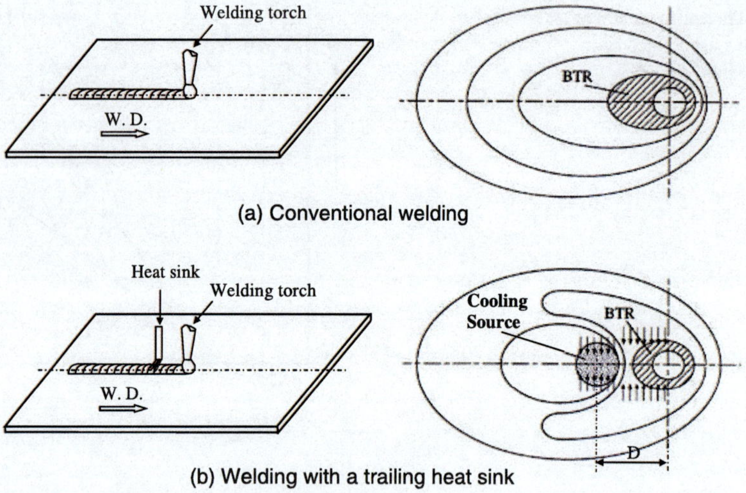

Fig. 3 Illustration of welding with a trailing heat sink

and Equation (1) takes the form of

$$\frac{d\varepsilon_w}{dT} + \frac{d\varepsilon_c}{dT} < CST \tag{4}$$

As a result, hot cracking can be prevented if the reduction of the strain rate with respect to temperature is sufficient within the BTR region.

Since the actual reduction of the strain rate (with respect to temperature) is dictated by the complex interactions between the instantaneous strain rate ($\partial\varepsilon/\partial t$) and cooling rate ($\partial T/\partial t$) during welding as indicated in Equation (2), advanced finite element procedures were used to quantify and optimize the effects of the trailing heat sink method on the dynamic thermomechanical interactions during welding.

Figure 4 shows the predicted and measured temperature distribution. The temperature distribution was predicted recently with commercial software ABAQUS. The measurement was done with a row of thermocouples which was installed in the holes drilled from the backside of the plate shown in Fig. 4 along the direction perpendicular to welding direction. Reference 5 gave a detailed experimental setup. With the cooling source shown in Fig. 3, the temperature cooling rate behind the weld pool has been significantly increased which was validated by the measurement. The cooling effect depends on the cooling media and distance (D) between the welding torch and the cooling source. For a given cooling media, the distance between welding torch and the cooling source is critical to control weld hot cracking.

Figure 5 shows the transverse (perpendicular to welding direction) strain evolution during cooling after solidification at a point near weld center line (X = 29.5 mm and Y = 0.43 mm referring to the coordinate system shown in the left picture of Fig. 4). The cooling media is liquid nitrogen with a flowing rate (V_c), 35 mL/min.

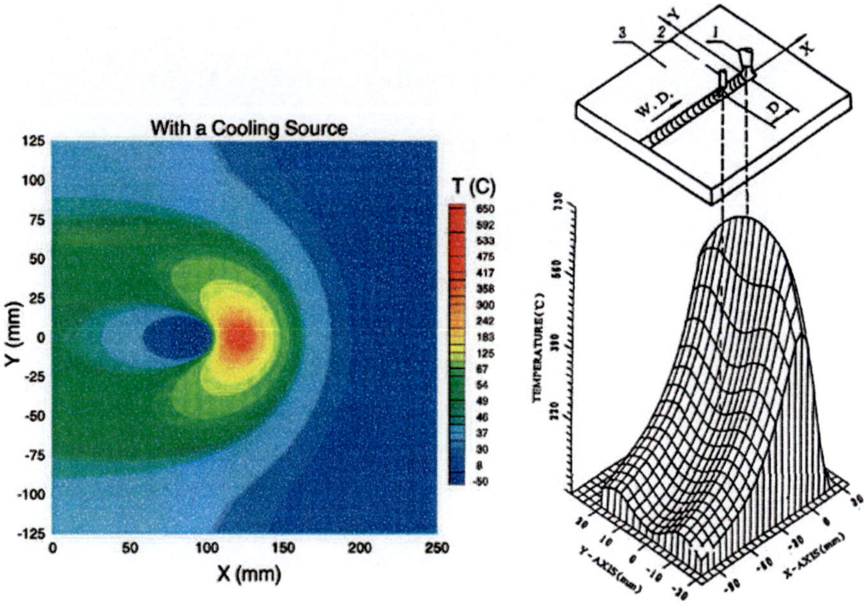

Fig. 4 Temperature distribution welding with a trailing heat sink

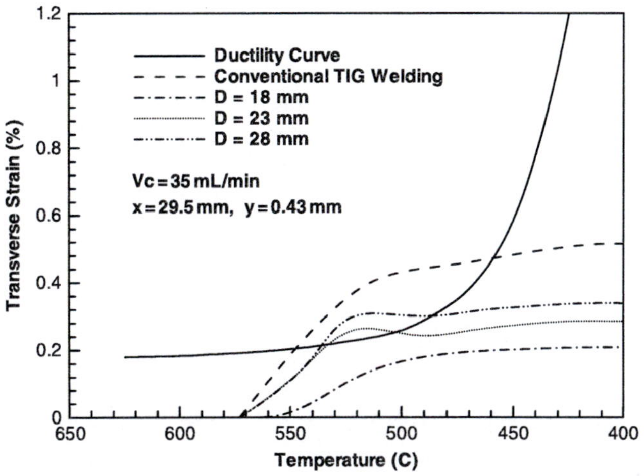

Fig. 5 Effects of cooling distance (D) on the evolution of transverse strain

The solid line is the ductility curve for the material. The dashed line is the transverse strain evolution for conventional welding. The remaining three lines are the transverse strain evolutions with trailing cooling for three trailing distances (*D*): 18, 23, and 28 mm, respectively. The trailing distance is the distance between the center of weld pool and the center of heat sink. The transverse plastic strain rate is reduced

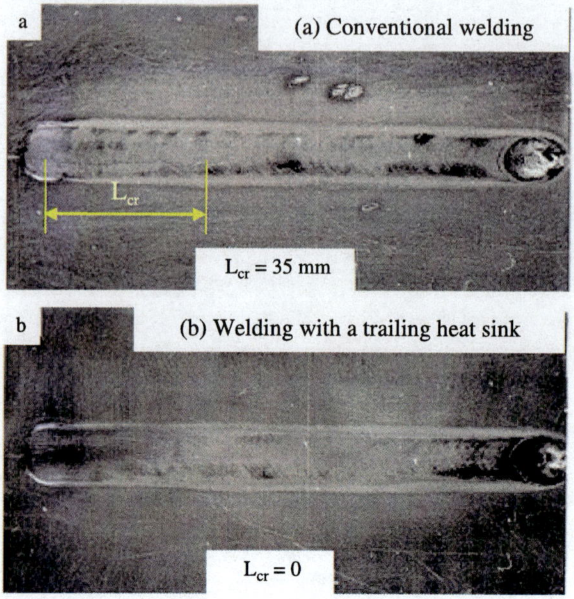

Fig. 6 Photographs of weld samples without and with a trailing heat sink

as the distance between welding torch and heat sink is reduced. When D is smaller than 23 mm, hot cracking can be eliminated.

The distance (D) between the welding torch and the trailing heat sink was an important parameter in reducing the propensity for hot cracking. With the optimal process conditions being approximately established from the detailed finite element simulations, welding trials were performed for various values of D. Experimental results showed that the smaller the cooling distance D, the shorter the hot cracking length [1]. As the cooling distance became shorter than 18.5 mm, hot cracking was completely eliminated. Figure 6 shows the photographs of two typical welding trial specimens without and with using the trailing heat sink method, respectively. The hot crack length under conventional GTA welding conditions was measured at 35 mm, while there was no indications of any cracks for specimens welded with the trailing heat sink (D = 18 mm).

Based on the results of this study, it was found that trailing cooling can be an effective method to control cracking for materials not sensitive to cold cracking such as aluminum alloys and austenitic stainless steel. The limitation of this method is that strong cooling can induce martensite for high strength steel and increase the risk of cold cracking.

Side Rolling

Side rolling [2] is another example to control hot cracking with mechanical methods. Sometimes, the trailing cooling technique cannot be used such as when welding high strength steel because trailing cooling can enhance the possibility of cold cracking.

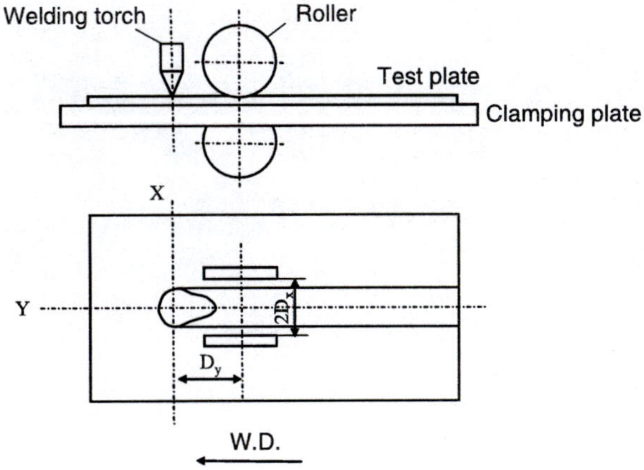

Fig. 7 Schematic diagram of rolling position

Therefore, side rolling was developed and the concept is shown in Fig. 7. During welding, two rollers amounted on both sides of weld move synchronously with the welding torch. By rolling both sides of the weld, a compressive strain ε_c can be induced to counterbalance the welding-induced tensile strain ε_w exerted on weld metal and HAZ experiencing BTR. Therefore, the resultant strain acting on the weld and HAZ within the BTR can be reduced to prevent hot cracking.

Finite element analysis was conducted to investigate the mechanism and optimize the technique parameters on aluminum 2024 alloy. The critical parameters are pressure, *Dy*, and *Dx* shown in Fig. 7. Figure 8 shows the finite element model and predicted temperature distribution. The model is a half model and the left side is a symmetric plane. Behind the weld torch and on the side, two rollers will move with

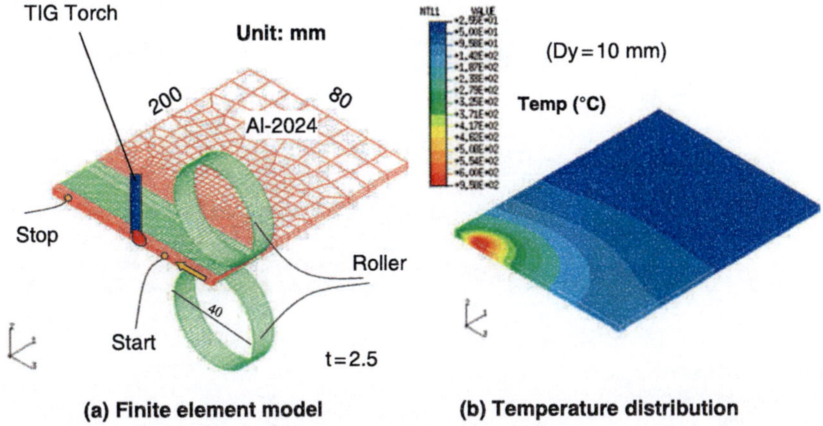

Fig. 8 Finite element model and predicted temperature distribution

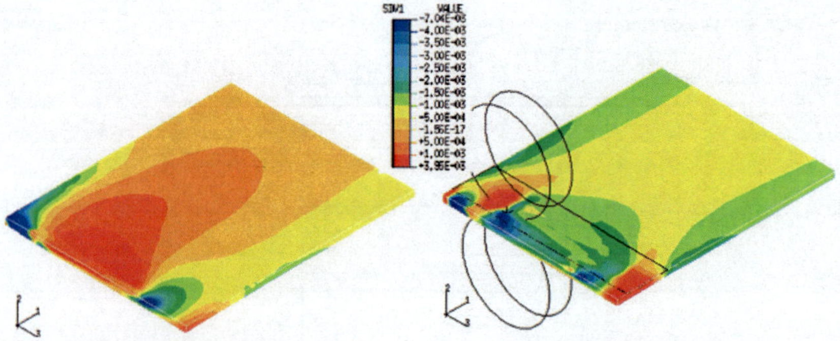

(a) Traditional TIG Welding (Dy = 10 mm Pd = 0.05 mm)

(b) Welding with Rollers (Dy = 10 mm Pd = 0.05 mm)

Fig. 9 Effect of rolling technique on transverse strain

welding torch and apply pressure on the plate. Contact between the roller and plate was modeled.

Figure 9 shows a comparison of transverse strain without and with side rolling when welding torch is moving to the end of the plate. Instead of applying pressure on the rollers, a 0.05-mm displacement (P_d) was applied on the top roller during finite element analysis. Without rolling, tensile strain is in the weld and HAZ. With

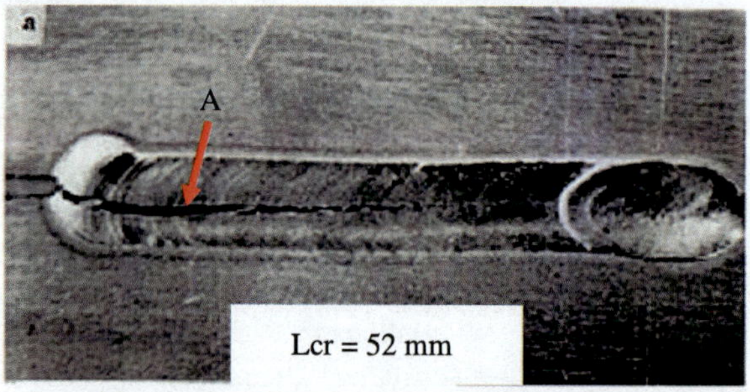

(a) Conventional welding

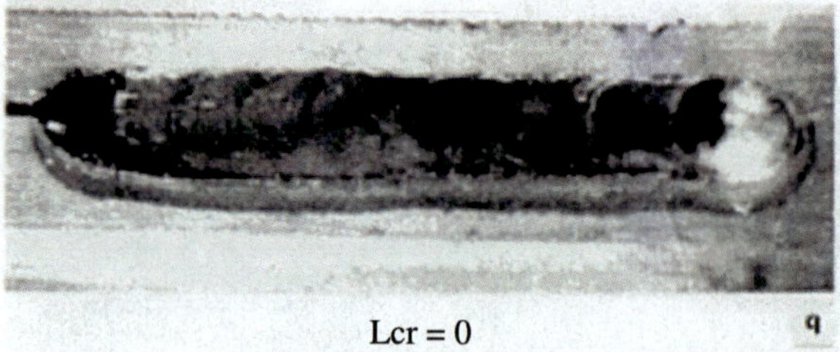

Lcr = 0

(b) Welding with side rolling

Fig. 10 Photographs of welded samples without and with side rolling side rolling, the tensile strain becomes compressive strain or small tensile strain. Therefore, hot cracking can be reduced or eliminated.

A finite element model was also used to optimize the side rolling technique parameters. With the optimized parameters, experiments were conducted without and with side rolling. Figure 10 shows that a 52 mm crack was found in the weld with conventional welding and there is no cracks found in the welding with side rolling. The effect of rolling parameters can be found in Ref. 2.

Based on the results of this study, it was found that side rolling cooling can be an effective method to control cracking. This method can be used with materials where trailing cooling cannot be applied. The limitation of this method is the complex tooling design and cannot only be used for straight or circular welds.

Controlling Welding Heat Input

Although dissimilar materials have been successfully welded for many applications in the nuclear industry, the cladding of low alloy steel pipes with Ni-Cr filler wire has remained a challenge. This challenge is related to ductility-dip cracking (DDC) and weld solidification cracking [3].

Ref. 3 investigated the DDC with both numerical modeling and experimental mockup. Based on the results of this study, it was found that reducing heat input is a valid method toward reducing the cracking tendency. The detailed results can be founded in Ref. 3.

Changing Welding Sequence

Changing welding sequence can change weld stress distributions which could be used in reducing cracking tendency. This example shown here is not for hot crack mitigation, but it serves as an example how welding sequence changes impact the weld stress distribution.

This study [4] was performed to help support the assessment of the cracking found in the 'A' reactor pressure vessel nozzle to hot-leg pipe bimetal weld (Fig. 11) in a US nuclear plant. The hot leg weld is a bimetallic weld joining a SA-508 (Class 2) reactor vessel nozzle with a Type 304 stainless steel pipe using an Inconel weld procedure (Fig. 12). The hot leg pipe carries reactor-heated water to the steam generator. It is then re-circulated by the pump back through the 'cold leg'. Both the hot and cold leg stainless steel pipes are joined to the reactor vessel nozzles via bimetallic welds. The cracking of concern occurs in the Inconel weld only on the inner surface. Tensile weld residual stresses, in addition to service loads, contribute to the cracking.

An axis-symmetric finite element model was used in the analysis (Fig. 12). The distance from the center of the weld to the pressure vessel is about 0.5 m and the distance from the weld center to the steam generator is about 7 m. The finite element

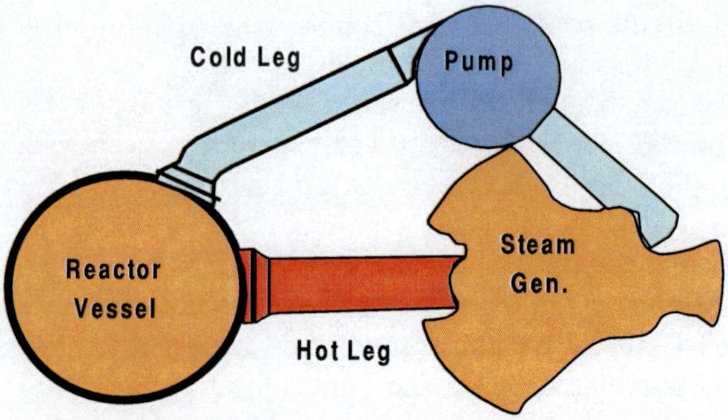

Fig. 11 Components in a nuclear reactor

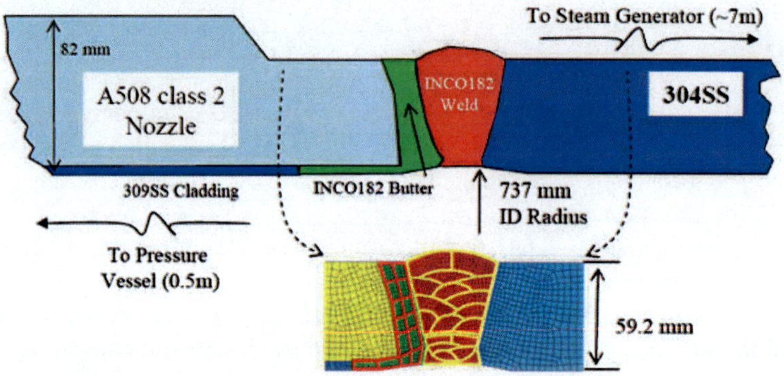

Fig. 12 Bimetallic welding geometry and finite element mesh near the weld

model included the entire 7.5 m of pipe system. The boundary conditions used in the finite element model included fixed displacements at the pressure vessel and steam generator. A lump-pass welding simulation technique was used in the analysis since more than 50 weld passes were used in the weld.

The numerical analysis involves several steps. Step 1 is to preheat nozzle to 66°C in preparation for cladding layer application. Step 2 is to model application of the cladding layer over the nozzle end with the model shown in Fig. 13. Step 3 is to conduct a local post weld heat treatment (PWHT) on the cladding area. Figure 14 shows the preheating temperature distributions (593°C) and heating length (254 mm). The heating lasts for 3 h which was modeled via creep analysis. Step 4 is to preheat nozzle and buttered region to 66°C in preparation for weld modeling. Step 5 is to deposit weld passes from inside to a depth of 17.8 mm as shown in upper picture of Fig. 15. At that time, defects were found in the weld and the weld had to be ground out. For holding the two parts together, a bridge was left as shown in the lower picture of Fig. 15.

There are two welding sequences which can be used to finish the hot-leg welding from the bridge. Sequence 1 is to weld inside and then weld outside. Sequence 2 is to weld outside and then weld inside. Finite element analyses were conducted to

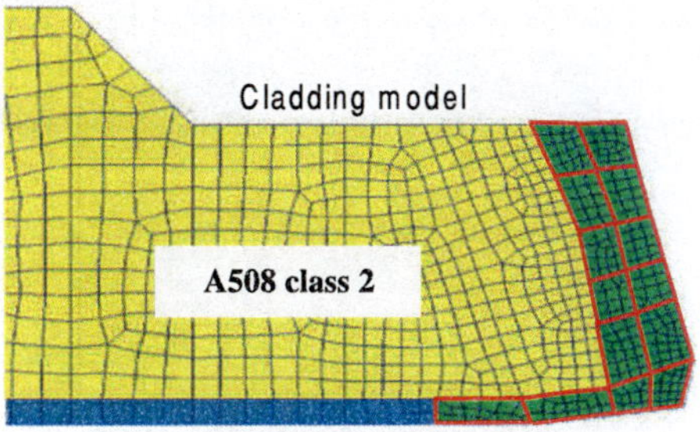

Fig. 13 Mesh for cladding on the carbon steel

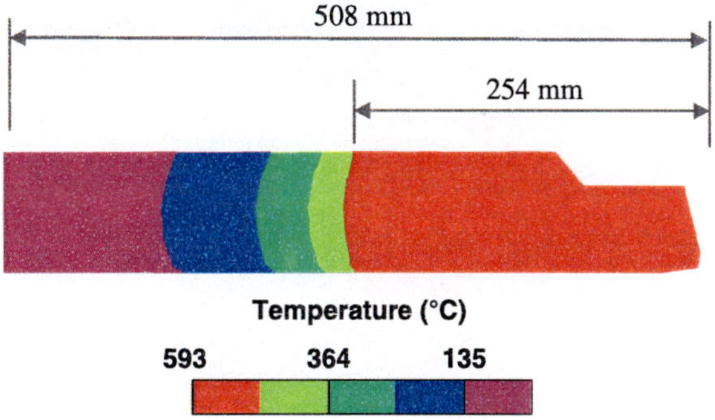

Fig. 14 Post-weld heat treatment modeling after cladding

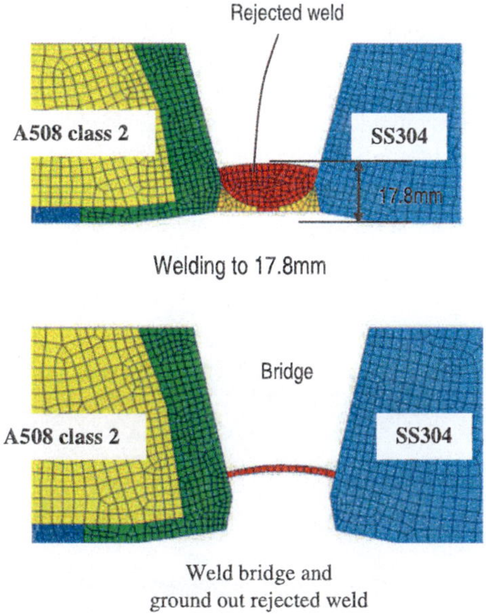

Fig. 15 Welding to 0.7 in. and ground out the defects

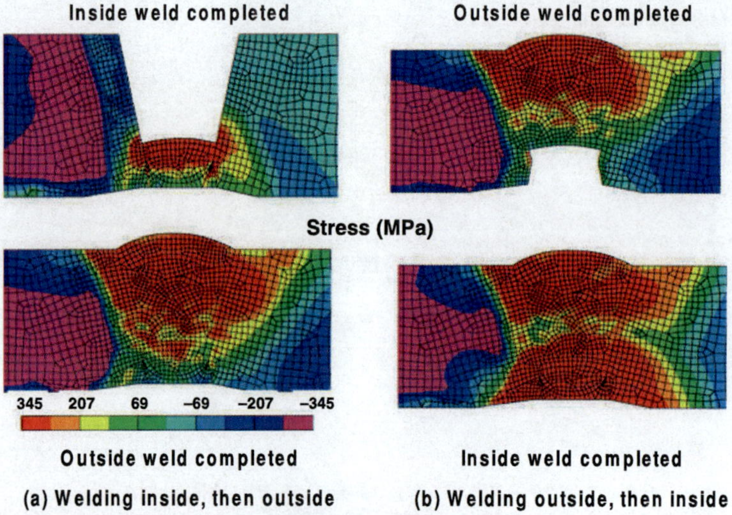

Fig. 16 Hoop stress comparison between two sequences

compare the stress distribution on the inner surface. Figure 16 shows the hoop stress comparison between the two welding sequences. The left-hand two figures show the hoop stress distributions after finishing inside and after completing the weld with sequence 1. The right-hand two figures show the hoop stress distributions after finishing outside and after completing the welding with sequence 2. The results show that sequence 1 has lower stress on the inner surface than sequence 2. This reason of lower stress in the sequence 1 is that the outside weld shrinkage in hoop direction applies compression on the finished inside weld. This results in high tensile stress outside and low tensile stress or compressive stress inside. The same principle should apply to sequence 2. But the inside weld is small compared to the outside weld and cannot produce enough shrinkage to reduce the tensile stress outside.

The shrinkage theory can also be used to explain the lower stress in sequence 1 than sequence 2 in the axial direction as shown in Fig. 17. Figure 17 shows the

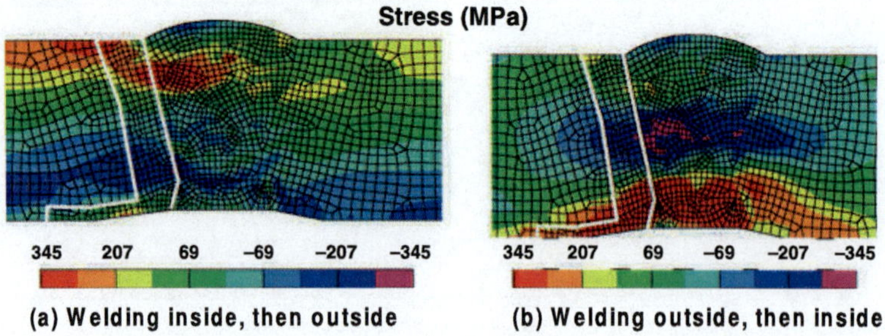

Fig. 17 Axial stress comparison between two sequences

final residual stress distribution for the two sequences. For the sequence 1, the outside weld shrinkage after welding applies compression on the inside weld in the axial direction. Therefore the stress on the inner surface has lower tensile stress or compressive stress.

Based on the results of this study, it can found that changing weld sequence can alter the stress distributions on the inner surface of a pipe component. This could be used in controlling stress distributions to prevent cracking during welding.

Summary

Stress and strain conditions associated with hot cracking were discussed and four thermomechanical methods were reviewed to demonstrate how to control hot cracking from a thermomechanical point of view. Experiment and numerical modeling tool was used in the development to understand the physics and optimize process parameters.

Trailing cooling and side rolling can be used for thin materials to prevent hot cracking. For materials not sensitive to cold cracking such as aluminum alloys and austenitic stainless steel, trailing cooling can be selected and side rolling can be used for most materials in principle, but it requires a complicated mechanical rolling equipment design. Reducing heat input and changing weld sequence can be practical methods to control hot cracking or other-type of cracking such as stress corrosion cracking Primary Water Stress Corrosion Cracking (PWSCC).

References

1. Yang YP, Dong P and Zhang J (2000) A Hot-cracking Mitigation Technique for Welding High-strength Aluminum Alloy. Welding Journal 79(1): 9-s–17-s.
2. Yang YP, Dong P and Tian XT (1998) Prevention of Welding Hot Cracking of High Strength Aluminum Alloys by Mechanical Rolling. Proceedings of the Conference on Trends in Welding Research, American Society for Metals, Pine Mountain, GA: 700–705.
3. Yang YP, Babu SS, Vaze S, Kikel JM and Dewees D (2008) Crack Mitigation During Buttering and Cladding of a Low Alloy Steel Pipe. Proceedings of the 8th International Conference on Trends in Welding Research, Pine Mountain, Georgia, June 2–6.
4. Brust FW, Scott PM and Yang YP (2003) Weld Residual Stresses and Crack Growth in Bimetallic Pipe Welds, Proceeding of 17th International Conference on Structural Mechanics in Reactor Technology (ed. S. Vejvoda), Praque, Czech Republic, August, 2003, Brno University of Technology, Brno, 2003, Section G, CR-ROM, http://www.iasmirt.org/SMiRT17/G08-1.pdf, Accessed 4 June 2010.
5. Yang YP (1995) Preventing Hot Cracking of High-strength Aluminum Alloys with Inverse Strain Methods. Ph.D. Dissertation, Harbin Institute of Technology, China.

Weld Cracking Susceptibility of Alloy C-22 Weld-Metal

M.L. Gallagher and John Lippold

Introduction

As the demand for materials with increased corrosion resistance and elevated temperature strength increases, the use of nickel-based alloys also increases. However, new materials can pose challenges joining, especially in welding. Alloys are often created without sufficient consideration given to the weldability of the material. As a result, welds are often the "weak-link" in a structure with respect to mechanical properties or corrosion resistance. One such weldability consideration is the formation of discontinuities, including the various forms of weld cracking. Discontinuities are especially problematic in critical applications where they can not only be costly and difficult to repair, but can also pose significant health and safety risks. The mechanisms behind the creation of such discontinuities must be well understood in order to avoid their formation, allowing these new materials to be used to their full potential.

One such critical application, in which Alloy C-22 is to be used, is the Yucca Mountain Project (YMP). Yucca Mountain is the proposed site for the first long-term geological repository of radioactive waste within the United States. This waste, from both energy and defense applications, will be stored in canisters within tunnels well below the surface of the mountain and well above the nearest water table. Due to its excellent corrosion resistance, the outer portion of these storage canisters is to be constructed of Alloy C-22. These storage canisters will require a 10,000 year compliance period, and will initially be exposed to high temperatures and levels of radiation [1]. In such a critical application, welds must be free of defects and have a minimum number of discontinuities. The prospect of weld-cracking during the fabrication of these storage canisters is unacceptable. It is therefore necessary to obtain a metallurgical understanding of the mechanisms that cause elevated temperature cracking within Alloy C-22.

M.L. Gallagher (✉)
Edison Welding Institute, Columbus, OH, USA
e-mail: mgallagher@ewi.org

J. Lippold et al. (eds.), *Hot Cracking Phenomena in Welds III*,
DOI 10.1007/978-3-642-16864-2_19,

Hastelloy alloys are nickel-based alloys known throughout the chemical processing industry as premier corrosion resistant alloys [2]. The "C-series" of Hastelloy alloys are derived from the Ni-Cr-Mo ternary system. Alloy C-22 is the third iteration in the C-series, following alloys "C", "C-276", and "C-4". Alloy C-22 contains low carbon levels to retard carbide formation, and levels of W and Mo that are a balance such that μ-phase formation during elevated temperature aging is retarded yet pitting and crevice corrosion resistance are not appreciably sacrificed. Alloy C-22 is suitable for most applications in the as-welded condition, resists the formation of grain-boundary precipitates, and is resistant to intergranular attack [2,3].

Solidification Cracking

In early work with Hastelloy X, Savage and Krantz [4] and Yeniscavich and Fox [5] observed hot-cracking to occur exclusively in areas affected by microsegregation. Local areas with higher solute concentrations have lower melting temperatures. When these local areas that are still solidifying are subjected to the shrinkage strains inherent to welding processes, cracks may form.

The mechanism by which hot-cracking occurs is most often explained using the modified-generalized theory [6]. In this theory, solidification is subdivided into four stages corresponding to decreasing temperature ranges. Stage-one, known as the mushy-stage, occurs at the beginning of solidification; during this stage too much liquid is present in the microstructure for cracking to occur. Stage-two is known as the coherency-stage. Cracking may occur in this stage; however, enough liquid is present in the microstructure in this temperature range such that all cracks that form will be healed. Stage-three is known as the critical-stage; in this stage, cracking occurs and there is insufficient residual liquid for crack healing. This stage is further subdivided into stages 3H and 3L, known as the film-stage and droplet-stage respectively. In the film-stage, cracks may both initiate and propagate; in the droplet-stage, cracks may only propagate from film-stage cracks. Stage-four corresponds to complete solidification, and hot-cracks neither initiate nor propagate in this temperature range.

Within the weld metal, hot cracking occurs at both grain boundaries and subgrain boundaries, which have high solute concentrations as a result of partitioning [7] during solidification. This phenomenon has been reported by numerous researchers, including Savage and Lundin who, during the development of the Varestraint test, reported that hot cracking initiated exclusively at sites of segregation [8]. Solidification cracking usually appears along the weld centerline, *i.e.* the last portion of the weld to solidify, in thick or highly restrained welds [9].

The wetting characteristics of the solidification grain boundaries and sub-grain boundaries also play an important role in the solidification cracking behavior of the alloy [10]. When a low fraction of liquid (less than 1%) with a high ratio of solid/solid to solid/liquid interfacial energies is present at the end of solidification, the tendency for the alloy to experience solidification cracking is remote. The liquid will likely exist in isolated pockets, where sufficient solid-solid interface exists to resist cracking. A moderate fraction of liquid (1–2%) with a low ratio of S/S to S/L interfacial energies at the end of solidification represents the worst case scenario for solidification cracking. This liquid will continuously wet the solidification boundaries, thereby creating a very susceptible structure for solidification cracking. If greater than 10% liquid is present at the end of solidification, the microstructure becomes less susceptible to solidification cracking because the large quantity of liquid at the end of solidification is able to "back-fill" any cracks that may form [7].

Cieslak et al. [11] investigated the relative weldability of alloys C-4, C-22, and C-276. It was discovered that C-22 and C-276 both formed non-equilibrium, low melting terminal solidification constituents. These terminal solidification constituents were seen to be significantly enriched in Mo and W, and depleted in Ni and Fe; Cr and Co were not observed to partition within these alloys. Differential Thermal Analysis (DTA) conducted on Alloy C-22 revealed a temperature range of approximately 1,400–1,360°C for primary austenitic solidification, while the σ-phase detected on the DTA curve did not solidify until approximately 1,290°C. Thin foil micrographs and EDS spectra revealed that Alloy C-22 contained three terminal solidification constituents: P-phase (orthorhombic), σ-phase (tetragonal), and μ-phase (hexagonal). As a whole, the terminal solidification constituent population consisted of approximately 80% P-phase, 20% σ-phase, and trace amounts of μ-phase.

Subsequent research on the solidification behavior of C-Series alloys was conducted by Perricone, Dupont, and Cieslak [12]. This investigation used computational thermodynamics to investigate the solidification sequences of alloys C-22 and C-276; the computational thermodynamic software allowed the researchers to account for alloying additions other than just Ni, Cr, and Mo, which had previously limited the analysis performed by Cieslak et al. [11]. The Mehrabian-Flemings model [13], which describes macrosegregation, was used to for the purpose of solidification path calculations. From this, the predicted solidification sequence of Alloy C-22 is:

$$L \rightarrow L + \gamma \rightarrow L + \gamma + P \rightarrow L + \gamma + P + \sigma \rightarrow \gamma + P + \sigma$$

Ductility-Dip Cracking

Collins, Lippold, and Ramirez [14] authored an extensive three part article on Ductility Dip Cracking (DDC) of nickel-based filler metals. DDC is an elevated temperature, solid-state cracking phenomenon occurring below the effective solidus temperature of an alloy, see Fig. 1. It is often observed in thick-section, multi-pass welds in austenitic materials. The large grain size and high restraint levels characteristic of these weldments contribute to the occurrence of DDC. Although DDC is relatively uncommon, it is seen most often in applications with low defect

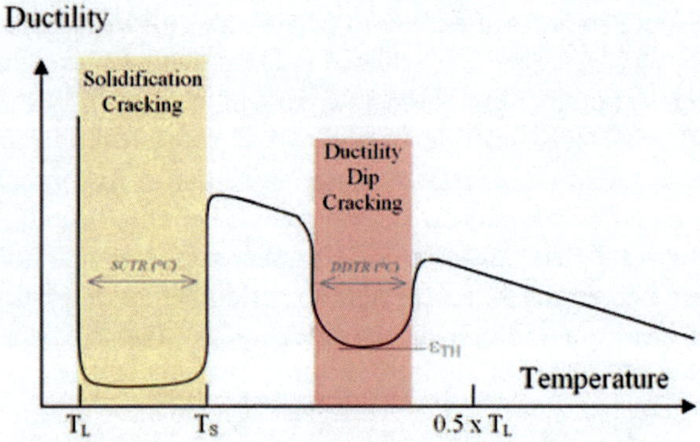

Fig. 1 Ductility vs. temperature schematic for a DDC susceptible alloy

tolerances, where cracking can be very costly (such as high-pressure steam, nuclear, and power generation applications).

DDC occurs along migrated grain boundaries (MGB), and becomes increasingly favorable when these boundaries are long and straight, as compared to tortuous. DDC has also been observed to initiate (and possibly terminate) at triple points, which have demonstrated to be microscopic stress concentrators. Unfortunately, the mechanisms behind DDC are currently not well understood, and preventative methods are limited.

Collins et al. [14] specifically investigated two nickel-based filler metals: Filler Metal 52 and Filler Metal 82. The STF test was selected as the method to quantify the relative susceptibility of the filler metals; the STF test is a Gleeble based test for DDC susceptibility. The development of the test is discussed by Nissley and Lippold [15] and is explained in subsequent sections of this paper. FM-52 was found to be considerably more susceptible to DDC than FM-82. Of the two filler metals, only FM-82 contained intentional Nb-additions. As a result, FM-82 forms both NbC and Laves-phase during solidification; FM-52 only forms the Laves-phase during solidification. The niobium-carbides within the FM-82 microstructure inhibits the migration of grain boundaries and produces tortuous grain boundaries more effectively than the secondary constituents in FM-52. As a result, FM-82 possesses only tortuous grain boundaries within its microstructure, whereas FM-52 exhibits a mixture of both straight (migrated) and tortuous grain boundaries. As mentioned previously, tortuous grain boundaries appear to inhibit DDC, especially at low strain levels.

Hydrogen and sulfur additions were observed to promote the occurrence of DDC, hydrogen having a much more pronounced effect. At the upper end of the Ductility-dip Temperature Range (DTR), dynamic recrystallization is observed along the edges of ductility dip cracks. This dynamic recrystallization, a function of temperature and deformation, suppresses both crack initiation and propagation, signaling ductility recovery.

Collins et al. [16] also investigated the strain distribution along grain boundaries using electron backscattered diffraction (EBSD), a technique that reveals crystallographic information on the sub-micron scale. The EBSD analysis revealed that

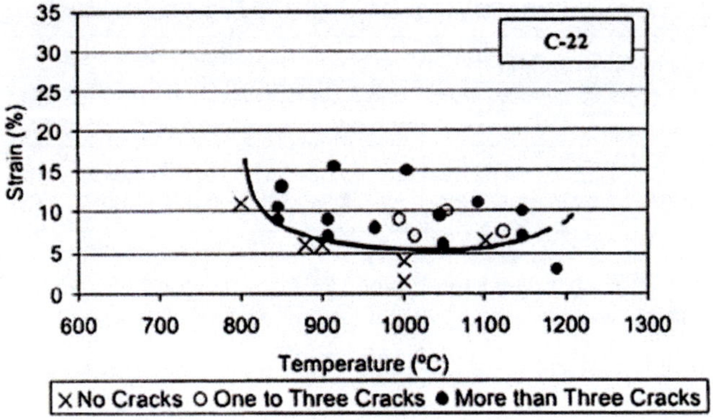

Fig. 2 Initial DDC evaluation of alloy C-22 using the STF test [17]

strain concentrates along grain boundaries, and especially at triple point junctions. Essentially no strain was seen within the grain interiors.

The nature of the boundary also influences the susceptibility to of the alloy to cracking. Low angle grain boundaries, *i.e.* boundaries where the lattice mismatch to its nearest neighbor is small, are known to have a high resistance to crack propagation. These grain boundaries include solidification subgrain boundaries and solidification grain boundaries where the high-angle boundary has migrated away. Collins et al. [16] observed DDC almost exclusively along high angle grain boundaries, specifically migrated grain boundaries. In multi-pass welding a higher degree of grain boundary migration occurs, due to the reheating of previously deposited passes. This boundary migration also results in further segregation of impurity elements to the migrating boundaries, due to a grain boundary "sweeping" effect. This impurity enrichment along the MGB may increase the susceptibility of an alloy to DDC.

Nissley et al. [17] performed work on the evaluation of DDC in austenitic stainless steels and nickel-base alloys. It was noted that the threshold strain (ϵ_{th}) and the DTR are the two primary quantitative measures of how susceptible an alloy is to DDC in the STF test, see Fig. 1. Additionally, it was noted that DDC is often observed in alloys free of second phases or precipitates. Nissley et al. [17] also provided initial results on the DDC susceptibility of Alloy C-22 in the STF test, see Fig. 2.

Experimental Procedure

Materials

Two separate commercial heats of 25.4-mm (0.25-in) thick Alloy C-22 plate were obtained from Haynes International. The compositions of these two heats are listed in Table 1; note the similarity of the compositions of the two heats. Small quantities

Table 1 Composition of the two heat of Alloy C-22 used for initial testing

	Chemical analysis (wt%)											
Heat number	*C*	*Co*	*Cr*	*Fe*	*Mn*	*Mo*	*Ni*	*P*	*S*	*Si*	*V*	*W*
2277-3-3114	0.0039	1.66	21.34	3.54	0.24	13.62	56.63	0.007	0.0037	0.025	0.12	2.91
2277-3-3119	0.0050	1.05	21.38	3.88	0.24	13.55	56.64	0.006	0.0057	0.028	0.14	3.07

of commercially pure metals were obtained in order to conduct the button-melting experiments. These metals included batches of nickel, chromium, molybdenum, tungsten, iron, and cobalt; all having a minimum purity of 99.90%.

Weldability Tests

The Transverse Varestraint Test

The Transverse Varestraint (TV) test was used to quantify the solidification cracking susceptibility of Alloy C-22. The Varestraint machine consists of a GTAW torch, power-supply, and a hydraulic system. Computer control of the X, Y, and Z-axis of the GTAW torch allows the torch to travel along the surface of the test specimen at a specified arc gap, travel speed, and distance. The hydraulic system of the Varestraint machine is used to conform the test plate around a die block of known radius, which induces augmented strain. Computer control of the hydraulic system, in coordination with the GTAW torch, allows the hydraulic system to bend the test plate around the die block using a specified stroke and stroke rate, at a user specified time in the GTAW weld sequence.

Six TV samples were cut from both heats of 25.4-mm (0.25-in) thick plate. Prior to testing, the surfaces of the samples were prepared by grinding, and cleaning with acetone. The parameters for TV testing were selected from the recommendations of Finton and Lippold [18]. A combination of current and travel speed was selected such that a weld bead with a width of approximately 10-mm (0.4-in) was produced; this weld bead width is standard for the Transvarestraint test, and permits comparison with other tested alloys.

After testing, cracking of the sample surfaces was evaluated using a low magnification stereomicroscope. The measurement of particular interest in assessing cracking susceptibility was the Maximum Cracking Distance (MCD). The MCD, or the longest crack in a direction perpendicular to the local solidification front at the instant of strain application, was measured using a filar scale within the ocular of the stereomicroscope.

The Strain-to-Fracture Test

The STF test was used to evaluate the DDC susceptibility of Alloy C-22; this is a Gleeble based test, which was performed on a model 3800 Gleeble® thermal-mechanical simulator. Twenty Gleeble "dog-bone" samples were sectioned

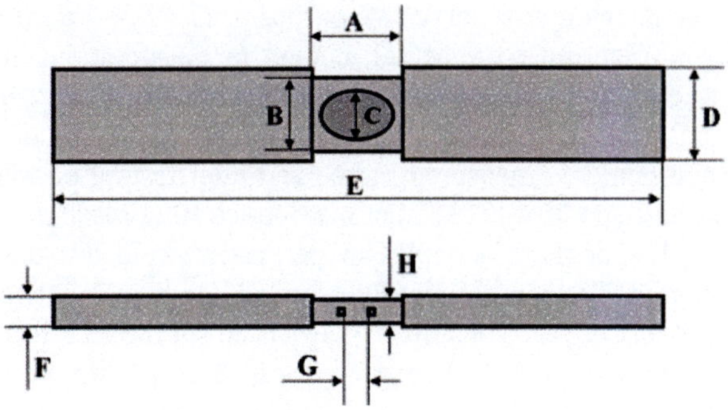

Fig. 3 Schematic of the strain-to-fracture test specimens. A = 19 mm (0.75 in), B = 15.3 mm (0.6 in), C = 10.2 mm (0.4 in), D = 19 mm (0.75 in), E = 127 mm (5 in), F = 6.4 mm (0.25 in), G = nominally 4 mm (0.16 in), and H = nominally 5.6 mm (0.22 in)

from heat 2277-3-3114 using standard machining techniques. Figure 3 displays the nominal dimensions of an STF sample.

A spot weld, approximately 10-mm (0.4-in) in diameter, was made on each side of the reduced section of the STF samples. The spot welds were autogenous GTA welds made using the parameters listed in Table 2. Copper chill blocks were used to fixture the samples during spot welding, ensuring that the spot welds would not grow larger in diameter than the width of the reduced section. The spot weld is made on the samples to create a reproducible microstructure that is susceptible to DDC; the slow radial solidification of the weld pool is controlled by the current down-slope time and produces large columnar grains that are oriented at all angles to the axis of tensile strain experienced by the samples during the STF test.

Making a spot-weld on the faces of the reduced sections creates an uneven surface, due to formation of a solidification crater. This surface is unfavorable for post-testing optical examinations. As a result, these faces are ground flat prior to testing. Grinding of the reduced section was accomplished using a 178-mm (7-in) diameter by 19-mm (0.75-in) wide aluminum oxide grinding wheel.

Gauge marks, located 4 mm (0.16 in) apart, are stamped on the top and bottom of the reduced section of the samples using diamond shape indenter and a special

Table 2 STF sample spot-weld welding parameters

Parameter	Time/setting	Level
Current up-slope time	None	–
Weld (constant)	30 s	145 A
Current down-slope time	12.7 s (standard)	11.4 A/s
Arc length	2 mm (0.079 inches)	10–12 V
Shielding gas pre-flow	10 s	Argon
Shielding gas post-flow	15 s	Argon
Gas flow rate	0.57 m^3/min	Argon

locator tool. These gauge marks serve two purposes: (1) they are measured before and after testing to determine the actual amount of plastic strain induced in the sample during testing, and (2) they serve as anchor points for the ceramic prongs of the longitudinal extensometer used in the Gleeble STF program. The distance between the gauge marks is measured before and after testing on the micrometer controlled stage of a Leco M-400-H1 microhardness testing machine.

Prior to testing, K-type thermocouples are percussion welded to the center of the reduced gauge section. The samples are then loaded into the Gleeble chamber with a free span of 50.8 mm (2 in). The Gleeble jaws used for the STF test are stainless steel "hot jaws" with hardened steel serrated inserts used to grip the sample. Once the sample has been loaded, the Gleeble chamber is closed, evacuated to a partial vacuum of 2.0×10^{-1} torr, and backfilled with argon gas. This procedure is repeated twice to decrease the level of sample oxidation during testing.

Initially the Gleeble heats the sample at 100°C/s (180°F/s) to the desired testing temperature; the temperature of the sample is then kept constant for 10 s as the gauge section is allowed to achieve an equilibrium temperature profile. During these initial steps, the mechanical system is under force control, and the sample experiences 0 kg (0 lbs) force. The mechanical system is next switched to stroke control mode in order to strain the sample. A control loop strains the sample at a constant stroke rate of 0.6-mm/s (0.024-in/s) until a specified extension of the gauge length, as measured by a longitudinal extensometer, is reached. After the desired gauge length extension is reached, the thermal and mechanical systems are turned off, and the sample is allowed to free cool under negligible tensile force.

The sample gauge marks are re-measured following STF testing. The difference between the initial and final gauge lengths is noted, and the plastic strain of the gauge section is calculated and recorded. Samples are next sectioned to separate the gauge section from the shoulders of the specimens. The gauge sections are mounted in Bakelite, ground flat, and polished to a 1 μm finish. This mirror surface is next examined under a stereomicroscope at magnifications between 10 and 80X; all cracks on the sample surface are counted and recorded. The samples are subsequently broken out of the Bakelite mold, and the opposite sample face is mounted, ground, and polished to count and record the number of cracks. The number of cracks reported per sample is the average of the number of cracks on the two faces. This cracking data is plotted on a strain versus temperature plot. Based on the data, a curve is drawn that separates cracked from un-cracked samples across the entire temperature field. This curve represents the threshold strain below which cracking will not occur. This threshold strain line defines the cracking susceptibility "envelope" for the alloy and can be used to compare behavior among alloys.

The Hot Ductility Test

Hot ductility testing was performed on Alloy C-22 base-metal and weld-metal using a Gleeble® 3800 thermo-mechanical simulator. All samples were machined from heat 2277-3-3114 of Alloy C-22. Base-metal samples were created by sectioning 100 mm × 6.4 mm × 6.4 mm (4 in × 0.25 in × 0.25 in) samples from 6.4 mm

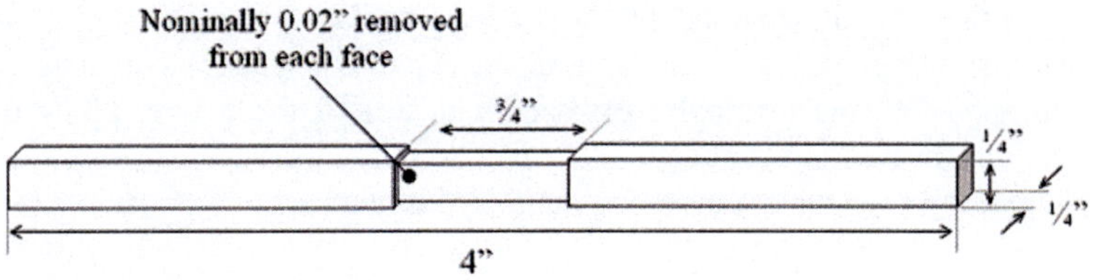

Fig. 4 Hot-ductility sample schematic

(0.25 in) thick Alloy C-22 plate. Weld metal samples were created in a similar fashion from plate with the center section consisting entirely of autogenous weld metal. The center sections of all samples were subsequently ground on all four faces to produce a 19 mm (0.75 in) gauge section that allowed for heat and strain concentration during Gleeble testing. Grinding was accomplished using a 178 mm (7 in) diameter by 19 mm (0.75 in) wide aluminum oxide grinding wheel. The nominal dimensions of the hot- ductility samples are displayed in Fig. 4. The autogenous weld-metal center sections of the weld-metal samples were created by first running a deep penetrating GTAW weld on the face of the Alloy C-22 plate. The plate was then flipped over, and an identical weld was made on the opposing face of the plate such that the second weld penetrated into the weld metal of the first pass, creating a center section of through-thickness weld-metal.

The Gleeble testing parameters for the hot-ductility test are listed in Table 3. The Nil-Strength-Temperature (NST) of the metal is determined by subjecting the sample to a small load (approximately 20 kg) and heating at a rate of 111°C/s (200°F/s) until the sample fails. The point where the recorded temperature begins to deviate from the programmed temperature is determined to be the NST. On-heating samples are heated at a rate of 111°C/s (200°F/s) to the testing temperature and immediately pulled to failure. On-cooling samples are heated at a rate of 111°C/s (200°F/s) to a temperature just below the NST. The samples are then free cooled, through conduction with the Gleeble jaws, down to the desired testing temperature, where they are immediately pulled to failure.

Table 3 Hot-ductility Gleeble testing parameters

Parameter	On-heating	On-cooling
Atmosphere	Vacuum	
Jaws	Stainless steel (with serrated grip inserts)	
Sample free-span	25.4-mm (1-in)	
Heating rate	111°C/s (200°F/s)	
Peak temperature	T_{test}	NDT < T_{test} < NST
Cooling rate	N/A	Free-cool
Hold time	0 s	
Stroke rate	50.8 mm/s (2 in/s)	

After testing, the dimensions of the fracture surface were measured with a micrometer, and the cross-sectional area of the fracture surface is calculated. This area is compared to the original cross-sectional area of the gauge, and the percent reduction in area is calculated. Plotting the reduction in area versus testing temperature allows for the creation of a ductility "signature".

Compositional Variation

Thermodynamic Simulations

Thermodynamic simulations were conducted using both the JMatPro v4.1 and Thermo-Calc Version-P thermodynamic simulation programs; both programs utilized the ThermoTech Ni-DATA V.7 thermodynamic database. These programs were used to create a number of diagrams used to understand the welding metallurgy and phase transformation sequence of Alloy C-22. The module that was most frequently used to understand the behavior of Alloy C-22 weld-metal was the Scheil-Gulliver non-equilibrium solidification module.

A response surface method (RSM) experimental approach was used to conduct the Alloy C-22 Scheil-Gulliver simulations in this investigation. The elements that are significantly enriched or depleted during solidification of Alloy C-22 were chosen as the factors to be varied. According to Cieslak et al. [11] the elements that are enriched or depleted in the terminal solidification constituents of Alloy C-22 are Mo, W, and Fe. For the remainder of the elements that comprise Alloy C-22, the levels of these elements in the Scheil-Gulliver simulations were chosen to be the average level from the two heats of Alloy C-22 used within this study. The low, intermediate, and high values for the RSM matrix were selected based upon the compositional limits of Alloy C-22. The RSM design was a three-factor EIMSE-Optimal RSM design [19]. For the Scheil-Gulliver simulations the following parameters were used: start temperature of 1,450°C, temperature step size of 1°C, and solidification cut-off of 0.02 fraction liquid.

Button-Melter Experiments

The button-melter constitutes a water cooled copper hearth that uses a GTAW torch as a heat source to melt and stir small charges of metal. A quartz cylinder and a brass lid, into which the GTAW torch enters, create an air-tight chamber through which argon flows to provide a protective atmosphere. Button melting experiments were performed to physically replicate the Scheil simulations performed in the thermodynamic simulation phase of the project.

The first step in this process was the creation of Ni-42Mo, Ni-25Fe, Ni-25 W, and Ni-25Co binary alloys. These binary alloys allowed for more precise addition of small quantities of alloying additions, and allowed for easier addition of high melting point elements such as molybdenum, utilizing the Ni-Mo eutectic composition. Once the compositions of the binary alloys were verified via Energy Dispersive

Spectrometry (EDS), precise quantities of the binaries were sectioned along with precise quantities of commercially pure nickel and chromium. These charges were melted together in the button-melter to form the desired compositions of the final buttons of the RSM matrix. Energy dispersive spectrometry, utilizing pure metal standards, was used to verify the final composition of the sixteen RSM buttons.

The volume fractions of secondary phases of the various alloys were determined through metallographic techniques. The samples were metallographically prepared to a finish of 0.05 μm, and examined in an SEM in the unetched condition. As a result of the high molybdenum and tungsten content of the secondary phases, these phases were clearly visible in the backscattered electron imaging mode. Using image analysis software, thirty scanning electron micrographs (taken at 2000X) of each composition were analyzed to determine the volume fraction of secondary phases.

Microstructural Preparation and Characterization

Samples to be examined using Light-Optical Microscopy (LOM), and samples that were to be electropolished were mounted in thermosetting compression molding compound. Samples that were to be examined using Scanning Electron Microscopy (SEM) were mounted in a conductive mounting compound. All metallographic samples were ground from 180 to 800-grit using progressively finer grits of SiC paper; intermediate grinding steps. Samples were next vibratory polished in a neutralized colloidal silica polishing suspension for approximately 3 h. Care was taken such that samples to be etched were not exposed to air after the final polish, as recommended by the alloy manufacturer [20].

Two etching methods were used to reveal the structure of the metal. A light etching of the structure was accomplished using a 10% chromic acid solution at an applied voltage of 2.5 V for approximately 30 s. A heavier etch was accomplished using a 95 mL HCl +5 g oxalic-acid solution at an applied voltage 6 V for 2–5 s. Samples to be examined using the Electron Backscattered Diffraction (EBSD) technique required electropolishing. These samples were electropolished for 10 s, at an applied potential of 22 V, using the Struers A2 solution at 20°C (70°F).

Optical microscopy was accomplished using a Nikon Epiphot microscope and a PAXcam 2MP digital camera using PAX-it™ software. Image analysis, of both optical and electron micrographs, was accomplished using ImageJ v1.37 software. Low magnification images were captured using a Nikon SMZ1000 stereoscope and a Nikon DS-2Mv 2MP digital camera. All electron microscopy was accomplished using a Quanta scanning electron microscope (SEM). The Quanta is a heated tungsten filament SEM equipped with a range of detectors including a secondary electron detector, a backscattered electron detector, and an internal CCD camera. The Quanta is also equipped with an Energy Dispersive Spectrometer (EDS) for chemical analysis using x-rays, and an Electron Backscattered Pattern (EBSP) detector for mapping and phase analysis.

Results

Solidification Cracking

Figure 5 displays the results from the Transvarestraint testing of Alloy C-22. The blue line represents heat 2277-3-3114 and the red line represents heat 2277-3-3119. One can see that strain saturates at approximately 2% augmented strain; that is, further increases in augmented strain does not lead to an increased cracking response.

Maximum Cracking Distance (MCD) is of particular interest because it is proportional to the temperature range over which solidification cracking occurs. The Solidification Cracking Temperature Range (SCTR) is given by the MCD (past the point of saturated strain) being multiplied by the cooling rate of the weld (as measured through thermocouple plunges), all divided by the travel speed of the GTAW torch during the Transvarestraint test. Using this formula, it was found that Alloy C-22 had an SCTR of approximately 50°C (90°F). Standard metallographic techniques found cracking to occur primarily along solidification grain boundaries (SGB), and to a lesser extent along solidification sub-grain boundaries (SSGB).

Strain-to-Fracture Test

Figure 6 displays the results of the strain-to-fracture test for Alloy C-22. Each circle represents an individual STF sample; the adjacent bracketed numbers are the number of cracks experienced by the corresponding sample. The dotted line represents the threshold level of strain to cause cracking. It is observed that Alloy C-22

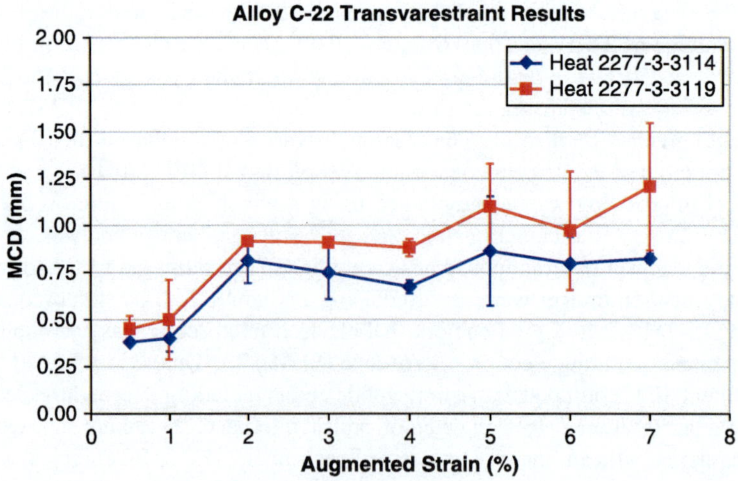

Fig. 5 Alloy C-22 Transvarestraint testing results

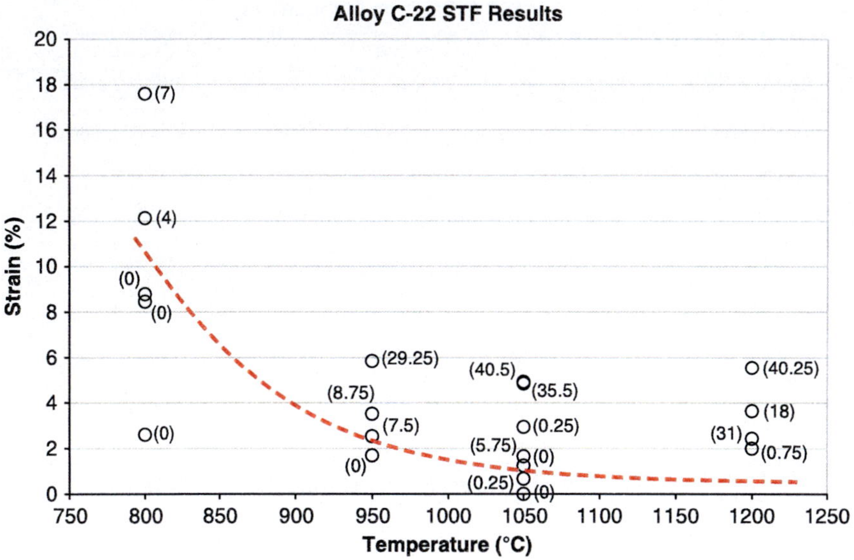

Fig. 6 Alloy C-22 strain-to-fracture test results; *red dotted line* represents the cracking threshold

has both a low cracking threshold, a wide temperature range over which cracking occurs, and no recovery of cracking resistance at the upper end of this susceptible temperature range. It is noted that only one heat of Alloy C-22 (heat 2277-3-3114) was used within this investigation. Earlier unpublished research demonstrated that the minor compositional differences between heats 2277-3-3114 and 2277-3-3119 were insufficient to cause observable differences in cracking response during the strain-to-fracture test.

In Fig. 7, the susceptibility of Alloy C-22 to ductility-dip cracking is compared to two other ductility-dip susceptible alloys that have been previously tested in the strain-to-fracture test: Filler-Metal 52 and Filler-Metal 82. The susceptibility of these three alloys is compared at 950°C (1,742°F). This temperature is used for baseline comparisons as it often corresponds to the minimum of the ductility trough for susceptible alloys. In the strain-to-fracture test, Alloy C-22 is observed to have a ductility-dip cracking susceptibility on the same order as that of FM-52.

Normally, alloys that are susceptible to ductility-dip cracking have three distinct regions on the strain-versus-temperature plot: (1) a low temperature region associated with good resistance to cracking, (2) an intermediate temperature range susceptible to ductility-dip cracking, and (3) a high temperature range where the resistance to cracking is restored due to the onset of recrystallization. However, Alloy C-22 does not recover its cracking resistance at the upper end of the cracking susceptible temperature range; instead, the intermediate ductility-dip temperature range of Alloy C-22 extends completely to the solidus temperature in the strain-to-fracture test.

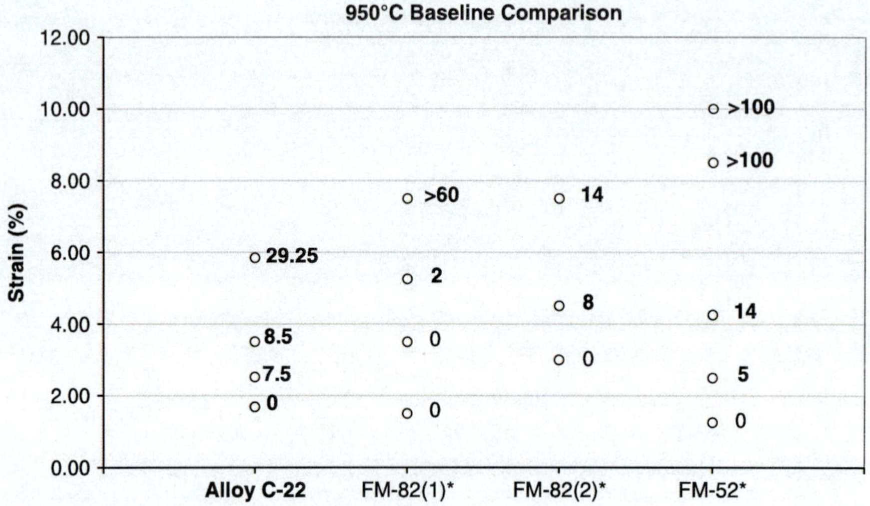

Fig. 7 Baseline comparison between the ductility-dip cracking susceptibility of alloy C-22, FM52, and FM82. Data provided courtesy of N.E. Nissley [21]

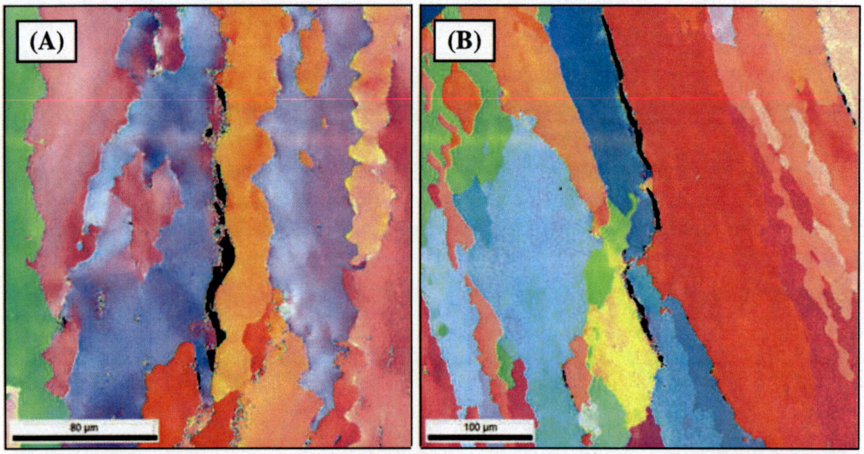

Fig. 8 Orientation imaging of strain-to-fracture samples; cracking is observed to occur along high-angle boundaries in both images. (**a**) STF sample subjected to 17% strain at 800°C (1,472°F), and (**b**) STF sample subjected to 3.6% strain at 1,200°C (2,192°F)

Optical, electron, and orientation imaging of ductility-dip cracks all found cracking to occur along high angle boundaries. Figure 8 displays two orientation images of ductility-dip cracks occurring along high-angle boundaries in strain-to-fracture specimens; the boundaries along which cracking has occurred are free of recrystallization. Metallographic examinations also determined that cracking was not associated with secondary phases, *i.e.* secondary phases did not initiate cracking.

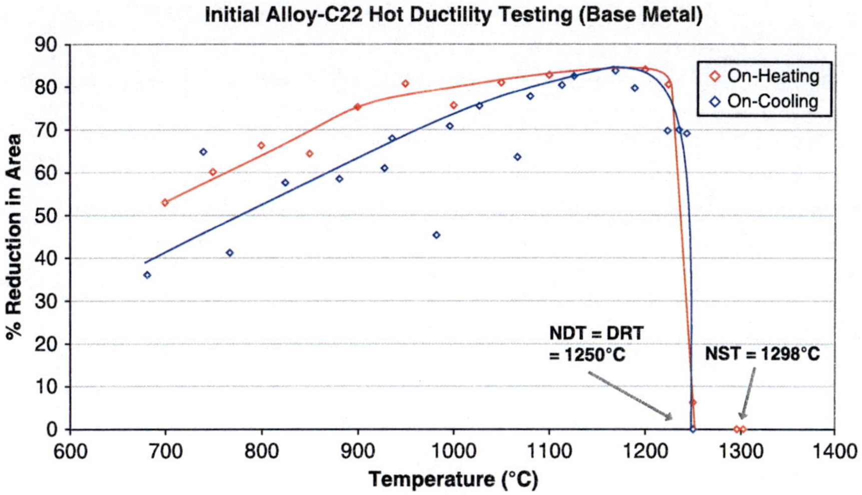

Fig. 9 Hot-ductility results for Alloy C-22 base metal

Hot-Ductility Test

Figure 9 displays the results from the hot-ductility testing of Alloy C-22 base-metal. Both the on-heating and on-cooling curves display excellent elevated temperature ductility, with no ductility-dip. The alloy is observed to have a narrow liquation cracking temperature range (LCTR), which indicates a low susceptibility towards HAZ liquation cracking. The LCTR is given by the difference between the nil-strength-temperature (NST) and the ductility-recovery-temperature (DRT), and represents the maximum temperature range over which liquid films will be present in the HAZ.

Figure 10 displays the results of the weld-metal hot ductility testing. The nil-strength-temperature, nil-ductility-temperature (NDT), and ductility-recovery-temperature are similar to those of the base-metal samples. However, a pronounced difference in the ductility signature of the weld-metal is seen as compared to the base-metal. The ductility of the weld-metal is lower than that of the base-metal throughout the entire range of test temperatures. More importantly, there are obvious ductility-dips in both the on-heating and on-cooling ductility curves.

Because of the difficulties associated with etching Alloy C-22 (namely, etching of high-angle boundaries), it is difficult to tell how cracks have propagated along the failure surface of the broken samples. However, all samples characterized had internal cracks. All internal cracks were noted to have occurred along high-angle boundaries. As with the strain-to-fracture test, cracking did not appear to be associated with secondary phases; *i.e.* cracks did not initiate at secondary phases.

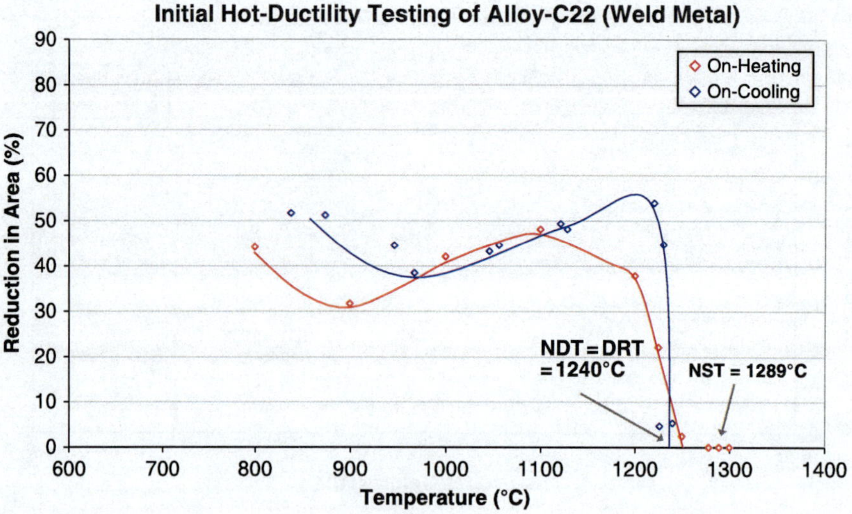

Fig. 10 Hot-ductility results for Alloy C-22 weld metal

Discussion

Solidification Cracking

As previously mentioned, Alloy C-22 was found to have an SCTR of approximately 50°C (90°F). An SCTR of approximately 50°C (90°F) demonstrates excellent resistance to solidification cracking; this SCTR ranks Alloy C-22 only slightly behind highly crack resistant alloys duplex SS 2205, austenitic SS 304 (FN6), and duplex SS 2507 in cracking susceptibility [22].

Examining Fig. 5, it is interesting to note that heat 2277-3-3119 has a higher cracking response than heat 2277-3-3114 throughout the entire range of augmented strain. However, the error bars in Fig. 5 indicate that this effect lies on what might be considered the border of statistical significance. That is, the standard deviation of one series often extends to the median of the other series, but rarely extends past the median. This is especially true when considering only the lower levels of augmented strain (below 5%), which are more representative of real weldments. Examining Table 1, one detects both heats to be extremely close in composition. Accordingly, Scheil simulations predict both heats to have virtually identical solidification temperature ranges. It is however noted that the Scheil simulations do not account for sulfur or phosphorus, and heat 2277-3-3119 has approximately 9% more (S + P) than heat 2277-3-3114. Perhaps this minor compositional difference, albeit between elements that are well known to aggravate hot cracking, explains the difference in the cracking response of the two heats. Lastly, the reader is cautioned that this discussion point is not intended to suggest that heat 2277-3-3119 is high in sulfur and phosphorus; both heats have extremely low levels of these two impurity elements.

Understanding how variations in composition effects the cracking susceptibility of an alloy is important with regards to heat-to-heat compositional variation. Every alloy is defined by a range of composition; as such, total confidence in the weldability of the alloy requires confidence over the entire compositional range. Small heat-to-heat variations in composition must not lead to significant increases in the cracking susceptibility of Alloy C-22.

The non-equilibrium solidification temperature range is proportional to the solidification cracking temperature range (SCTR) [23]; therefore, the SCTR is an indicator of the susceptibility of the alloy to solidification cracking. A response-surface-method (RSM) experimental design was used to examine the individual effects of Mo, W, and Fe on the non-equilibrium solidification temperature range (T_L-T_S) in Alloy C-22. Regression analysis of the factors and responses from the experimental design matrix was performed using the software package Minitab-14. The results presented are in the form of main effects plots; these plots use the mean value of all but one factor to show the individual effect of the remaining factor on the response.

Figure 11 displays the predicted (Scheil simulations) relationships for the effects of the Mo, W, and Fe on the non-equilibrium solidification temperature range. Examining these results, it is unexpectedly found that the predicted solidification temperature ranges are *decreased* as the alloy becomes richer in Mo, W, and Fe; with Mo having the most pronounced effect. However, examining the scale of the Y-axis in Fig. 11, it is observed that over the Mo, W, and Fe composition ranges that define Alloy C-22, the solidification temperature range is only affected by a few degrees Celsius. This level of response to the compositional variation of Alloy C-22 would not be expected to significantly effect the solidification cracking susceptibility of the alloy.

STF vs. Weld-Metal Hot-Ductility Signatures for Alloy C-22

The most striking difference between the results from the strain-to-fracture test (Fig. 6) and the hot-ductility test (Fig. 10) is the high temperature behavior of the alloy. In the hot-ductility test there is a recovery of ductility at the upper end of the testing temperature range; whereas, there is no recovery of cracking resistance in this temperature range in the strain-to-fracture test. This phenomenon is in large part explained by the recrystallization behavior of the alloy. Figure 12 displays orientation images of an internal crack from a weld-metal hot-ductility specimen tested at 1,100°C (2,012°F) in the on-heating condition. The boundaries of this sample are decorated with a vigorous amount of recrystallization. The recrystallized grains are heavily twinned equiaxed grains with an average diameter of approximately 5-μm (Fig. 12B).

This microstructure is compared to that of the 1,200°C (2,192°F) strain-to-fracture sample (Fig. 8b), where the boundaries are completely absent of recrystallization. In the hot-ductility test, recrystallization has worked to dynamically annihilate internal strain, allowing for higher levels of deformation before the

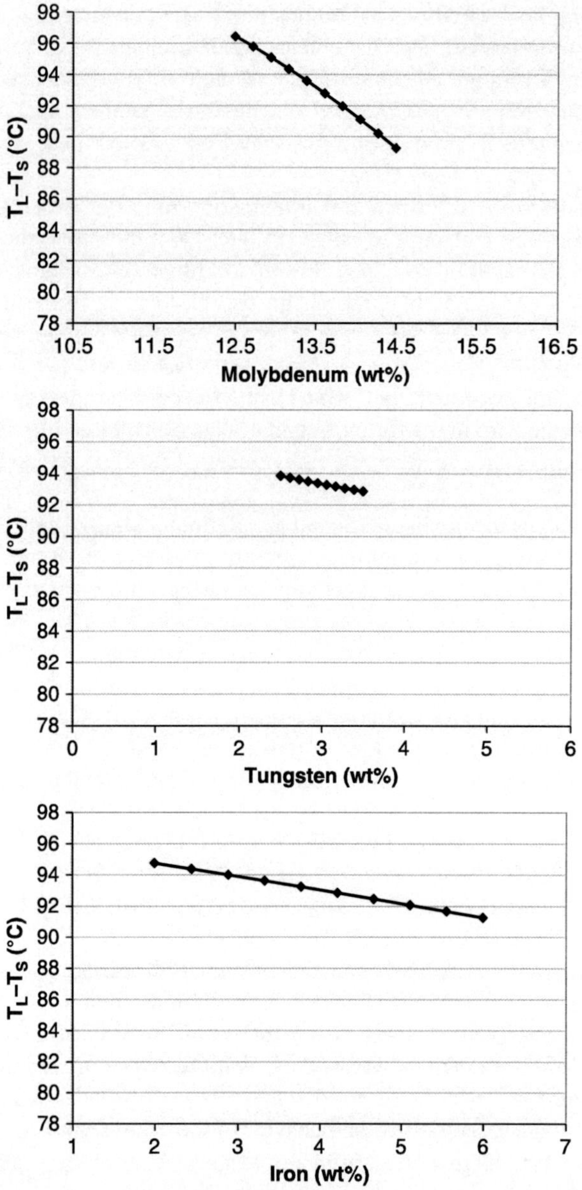

Fig. 11 Predicted ranges of the non-equilibrium solidification temperature ranges of the Alloy C-22 variants tested in the response-surface-methods trials

eventual failure. This condition produces the recovery of ductility at the high end of the testing temperature range. This is not the case during the strain-to-fracture test; recrystallization is not seen within the microstructure at the upper end of the testing temperature range, and ductility recovery does not occur.

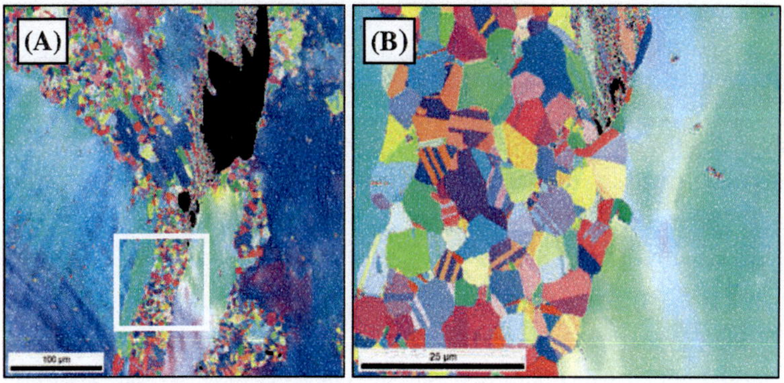

Fig. 12 (**a**) Orientation imaging of an internal crack in a weld-metal hot-ductility specimen tested at 1,100°C (2,012°F) in the on-heating condition, and (**b**) a zoomed in view of the region within the white-box in "A". Noted is the vigorous recrystallization along the boundaries

Alloy C-22 vs. Previously Tested Alloys in the STF Test

The fact that Alloy C-22 does not experience a recovery of cracking resistance at the upper end of the testing temperature range is to date unique amongst alloys tested in the STF test. The reason for this unique behavior also relates to the recrystallization behavior of the alloy. Reed-Hill [24] states that "*Dynamic recovery occurs most strongly in metals of high stacking-fault energies and is not readily observed in metals of very low stacking-fault energies*", and that "*Alloying normally reduces the stacking-fault energy of a metal.*" Stacking-Fault-Energy (SFE) and alloying content were examined and compared for Alloy C-22 and six other austenitic alloys previously tested in the strain-to-fracture test [17]. These alloys included austenitic stainless steels 304SS, 310SS, and AL-6XN, as well as nickel-base alloys 690, FM52 and FM82.

Of the nickel-base alloys within this group of materials, Alloy C-22 is the most highly alloyed, and therefore the most likely to have a low SFE and to be resistant to recrystallization. The SFE of these seven alloys was calculated using JMatPro (v4.1). The software calculates SFE by determining the difference between the Gibbs free energy of the FCC and HCP crystal structures of the alloy (a stacking-fault in austenitic alloys may be visualized as, or understood to be, a thin layer of HCP within the FCC matrix). At a temperature of 1,200°C (2,192°F), the SFE of these seven alloys rank from highest to lowest as follows: FM82, 304SS, 310SS, Alloy 690, AL-6XN, FM52, and Alloy C-22. Of these seven alloys, Alloy C-22 has the lowest SFE, indicating that it is the most resistant to recrystallization. This explains why Alloy C-22 does not experience a recovery of cracking resistance at the upper end of the testing temperature range, where other alloys do.

Secondary Phases and Elevated Temperature Ductility

The results above, specifically the STF results, indicate that Alloy C-22 is susceptible to DDC. What then can be done to improve the cracking resistance of the alloy? To answer this question, consider what microstructures are considered susceptible and resistant to DDC. Susceptible microstructures are usually single phase austenitic microstructures with long straight weld-metal boundaries. Resistant microstructures have secondary phases that inhibit grain boundary migration at elevated temperatures, which results in winding, tortuous grain boundary morphologies. With that in mind, would increasing the volume percent of secondary phases in Alloy C-22 increase the resistance of the alloy to DDC?

A small set of experiments were undertaken in order to verify the hypothesis that an increasing the volume fraction of secondary phases will decrease the susceptibility of the alloy to DDC. Several weld-metal hot-ductility and strain-to-fracture samples were heat treated to purposefully form secondary phases. These samples were heat treated, under an argon atmosphere, for 12 h at 990°C (1,814°F); this time and temperature corresponded to the nose of the time-temperature-transformation diagram for Alloy C-22, as predicted by JMatPro v4.1. The fraction-phase *vs.* temperature plot (also created using JMatPro v4.1) predicted that the secondary phases formed during this heat treatment should be a mixture of predominately P-phase with a very small amount of M_6C. Figure 13 displays the quantity and morphology of the secondary phases in the weld-metal before and after this heat treatment. The respective volume fractions of secondary phases in the two microstructures was measured to be 0.19% and 2.19% for the as-welded and heat-treated microstructures, respectively. Figure 13b demonstrates that the formation of secondary phases first occurred along boundaries within in the weld-metal.

Figure 14 displays the results of the hot-ductility testing of the heat-treated versus as-welded weld-metal samples, both in the on-heating condition. It is observed

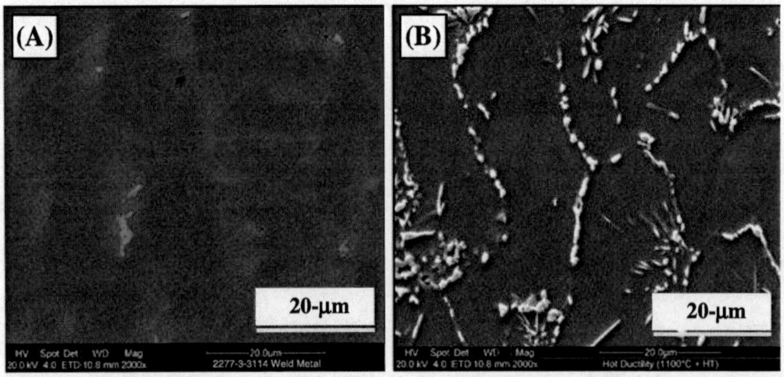

Fig. 13 (**a**) Backscattered electron imaging of Alloy C-22 weld-metal in the unetched condition, and (**b**) Backscattered electron imaging of Alloy C-22 weld-metal that has been heat-treated at 990°C (1,814°F) for 12 h, in the etched condition

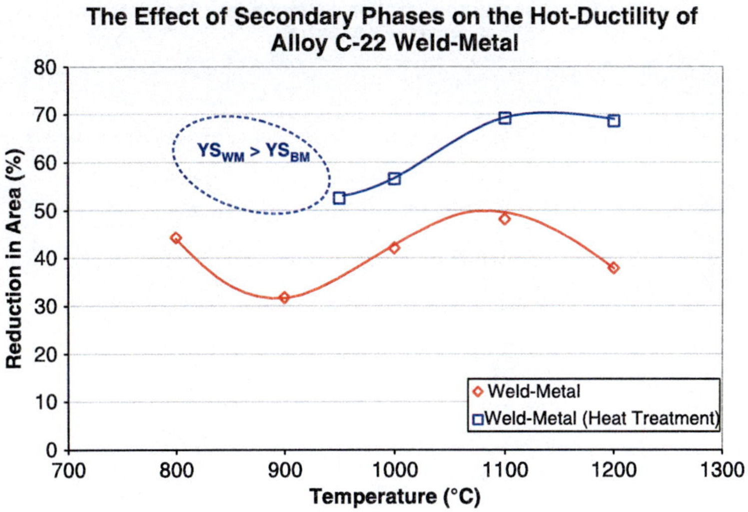

Fig. 14 A comparison between the elevated temperature ductility of as-welded Alloy C-22 weld-metal and Alloy C-22 weld-metal subjected to a heat-treatment of 990°C (1,814°F) for 12 h

that the samples with the higher volume fraction of secondary phases are significantly more ductile over the testing temperature range. Reduction in area data is not plotted for the 800°C (1,472°F) and 900°C (1,652°F) heat-treated weld-metal samples because necking and failure occurred in the base-metal of these samples. The heat-treatment had actually strengthened the weld-metal beyond that of the (cooler) surrounding base-metal during the Gleeble hot-ductility test.

The formation of secondary phases also improved the performance of the alloy in the strain-to-fracture test. Figure 15 displays the results from the strain-to-fracture testing of as-welded and heat-treated Alloy C-22 samples, at a testing temperature of 1,200°C (2,192°F). A significant improvement in the cracking resistance of the heat-treated samples is noted; the cracking threshold has climbed from below 2% to above 4% strain. Additionally, the rise in the rate of crack initiation appears to be lower with increasing levels of applied strain. Optical micrographs revealed recrystallization along the weld metal grain boundaries of the heat treated STF samples; recrystallization was not observed in the as-welded strain-to-fracture samples.

Similar to the hot-ductility tests of heat-treated weld-metal, the secondary phases present in the microstructure appear to have significantly strengthened the boundaries of the metal. Aside from boundary strengthening, a more subtle mechanism may have played a role in the increased ductility and cracking-resistance of the heat-treated Alloy C-22 samples with the higher volume-fraction of secondary phases. This mechanism is known as Particle Stimulated Nucleation (PSN), and has been widely studied in aluminum alloys [25]. This phenomenon occurs in alloys that contain coarse non-deformable particles; when such alloys are subjected to strain, from cold- or hot-working, a highly deformed region forms around these particles. In subsequent annealing, recrystallization occurs preferentially in

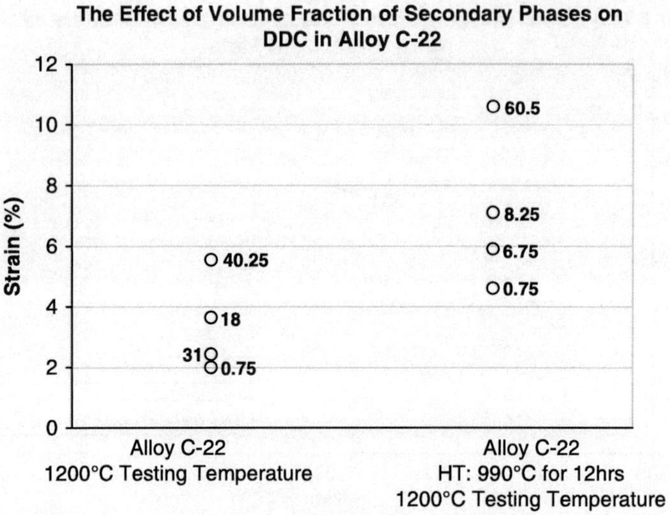

Fig. 15 Comparison of the behavior of as-welded and heat-treated Alloy C-22 weld-metal in the strain-to-fracture test at a testing temperature of 1,200°C (2,192°F)

the highly strained immediate vicinity of these particles; this phenomenon can presumably be extended to participate in dynamic recrystallization. Additionally, aside from producing recrystallization in situations where it may not be expected, particle-stimulated-nucleation also increases the kinetics of the recrystallization process.

If the P-phase of the heat-treated Alloy C-22 samples is in fact non-deformable, this mechanism may have occurred during the hot-ductility and strain-to-fracture testing of the heat-treated samples. This mechanism would have lessened the barrier to recrystallization of the alloy and allowed for higher strain levels to be achieved before the initiation of cracking in the strain-to-fracture test, or before failure in the hot-ductility test. This mechanism would of course be more active in the upper region of the testing temperature range, where dynamic recrystallization becomes possible.

Finally, it should be mentioned that these experiments involving the formation of secondary phases in the weld-metal have been conducted merely for demonstrative purposes. Secondary phase formation in Alloy C-22 weld-metal is known to be detrimental to corrosion resistance, and therefore discouraged. Corrosion considerations aside, it would be impossible to use heat-treatment in practical welding situations to form secondary phases for the purpose of improving the elevated temperature ductility and DDC resistance of a material. However, it would be very easy to form secondary phases as solidification constituents by varying the composition of the weld wire.

Unpublished data from button-melting experiments indicated that the formation of secondary phases during solidification is dependant more upon the total amount

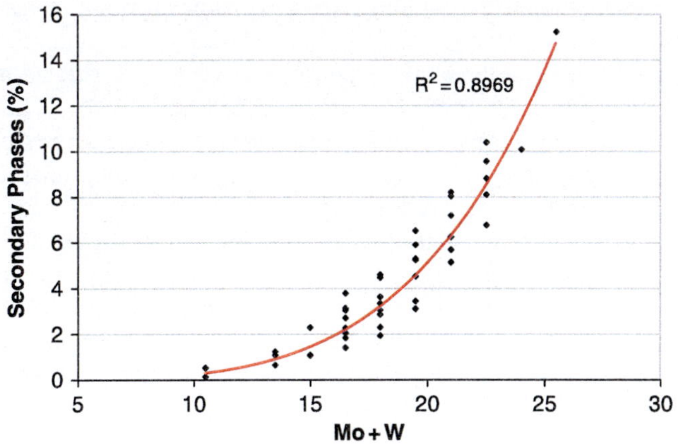

Fig. 16 The predicted effect (Scheil simulations) of molybdenum, tungsten, and iron on the volume fraction of secondary phases in the expanded fractional-factorial composition range

of solute than on the individual contributions of molybdenum, tungsten, or iron. An attempt was made to develop further understanding of this phenomenon; over fifty Scheil simulations were conducted using the nominal compositions of Alloy C-22, plus randomized levels of molybdenum, tungsten and iron. Molybdenum was varied from 10.5 to 18 wt%, tungsten varied from 0 to 7.5 wt%, and iron varied from 1 to 8.5 wt%.

Figure 16 displays the results from these Scheil simulations. It appears as if the formation of secondary phases over this composition range is strongly dependant upon the combined level of molybdenum and tungsten; the volume-percent of secondary phases increases exponentially with an increase in (Mo + W). Iron, by comparison, was found to have a much less pronounced effect on the secondary phase formation. The assumption from these findings is therefore that specifically increasing the combined Mo + W levels of Alloy C-22 should increase the elevated temperature ductility and solid cracking resistance of the alloy.

Conclusions

1. Alloy C-22 was found to have a solidification cracking temperature range of approximately 50°C (90°F); this indicates that the alloy is resistant to solidification cracking. The susceptibility of Alloy C-22 to solidification cracking is only slightly higher than very crack resistant alloys such as DSS2205, SS304 (FN6), and DSS2507.
2. Scheil simulations revealed that increasing the amount of Mo, W, and Fe (within the compositional limits of Alloy C-22) decreased the solidification temperature range. However, this decrease was less than ten degrees Celsius, and therefore would not significantly affect the cracking susceptibility of the alloy.

3. In the STF test, Alloy C-22 demonstrated a very low cracking threshold strain. The threshold strain for cracking at 950°C (1,742°F) was roughly 2%; this is approximately the same threshold strain for cracking demonstrated by FM-52, which is known to be susceptible to ductility-dip cracking. Alloy C-22 also has a wide temperature range over which ductility-dip cracking occurs in the strain-to-fracture test, and Alloy C-22 displays no recovery of cracking-resistance at the upper end of the testing temperature range.
4. In the hot-ductility test the on-heating and on-cooling ductility signatures for the weld-metal had pronounced intermediate temperature ductility-dips. The ductility minimum of the on-heating signature occurred at 900°C (1,652°F), while the ductility minimum of the on-cooling signature occurred at 1,000°C (1,832°F). Recovery of ductility occurred at the upper end of the testing temperature range in both the on-heating and on-cooling conditions.
5. Alloy C-22 is highly alloyed, has a low SFE, and is resistant to recrystallization. For this reason, Alloy C-22 samples do not experience dynamic recrystallization or a recovery of cracking resistance at the upper end of the testing temperature range in the STF test.
6. Increasing the volume fraction of secondary phases within Alloy C-22 was demonstrated to increase the elevated temperature ductility and decrease the susceptibility of the alloy to DDC. The volume fraction of secondary phases that form as solidification constituents can be increased exponentially by increasing the combined amount of Mo + W in the weld metal.

Acknowledgments The authors would like to thank AWS and The Ohio State University for providing financial support of this project. Additionally, gratitude is extended to all members of the Welding and Joining Metallurgy Group and the faculty of the Welding Engineering department at OSU who helped during the course of this work. Thanks also to Haynes International for provision of the base-material used within this investigation.

References

1. Lawrence Livermore National Laboratory (2003) Yucca Mountain Project. http://ymp.llnl.gov. Accessed 1 December 2007.
2. Haynes International (2002) HASTELLOY® C-22® Alloy Brochure. http://www.haynesintl.com/pdf/h2019.pdf. Accessed 27 September 2010.
3. Haynes International (2009) HASTELLOY® C-22® Alloy Welding Data. http://www.haynesintl.com/pdf/h2066.pdf. Accessed 27 September 2010.
4. Savage W and Krantz B (1966) An Investigation of Hot Cracking in Hastelloy X. Welding Journal 1:13S–25S.
5. Yeniscavich W and Fox C (1969) Effects of Minor Elements on the Weldability of "Hastelloy" Alloy X. Proceedings of a Symposium Sponsored by the Welding Research Council pp. 24–35.
6. Matsuda F (1990) Hot crack susceptibility of weld metal. Advances in Welding Metallurgy - 1st US-Japan Symposium pp. 19–35.
7. Savage W, Nippes E, Miller T (1976) Microsegregation in 70Cu-30Ni Weld Metal. Welding Journal 6:165s–173s.
8. Savage W and Lundin C (1965) The Varestraint Test. Welding Journal 10:433s–442s.

9. Rowe M, Crook P, Hoback G (2003) Weldability of a Corrosion-Resistant Ni-Cr-Mo-Cu Alloy. Welding Journal 11:313s–320s.
10. Clyne T and Kurz W (1982) The Effect of Melt Composition on Solidification Cracking of Steel. Metallurgical Transactions B 13A:259–266.
11. Cieslak M, Headley T, Romig A (1986) The Welding Metallurgy of HASTELLOY Alloys C-4, C-22, and C-276. Metallurgical Transactions A, 11:2035–2047.
12. Perricone M, Dupont J, Cieslak M (2003) Solidification of Hastelloy Alloys - an Alternative Interpretation. Metallurgical Transactions A 5:1127–1132.
13. R. Mehrabian R and M.C. Flemings M (1970) Metallurgical Transactions A 1:455–464.
14. Collins M and Lippold J (2003) An Investigation of Ductility Dip Cracking in Nickel-Based Filler Materials – Part I. Welding Journal 10:288s–295s.
15. Nissley N and Lippold J (2003) Development of the Strain-to-Fracture Test. Welding Journal 12:355s–364s.
16. Collins M, Ramirez A, Lippold J (2004) An Investigation of Ductility Dip Cracking in Nickel-Based Weld Metals – Part III. Welding Journal 2:39s–49s.
17. Nissley N, Collins M, Guaytima G, Lippold, J (2002) Development of the Strain to Fracture Test for Evaluating Ductility-Dip-Cracking in Austenitic Stainless Steel and Ni-Base Alloys. Welding in the World 7–8:32–40.
18. Finton T, Lippold J (2004) Standardization of the Transvarestraint Test - A Statistical Study using Austenitic Stainless Steels and Nickel-Base Alloys. Edison Welding Institute Summary Report SR0402.
19. Allen T (2006) Introduction to Engineering Statistics and Six Sigma. Springer-Verlag, London.
20. Haynes International (2003) Metallographic Preparation of HASTELLOY® and HAYNES® Alloys. http://www.haynesintl.com/pdf/h1107.pdf. Accessed 27 September 2010.
21. Nissley N and Lippold J (2008) Ductility-Dip Cracking Susceptibility of Nickel-Based Weld Metals Part 1 - Strain-to-Fracture Testing. Welding Journal 10:257s–264s.
22. Lippold J, Kotecki D (2005) Welding Metallurgy and Weldability of Stainless Steels. Wiley, Hoboken, NJ.
23. Dupont J, Michael J, Newbury B (1999) Welding Metallurgy of Alloy HR-160®. Welding Journal 12:408s–415s.
24. Reed-Hill R, Abbaschian R (1992) Physcial Metallurgy Principles. PWS-Kent Publishing, Boston.
25. Radhakrishnan B and Sarma G. (2003) The Effect of Coarse Non-deformable Particles on the Deformation and Static Recrystallization of Aluminium Alloys. Philosophical Magazine 22:2341–2366.

Weldability of a New Ni-Cu Welding Consumable for Joining Austenitic Stainless Steels

J.W. Sowards, B.T. Alexandrov, John Lippold, and G.S. Frankel

Acronyms

CPTT	Cast pin tear test
CSR	Crack susceptible region
DDC	Ductility-dip cracking
DDTR	Ductility-dip temperature range
DRT	Ductility recovery temperature
EDS	Energy dispersive spectroscopy
GTAW	Gas tungsten arc welding
MCD	Maximum crack distance
MCL	Maximum crack length
MGB	Migrated grain boundary
NDT	Nil ductility temperature
NST	Nil strength temperature
SCTR	Solidification cracking temperature range
SEM	Scanning electron microscopy
SGB	Solidification grain boundary
SMAW	Shielded metal arc welding
SS DTA	Single sensor differential thermal analysis
SSGB	Solidification subgrain boundary
STF	Strain-to-fracture

Introduction

Hexavalent chromium (Cr^{6+}) contained in fume produced during welding of stainless steels can pose a significant health risk to the safety of welding related personnel. Current methods to control fume exposure, such as ventilation, fume

J.W. Sowards (✉)
The Ohio State University, Columbus, OH, USA; National Institute of Standards and Technology, Boulder, CO, USA
e-mail: jeffrey.sowards@nist.gov

J. Lippold et al. (eds.), *Hot Cracking Phenomena in Welds III*,
DOI 10.1007/978-3-642-16864-2_20,

extraction, and respirator systems, may not reduce exposure levels to current government standards. This is especially true of stainless steel fabrication or repair in tightly enclosed spaces such as inside pressure vessels, piping, ship hulls, etc. An alternative approach to controlling Cr^{6+} exposure was investigated where chromium was eliminated from the consumable by utilizing the Ni-Cu system [1,2]. The new consumable produced welds with similar corrosion resistance and mechanical properties to welds deposited with conventional stainless steel consumables (ER308L). The Ni-Cu system was selected to avoid galvanic corrosion since it is nobler than austenitic stainless steels on the galvanic series. Resistance to localized corrosion such as pitting and crevice corrosion was improved by the addition of small amounts (< 1 wt-%) of the noble alloying elements Pd [2,3] and Ru [4] to the consumable composition. Mechanical properties including tensile strength and elongation exceeded minimum values required of Type 304L stainless steel. Elimination of Cr from the consumable reduced Cr^{6+} levels in the fume produced by the Ni-Cu consumable by two orders of magnitude.

Substituting a Ni-Cu consumable for a conventional stainless steel welding consumable, such as those based on 308L, poses several issues related to weldability. Welds deposits made with the 308L composition are resistant to solidification, liquation, and ductility-dip (DDC) cracking under most arc welding conditions since they solidify under the primary ferrite-austenite (FA) solidification mode [5]. However, a simple weld dilution analysis with the Schaeffler diagram [6] shows that the use of a Ni-base welding consumable to weld austenitic stainless steel alloys (such as 304L) will shift solidification to the fully austenitic mode (type A) due to an increase in the Ni-equivalency of the weld metal. Fully austenitic weld deposits generally have a higher likelihood to experience these weldability issues during fabrication and repair [5,7]. Segregation is more pronounced during austenitic (type A) solidification, which promotes liquid film formation along grain boundaries at the end of solidification. The films more easily wet austenite/austenite grain boundaries than austenite/ferrite boundaries in FA weld deposits, which increases the likelihood of solidification cracking given that sufficient restraint levels are present [5]. Weld metal liquation cracking susceptibility of fully austenitic deposits may also increase compared to FA deposits during reheating in multipass welding for similar reasons. DDC susceptibility of fully austenitic weld deposits is also increased since they lack the "tortuous" grain boundaries of FA weld deposits which act to limit grain boundary sliding and are more resistant to crack initiation and propagation.

The use of a welding consumable resulting in fully austenitic weld deposits necessitates weldability testing to evaluate cracking susceptibility. The current study was performed to evaluate the weldability of the new Ni-Cu consumables during welding of Type 304L stainless steel. Weldability testing including the Transvarestraint, hot ductility, and strain-to-fracture tests were performed to assess weldability of the deposits and compare with conventional stainless steels and other Ni-base alloys.

Experimental Procedures

Materials and Sample Preparation

The welding consumables used in this investigation were custom made in the form of covered electrodes for shielded metal arc welding (SMAW) and bare wire electrodes for gas tungsten arc welding (GTAW). Four "generations" of Ni-Cu welding electrodes were developed during the progression of this study as modifications were made to obtain optimal weld deposit compositions. Generation I electrodes were fabricated by adding Cu and Pd metal powders to Ni tubular wire for intended use with GTA welding. These electrodes were used to make weld deposits for an initial evaluation study of mechanical and corrosion properties. However, the intended use of the Ni-Cu consumables was with the SMAW process since it is both highly versatile and portable. Therefore, extensive weldability testing was not performed with Generation I consumables and results are not reported here. Generation II consumables were made in the form of SMAW consumables and consisted of Ni-209 core wire (3.2 mm diameter). Generation II-B and II-C coatings contained additions of Cu, and Cu-Pd, respectively. The Cu and Pd were alloyed into the weld deposit by transfer across the arc from the coating. However, the transfer efficiency with this method was not sufficient to achieve desired levels of Cu and Pd in weld deposits. Therefore, Generation III and IV consumables were made by melting custom heats of material containing sufficient additions to achieve a minimum of 7.5Cu and 1Ru in an all-weld-metal deposit. Ru replaced Pd as the noble alloying addition since it provided higher corrosion resistance [4]. The Generation III heat was extruded into 1.14 mm wire for GTA welding and the Generation IV heat was extruded into 3.2 mm wire and then coated for SMA welding. Compositions of Generation II all-weld-metal deposits, and Generation III and IV core wires are shown in Table 1. Compositions were measured with mass spectrometry and Leco® interstitial analysis. Type 304L base metal composition, as provided by the manufacturer, is also shown in Table 1.

Since this study focused on welding of stainless steel, Type 304L was utilized as the base material and Ni-Cu consumables were used to deposit weld metal into grooves machined into the base plate. A variety of tests were performed that required the use of different specimen geometries. SMA welds were produced using Generation II and IV welding electrodes on grooves machined in Type 304L base metal. SMA welding was performed in the DC electrode positive configuration. Current was maintained in the range of 120–130 A at a potential between 24 and 25 V, and travel speed was approximately 2.5 mm s^{-1} for all welded samples produced with the SMAW consumables. During GTA welding with Generation III bare wire consumable, the weld deposits were made with a current of 200 A at a potential of 13.5 V, travel speed of 2.1 mm s^{-1}, wire feed speed of 25.4 mm s^{-1}, and high-purity Ar was used as the shielding gas. GTA welding was performed on a Jetline sidebeam carriage with automatic arc voltage control (set to 10% sensitivity)

Table 1 Chemical composition (wt-%) of welding consumables and base metal

Element	Gen. II-B	Gen. II-C	Gen. III	Gen. IV	304L 6.4 mm	304L 9.5 mm
C	0.022	0.016	0.014	0.019	0.02	0.01
Cu	8.01	4.94	8.2	7.78	–	0.41
Fe	0.09	0.08	–	–	Balance	Balance
Mn	0.32	0.147	–	–	1.76	1.78
Mo	–	–	–	–	–	0.28
Ni	Balance	Balance	Balance	Balance	8.11	8.36
Cr	0.01	0.04	–	–	18.1	18.3
Al	0.07	0.05	0.56	0.83	–	–
Ti	1.56	0.97	0.53	4.31	–	–
Ru	–	–	1.36	1.11	–	–
Pd	–	0.24	–	–	–	–
Si	0.70	1.08	–	0.1	0.36	0.29
N	NA	NA	<0.001	<0.001	0.08	0.1
P	0.003	0.005	<0.005	<0.002	0.03	0.03
S	0.001	0.005	<0.001	<0.002	0.003	0.003

using a Miller Dynasty 300 LX power supply. Weldability test specimens were machined from the weld deposits in the transverse orientation. Hot ductility and Transvarestraint test specimens were machined from the 6.4 mm thick plate and strain-to-fracture specimens were machined from the 9.5 mm thick plate. Weld dilution levels were approximately 10–15% for the Transvarestraint specimens, 25–40% for the hot-ductility specimens, and 25% for the strain-to-fracture specimens.

Weldability Testing

Several weldability tests were performed to assess cracking susceptibility of weld deposits to solidification, weld metal liquation, and ductility-dip cracking. Brief descriptions of each test, and relevant testing parameters, are provided below. A complete list of important acronyms is also defined at the end of this document.

Solidification Assessment

Buttons were cast on a copper hearth with a GTA heat source under Ar shielding gas. The entire button was made with desired dilution level by mixing appropriate masses of Generation III wire and Type 304L. A type-C thermocouple was inserted into the bottom of each button into a predrilled hole. An arc weld was made on top of the button with a current of 300 A and constant arc length of 3.8 mm for 5–7 s. Time-temperature data was acquired from the weld pool and then processed with single sensor differential thermal analysis (SS DTA^{TM}) software [8] to determine phase transformations associated with solidification, i.e., the liquidus (T_L), solidus (T_S), and eutectic (T_E) temperatures.

The CALPHAD-based Thermo-Calc software was utilized in conjunction with the Thermotech TTNI7 database to perform Scheil-Gulliver simulations with various weld deposit compositions including the compositions of the buttons analyzed with SS DTA™. Liquid and FCC (γ and MC) phases were included in the calculations.

Transvarestraint Test

The Transvarestraint test was performed to assess solidification cracking susceptibility of Generations II and III weld deposits on Type 304L. This test was performed by transverse bending of the test coupon over a die block while moving an autogenous GTA weld over the deposited weld metal. The die block had a curved radius, which was varied to change the strain level applied to the sample. The maximum crack distance (MCD) was measured on the surface of the sample as an indicator of cracking susceptibility. The MCD is the longest crack length measured perpendicular to the instantaneous solid/liquid interface at the moment of bending. Welding and test parameters, as well as the equation for calculating strain are shown in Table 2. The solidification cracking temperature range (SCTR) was also used as an indication of cracking susceptibility. The equation for calculating SCTR is shown in Table 2, where cooling rate was measured during testing by plunging a type-C thermocouple into the weld pool and acquiring time-temperature data.

Cast Pin Tear Test

The cast pin tear test (CPTT) was designed as an economical method of evaluating solidification cracking susceptibility alloys when modification of composition is of

Table 2 Transvarestraint test parameters and equations for calculating strain and SCTR

Test parameter	Transverse varestraint
Current	180 A
Voltage	~10 V (2 mm arc gap)
Travel speed	2.1 mm s^{-1}
Ram speed	152.4 mm s^{-1}
Total weld length	50.4 mm s^{-1}
Weld length at bending	38.1 mm s^{-1}
$\varepsilon = \frac{t}{2R+t}$	ε, augmented strain *t*, plate thickness *R*, die block radius
$SCTR = CR\frac{MCD}{V} = \frac{dT}{dt}\frac{dt}{dx}MCD$	*SCTR*, solidification cracking temperature range *CR*, cooling rate in solidification temperature range $(dT/dt)_{L+S}$ *MCD*, maximum crack distance *V*, torch travel speed *(dx/dt)*

Table 3 Summary of cast pin tear test information including mold lengths, mass required to fill molds, number of test replications, and equation for calculating crack susceptibility

Mold length (mm)	Mass of button (grams)	Dilution level (%) 0	25	50
12.7	10	–	3	3
15.9	10.5	–	–	3
19.1	11	–	3	3
22.2	11.5	–	–	3
25.4	12	3	3	3
28.6	12.5	–	–	3
31.8	13	3	3	3
34.9	13.5	–	–	–
38.1	14	3	3	3
44.5	15	3	3	–
50.8	16	3	3	–

$$C = \frac{L_{total}}{360^{\circ}} \cdot 100$$

C, percent cracking
L_{total}, total arc length of all cracks measured on surface (degrees)

interest [9]. CPTT was selected for the current weldability evaluation to determine dilution effects of the filler metal by Type 304L on solidification cracking susceptibility. The current method of CPTT has been modified extensively [10] from the original test. Generation III wire was melted into buttons to simulate undiluted weld metal. Buttons were also made to simulate two different dilution levels (25 and 50%) by arc melting Generation III wire and chips of Type 304L (6.4 mm thickness). The buttons were re-melted three times to ensure sufficient mixing and were cast into the pin molds after the final melt. Cracking was induced along the pin near the shoulder location when solidification induced shrinkage strain exceeded a threshold cracking level [10]. Circumferential cracking was measured around the pin diameters. Longer mold lengths produce higher strain levels; therefore the pin length was varied to create a cracking envelope for each of the three compositions. Table 3 contains a list of the mold lengths utilized, the mass of material required to fill each mold (appropriate for Ni-base alloys or other alloys with similar density), the number of replications at each mold length, and the formula used to calculate circumferential cracking.

Hot-Ductility Test

Hot-ductility curves provide a measure of a material's ductility as a function of temperature [11]. Weldability failures are typically associated with an exhaustion of ductility, thus weld ductility during heating and cooling was determined. Transverse weld metal tensile specimens were machined from Type 304L plates and tested in a

Table 4 Hot-ductility test parameters for on-cooling and on-heating tests

Test parameter	On-heating hot ductility curve	On-cooling hot ductility curve
Sample free span	25.4 mm	25.4 mm
Heating rate	111°C s^{-1}	111°C s^{-1}
Cooling rate	Not applicable	Free cooled (~30°C s^{-1})
Displacement rate	50.8 mm s^{-1}	50.8 mm s^{-1}

Gleeble 3800 thermo-mechanical simulator to determine on-heating and on-cooling hot ductility curves. Test parameters for on-heating and on-cooling portions of the test are shown in Table 4. Reduction in area of specimen cross-sections was used as a measure of ductility. Nil-strength temperatures were determined by placing a small tensile load (approximately 25 kg) on specimens then heating at a rate of 111°C s^{-1} until failure occurred.

Strain-to-Fracture Test

Transvarestraint and hot-ductility testing revealed an apparent decrease in ductility at intermediate temperatures. For these tests, this is often an indicator of susceptibility to ductility dip cracking (DDC). The Strain-to-fracture (STF) test was used to determine the threshold strain levels necessary to induce DDC. The test consisted of fabricating dogbone-type samples with Generation II and III weld metal at the reduced section, and placing an autogenous GTA spot weld within the weld deposit [12]. GTA spot welds were placed at the reduced section with an arc gap of 2 mm, welding current of 125 A, under Ar shielding gas flowing at ~14 liter min^{-1}. Specimens were electropolished after placement of the spot weld region to aid in crack identification. The samples were heated in a partial vacuum to an elevated temperature at a rate of 100°C s^{-1}, held for a 10 s soak, and then loaded in tension to a predetermined amount of strain at a constant displacement rate of 0.06 cm s^{-1}. After straining, the samples were free-cooled under vacuum and cracks were counted in the spot weld. Cracking threshold strains were then plotted as a function of temperature to provide the DDC envelope.

Microstructure Characterization

Weld microstructure analysis was performed by preparing metallographic mounts, polishing with diamond suspension and vibratory polishing, followed by etching with Marbles Reagent (10 g $CuSO_4$, 50 mL HCl, 50 mL H_2O) for 2–10 s. Scanning electron microscopy (SEM) and energy dispersive spectroscopy (EDS) chemical analysis were performed with a field emission SEM at 15–30 kV.

Results and Discussion

Microstructure Evolution

As-deposited weld metal microstructures representative of weldability test samples are shown in Fig. 1. Note that weld metals produced with all consumables consist of a columnar austenitic (fcc) grain structure decorated with secondary phases identified as Ti(C,N). A cellular sub-structure is evident along with secondary phase particles which resided along solidification grain (SGB) and solidification subgrain boundaries (SSGB). Generation II-B deposits had a higher volume fraction of TiC particles than Generation II-C deposits as a consequence of the lower Ti content of the consumable. Generation III welds had the lowest observable volume fraction of Ti(C,N) particles due to the lowest Ti content (~0.5 wt-%). Generation IV welds contained a similar Ti concentration as Generation II and exhibited a comparable distribution of Ti(C,N) in the microstructure.

Button melting experiments were performed to evaluate the microstructural evolution during solidification by determining the phase transformation behavior with SS DTA^{TM}. The experimental setup and acquired time-temperature data for three different button compositions (Generation III filler metal, 25% dilution, and 50%

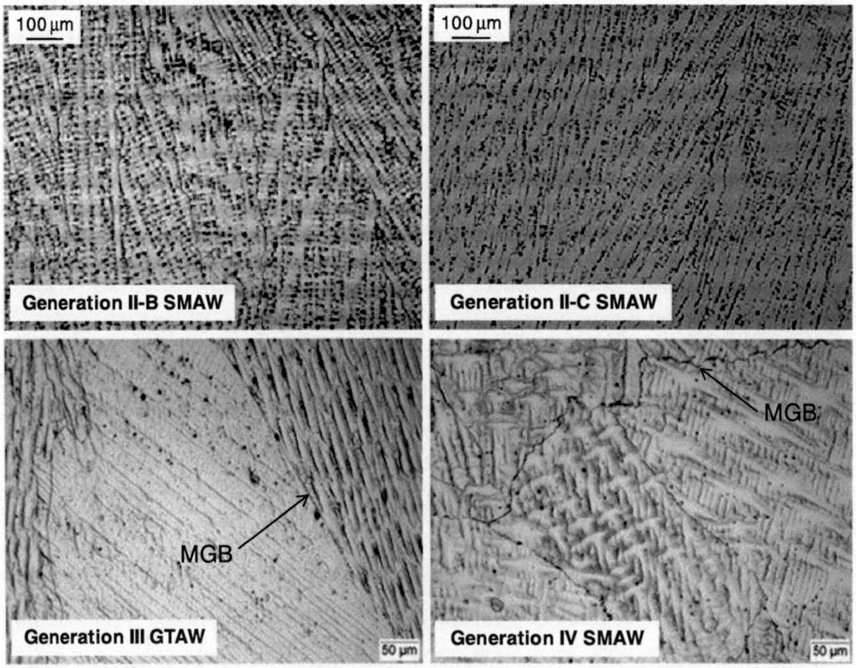

Fig. 1 Representative microstructures of Generation II, III and IV weld deposits; note different magnifications. *Arrows* indicate examples of a migrated grain boundary (MGB)

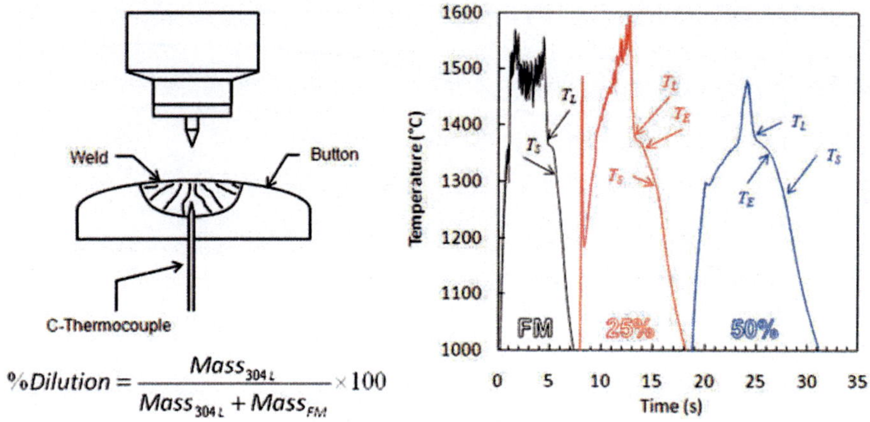

Fig. 2 Experimental setup and time-temperature data acquired for solidification analysis of Generation III filler metal and two dilution levels

dilution) are shown in Fig. 2. Inflection points in the thermal histories were noted during heating and cooling. The on-heating inflections, which are indicated by a decrease in the heating rate, coincide with melting and the absorption of the latent heat necessary to melt the weld pool. After melting there was some fluctuation in the temperature signal as a result of either weld pool convection or electromagnetic arc noise. After the arc was extinguished, cooling of the molten pool commenced. Inflections in the thermal histories associated with the liquidus, eutectic, and solidus temperatures were observed in the 25 and 50% dilution button compositions during analysis with SS DTA. There was no indication of a eutectic reaction during solidification of the filler metal (FM) composition. Note that the measured solidus temperature decreased with dilution level, effectively expanding the solidification temperature range.

Scheil-Gulliver simulations were performed over the range of undiluted filler metal up to 50% dilution by Type 304L. Results of the calculations are shown in Fig. 3 where transformation temperatures are plotted as a function of the amount of dilution. The diagram contains solid lines that indicate predicted phase transformation temperatures during solidification under Scheil conditions (non-equilibrium solidification where $D_S = 0$ and $D_L \to \infty$). As expected, austenitic solidification occurred for all levels of dilution since the Ni-equivalency of buttons was sufficiently high. Simulations predicted no eutectic reaction at the end of solidification of the filler metal composition although TiN was predicted to be stable in the liquid and solid phases at all compositions that had some dilution by Type 304L. This is explained by examining the compositions listed in Table 1, which show that the Generation III filler metal contained <0.001 wt-% nitrogen whereas the stainless steel contained 0.08 wt-% nitrogen. However, calculations do not consider that some nitrogen pickup would be expected during actual GTA welding. It is expected that welds with dilution levels approaching 0% would be expected to contain some TiN because of this. At dilution levels above ~8%, a TiC eutectic reaction was predicted

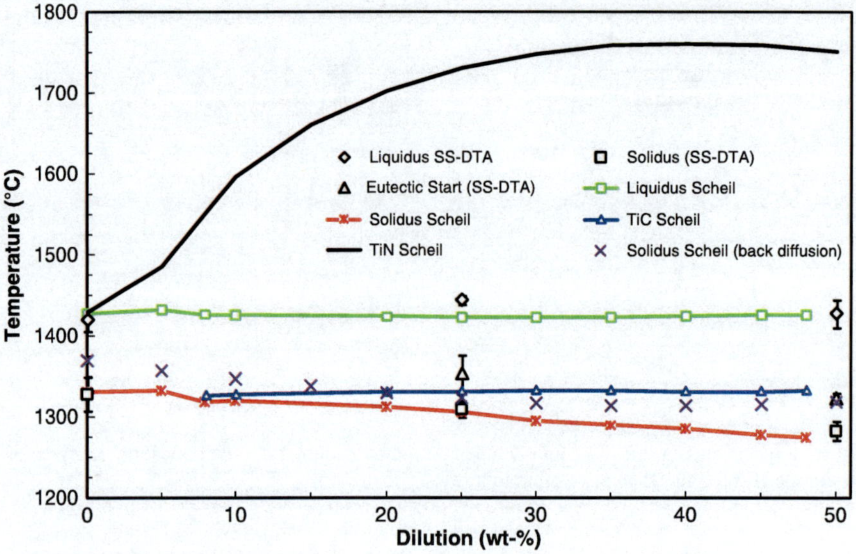

Fig. 3 Solidification diagram for Generation III consumable diluted by Type 304L

Table 5 Predicted solidification sequences for Ni-Cu consumables diluted by Type 304L

Dilution range	Solidification sequence
0–8%	$L \rightarrow L + TiN \rightarrow L + TiN + \gamma \rightarrow TiN + \gamma$
8–50%	$L \rightarrow L + TiN \rightarrow L + TiN + \gamma \rightarrow L + \gamma + Ti(C,N) + [TiN \rightarrow Ti(C,N)] \rightarrow \gamma + Ti(C,N)$

at the end of solidification. This reaction suppressed the solidus temperature and was more pronounced with increasing dilution level. Thus the total solidification temperature range widened with dilution by the 304L base metal. This is expected to play a role with respect to solidification cracking susceptibility and is discussed below. Experimentally determined phase transformation temperatures are indicated on the plot as discrete data points. Note that the Scheil-Gulliver model showed good agreement with experimental results.

Based on the observation of weld microstructures and solidification analysis, the microstructure evolution of weld deposits in this study can be described with the sequences listed in Table 5. The effect of Ti(C,N) in the microstructure is related to weldability behavior below.

Solidification Cracking

Two tests were performed to assess solidification cracking susceptibility including the Transvarestraint and CPT tests. Transvarestraint testing was performed on welds deposited with the SMA (Generation II-B and II-C) and bare wire GTA (Generation III) welding processes on Type 304L base metal. CPTT was performed to evaluate the effect of weld dilution (by Type 304L) on solidification cracking since it was more economical and easier to control dilution than Varestraint sample preparation. These tests were designed to induce solidification cracking so that susceptibility could be compared with other alloy systems under controlled cracking conditions.

Generation II-B, II-C, and III were subject to at least three strain levels during Transvarestraint testing. Two distinct regions of cracking were observed in all three weld metals as indicated in the Generation II-B Transvarestraint test weld depicted in Fig. 4a. The first region was directly behind the instantaneous liquid-solid interface at the moment strain was applied. Solidification cracking is associated with

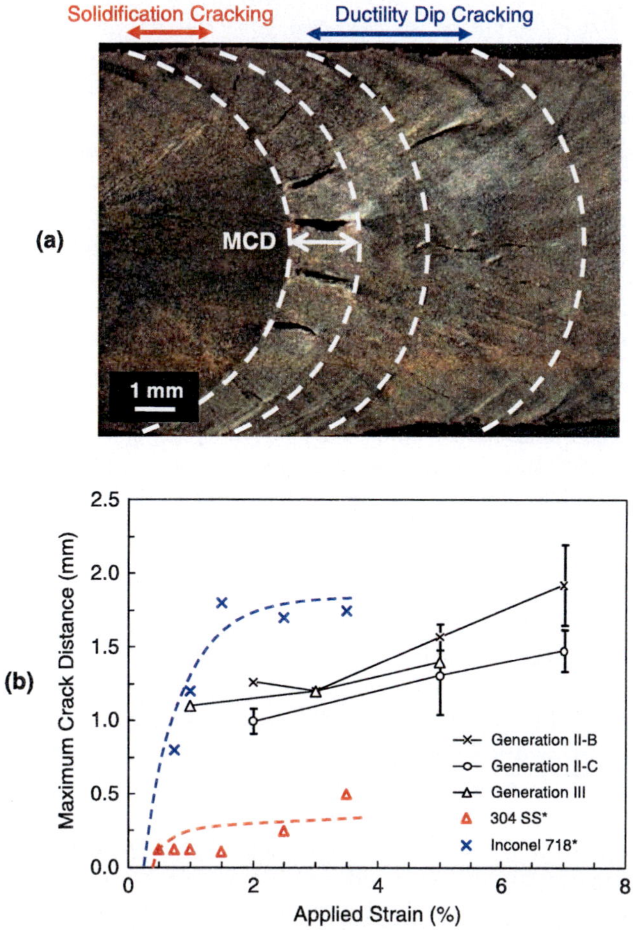

Fig. 4 Solidification cracking results showing (**a**) plan view of Transvarestraint specimen, and (**b**) cracking envelope of Generation II and III weld deposits (~10–15% dilution by Type 304L) determined with Transvarestraint testing – other alloys were adapted from [13]

this region and the maximum crack distance (MCD) values were determined. The second region was remote to the solidification cracks, and was typically observed several millimeters behind the weld region where liquid films would not be present. This second region of cracking presumably occurred in the solid state which is indicative of ductility dip cracking. Note that since a gap of over 1 mm exists between the two regions of cracking, the mechanism of each form of cracking is associated with a separate, distinct temperature range.

MCD is plotted as a function of applied strain in Fig. 4b for each of the three weld deposits. Error bars indicated standard deviation of three measurements. For comparative purposes, maximum crack lengths (MCL) of Type 304SS and Alloy 718 alloys have been included on the plot as determined by Varestraint testing performed in another study [13]. Type 304 has a low susceptibility to solidification cracking whereas Alloy 718 has a moderate to high solidification cracking susceptibility. The MCD values of Ni-Cu welds deposited on 304L are between those of the other two alloy types. MCD and MCL are not equivalent, but are typically close enough in value to provide a useful comparison. Therefore the Ni-Cu consumables appear to have a moderate susceptibility to solidification cracking based on the crack distances at strain levels between 1 and 7%. Note that threshold cracking strains were not measured as shown for the 718 and 304 alloys. Threshold strain is the minimum strain to cause solidification cracking and appears to be less than 0.5% for 718 and Type 304. The II-B deposits appear to have the highest cracking susceptibility of the three types of Ni-Cu weld deposits, which is a result of the higher Ti content of the welds. The II-C and III deposits appear to have similar cracking susceptibility but based on the error bars of the II-C deposit, and the lack of replications of the Generation III deposit, this similarity may not be a statistically significant.

Metallographic analysis performed on plan view sections of cracked regions showed that solidification cracking occurred predominantly on SGBs though some cracking was observed along SSGBs. Figure 5 shows an SEM plan view of a solidification crack observed after Varestraint testing of Generation II at 5% strain. The welding direction is from right to left on the figure. There was evidence of some backfilling of the cracks near the instantaneous solid/liquid interface. These backfilled regions were not observed until the samples were polished and etched, therefore they were not included in measurements of MCD. EDS analysis was performed on the leading crack edges and showed that SGBs were enriched in Ti and Si and depleted in Fe and Ni. The leading edge of solidification cracks observed in Generation II-B and II-C weld deposits were decorated with secondary phase particles rich in Ti. Based on solidification analysis and modeling, these particles were likely Ti(C,N) though diffraction analysis was not performed to positively identify the phases in these particular weld deposits. These particles were also observed in interdendritic regions surrounding the cracks and throughout the weld deposit microstructure. Generation II-B deposit contained a higher volume fraction of Ti(C,N) which suggests that there was a higher fraction of eutectic at the end of solidification. This likely exacerbated solidification cracking and resulted in increased susceptibility during Transvarestraint testing.

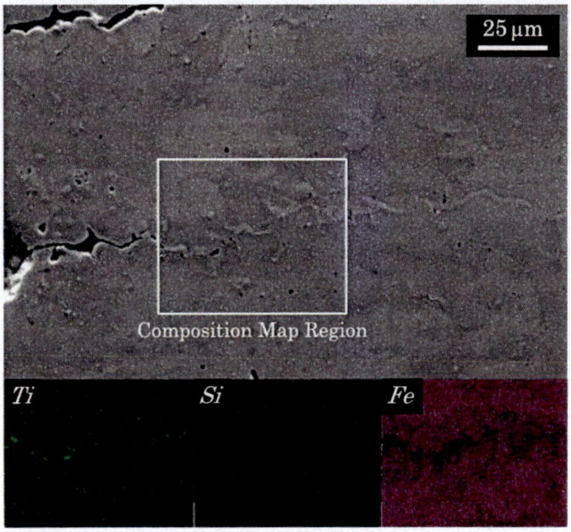

Fig. 5 Secondary electron plan view image of Generation II Varestraint weld region showing composition maps along solidification grain boundary

SCTR values were calculated with the equation and travel speed listed in the procedures (Table 2). Cooling rates in the two-phase region (L + S) of weld pools were found to vary between approximately 170–200°C s^{-1} in Transvarestraint specimens. For the purpose of SCTR calculations, the MCD is typically taken from the saturated strain level – the strain level where crack distances will not see further increases with an increase in strain. It is not quite clear from Fig. 4 that saturated strain was determined. Therefore, MCD values used during SCTR calculations were taken from the samples tested at 7% strain. Calculated values for the current study are shown in Table 6 along with values obtained for several other stainless steel alloys reported elsewhere [14]. Note that SCTR values of Generation II and III deposits are higher than Type 304L and 316L indicating welds are more prone to solidification cracking. However, SCTR values are less than those determined for other fully austenitic stainless steels that are readily fabricated with fusion welding such as Al6XN and Type 310. Special precautions are typically necessary for welding with consumables that undergo fully austenitic solidification, in particular control of impurities such as sulfur and phosphorus [5]. Such precautions would also apply to welding with the

Table 6 SCTR and DDC-TR values measured from generation II and III varestraint deposits; SCTR values for stainless steel alloys were adapted from [14]

Material	SCTR (°C)	DDC-TR (°C)
Type 304L SS, FN 6	31	–
Type 316L SS, FN 4	49	–
Generation II-C	109 ± 6	770–1,140
Generation II-B	111 ± 7	855–1,055
Generation III	113	920–1,130
Al6XN Superaustentic SS	115	–
Type 310 SS	139	–

consumables of the current study to minimize solidification cracking. The temperature range where DDC occurred is also included in Table 6, and will be discussed below.

Evaluation of solidification cracking was also performed with the CPTT since thermal analysis and modeling work revealed an expansion of the solidification temperature range with increasing levels of dilution. The CPTT was ideally suited for such an evaluation since the button melting sample preparation results in precisely controlled compositions. CPTT testing was performed with three button compositions: Generation III filler metal, 25% dilution, and 50% dilution, i.e., the same compositions evaluated during solidification analysis. Results of the test are shown in Fig. 6 where circumferential cracking around the cast pins are plotted as a function of pin length for each button composition. The error bars represents one standard deviation of at least three measurements.

The lowest crack susceptibility was observed in the pins cast with undiluted Generation III filler metal. Approximately 20% circumferential cracking was observed in pins cast in the maximum available mold lengths of 50.8 mm. Threshold cracking length was between the 38.1 and 44.5 mm mold lengths. These values of crack susceptibility were similar to those reported for Alloy 600 [10], a solid-solution strengthened nickel-base alloy, which is fairly resistant to solidification cracking. The pins cast from 25% dilution buttons had a cracking threshold length between the 25.4 and 31.8 mm molds and complete circumferential cracking (100%) occurred in the 50.4 mm mold length. Therefore the 25% dilution composition had higher cracking susceptibility than undiluted filler metal and susceptibility fell between Alloys 718 and 600. These data show good agreement with Transvarestraint results in Fig. 4 where cracking susceptibility of Generation III deposit (~15% dilution by 304L) was lower than Alloy 718 determined elsewhere

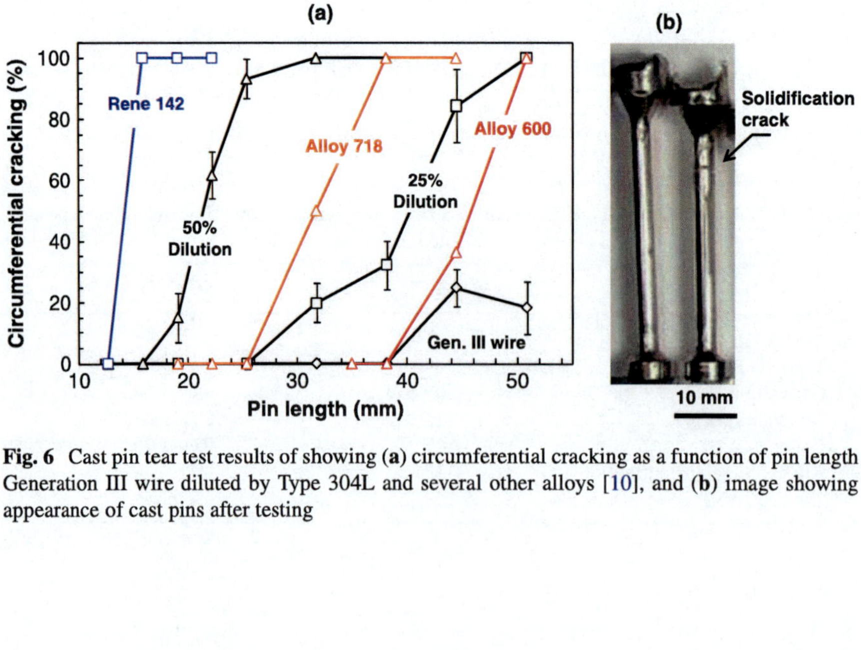

Fig. 6 Cast pin tear test results of showing (**a**) circumferential cracking as a function of pin length Generation III wire diluted by Type 304L and several other alloys [10], and (**b**) image showing appearance of cast pins after testing

[13]. The 50% dilution composition had the highest cracking susceptibility of the three compositions evaluated. This is indicated by both the threshold and complete (100%) cracking occurring at shorter mold lengths than the buttons cast from 25% dilution and Generation III wire compositions.

Increasing weld dilution by Type 304 increases solidification cracking susceptibility of the Generation III consumable. The solidification assessment and Scheil simulations revealed that solidification temperature range increased with dilution and stabilized a Ti(C,N) eutectic reaction at the end of solidification. CPTT revealed that the increase in solidification temperature range also increases solidification cracking susceptibility. It is unclear how the CPTT results would translate to welding conditions but other work has shown a direct correlation between experimentally determined solidification crack lengths and solidification temperature range [13]. Therefore, in production situations it may be necessary to control dilution to mitigate solidification cracking during fabrication of highly restrained welds. Generation III wire has a different composition from Generation II and IV, though the weld deposits made with all three consumables were nominally the same. Therefore, the solidification cracking susceptibility and dilution effects of all consumables are expected to be similar.

Weld Metal Hot Ductility

On-heating and on-cooling hot ductility curves were determined with the hot ductility test for Generation II-C and III weld deposits on Type 304L. The curves are shown in Fig. 7 where reduction in area (%RA) is plotted as a function of test temperature. Nil-strength temperatures (NST) of both weld deposits are also indicated on the figure. A hot ductility curve of Type 304 stainless steel tested in tension at a strain rate of 6 s^{-1} is shown for comparison as reported elsewhere [15]. Overall, the hot ductility of Generation III welds was higher than Generation II-C welds. Generation III welds were deposited with the GTAW process whereas Generation II-C welds were deposited with SMAW. Higher inclusion densities in the microstructure of SMAW deposits likely promoted the observed strengthening and loss of ductility. Both weld deposits exhibited a ductility dip below the solidus temperature. This is discussed below and referred to as the ductility dip temperature range (DDTR).

On-heating curves reached a point of zero ductility, termed the nil ductility temperature (NDT), where necking of test specimens approached zero reduction in area (RA). This occurred at 1,300°C and 1,360°C for the Generation II-C and III weld deposits, respectively. At the NDT, weld metal grain boundaries are covered with a continuous liquid films resulting in a complete ductility loss since liquid will accommodate very little strain. However, some strength still remains at the grain boundaries. The NST represents the temperature at which sufficient liquid has formed at grain boundaries to allow failure at very low loads. The NST temperatures for the Generation II-C and III weld deposits were 1,333°C and 1,374°C, respectively.

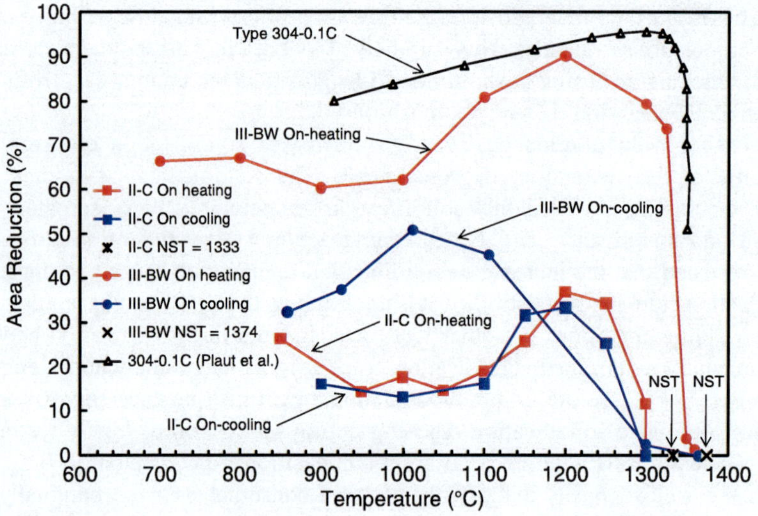

Fig. 7 Hot ductility curves of generation II-C and III weld deposits on Type 304L

On-cooling hot ductility tests were performed by heating the samples to a temperature midway between NDT and NST (II-C samples were heated to 1,315°C and III-BW samples were heated to 1,365°C) then cooling to a predetermined temperature before tensile loading the sample until failure. The temperature at which the sample recovers some ductility on-cooling is termed the ductility recovery temperature (DRT). DRT values were determined to be approximately 1,300°C and 1,320°C for the Generation II-C and III weld deposits, respectively. The ductility of both deposits appeared to recover at similar rates upon decreasing from the DRT. The Generation III deposit recovered more overall ductility upon cooling, implying that it has a higher resistance to solid state cracking below approximately 1,150°C.

The difference between the NST and DRT is considered the crack susceptible region (CSR). Ductility in a weld HAZ is essentially zero within this temperature range since liquid films are present along the grain boundaries adjacent to the weld pool. If a weld experiences a prolonged time duration in this region, stress will be accumulated during cooling of the weld increasing the likelihood of cracking [16]. The Generation II-C weld deposits had a CSR of approximately 33°C which represents the temperature range in which the weld metal may be susceptible to cracking, such as weld metal liquation cracking in the HAZ of a multipass weld. The Generation III deposit had a CSR of approximately 54°C, which was larger than the CSR of the II-C welds. However, the CSR values of both weld deposits are somewhat narrow suggesting that the welds have good resistance to weld metal liquation cracking in multipass weld deposits. In single pass welds, the liquation cracking susceptibility of the HAZ is determined by the hot ductility behavior of the austenitic stainless steel base metal.

Table 7 Summary of critical temperatures (reported in °C) from hot ductility experiments of Generation II-C and III welds on Type 304L

Hot ductility critical temperature	Generation II-C SMAW (~40% dilution)	Generation III GTAW (~20% dilution)
NDT	1,300	1,360
NST	1,333	1,374
DRT	1,300	1,320
CSR (NST-DRT)	33	54
DDTR	850–1,150	800–1,100
Eutectic temperature range – Ti(C,N)	1,230–1,340 (42% dilution)	1,310–1,354 (25% Dilution)
Solidus temperature (from SS DTA)	1,340 (42% dilution)	1,354 (25% Dilution)

Important hot ductility parameters including NDT, NST, DRT, CSR, and DDTR are summarized in Table 7 along with the experimentally determined eutectic temperature ranges of welds of similar dilution levels. The Generation II-C hot ductility deposit had a higher dilution of Type 304 (approximately 40% dilution) than the III-BW deposit (approximately 20% dilution). Lower dilutions would be expected in the GTAW deposits of III-BW since that welding process produced lower penetration welds. The listed eutectic temperature ranges coincide with the region of the hot ductility curves where on-heating ductility is reduced from peak ductility values to the nil ductility values. It is probable that eutectic Ti(C,N) present along SGBs and SSGBs undergoes melting and provides a network of liquid between solid grains that would result in preferred failure paths when sufficient deformation is applied to the test sample. The lower eutectic temperature range of the II-C deposit results in a decrease in ductility at lower temperatures than the III-BW deposit. Solidus temperatures are also shown as determined with SS DTA™. The values represent the temperature at which solidification was mostly complete and correlate reasonably well with the NST values. Generation III deposits had a higher solidus corresponding to a higher NST. Phase transformation temperatures were measured on-cooling and may differ from on-heating transformations depending on kinetic rate effects. Since dilution was generally found to widen the solidification temperature range, the higher dilution of the II-C hot ductility deposits reduced the NST compared to the lower dilution of the III weld deposit.

Solid-State Cracking

The solid state cracking phenomenon known as DDC was observed during Varestraint testing of weld metal and was associated with sub-solidus temperatures as shown in Fig. 4a. Evidence of ductility dips were observed in similar temperature ranges during hot ductility testing as shown in the on-heating curves in Fig. 7 and reported in Table 7 (DDTR). The observed hot ductility loss is believed to be associated with solid-state weld cracking. However, the Varestraint and hot ductility tests were not designed to specifically characterize weld metal susceptibility to DDC. Therefore, the STF test was utilized for this purpose. Generation II and III weld deposits on 304L were evaluated with this method and the results of the three tests are compared below.

The temperature range over which DDC occurred (DDC-TR) during Varestraint testing was determined with a similar method as SCTR and is summarized in Table 6. Cracking threshold strains were below 2% strain for the Generation II-B and II-C deposits and below 3% for the Generation III-BW deposit. It was not clear whether cracks initiated subsurface and propagated towards sample surfaces, or initiated on the sample surfaces. Strain is highest at the surface of the Transvarestraint specimens during testing, which would make initiation most likely near surface locations. As strain was increased by bending samples over blocks with smaller radii, the total number of cracks increased in a near linear fashion up to 7% strain. Generation II-C deposits exhibited the greatest number of cracks at all strain levels suggesting it has the highest susceptibility to DDC. The Generation II-B and III deposits had similar cracking behavior in the range of 3–5% strain based on the number of cracks counted although the DDC-TR was narrower in the II-B deposit, which could be a result of the higher volume fraction of TiC. Favorable distributions of TiC have been shown to promote DDC resistance in nickel-base weld deposits [17].

Note that the Generation II-C on-heating hot ductility curve in Fig. 7 shows a ductility decrease of approximately 9% in the temperature range of 850° to 1,150°C. A dip was also apparent in the on-heating curve of the Generation III deposit in the range of 800°C–1,100°C where ductility decreased by approximately 7%. While these values are similar, the dip in the Generation II-C deposit was a significantly greater portion of the maximum ductility. The temperature ranges where the dip occurred and the magnitude of ductility reduction correspond to the solid-state cracking ranges found during Varestraint testing.

To further examine this solid-state cracking ductility dip, samples of II-C and III-BW weld metal were subjected to the Strain-to-fracture test. A comparison of cracking susceptibility of Generation II-C and III deposits, Type 304L, and several other Ni-base filler metals [17] at 950°C is shown in Fig. 8. The temperature 950°C typically corresponds to that of minimum ductility for stainless steel and Ni-base alloys that are susceptible to DDC. The number of cracks observed in each specimen is listed and the plot shows that Generation III weld deposits had a higher threshold strain than II-C deposits, and a comparable threshold to FM-82, which is moderately resistant to DDC. The II-C deposits have a cracking susceptibility similar to FM-52, an alloy that is considerably prone to DDC in highly restrained multipass weld deposits. In addition to the threshold cracking strain, the transition to "massive cracking" has been established as an important indicator of cracking susceptibility [17]. Weld metals that transition from the threshold strain to massive cracking over a narrow interval are typically more susceptible to DDC. This transition was described as the strain where the number of cracks exceeded 50 in the spot weld or when they became too numerous to accurately count. Based on that

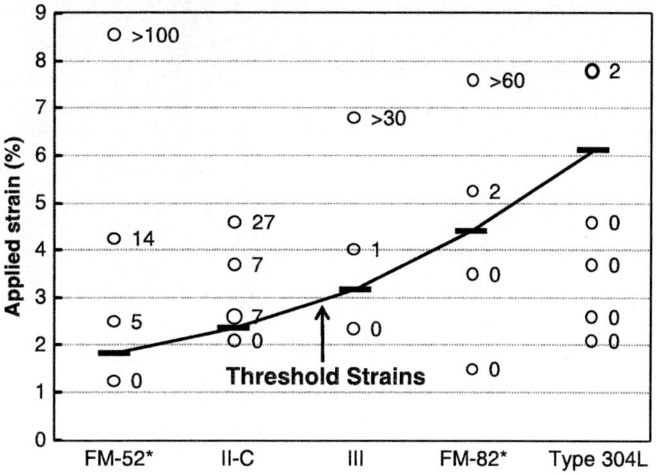

Fig. 8 Determination of threshold strains to induce ductility dip cracking during strain-to-fracture testing at 950°C. Data is included for several other nickel-base weld metals FM-52 and FM-82 [17]

criterion, the transition to massive cracking was not observed in Generation II-C or III deposits within the strain levels evaluated (less than 5% in II-C and less than 7% in III). The Generation IV weld deposits exhibited higher grain boundary tortuosity due to extensive grain boundary pinning. It is expected that future evaluations of those weld deposits will yield improved resistance to DDC since boundary pinning prevents the grain boundary sliding that leads to cracking.

Conclusions

1. Weld deposits on Type 304L with the developed consumables were austenitic (fcc) and exhibited a Ti(C,N) eutectic reaction at the end of solidification over dilution ranges typical of most arc welding processes.
2. Solidification modeling and button melting experiments revealed that increasing dilution by Type 304L expanded the solidification temperature range and promoted the eutectic reaction at the end of solidification.
3. Cast pin tear testing revealed that increasing dilution by Type 304L increased the solidification cracking susceptibility of Generation III weld metal as a result of the expanded solidification temperature range.
4. Solidification cracking susceptibility of Generation II and III weld deposits was higher than Type 304L yet comparable to nickel-base alloys and fully austenitic stainless steels based on Transvarestraint testing and the experimentally determined solidification cracking temperature range (SCTR) values.
5. Generation II and III weld deposits had a narrow crack susceptible region during hot ductility testing indicating weld metal liquation cracking susceptibility is low.

6. Hot ductility of the Generation III GTA welds was higher than Generation II SMA welds and the Generation II welds exhibited a more pronounced intermediate temperature ductility dip.
7. Ductility-dip cracking was observed in Transvarestraint test coupons at levels of 2–3% strain. The temperature range of cracking was estimated using an approach similar to SCTR and showed good correlation with dips in on-heating hot ductility of weld deposits.
8. Strain-to-fracture testing revealed DDC threshold strains of approximately 2 and 3% for Generation II and III weld metals, respectively. These are lower than Type 304 but similar to other Ni-base filler metals that have a moderate susceptibility to DDC in highly restrained welds.
9. Based on this study, the newly developed welding consumables have similar weldability to fully austenitic stainless steels and nickel-base alloys with moderate to good weldability.

Acknowledgements The authors wish to acknowledge the financial support of this work from the Strategic Environmental Research and Development Program (SERDP). Consumable development was greatly aided by Special Metals and Electrode Engineering. The lab assistance of Adam Hope, Pat Varga, Kenny Izor, and Hadley Cluxton is also appreciated.

References

1. Kim Y-H, Frankel GS, Lippold JC and Guaytima G (2006) Development of a chromium-free consumable for austenitic stainless steels – part 1: monel (alloy 400) filler metal. Corrosion 62:44–53.
2. Kim Y-H, Frankel GS and Lippold JC (2006) Development of a chromium-free consumable for austenitic stainless steels – part 2: optimization of alloy composition based on corrosion behavior. Corrosion 62:109–120.
3. Liang D, Sowards JW, Frankel GS, Alexandrov BT and Lippold JC (2010) A corrosion study of nickel-copper and nickel-copper-palladium welding filler metals. Mater Corr doi: 10.1002/maco.200905583.
4. Liang D, Sowards JW, Frankel GS, Alexandrov BT and Lippold JC (2010) Corrosion resistance of welds in type 304L stainless steel made with a nickel-copper-ruthenium welding consumable. Corrosion Science 52:2439–2451.
5. Lippold JC and Kotecki DJ (2005) Welding metallurgy and weldability of stainless steels. Wiley, Hoboken.
6. Schaeffler AL (1948) Welding dissimilar metals with stainless electrodes. Iron Age 162:72.
7. DuPont JN, Lippold JC and Kiser SD (2009) Welding metallurgy and weldability of nickel-base alloys. Wiley, Hoboken.
8. Alexandrov BT and Lippold JC (2007) Single sensor differential thermal analysis of phase transformations and structural changes during welding and postweld heat treatment. Welding World 51:48–59.
9. Hull FC (1959) Cast-pin tear test for susceptibility to hot cracking. Welding Journal 38: 176s–181s.
10. Alexandrov BT, Lippold JC and Nissley NE (2008) Evaluation of weld solidification cracking in Ni-base superalloys using the cast pin tear test. In: Bollinghaus T, Herold H, Cross CE, Lippold JC (Ed) Hot cracking phenomena in welds II. Springer, Berlin.
11. Nippes EF, Savage WF, Bastian BJ, Mason HF and Curran RM (1955) An investigation of the hot ductility of high temperature alloys. Welding Journal 34:183s–196s.

12. Nissley NE and Lippold JC (2003) Development of the strain-to-fracture test. Welding Journal 82:355s–364s.
13. DuPont JN, Michael JR and Newbury BD (1999) Welding metallurgy of alloy HR-160. Welding Journal 78:408s–415s.
14. Lippold JC (2005) Recent developments in weldability testing. In: Bollinghaus T, Herold H (Ed) Hot cracking phenomena in welds. Springer, Berlin.
15. Plaut RL, Herrera C, Excriba DM, Rios PR and Padihla AF (2007) A short review on wrought austenitic stainless steels at high temperatures: processing, microstructure, properties and performance. Mater Res 10:453–460.
16. Lin W, Baeslack WA and Lippold JC (1990) Hot ductility testing of high-strength, low-expansion superalloys. In: David SA, Vitek SM (Ed) Recent trends in welding science and technology. ASM International, Metals Park, OH.
17. Nissley NE (2006) Intermediate temperature grain boundary embrittlement in nickel-base weld metals, PhD Dissertation, The Ohio State University, Columbus.

Metallurgical Response of Electron Beam Welded Allvac® 718Plus™

J. Andersson, G. Sjöberg, and H. Hänninen

Introduction

Ni-based superalloys are generally composed of an austenitic face-centered cubic (FCC) matrix with additional secondary phases. They are frequently used at high temperatures in hostile corrosive environments and highly stressed [1].

Allvac® 718Plus™ is a newly developed superalloy intended for use up to ~700°C, which is 50°C higher than that for the dominating Alloy 718 [2]. Part of the success of Alloy 718 is due to its versatility as available in many shapes as forged products, sheet metal and as castings, whereas the new 718Plus alloy is still limited to wrought products (AMS 5441). However, there are different ongoing programs for standardizing both sheet and cast materials.

Large assembly structural parts belong to a new concept where weight optimization is acquired by joining different shapes (wrought and cast) by welding [3]. As a candidate material for such large structural aircraft engine components it goes without saying that the weldability is the key property not only for Allvac 718Plus but for any material chosen [4].

The chemical composition of this new alloy is close to that of Alloy 718 as shown in Table 1. The most important differences are the Ti and Al contents, which have been modified, which alters the primary strengthening phase from a γ'' (Ni_3Nb, Al, Ti) body-centered tetragonal (BCT) phase, which governs the strength in Alloy 718, to a γ' (Ni_3Al, Ti, Nb) FCC phase, which is the main reason for achieving the higher thermal stability of 718Plus alloy, since the γ'' phase degrades fast above 650°C [5, 6, 7]. γ'' in Alloy 718 is a metastable phase and transforms to the orthorhombic δ-phase (Ni_3Nb) after long time exposure at high temperature [6].

In addition, Fe has been reduced, whereas Co and W are added to improve the thermal stability of the alloy. The stable δ-phase which is a constituent of 718Plus

J. Andersson (✉)
Volvo Aero Corporation, Trollhättan, Sweden; Chalmers University of Technology, Gothenburg, Sweden
e-mail: joel.andersson@volvo.com

J. Lippold et al. (eds.), *Hot Cracking Phenomena in Welds III*,
DOI 10.1007/978-3-642-16864-2_21, © Springer-Verlag Berlin Heidelberg 2011

Table 1 Chemical composition (wt%) of Ni-base superalloys

Alloy/Elements	Ni	Cr	Fe	Co	Mo	Nb
Allvac 718Plus	BAL.	18.0	10.0	9.0	2.8	5.45
Alloy 718	BAL.	18.1	18.0	–	2.9	5.40
Alloy/Elements	Al	Ti	W	C	P	B
Allvac 718Plus	1.45	0.7	1.0	0.025	0.014	0.004
Alloy 718	0.45	1.0	–	0.025	0.007	0.006

alloy is used, as in Alloy 718, for controlling the grain size both during processing and heat treatment [8]. Some studies [9, 10] indicate that δ-phase increases the susceptibility to heat affected zone (HAZ) liquation cracks, whereas other studies indicate the contrary [11]. Apart from the solid state reactions also MC-carbide (NbC and TiC) and Laves (Ni_2Nb, hexagonal close packed) phase form by eutectic reactions both in Alloy 718 and 718Plus [1, 12].

When deep penetration welds are required electron beam welding (EBW) is the preferred method for aircraft engine parts. In EBW an energy dense beam of electrons penetrates the joint between the two work pieces to be welded by melting. EBW is a method by which one can achieve a high depth-to-width ratio weld, low distortion, an oxidation free weld at a high welding speed. It is also important to mention that due to the low heat input, compared with the other welding methods, the HAZ is shallow which is beneficial not least from the mechanical point of view. EBW is a method which is used when conventional methods, as TIG or plasma welding, are insufficient but it suffers from the fact that it is an expensive method; the equipment is costly and the welding needs to be carried out in vacuum with careful preparation of the joint which increases the lead times.

Experimental

Forged 718Plus rings with the chemical composition shown in Table 2 were used in the EB-welding trials. The rings were delivered in two different grain sizes, ASTM 5 and 8, due to super- (above δ-solvus) and sub-solvus (below δ-solvus) forging temperatures, respectively.

Table 2 Chemical composition of Allvac 718Plus

	Ni	Cr	Fe	Co	Mo	Nb	Al	Ti
HeatX06/elements	BAL.	17.85	9.54	9.04	2.67	5.51	1.43	0.75
	W	Mn	Si	C	P	V	B	
	1.01	0.04	0.04	0.018	0.010	0.02	0.004	

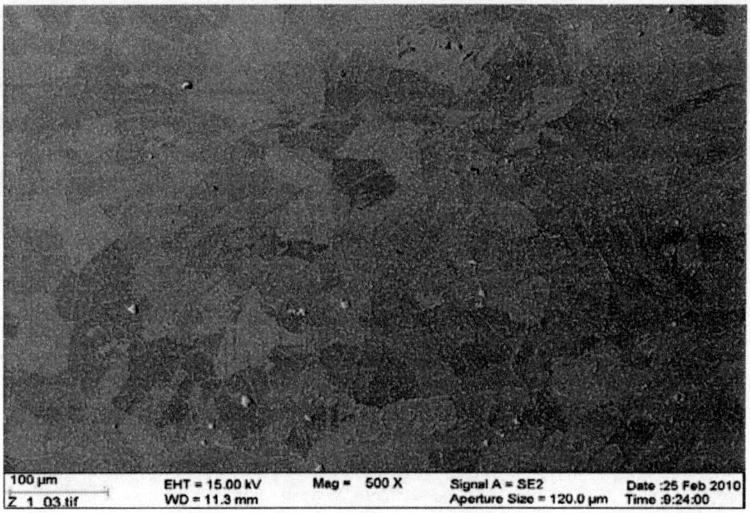

Fig. 1 The microstructure of the as-received Allvac 718Plus. Note that the major part of the δ-phase pattern does not coincide with the present grain boundaries

The as-received microstructure from one of the rings with ASTM 8 grain size is shown in Fig. 1, which reveals a uniform grain structure, where new recrystallized grains are evident and where the old grain boundaries may be delineated by the patterns of the prior grain boundary δ-phase. Some dispersed carbides are also present.

The material was open-die forged and ring rolled to produce two large rings, Fig. 2. Water-jet machining was used to extract thinner rings out of the two large ones and severe distortion was evident. Before final machining, stress relief heat treatments in three steps were carried out at the recommended solution heat treatment temperature, 954°C [8]. Total exposure time at solution treatment temperature was 1.5 h (3 × 0.5 h) prior the final heat treatment which adds another hour. All heat treatments were carried out in a vacuum furnace using forced argon convection to achieve cooling faster than 20°C/min down to 500°C to minimize precipitation. As indicated in Fig. 2 a number of specimens were excised for further metallographic cross-sectioning.

Rings with three different thicknesses (6, 12 and 20 mm) were produced and welded using a 30 kW low voltage (60 kV) Sciaky EB welding machine. First, optimization of weld parameters (here called "nominal") to minimize weld thickness to at most 2 mm was done. Variation of the heat input by plus/minus 10% was carried out through adjustments of the beam current (mA), the accelerating voltage (kV) as well as by adjustment of the welding speed (mm/min), alternatively. These adjustments of specific weld parameter were altered by ramping up from minimum heat input to maximum heat input (e.g. 40–60 kV) over a welding distance of 300 mm. The microstructure response of a cosmetic weld pass (~35% heat input relative to "nominal" heat input) performed onto a "nominal" weld pass was also examined.

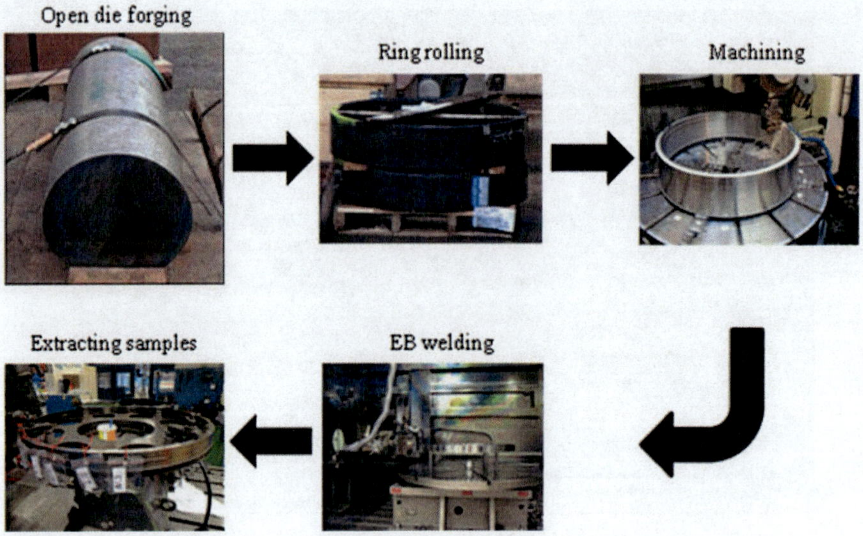

Fig. 2 Schematic illustration of the processing of Allvac 718Plus forged rings

Circular oscillation (frequency of 1,000 Hz) and top surface focal point was used throughout the whole study except for the cosmetic weld pass where a focal point above the surface was used in order to broaden the weld bead. The complete series of parameters can be seen in Table 3. Weld bead cross-sections were examined using both light optical microscopy (LOM) and scanning electron microscopy (SEM) equipped with a backscattered electron detector (BSE). Metallographic work was

Table 3 EB welding process parameters in respect of kV, mA and velocity are presented together with the respective thickness and grain size

	kV	mA	Velocity, mm/min	Thickness, mm	Grain size, ASTM
Nominal	50	66	440	6	5 & 8
kV	40–60	66	440	6	5 & 8
mA	50	53–79	440	6	5 & 8
Velocity	50	66	528–352	6	5 & 8
Cosmetic weld pass	35	30	400	6	5 & 8
Nominal	50	112	400	12	5 & 8
kV	40–60	112	400	12	5 & 8
mA	50	90–134	400	12	5 & 8
Velocity	50	196	480–320	20	8
Cosmetic weld pass	35	30	400	12	5 & 8
Nominal	50	196	400	20	8
kV	40–60	196	400	20	8
mA	50	157–235	400	20	8
Velocity	50	196	480–320	20	8
Cosmetic weld pass	35	30	400	20	8

carried out using standard metallographic procedures. Etching of the weld bead cross-sections and parent metal was performed electrolytically using oxalic acid and 4–6 V for ~10 s.

Results

No cracks could be detected in the weld bead cross-sections (117 cross-sections in total) in this study. Typical appearances of weld bead cross-sections are shown in Fig. 3 for the ASTM 8 ring material.

On closer examination of the weld cross-sections, irrespective of thickness and grain size, some well-defined healed cracks were seen in the top part of the weld bead cross-section in the HAZ with evident liquation of carbides (MC), Figs. 4 and 5. Healed cracks that did not coincide with the carbides were also seen in the HAZ, Fig. 4.

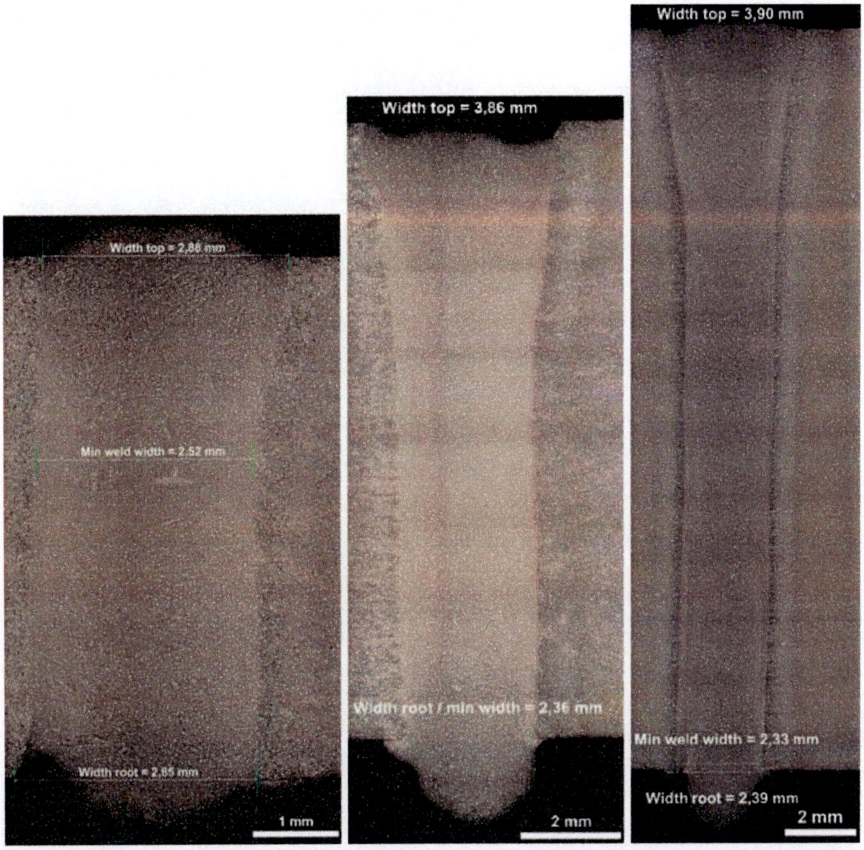

Fig. 3 Weld cross-sections for "nominal" parameters on the 6, 12 and 20 mm thicknesses, respectively

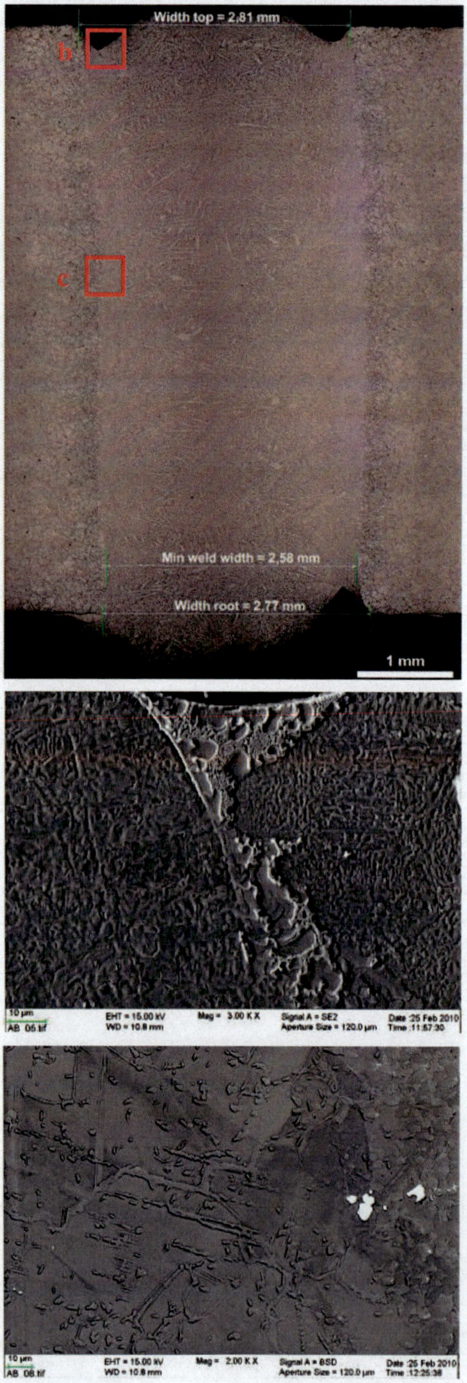

a b c

Fig. 4 Metallographic microstructures of a 6 mm thick weld cross-section for "nominal" parameters using LOM (**a**) and SEM photos taken within the boxed areas of the LOM picture

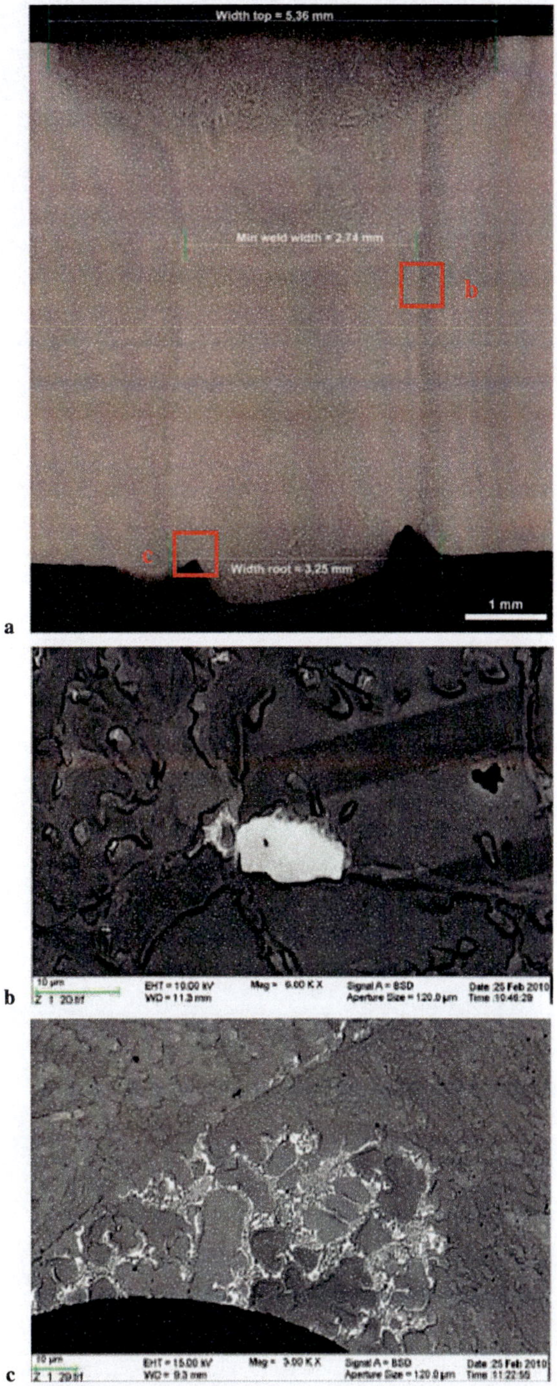

Fig. 5 Close-up SEM pictures (**b** and **c**) taken from the boxed areas within the LOM picture (**a**) of a 6 mm thick weld cross-section for the "nominal" and "cosmetic weld pass" weld

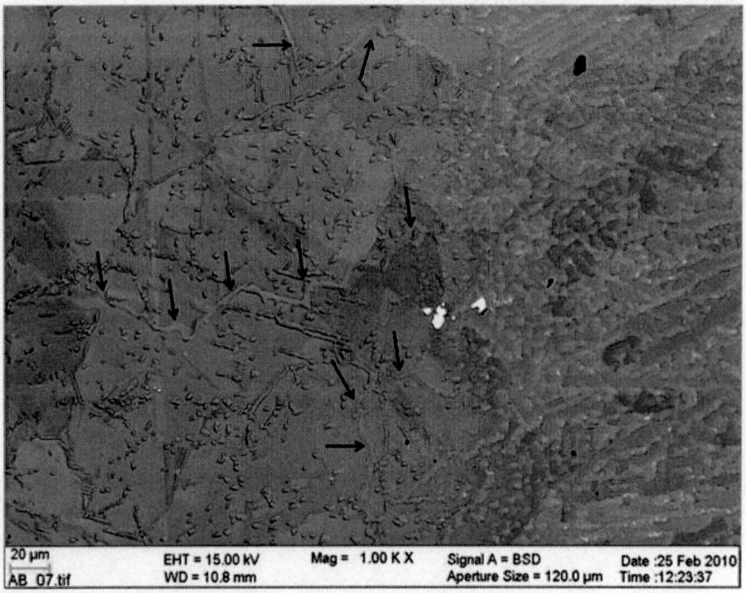

Fig. 6 SEM micrograph revealing healed cracks in the HAZ (fusion zone to the right)

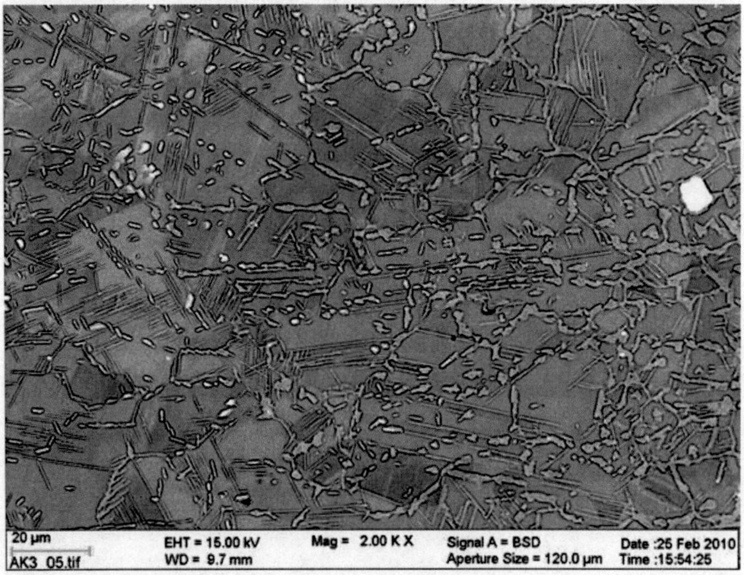

Fig. 7 SEM micrograph revealing how δ-phase constituents have broken up and reshaped as compared to its original shape indicating liquation (fusion zone to the right)

In the "cosmetic weld pass" cross-sections no healed cracks were seen in the top side of the cross-section, Fig. 5. However, healed cracks in the HAZ and root side were still there.

In Figs. 6, 7 and 8 healed cracks are seen and the δ-phase lamellas have partly broken which indicates a liquation reaction.

Figure 9 convincingly reveal liquated areas with healed cracks in the HAZ with a phase with script morphology indicating the presence of the Laves phase with high

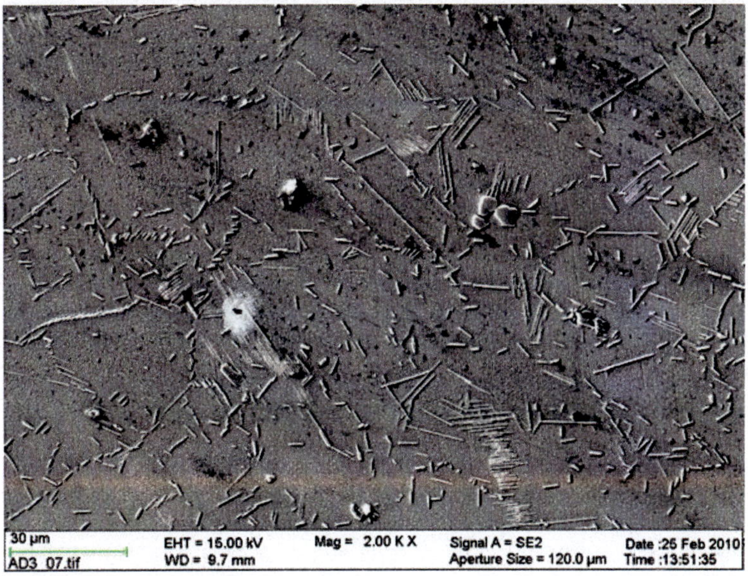

Fig. 8 General appearance and morphology of δ-phase lamellas in the forged and solution heat treated Allvac 718Plus

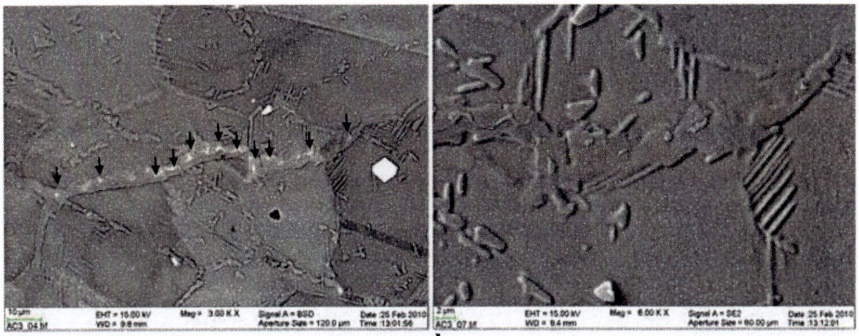

a b

Fig. 9 SEM micrograph of low and high magnification using BSE, which enhances high atomic number elements such as Nb, convincingly reveals a healed crack including a phase constituent of script morphology (indicated by the *arrows* in **a**) together with liquated δ-phase needles in EB-welded Allvac 718Plus. (**a**) Low magnification. (**b**) High magnification niobium content [1]. The rounded shape of the δ-phase also indicates that a liquation reaction has occurred.

Discussion

By considering the lattice diffusion of Nb in Ni it is possible to estimate the time for dissolution of the δ-phase. Patil et al. showed that Nb diffusion in Ni can be described by the standard Arrhenius expression [13].

$$D_{Nb} = D_0 \cdot \exp^{(Q/RT)} \tag{1}$$

where:

D_{Nb} = diffusion coefficient of Nb in Ni matrix (m^2/s)
$D_0 = 1.04\ (+0.87/-0.47) \cdot 10^{-6} (m^2/s)$
$Q = -202.59\ (+0.47/-0.47)$ (kJ/mol)
$R = 8.31451$ (J/molK)
T = absolute temperature (K)

In Table 4, diffusion values using Eq. (1) are calculated for 4 different temperatures.

Diffusion coefficients of Nb in Ni in Table 4 can be used together with Einstein theorem of random walk, Eq. 2, and simple estimates can be obtained of the diffusion distances, Table 5.

$$1 = \sqrt{(2D \cdot t)} \tag{2}$$

where:

1 = diffusion distance (m)
D = diffusion coefficient (m^2/s)
t = time for diffusion (s)

Table 4 Diffusion coefficients of Nb in Ni at 1,000°C through 1,300°C in steps of 100°C

Diffusion coefficient of Nb in Ni (m^2/s)	Temperature (°C)
9.9×10^{-16}	900
5.1×10^{-15}	1,000
2.0×10^{-14}	1,100
6.8×10^{-14}	1,200
2.0×10^{-13}	1,300

Table 5 Diffusion time of Nb in Ni at 1,000°C through 1,300°C in steps of 100°C for a typical diffusion distance of 1 μm

Diffusion distance (μm)	Temperature (°C)	Diffusion time (s)
1	900	505
1	1,000	98
1	1,100	25
1	1,200	7
1	1,300	3

When considering the HAZ weld thermal cycle of metallurgical significance (i.e., above 900°C) with heating rate of 100°C/s and cooling rate of 50°C/s common to arc welding processes it will roughly give total time of ~12 s in this temperature range. One should keep in mind that the dissolution of δ-phase usually takes place by solid state diffusion in the 1,010–1,040°C range [8]. Having the above rationale in mind the solid state dissolution should then take place within ~3 s with a heating rate of 100°C/s at the warmest point in the HAZ, which reaches ~1,300°C as a peak temperature. Further away from the fusion zone boundary the peak temperature decreases and also the time available for the solid state diffusion to take place.

With this explanation it seems reasonable that the dissolution through solid state diffusion is insignificant due to the fact that heating and cooling rates in a EB weld such as the ones performed in this study are significantly higher than the ones used for the time estimates above since the power densities of the electron beam is very high and can melt up to ~50 mm thick steel alloys within just a second [14]. Also the actual heat input in the EB-process per unit length is also in comparison lower and much less heat is to be dissipated with corresponding shorter dwell times both during the heating and cooling. This advocates that another type of non-equilibrium transformation mechanism is governing the δ-phase dissolution when rapidly heated above its solvus temperature, specifically for processes where high power densities are involved.

Andersson et al. performed solidification experiments on Allvac 718Plus together with EDX phase mapping [12]. Selected results of these analyses are shown in Table 6 and Fig. 10.

The DSC and EDX results can be used to produce a pseudo-binary phase diagram as shown in Fig. 11, which is similar to the Alloy 718 diagram of Knorovsky et al. [15]. C_0 is the Nb content of the 718Plus material used in DSC test, C_m is the maximum solubility of Nb in the gamma matrix phase ("Dark grey area" in phase field map of Fig. 10), $C_{Eut.}$ is the Nb concentration of the gamma/Laves

Table 6 Results from differential scanning calorimetric (DSC) analysis of Allvac 718Plus showing solidification reaction temperatures

Cooling rate	$T_{liquidus}$ °C	$T_{solidus}$ °C	Laves eutectic °C
20K/min	1,340	1,300	1,160

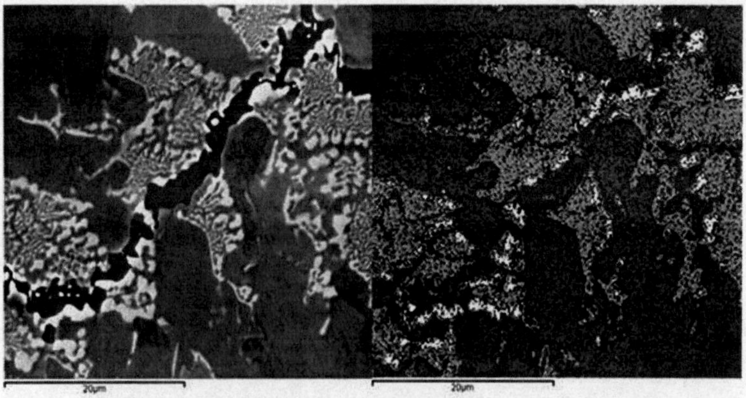

Average	Dark grey area	Light grey area	White area
Nb 5.47	8.64	20.04	28.3

Fig. 10 SEM backscatter electron diffraction mode to the *top left* and corresponding EDX phase map to the *right*. The EDX phase mapping shown in the table below the SEM pictures reveals the Nb content in three different distinct areas (wt-%)

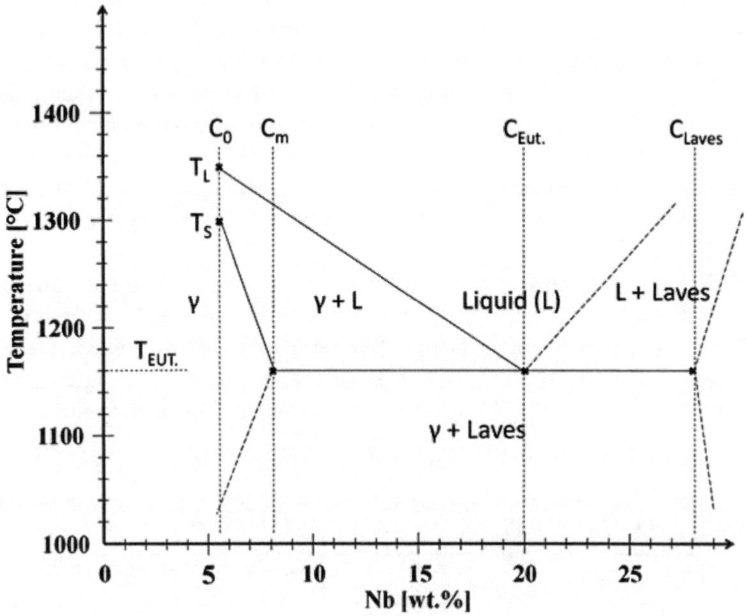

Fig. 11 Pseudo-binary phase diagram for Allvac 718Plus

eutectic ("Light grey area"), whereas C_{Laves} is the concentration of Nb in Laves phase ("White area").

The DSC analysis and phase diagram reveal that a eutectic reaction of gamma/Laves takes place as the final step of solidification at approximately 1,160°C. The microstructure of a cast material differs significantly compared with that of the wrought material not least due to the segregation of various elements. However, as revealed by the as-received microstructure in Figs. 1 and 8 the 718Plus wrought material used in this study contained significant amount of Nb-rich δ-phase (Ni_3Nb). With that in mind, together with what can be extracted from Figs. 6, 7, 8, 9, and 10, the following rationale can be deduced. As rapid heating begins and due to the limited time for the solid state diffusion the δ-phase particles can remain solid to the $T_{Eut.}$ with the amount of Nb in the δ-phase high enough ($>C_m$) to induce eutectic liquation which is supported by the appearance of the δ-phase lamellas in Figs. 7, 9 and 10. Also, the C_m composition in the pseudo-binary phase diagram is thermodynamically in equilibrium with the eutectic composition $C_{Eut.}$.

Interestingly, no open cracks were seen in any of the >100 cross-sections examined while healed cracks were seen in most of them. Generally eutectic backfilling occurs when a crack is opened in the solidifying fusion zone and the liquid from the fusion zone starts to penetrate the crack by capillary effect. Evidently the δ-phase undergoes constitutional liquation and lowers the ductility of the material with cracks in the HAZ, but due to the large amount of eutectic liquid available as a result of the same liquation reaction healing of the opened cracks takes place at a later stage. One should therefore be careful when interpreting a material's weldability since the lack of open cracks not necessarily means that no degrading features have been encountered. The presence of a healed crack through the eutectic liquid and its influence on the mechanical properties is something which should be investigated further.

Conclusions

1. No "open" cracks were found.
2. Backfilled cracks were found at the top, root and in the heat affected zone.
3. δ-phase undergoes constitutional liquation rather than dissolution through the solid state diffusion in the heat affected zone.
4. δ-phase assists healing of cracks through liquation, in the heat affected zone.

Acknowledgements Mikael Fransson and Joakim Carlsson at *the welding department* (Volvo Aero) are greatly acknowledged for their help with EB welding operation. We also thank Anders Oskarsson at *the department of Materials Technology* (Volvo Aero) for helping out with metallographic preparation of 718Plus. Tapio Saukkonen at *the department of Engineering Design and Production, Engineering Materials* (Aalto University School of Science and Technology) is greatly acknowledged for his help and assistance in SEM analyses.

The work presented in this paper is partly financed by VITAL, which is a new collaborative research project, running for 4 years, which aims to significantly reduce aircraft engine noise and CO_2 emissions. Snecma leads a consortium of 53 partners gathering all major European engine manufacturers: Rolls-Royce Plc, MTU Aero Engines, Avio, Volvo Aero, Techspace Aero, Rolls-Royce Deutschland and ITP, and the airframer Airbus. The work in this paper has been performed under WP 4.4 Advanced Hot Structures/4.4.5 Evaluation of Alternative Alloy with Temperature Capability over IN 718, where Volvo Aero specifically contributed to the work presented here.

References

1. Sims CT, Stoloff NS and Hagel WC (1987) Superalloys II - high temperature materials for aerospace and industrial power. Wiley, New York, NY.
2. Cao WD and Kennedy R (2004) Superalloys, pp. 91–99.
3. Sjöberg G (2010) Casting Superalloys for Structural Applications. To be presented at the 7th International Symposium on Superalloy 718 and Derivatives.
4. Sjöberg G, Andersson J and Sjunnesson A (2009) New Materials in the Design and Manufacturing of Hot Structures for Aircraft Engines – ALLVAC 718PLUS. *ISABE-2009-1286*.
5. Merrick HF (1978) Proceedings of Symposium Precipitation Process in Solids, TMS-AIME, Warrendale, PA, pp. 161–190.
6. Sundararaman M, Mukhopadhyay P and Banerjee S (1988). Metallurgical Transactions A 19A:453–465.
7. Sundararaman M, Mukhopadhyay P and Banerjee S (1992). Metallurgical Transactions A 23A:2015–2028.
8. Cao WD and Kennedy RL (2006) Recommendations for Heat Treating Allvac® 718Plus® Alloy Parts. *ATI Allvac* (www.allvac.com/718Plus), Feb. 24 (2006).
9. Qian MJ and Lippold JC (2003). Acta Mater 51:3351–3361.
10. Hong JK, Park JH, Park NK, Eom IS, Kim MB and Kang CY (2008). Journal of Materials Processing Technology 201:515–520.
11. Vincent ER (1980) The Microstructure of Welds in Inco 718. University of Bristol, UK.
12. Andersson J, Sjöberg G, Viskari L, Brederholm A, Hänninen H and Knee CS (2008) Proceedings of the 47th Conference of Metallurgists, pp. 401–413.
13. Patil RV, Kale GB (1996). Journal of Nuclear Materials 230:57–60.
14. Linnert GE (1994) Welding Metallurgy - Carbon and Alloy Steels 4th edition 1. American Welding Society, Miami, pp. 459–460.
15. Knorovsky GA, Cieslak MJ, Headley TJ, Romig AD Jr. and Hammetter WF (1989). Metallurgical Transaction 20A:2149–2158.

Stray Grain Formation and Solidification Cracking Susceptibility of Single Crystal Ni-Base Superalloy CMSX-4

John N. DuPont and T.D. Anderson

Introduction

Ni based single crystal superalloys are used for turbine blades due to their excellent creep resistance. These materials are very expensive due to the complex fabrication conditions required to maintain the single crystal structure during casting. Some blades must be scrapped after casting due to the formation of defective "stray" grains, while other blades may need to be replaced after finite service exposures due to damage associated with wear, fatigue, or creep. Thus, there is a need to develop reliable welding techniques in order to rejuvenate damaged blades or repair blades with casting defects. This involves the establishment of welding parameters that will avoid the formation of stray grains (SG) in the weld. The stray grain formation tendency is a strong function of the local solidification conditions, which are controlled by the welding parameters. Another type of welding defect which is commonly observed in this class of alloys is solidification cracks. Solidification cracks are commonly associated with SG formation, but the particular range of welding parameters over which cracking can occur is not typically well established. The complex relations between welding parameters, solidification parameters, resultant SG formation, and solidification cracking susceptibility make the development of reliable weld repair strategies difficult.

Stray Grain Development

Stray grain formation is known to be a result of constitutional supercooling (CS), which is controlled by temperature gradient (G) in the liquid directly ahead of the solid/liquid interface and the growth rate (V). Research has shown that a low G/V ratio will promote the nucleation of stray grains by introducing excessive liquid undercooling ahead of the solidifying columnar dendrite front [1]. There exists

J.N. DuPont (✉)
Department of Materials Science and Engineering, Lehigh University,
Bethlehem, PA, USA
e-mail: jnd1@lehigh.edu

J. Lippold et al. (eds.), *Hot Cracking Phenomena in Welds III*,
DOI 10.1007/978-3-642-16864-2_22,

a wide range of G and V across the solidification interface of a weld pool. In a polycrystalline material, dendrite growth will generally be opposite to the direction of heat flow, so G and V are considered normal to the solid/liquid interface. For the case of a SX material, the growth directions are limited to one of the six crystallographic <100> "easy" growth vectors. The relevant G and V for SX solidification are those parallel to the local dendrite growth directions, G_{hkl} and V_{hkl}. These can be calculated using a geometric model [2–4]) if the weld pool shape and SX substrate orientation are known. The dendrite tip velocity V_{hkl} is calculated with the relationship:

$$V_{hkl} = S\frac{\cos\theta}{\cos\psi} \tag{1}$$

where S = travel speed, θ = the angle between S and the interface normal, and ψ = the angle between the interface normal and the active dendrite growth direction.

Current approaches to predict SG formation in weld structures stem from models originally developed to describe the Columnar-to-Equiaxed Transition (CET) in castings [2]. The model has been applied to the case of fusion welding, and the equation used to calculate the SG area fraction (ϕ) is given by [3]

$$\phi = 1 - \exp\left(-\frac{4\pi N_o}{3}\left(\frac{1}{(n+1)\,(G^n/aV)^{1/n}}\right)^3\right) \tag{2}$$

where N_o = the nuclei density and both **a** and **n** are material constants. The nuclei density is a critical factor because stray grains nucleate independently in the liquid. Estimates of N_o have been made based on SG measurements [3]. The incidence of SG formation has also been related to the presence of a wide solidification temperature range, ΔT_s [4]. Recent solidification studies of alloy CMSX-4 have shown that it exhibits a relatively wide solidification temperature range of ~200°C (360°F) and therefore may be quite susceptible to stray grain formation [5].

Solidification Cracking

Weld solidification cracking in the fusion zone of Ni-base alloys has been the subject of considerable investigation and the mechanism is generally well understood. As is characteristic of weld solidification cracking in other systems, cracks form during the terminal stages of solidification when liquid films are distributed along solidification grain boundaries and, in some cases, interdendritic sites. At this stage, shrinkage strains across the partially solidified boundaries can become appreciable. If the terminal liquid is distributed along the boundaries as a continuous film, the strains cannot be accommodated and the boundaries separate to form a crack.

Susceptibility to weld solidification cracking is a function of both metallurgical factors and the level of local strain present at the end of solidification. In terms of metallurgical factors, it is well established that the solidification temperature range as well as the amount and distribution of the interfacial terminal liquid are the primary factors that control solidification cracking susceptibility of Ni-base alloys [6–8]. Solute redistribution plays an important role in solidification cracking as it affects the solidification temperature range and amount of terminal liquid.

The effect of the solidification temperature range can be understood in simplified terms by considering its influence on the size of the solid + liquid (mushy) zone. During welding, the mushy zone trails behind the liquid weld pool. It is this mushy region which is susceptible to cracking under the influence of shrinkage strain and external restraint. For a fixed temperature gradient in the mushy zone (constant processing parameters), alloys with relatively high solidification temperature ranges can be susceptible to cracking due to a rather large crack-susceptible mushy zone.

The actual distance a solidification crack propagates through the mushy zone depends on the distribution of terminal liquid that exists near the end of the solid + liquid region [8] and the level of local strain present. When the amount of terminal liquid is moderate (typically between approximately 1–10 volume percent [8]) and/or the surface tension is low, the liquid tends to wet the boundary and forms a continuous film. This type of morphology is most detrimental as it interferes with the formation of solid/solid boundaries, thus reducing the ability of the material to accommodate strain. When the amount of terminal liquid is relatively high, (greater than approximately 10 volume percent), it can often flow into the cracks and provide a "crack healing" effect [8].

Since solidification cracking in superalloys is primarily associated with grain boundaries, SX weld zones have typically been observed to crack when SG formation has introduced grain boundaries into the weld metal. The large ΔT_S combined with the relatively small grain boundary surface area make these areas especially susceptible to cracking. Moreover, the character of the grain boundary itself can contribute towards cracking susceptibility. High-angle boundaries are more likely to crack than low angle boundaries due to the prolonged time interval that these boundaries require to coalesce [9]. While substrate preheating can be used help reduce the restraint and possibly reduce cracking tendency [10], it also reduces the overall G/V ratio in the weld zone, making a Columnar-to-Equiaxed Transition (CET) far more likely to occur. The resultant nucleation and growth of equiaxed grains will significantly increase the solidification cracking susceptibility. Since solidification cracks have never been observed without stray grains, the avoidance of such cracks can be achieved by reducing or eliminating SG formation. Past studies [11] have shown that weld parameters which produce higher G/V ratios can effectively eliminate solidification cracking. The objective of this study is to evaluate the stray grain formation and resultant cracking tendency of alloy CMSX-4 over a wide range of processing conditions. More detailed results from this study on the use of heat and fluid flow modeling to control stray grain formation has been published in separate articles [12,13]. In this article, experimental results are presented that correlate the process parameters and process type to the stray grain formation and cracking tendency. The results of this research can be used as the basis for designing processing strategies for successful single crystal welds repairs.

Experimental Procedure

A series of autogenous welds were prepared in order to study the SG formation behavior and cracking susceptibility as a function of welding parameters and welding process. The Ni-base superalloy CMSX-4 was selected as a representative alloy used in SX applications for its widespread use in the industry. The alloy composition, as measured through wet-chemical techniques, is given in Table 1. Substrates with dimensions 155 × 80 × 6 mm (6.1 × 3.1 × 0.25") were cast such that the (001) crystal plane was parallel to the sample surface. The substrates were solution heat-treated with a schedule used in industrial practice for alloy CMSX-4 (heated to 1,310°C [2,390°F] for 7 h while under vacuum). Electron Beam (EB) welds were performed on the CMSX-4 substrates at beam powers up to 1,500 W and travel speeds up to 95 mm/s using a large-chamber Leybold-Heraeus EB welding apparatus. The absorbed power of these welds was taken to be equal to the transmitted power by assuming an absorption coefficient of 1.0, since the transfer efficiency of the EB weld process is known to be very high. For comparison, a smaller set of autogenous welds were also conducted using the gas-tungsten arc welding (GTAW) process. The particular powers were selected such that the absorbed powers would overlap some of the EB welds after taking into account the arc transfer efficiency of ~0.7 for the GTAW process [14]. (All power values cited in this work are *absorbed* powers.) The torch travel speed also ranged from 1 to 100 mm/s (2.4–240 ipm), similar to the EB welds. All welds described in this article were produced in the [100] direction on the (100) plane.

Cross-sections from the welds were prepared using standard metallographic techniques. Orientation Imaging Microscopy (OIM) analysis was conducted using an electron-backscattered diffraction (EBSD) camera on a Hitachi 4300 field-emission-gun scanning electron microscope (FEG-SEM) in order to determine the area of stray grains. The weld structures were then revealed by immersing the specimens for 5 s in a reagent consisting of 50 mL of HCl, 50 mL of H_2O, and 2.5 g of $CuCl_2$. Light optical microscopy (LOM) photomicrographs of the weld cross-sections were used to measure the weld pool dimensions and the fusion zone area. The SG area fraction was determined by dividing the SG area (determined via OIM) by the fusion area (determined by LOM). The average of three different cross-sections was acquired for each weld condition. A more thorough description of the experimental weld trials and analysis is provided in a separate articles [12,13].

Table 1 Chemical composition of Alloy CMSX-4. All values in weight percent

Ni: Bal	Mo: 0.63	Al: 5.62
C: 0.002	W: 6.34	Hf: 0.10
Cr: 6.36	Ta: 6.52	Re: 2.87
Co: 9.68	Ti: 1.0	

Results and Discussion

Figure 1 contains typical examples of OIM-generated grain maps superimposed upon the LOM photomicrographs of EB weld microstructures from which they were collected. In these figures, stray grains are identified by the color pixels. The weld structure shown in Fig. 1a resulted from a relatively low heat input (500 W, 95 mm/s) and represents a crack-free weld with essentially no stray grains. By comparison, a high heat input (1,500 W, 25 mm/s) EB weld is shown in Fig. 1b. This weld exhibited a high stray grain content of 70%. The presence of a solidification crack is illustrated by a line of black pixels and is bordered on both sides by stray grains.

While stray grain formation behavior is directly a function of local solidification parameters, these values are functions of the overall welding parameters. Figure 2 shows the overall stray grain area fraction as a function of beam power and travel speed for the EB welds. Figure 2a provides a summary of all the results, and the data is repeated in Fig. 2b at a reduced range of stray grain area fraction values. The maximum stray grain area fraction is reached at an intermediate travel speed of ~ 6 mm/s. Beyond this value, the stray grain content decreases with increasing travel speed. This variation in stray grain content with travel speed has been observed in other work [4] and can be explained based on the relative increases in temperature gradient and growth rate with changes in travel speed. When the travel speed is low, initial increases in the speed will cause an increase in the growth rate with only minor changes in the temperature gradient. As a result, the G/V ratio generally decreases, and the amount of stray grains will therefore increase. Further increases

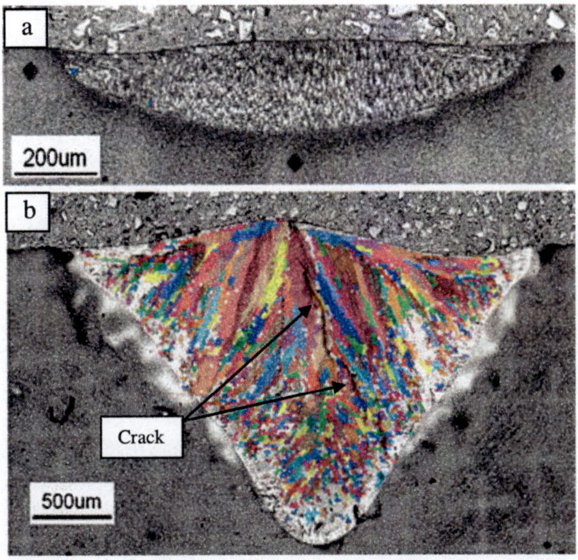

Fig. 1 Two EB weld microstructures with superimposed OIM maps: (**a**) 500 W, 95 mm/s; (**b**) 1,500 W, 25 mm/s

Fig. 2 The effect of welding parameters on stray grain area fraction within the electron-beam weld structures for the [100]||(001) substrate orientation

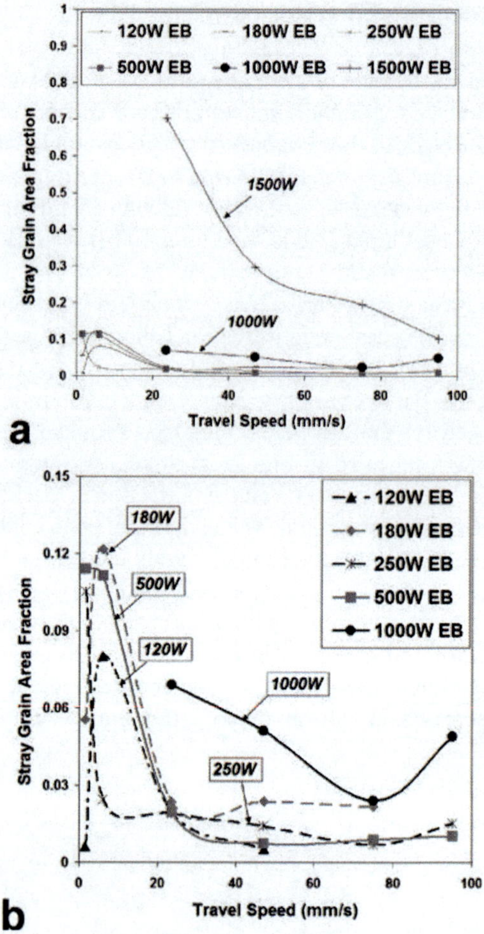

in the travel speed will induce larger increases in the temperature gradient, and, according to equation (2), G has a larger effect on stray grain formation than V. Thus, stray grain formation will subsequently decrease with further increases in the travel speed. The negative influence of increasing weld power on stray grain formation can be understood by considering its effect on the temperature gradient. An increase in the power will produce an decrease in the temperature gradient, thus promoting more stray grains to form in the weld.

A limited set of stray grain measurements were performed on the GTA weld structures. Those results are shown in Fig. 3 along with data from welds conducted using the EB processes at an equivalent absorbed power of 180 W. Data for several laser welds made with an Nd-YAG laser at an equivalent absorbed power are also shown for comparison [12,13]. Note that the GTA welds always exhibit more stray grains than the EB welds, and the laser welds are intermediate to these two cases.

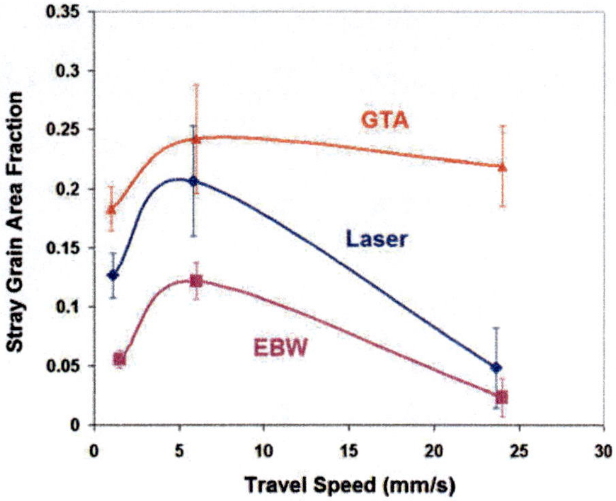

Fig. 3 The effect of welding process on stray grain area fraction for an equivalent absorbed power over a range of travel speeds

It is interesting to note that the trend in SG content between the three processes correlates to the differences in energy density. The energy density of the heat source influences the temperature gradient in the weld pool, where welds produced with higher energy density processes will experience higher temperature gradients. Thus, welds produced with higher energy density processes are expected to exhibit lower SG contents than welds made from lower energy density processes at equivalent levels of input power and travel speed. This accounts for the relatively high stray grain content of the GTA welds. The influence of welding parameters on the local solidification conditions and resultant stray grain formation behavior of these welds has been investigated in detail using heat an fluid flow modeling and is described elsewhere [12,13].

Figure 4 summarizes the cracking susceptibility of all the EB (Fig. 4a) and GTA (Fig. 4b) welds as a function of absorbed power and travel speed. These results clearly show that crack-free welds are promoted by low heat inputs (i.e., low power and high travel speed). This result is not surprising considering the influence of processing parameters on stray grain formation, and the link between stray grain formation and cracking susceptibility. Since stray grains can generally be reduced under low heat input conditions, the cracking susceptibility will also be reduced as the heat input is decreased. The reduced heat input may also be beneficial due its effect on solidification shrinkage and size of the crack-susceptible mushy zone. The smaller welds produced under lower heat input conditions will exhibit reduced strain from solidification shrinkage along with a smaller crack-susceptible mushy zone, and these factors may also contribute to the reduced cracking susceptibility.

Careful examination of Fig. 4 indicates there is a significant difference in the range of processing parameters between the two processes that can be used to produce crack-free welds. This is shown in Fig. 5, which compares the position of the crack/crack-free boundary for each process. Although the position of these boundaries are only approximate and apply only to the conditions used in this

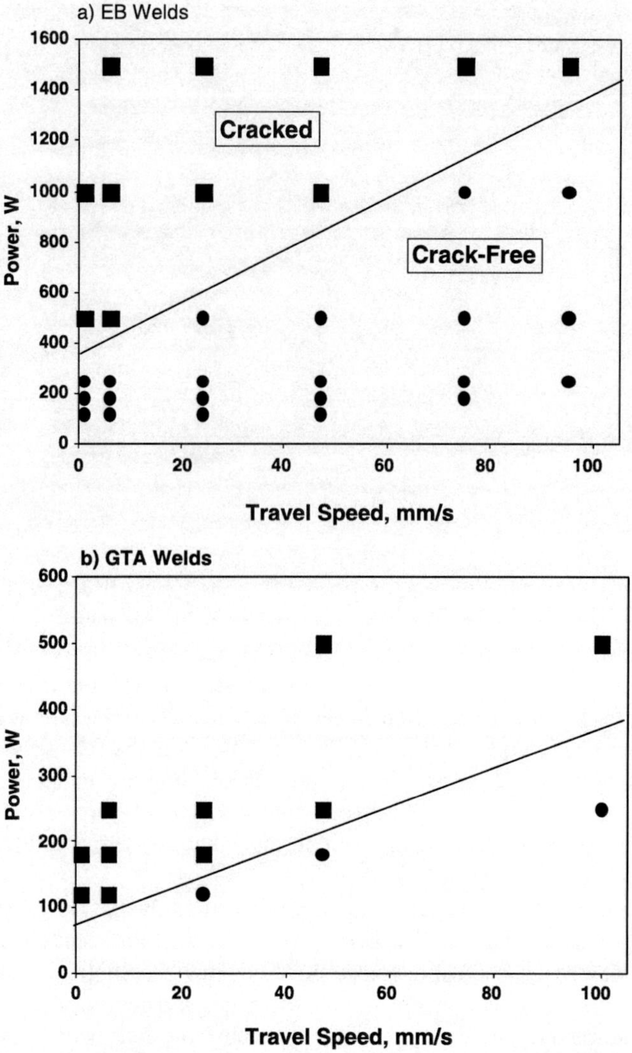

Fig. 4 Solidification cracking susceptibility in the (**a**) EB welds and (**b**) GTA welds as a function of absorbed beam power and travel speed

investigation, the results clearly demonstrate the beneficial effect of the EB process over the GTA process. Reference to Fig. 3 indicates this can likely be attributed to differences in power density and resultant temperature gradient. The higher power density and concomitant temperature gradient of the EB process reduces the stray grain content and, therefore, helps reduce the incidence of cracking.

As previously mentioned, successful weld repair of single crystal turbine blades requires minimizing both the amount of stray grains and solidification cracks.

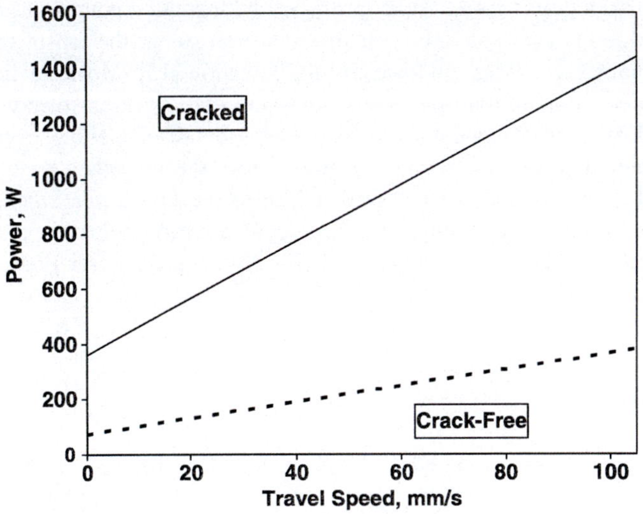

Fig. 5 Comparison of crack/crack-free processing regimes for the EB (*solid line*) and GTA welds (*dotted line*)

Fortunately, a reduction of the stray grain content typically leads to crack-free welds, and each defect, in turn, can be minimized by reductions in the heat input. In view of this, Fig. 6 summarizes the influence of heat input on stray grain area fraction and cracking susceptibility for the EB welds. These results show that, for the current conditions, there is a critical heat input of ~13 J/mm. Welds made below this heat input level are consistently crack-free with very low stray grain contents (<5%). It is important to note that effective weld repairs can still be accomplished when small

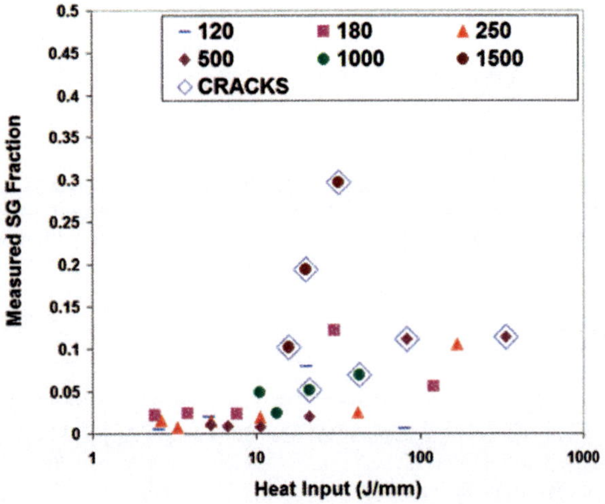

Fig. 6 Effect of heat input on the stray grain area fraction and solidification cracking susceptibility in EB welds. Legend values are power setting (in Watts)

amounts of stray grains form. Welds produced with small amounts of stray grains typically exhibit a very shallow layer of stray grains near the top of the weld at the centerline, since G/V is the lowest in that location [15]. Most practical repairs require the deposition of multiple layers. Thus, the shallow layer of stray grains can be removed by subsequent passes as long as the depth of melting from the next pass is greater than the depth of stray grains from the preceding pass. With this approach, the stray grains can be "pushed" up to the final layer, where they can be easily removed by machining. This approach has recently been applied to successfully prepare a 12 layer single crystal deposit using the Laser Engineered Net Shaping process [15].

Conclusions

Autogenous welds were prepared on the single crystal CMSX-4 using the EB and GTA welding processes. The stray grain area fraction and cracking susceptibility were determined and correlated to the processing parameters and process type. The following conclusions can be drawn from this work.

1. The stray grain content initially increases and then decreases with increasing travel speed. This effect is attributed to the complex relationship of travel speed on the temperature gradient and growth rate. Stray grain content decreases with decreasing weld power. In general, the stray grain content and cracking susceptibility can be reduced by welding at low heat inputs.
2. The stray grain content and corresponding cracking susceptibility was higher for welds prepared with the GTA process compared to the EB process. This difference is attributed to differences in power density and concomitant temperature gradient, where the EB process produces a higher temperature gradient that leads to reduced stray grains and less solidification cracking.
3. For the conditions evaluated in this work, EB welds produced at heat inputs below ~13 J/mm produced welds that were crack-free with very low stray grain contents (<5%).

Acknowledgements The authors would like to acknowledge the financial support of the National Science Foundation through grant # 0500254. The SX substrates used for welding trials were cast by the Howmet, a division of Alcoa, Inc. The EB welding experiments were made possible by Thomas Lienert and Paul Burgardt of Los Alamos National Laboratories.

References

1. Park JW, Babu SS, Vitek JM, Kenik EA and David SA (2003) Stray Grain Formation in Single Crystal Ni-base Superalloy Welds. Journal of Applied Physics 94:4203–4209.
2. Hunt J (1984) Steady State Columnar and Equiaxed Growth of Dendrites and Eutectic. Materials Science and Engineering 65:75–83.

3. Gaumann M, Bezencon C, Canalis P and Kurz W (2001) Single-Crystal Laser Deposition of Superalloys: Processing - Microstructure Maps. Acta Materialia 49:1051–1062.
4. Vitek JM, Babu SS, David SA and Park JW (2003) Microstructure Development in Single Crystal Welds. Materials Science Forum 426-432:4123–4128.
5. Walter C, Hallstedt B and Warnken N (2005) Simulation of the Solidification of CMSX-4. Materials Science and Engineering A 397:385–390.
6. Cieslak MJ (1991) The Welding and Solidification Metallurgy of Alloy 625. Welding Journal 70:49–56.
7. DuPont JN, Michael JR and Newbury BD (1999) Welding Metallurgy of Alloy HR-160. Welding Journal 78:408–414.
8. DuPont JN, Robino CV and Marder AR (1998) Solidification and Weldability of Nb-bearing Superalloys. Welding Journal 77:417–431.
9. Wang N, Mokadem S, Rappaz M and Kurz W (2004) Solidification Cracking of Superalloy Single- and Bi-Crystals. Acta Materialia 52:3173–3182.
10. Kou S (2003) Welding Metallurgy. Wiley, Hoboken.
11. Vitek JM, Babu SS, David SA, Park JW, Hu Y and Hehmann W. Cracking Behavior in Nickel-Based Single Crystal Superalloy Welds. 2005. Proceedings of the 7th International Conference on Trends in Welding Research. 5-16-2005. Ref Type: Conference Proceeding
12 Anderson TD, DuPont JN, and DebRoy T, Anderson TD, DuPont JN and DebRoy T (2010) Origin of Stray Grain Formation in SX Superalloy Weld Pools from Heat Transfer and Fluid Flow Modeling. Acta Materialia 58:1441–1454.
13. Anderson TD, DuPont JN and DebRoy T (2010) Stray Grain Formation in Welds of Single Crystal Ni-base Superalloy CMSX-4. Metallurgical and Materials Transactions A 41(1):181.
14. DuPont JN and Marder AR (1995) Thermal Efficiency of Arc Welding Processes. Welding Journal 74:406–416.
15. Liu W and DuPont JN (2005) Direct Laser Deposition of a Single-Crystal Ni3Al-Based IC221W Alloy. Metallurgical and Materials Transactions A 36A(12):3397–3406A. Dec. 2005.